Questions for Reviewing a Paper

For a fuller discussion, see Part III, Revising and Editing, pages 597–669 and the Checklist for Common Problems in Revising a Paper, back endpapers.

Title

1. Is the title clearly stated, in a phrase?
2. Does the title indicate precisely and accurately the focus of the paper?

Organization

1. Does the introduction give needed background, overview, and identification of the subject?
2. Does the introduction include a lead-in or forecasting statement that points the way for the remainder of the paper?
3. Does the body of the paper adequately develop the lead-in or forecasting statement?
4. Does each section of the paper have a topic statement that points back to the lead-in or forecasting statement?
5. Is the information in the body of the paper arranged in a logical sequence?
6. Is the closing of the paper compatible with the purpose and emphasis of the paper?

Content

1. Is the content of the paper suitable for the intended audience and purpose?
2. Is the content accurate, clear, complete, concise, and coherent?
3. Is all the information directly related to the topic?
4. Is there a sense of flow between sentences, between paragraphs, and between sections of the paper?

Mechanics and Usage

1. Do punctuation and spelling follow standard practice?
2. Is word choice accurate and precise?
3. Does grammatical usage follow standard practice?
4. Is each sentence effective in structure and emphasis?
5. Is the paper free of typos and careless errors?

Visuals

1. Do the visuals convey information and create interest?
2. Is each visual numbered, captioned, and referred to in the text of the paper?

Headings

1. Is the organization of the paper apparent through the headings
2. Are headings of equal value similarly positioned on the page?

Layout and Design

1. Is the material on each page arranged pleasingly with plenty of white space?
2. Does the entire paper (document) invite a positive response from the reader?

Technical English

Technical English

Writing, Reading, and Speaking

SEVENTH EDITION

Nell Ann Pickett
Hinds Community College

Ann A. Laster
Hinds Community College

■ HarperCollinsCollegePublishers

**For the Hinds Community College English Faculty,
our colleagues and friends**

Executive Editor: Anne Elizabeth Smith
Developmental Editor: Susan Messer
Project Coordination and Text Design: Ruttle, Shaw & Wetherill, Inc.
Cover Designer: Rubina Yeh
Electronic Production Manager: Christine Pearson
Manufacturing Manager: Helene G. Landers
Electronic Page Makeup: Ruttle, Shaw & Wetherill, Inc.
Printer and Binder: R.R. Donnelley and Sons Company
Cover Printer: Phoenix Color Corp.

For permission to use copyrighted material, grateful acknowledgment is made to the copyright holders on pp. 671–672, which are hereby made part of this copyright page.

Technical English: Writing, Reading, and Speaking, Seventh Edition

Library of Congress Cataloging-in-Publication Data
Pickett, Nell Ann.
    Technical English : writing, reading, and speaking / Nell Ann
Pickett, Ann A. Laster. — 7th ed.
        p.   cm.
    Includes index.
    ISBN 0-673-99794-4 (pbk.) (student edition)—ISBN 0-673-99795-2 (pbk.) (instructor's edition)
    1. Readers—Technology.   2. English language—Technical English.
3. Technical writing.   I. Laster, Ann A.   II. Title.
PE1122.T37P5   1996
808'.0666—dc20   95–12806
                                                                          CIP

96 97 98 99 9 8 7 6 5 4 3 2 1

Contents

Part III
Revising and Editing

Preface

This seventh edition of *Technical English* is another milestone on a journey that began some twenty-eight years ago. The first edition was the result of our frustration with the lack of available textbooks for technical communication. That first edition resulted from our asking several questions. What communication skills do students in such majors as drafting, computer science, health sciences, commercial art and design, marketing, agriculture, electronics, and horticulture need? What is the most effective way to teach these skills to a heterogeneous group? How can we present information that students can read, identify with, and learn from?

To find answers to these questions, we researched state departments of education, vocational and technical schools, and personnel offices in business and industry. We enrolled in several technical courses; we worked closely with technical instructors—sitting in on their classes, attending their departmental meetings, attending their advisory committee meetings, inviting them to help us plan class content and assignments. We interviewed technicians on the job. We collected samples of technical writing wherever we went, from the state fair to the county health office to home improvement stores. We read, we assimilated, and we read some more, and we talked with each other and with other professionals in the field.

We identified a body of communication skills that students need to be good communicators in their workplaces. We decided the most effective way for students to develop these skills is having the students plan and write documents and give oral presentations on subjects related to their career choices. Not only would this approach develop communication skills for the students but also this approach would reinforce and enhance the information learned in their major classes.

With this approach in mind, we prepared handouts for class assignments; the handouts grew into teaching units on activities such as explaining procedures, describing mechanisms, analyzing effects and causes, giving reports, and writing letters. The teaching units grew into a spiral-bound book, with photocopied units, sold in our college bookstore. And then came the contract with a publishing company.

Out goal then remains our goal now: to find successful ways to prepare technical students for their communication needs when they enter the job market and to upgrade communication skills for workers already in the job market. Over the past twenty-eight years we have continued to research needed communication skills for workers, to listen to students in technical communication classes, to seek direction from other instructors of technical communication, and to read professional literature. All of the gathered body of knowledge is the basis for this seventh edition.

We are humbled and pleased by the continued interest in the textbook and the comments we receive from students, instructors, and persons in the workplace.

Purpose of This Textbook

In the seven editions, we have included in each preface our purpose for this textbook: to provide a beginning textbook in technical communication, written on an appropriate level, for students and teachers in two-year and four-year colleges and universities, technical schools, and in-service short courses. We believe that effective communication skills are mandatory for success in the workplace. We believe this textbook helps those who use it to develop effective communication skills.

Organization of This Textbook

This seventh edition retains much of the familiar organization and content of earlier editions. This edition has a three-part division. Part I presents the basic principles and forms of communication that any student or person on the job needs to know, but with emphasis on technical communication demands. It emphasizes audience and purpose and layout and design; it includes chapters on instructions and process explanation, description of a mechanism, definition, analysis through classification and partition, analysis through effect-cause and comparison-contrast, summary, memorandums and letters, reports, the research report, oral communication, and visuals. Part II, Selected Readings, retains selections that students and instructors have found useful; it also includes new selections. Part III focuses on revising and editing, an emphasis new to this edition.

Part I Forms of Communication

Each chapter in Part I is an independent chapter. The chapters can be taught sequentially or in any other order. Each chapter opens with clearly stated outcomes so that students know exactly what is expected of them. Each chapter then has the following organization. The chapter topic is developed through discussion and examples, an explanation of principles, and an outline of a step-by-step procedure for writing and speaking. Plan Sheets, completed to serve as models, reinforce the need for careful thinking and planning before preparing a preliminary draft. A sample blank Plan Sheet corresponding to end-of-chapter assignments provides a guide for students to fill in as preparation for a written or oral presentation. Applications at the end of each chapter suggest ways the students can practice a variety of technical communication activities. Applications also suggest ways to relate chapter content to selected readings in Part II.

Part II Selected Readings

Part II begins with introductory material that suggests a method for selective reading. The readings following illustrate and amplify the assignments in Part I. At the end of each selection, questions stimulate thought and initiate responses. The readings were chosen for their inherent interest and pedagogical utility.

Part III Revising and Editing

Part III is a resource for instructors and students to use as a reference for revising and editing according to individual needs and preferences.

Appreciation

The earlier editions of *Technical English* and this seventh edition have been influenced and enhanced by many individuals. One of the rewarding benefits of writing a textbook is the people you meet and learn from. There have been so many—students and fellow teachers, who have been generous and helpful with their comments and suggestions. We are grateful to the manuscript reviewers for substantive suggestions for revision: Reva Daniel, Dynamic Business Writing; Faye Angelo and Jerry Carr, both of Hinds Community College; Jeannie Dobson, Greenville Technical College; Jeanny Pontrelli, Truckee Meadows Community College; Deborah Barberousse, Cochre College; James Kennedy, Richard J. Daley Community College; Rebecca Kamm, Northeast Iowa Community College. And of course our colleagues, friends, and students at Hinds Community College offer immeasurable help. Specifically we thank Nancy Tenhet and Curtis Kynerd and other staff at Hinds Community College Learning Resources Center; colleagues in the English Department, especially Jerry Carr; Mary Ellen Powell; Russell Porrier and Jim Porch, Hinds Community College Printing Department; and Mike Hataway, Commercial Art and Design Department.

Nell Ann Pickett
Ann Laster

Technical English

Technical Communication: Getting Started

The Communication Process

Communication is an active process with at least five parts: a sender, a message, a reason (purpose) for sending the message, a way to send the message, and a receiver (audience: reader, listener) to comprehend, analyze, and respond or react to the message. Technical communication involves all of these because technical communication involves people who generate and use information.

As a future communicator on the job, you must be very aware of the communication process, for you may be the person with a message or the person to receive a message. Making a sale, convincing management of needed changes, giving employees instructions for operating new equipment, preparing a report on work in progress, proposing the purchase of new materials or equipment—a worker daily sends and receives many different types of messages to and from many different audiences and for many different purposes.

This Text, You, and the Communication Process

Basically this text assumes that you are the sender in the communication process. You are introduced to some common types of messages—instructions and process explanations, descriptions of mechanisms, definitions, classification and partition analyses, effect-cause and comparison-contrast analyses, memorandums and letters, summaries, and reports—with explanations and suggestions on how to present these messages in writing and orally, verbally and visually. Then you are asked to plan and present similar messages, making choices about subject, audience, purpose, content, page layout, document design, and visuals.

For instance, you might be assigned the following. You (sender) write (method) a set of instructions (message) for consumers (receivers) who will purchase a clock radio and must have the instructions to know how to use the radio (purpose for writing the instructions). As you study this textbook, you will learn to ask such questions as the following: Do I know enough about the subject to write a clear set of instructions for buyers of the clock radio? If not, where can I find needed information? Should I practice using the clock radio before I try to tell consumers how to use the clock radio? How can I write the instructions to serve an obviously varied reading audience? How should I arrange the instructions so that consumers will react positively to the appearance of the instructions and follow them easily and accurately? How would visuals help? What kind of visuals?

As these questions suggest, every decision you the sender of the message make about the message depends upon at least four factors:

- *Who* is to receive the message (the audience: the reader, the listener, the viewer)
- *Why* you are preparing the message (the purpose)
- *How* you can best send the message (written or oral; visuals; page layout and document design)
- *What* you want the receiver to do with the message (the desired repsonse)

As you study the chapters within this text, you will be reminded often to ask and answer these four questions before you begin any technical communication.

Of course, the way an audience reads or listens, comprehends, and responds to communication depends to a great extent on the appearance and the content of the message. It also depends on such other factors as the mental state, the background, and the special skills of the reader or listener as well as the reason for reading or listening. The writer or speaker can use various visual and verbal techniques to make a message easier to comprehend. Since these techniques may be new to you, the following pages introduce you to frequently used visual and verbal techniques.

As you proceed through the various chapters and prepare the related assignments, you will select and adapt techniques to enhance a particular message for a particular audience for a particular purpose.

Audience

A concept that is perhaps new to you is the idea of audience. Who will read what you write? Or listen to what you say? What does that person already know about the subject? What attitude does the person likely have toward the subject? What background (education, culture, experience) does he/she have? Why does the reader or listener need the information? What must the reader or listener do once he/she has the information?

All technical communication has an intended audience: a reader or a listener. Different classification systems exist for such audiences. At best any system simply gives a workable way to talk about different audiences. Often an audience may fit into more than one category. A suggested system follows.

Category	Characteristics
Experts	Advanced in their field. Understand technical information within that field. Handle theory and practical application with ease.
Technicians	Understand technical information within their field. Handle practical application with ease.
Professionals (nonexperts)	College-level education
Lay (general) audience	Population at large

Analysis of the intended audience determines first of all the form, either oral or written, for presenting material. Such an analysis then determines length, content, and style. For written communication, audience analysis also determines the kind of document needed. Maybe the document should be a business letter, a memorandum, a set of instructions, a proposal, or a feasibility report. Further, audience analysis affects decisions about page layout and document design. Indeed, audience analysis determines every decision you must make about a message.

Primary and Secondary Audiences

In many instances, the audience for technical communication will be several persons. You as a communicator of information, therefore, have to make decisions about page layout and document design, organization and development, degree of complexity, and word choice, all the while keeping the several audiences in mind. Each audience needs the information for a specific reason, and you must present the material to meet those needs.

Suppose, for instance, that workers in a clerk/keyboard entry pool decide they need new computers. After much discussion among themselves, one of the workers, the senior worker, writes a proposal suggesting the purchase of the new computers. The primary readers of the proposal are the supervisor of the pool who will recommend the purchase and the department head in charge of the clerk/keyboard entry pool who can approve or deny the proposal. Secondary readers are the workers in the pool who supply data and information for the proposal and whose daily activities will be affected by the decision.

Primary readers are usually those readers who have requested a document or who will use the document as the basis for making a decision or taking action. Secondary readers are usually those whose activities will be affected by the decision; these readers may help carry out an approved project or may supply needed information to the writer or to the primary readers. Secondary readers may read the entire document or they may read only the portion that directly affects their area of work or responsibility.

Subject and Authorship

In technical communication, the subject of a presentation or the type of presentation, whether oral or written, usually is predetermined. That is, you will be assigned a document to prepare or a subject to research; perhaps the need for a document will grow out of conditions or circumstances within your department or company. Nobody has to tell you the document is needed. For instance, you may be asked to analyze an existing process to determine if a change in the process will increase productivity. You may see the need for new equipment and write a proposal recommending the purchase.

No one jumps out of bed on a sunny day and says, "I believe I'll write a report today." You might, however, arrive at work on that sunny day and your supervisor says to you, "Sales in the Southeast are down for the last quarter. Contact all the sales managers in this area to come up with an explanation for the decrease in sales and recommendations for solving the problem. I want a written report on my desk by Friday."

The report assignment described above implies another possible aspect of technical communication. You may be only one of several communicators gathering, analyzing, and presenting needed information. As a member of a team—collaborative group—you may be responsible for the final report, but the information for the report may come from a number of persons. In other instances, you may be responsible for only a section of a final report.

Collaboration

In collaborative work, you will likely be guided by a group leader who may give direction and make decisions so that the finished presentation will accomplish its purpose. Or you may be the group leader. The preparation may be carried out through group meetings, telephone conferences, or computer networking. Computer networking has decided advantages, allowing a project's team members to see on screen the work of each member and to interact with each other during revision.

As with any group work, some persons may find themselves doing "extra" work and some persons may not fulfill their responsibilities. The bottom line, nevertheless, is that the job must be completed and on time.

Peer Review and Peer Editing

To give you a "feel" for working with others, you will find throughout the writing assignments in this textbook a suggestion that you work in peer review and peer editing groups. Such group work allows you to benefit from suggestions from other students to improve your writing assignments. All students will have had an opportunity to study the assigned material and listen to class discussion. You can then help each other by reviewing each other's work and making suggestions for positive change. As you look at preliminary drafts, you can ask such questions as the following:

- Does the content meet the need of the intended audience?
- Does the content accomplish the stated purpose?
- Is the format appropriate?
- Does the presentation follow accepted standards for the type of communication?
- Are page layout and document design effective?
- Do headings signal sections of material?
- Are levels of headings clearly indicated?
- Are appropriate visuals included? Placed effectively? Labeled and captioned (titled)?
- Are grammatical usage and mechanics acceptable?
- Is the overall tone of the presentation appropriate?

Following the advice from a careful peer editor can result in a better final presentation. Rarely does even the most experienced writer complete a document in one attempt. Good writing takes hard work and revision.

Planning a Presentation

As you know, planning a written or an oral presentation takes time. You have to gather needed materials, then analyze and organize those materials in an effective presentation for the intended audience and purpose.

Upon receiving a writing assignment, you may be tempted to dash off the paper immediately. Resist such temptation; spend some time "getting ready."

The Plan Sheet

This textbook suggests the use of a Plan Sheet to help you think through aspects of a presentation. The Plan Sheet is a kind of blueprint; it helps you get ready to write or speak.

Filling in the Plan Sheet before beginning an actual presentation helps you in several ways:

- Reminds you to identify a specific audience and purpose
- Directs you as you gather, think through, and analyze information

- Helps you select appropriate material for the specific audience and purpose
- Provides an informal outline to help you organize material
- Enables you to get right to work without spending time staring at a blank sheet of paper or wondering where to begin

Spending a little extra time on the planning of a presentation will undoubtedly result in a more effective final presentation.

Page Layout

One of the simplest ways for a writer to make a message more readable is page layout—that is, placement of words and sentences, blocks of information, paragraphs, lists, tables, graphs, and the like on the page. Page layout makes an important visual impression on the reader.

For many years writers followed the centuries-old essay form, paragraph after paragraph after paragraph with only an occasional indentation to mark the beginning of a new paragraph. Sometimes no indentation appeared on an entire page; the page was literally filled with words with space only at the outer edges of the page. Look at Figure 1, Example A on page 7.

Today's writer uses page layout to enhance writing and reading. Look at Example B in Figure 1. This example illustrates techniques for making a page "come alive." Not only is the material easier to read when arranged similarly but also such a page has a positive psychological effect. The reader is not overwhelmed by the appearance of the page and does not dread reading it. White space offers places to stop, get a breath, and absorb—before reading further.

Basic page layout techniques can be grouped into two general categories:

- White space providers
- Emphasis markers

White Space Providers

White space providers yield an uncluttered, easily readable, inviting page. Indenting for paragraphs, double spacing, and allowing ample margins ($1^{1}/_{2}$ inches minimum on all four perimeters) are the most typical ways of allowing for plenty of white space. Additional white space providers include headings (with triple spacing before a major heading and double spacing after the heading), vertical listing (placing the items up and down rather than across the page), and columns (setting up the page with several columns of short lines, as in a newspaper, rather than a page of long lines).

Emphasis Markers

You can use a number of techniques to draw attention, lend variety, or give emphasis to material. For example, you can use different sizes and styles of typefaces. With the ready availability of computers and software programs, sizes of typefaces may vary from 7 point to 72 point.

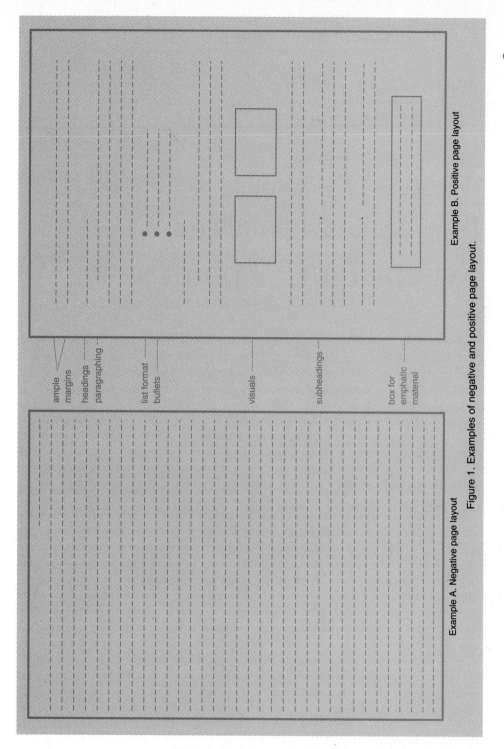

ample
margins

headings

paragraphing

list format
bullets

visuals

subheadings

box for
emphatic
material

Example B. Positive page layout

Example A. Negative page layout

Figure 1. Examples of negative and positive page layout.

EXAMPLES OF VARIOUS POINT SIZES

Typeface in 7 point size

Typeface in 10 point size

Typeface in 12 point size

Typeface in 14 point size

Typeface in 18 point size

Typeface in 24 point size

Typeface in 30 point size

Choose point sizes that balance each other. Avoid selecting an extremely large point size to use with an extremely small point size. Also, use varying point sizes to show levels of headings. The larger the size, the more important the heading; as the levels of headings decrease, the size of the headings decreases accordingly. (See Headings, pages 9–11, 311–312.)

Computers also make possible many styles of typefaces, such as **bold,** *italic,* Outline, and Shadow, in addition to underlining and CAPITALIZATION. Such styles can be used effectively for emphasis. A caution, however: Since such styles give emphasis, use them sparingly. An entire report or even an entire section of a report printed in bold, italic, outline, shadow, capital letters, or underlined would be very unappealing to look at and difficult to read.

Other ways to add emphasis include boxes to set off material, such as a caution in a set of instructions, or to enclose an entire page. Insets—small boxes that set off material—may provide a legend or an explanation of symbols for such visuals as maps, charts, and drawings. Insets are often used to set off key thoughts, a summary, or supplementary material. Bars (heavy straight lines) typically separate a table or other visual from the text. Also you can employ multiple colors (for contrast), shading (for depth), and borders (particularly in a paper requiring a title page). Various computer programs can be used to create any of the emphasis markers.

You can also use symbols as emphasis markers. Frequently used are the bullet (○), the square or box (□), other geometric shapes such as a triangle or diamond (△◇), the dash (—), the asterisk (*), and arrows (↓ →). The various geometric shapes can be solid (●■▲) or open (○□△). Especially easy to make is the bullet. You can make it on a keyboard by using the letter "o," or the period, or special keyboard commands. You can make it by hand easily with a pen or pencil.

In fact, all of these emphasis markers are fairly easy to produce. Readily available commercially prepared aids such as templates, transfer lettering sheets, transfer shading sheets, and charting tape can give your writing a professional look and enhance readability and comprehension. And, of course, with a computer, you have access to a variety of symbols.

Document Design

Design concerns individual pages as well as the document as a whole. In designing a document, the writer makes careful, conscious decisions concerning matters that affect the appearance and the impact of the total document.

Design involves the elements of composition: contrast, balance, proportion, dominance, harmony, opposition, unity. These elements direct the reader's attention and establish hierarchies of emphasis.

Among the design tools are color, format, column width, visuals (including kind, size, and placement), spacing, texture, geometric shapes, and sizes and styles of typefaces (for text, headings, and special materials). Through the use of these tools, you can consciously make choices to achieve the desired intellectual and psychological impact on the reader.

Visuals

Visuals—drawings, photographs, graphs, charts, tables, and the like—are an integral part of technical communication. Although visuals may not have been important in your previous writing, you will discover that in technical writing and speaking, visuals are essential.

Visuals can

- simplify or considerably reduce textual explanation. Consider, for instance, trying to assemble a prepackaged entertainment center without the drawings. The drawings show the "how to" of the assembly steps.
- convey some kinds of information better than words can. A physician's assistant circles the fracture on an X-ray of the femur. The patient sees exactly the site of the fracture.
- focus attention. In a proposal for replacing machines that grade the size of apples, an orchard supervisor highlights the efficiency of new machines. Through a multiple-line graph she shows the lower productivity of the old machines.
- add interest. Visuals break up lines of text, inviting the reader in.

Headings

A heading identifies the topic or subtopic for a section or a block of information. Each heading reflects content, that is, what the information is about. The heading helps the writer to stay on the subject; the heading plus the white space marking the block of information helps the reader by making the pages more inviting. Even a glance at a page of print without headings and one with headings (such as this page) will attest to the importance of headings. Headings give the reader a visual impression of major and minor topics and their relation to one another; headings reflect the organization of the material. Headings remind the reader of movement from one point to another. Headings help the reader interested only in particular sections, not the whole document, to locate these sections. And headings give life and interest to what otherwise would be a solid page of unbroken print.

Levels of Headings

Headings show organization. The form and placement of each heading indicates its rank or level. Uppercase letters, underlining, point size, and position on the page help to differentiate rank or level. Headings that identify higher levels of material should *look* more important than those that identify lower levels.

You need to work out a system of headings to reflect the major points of a document and their supporting points. (Should a document contain an outline or table of contents, the headings in them should correspond exactly to the headings in the body.)

Following are suggested systems of headings.

2 Levels of Headings (the underlining may be omitted)

MAJOR HEADING (this heading may be centered)
Division of Major Heading

or

MAJOR HEADING
Division of Major Heading

3 Levels of Headings (the underlining may be omitted)

MAJOR HEADING (this heading may be centered)
Division of Major Heading
Subdivision of division of major heading. The paragraph begins here.

4 Levels of Headings

MAJOR HEADING (this heading may be centered)
DIVISION OF MAJOR HEADING
Subdivision of Division of Major Heading
Sub-subdivision of division of major heading. The paragraph begins here.

With a computer, headings can be distinguished with different point size and style such as bold, outline, or color. While most text is printed in 10-point or 12-point type, headings can be printed in larger point sizes to make the headings easier to see and read.

2 Levels of Headings with Different Point Size

Major Heading (14 point)

Division of Major Heading (12 point)

A guide in using the computer to select levels of headings: Select a point size in proportion to the size chosen for the text of the document. As you move to divisions and subdivisions of headings, use decreasing point size. In other words, a division or a subdivision of a heading should be printed in a smaller point size than the immediately preceding level of heading. Be careful in using different fonts and styles, too. Choose fonts that are easy to read. Use styles such as **bold,** *italic,* and underlining to call attention to material.

Make headings obvious. Leave plenty of white space. As a general rule, triple-space above a heading and double-space below a heading (unless the heading is indented as a part of the paragraph).

Titles

The title of a document is the first thing a reader sees. It should identify the topic, suggest the writer's approach and extent of coverage, and interest the reader. A title should indicate to the reader what to expect in the content.

A title such as "Rotary Tillers" is vague and broad, giving a reader no idea of the writer's approach to the topic or the coverage. A title such as "Classification of Rotary Tillers According to Tines: Position, Type, and Use" clearly reflects the approach and the extent of coverage.

Be honest with your reader. A report entitled "Customizing a Van at Home" suggests the report covers everything an individual would need to know to customize a van. If the report discusses only special tools needed to customize a van at home, then the writer has misled the reader. A clearly focused title is "Special Tools Needed for At-Home Van Customizing."

Titles may include subtitles. For example, a proposal to employ a commercial service for periodic cleaning of a swimming pool might be entitled "Employing Hill Brothers Pool Service: A Proposal." Such a title might also be worded "A Proposal to Employ Hill Brothers Pool Service." Or a report on the advantages of using a commercial service might be entitled "Advantages of a Commercial Pool Cleaning Service: A Report" or "A Report on the Advantages of a Commerical Pool Cleaning Service."

A title is usually a phrase rather than just one word or a complete sentence.

Word Choice and Clear Sentences

As a writer or speaker, you must be very aware of the words you select to convey a message to an audience. The English language is filled with a rich, diverse vocabulary. Thus, you have a wide choice of words to express an idea. To select the words that best convey the intended meaning to an intended audience, use denotative and connotative words as needed, choose between specific and general words, avoid inappropriate jargon, practice conciseness where desirable, and select the appropriate word for the specific communication need.

Denotation and Connotation

Language could be used more easily and communication would be much simpler if words meant the same things to all people at all times. Unfortunately they do not. Meanings of words shift with the user, the situation, the section of the country, and the context (all the other words surrounding a particular word). Every word has at least two areas of meaning: denotation and connotation.

The denotative meaning is the physical referent the word identifies, that is, the thing or the concept; it is the dictionary definition. Words like *computer terminal, mannequin, T square, management, tractor, desk,* and *thermometer* have physical referents; *scheduling, production control, courage, recovery,* and *fear* refer to qualities or concepts.

The connotative meaning of a word is what an individual feels about that word because of past experiences in using, hearing, or seeing the word. Each person develops attitudes toward words because certain associations cause the words to suggest qualities either good or bad. For some people, such words as *abortion, liberal, leftist, Democrat, Republican,* or *right-wing* are favorable; for others, they are not. The effects of words depend on the emotional reactions and attitudes that the words evoke.

Consider the words *fat, large, portly, obese, plump, corpulent, stout, chubby,* and *fleshy.* Each of these could be used to describe a person's size, some with stronger connotation than others. Some people might not object too much to being described as *plump,* but they might object strongly to being described as *fat.* On the other hand, they proabably would not object to receiving a *fat* paycheck. So, some words may evoke an unfavorable attitude in one context, "*fat* person," but a favorable attitude in another, "*fat* paycheck."

Some words always seem to have pleasant connotations, for example, *integrity, success, bravery, happiness, honor, intelligence,* and *beauty.* Other words, such as *defect, hate, spite, insanity, disease, rats, poverty,* and *evil,* usually have unpleasant connotations.

You should choose words that have the right denotations and the desired connotation to clarify your meaning and evoke the response you want from your reader. Compare the following sentences. The first illustrates a kind of writing in which words point to things (denotation) rather than attitudes; the words themselves call for no emotional response, favorable or unfavorable.

> Born in the Fourth Ward with its prevailing environment, John was separated from his working mother when he was one year of age.

The second sentence, through the use of words with strong connotative meaning, calls for an emotional response—an unfavorable one.

> Born in the squalor of a Fourth Ward ghetto, John was abandoned by his barroom-entertainer mother while he was still in diapers.

Specific and General Words

A specific word identifies a particular person, object, place, quality, or occurrence; a general word identifies a group or a class. For example, *lieutenant* indicates a person of a particular rank; *officers* indicates a group. For any general word there are numerous specific words; the group identified by *officers* includes such specific terms as *lieutenant, general, colonel,* and *admiral.*

There are any number of levels of words that can take you from the general to the specific:

GENERAL	computer
SPECIFIC	Power Macintosh 5200/75 LC
GENERAL	drafter drafting pool architectural drafting pool
SPECIFIC	Michael Baker, Inc.

The more specific the word choice, the easier it is for the reader to know exactly the intended meaning.

GENERAL	The computer was expensive.
SPECIFIC	The Power Macintosh 5200/75 LC cost $3200.
GENERAL	The instrument was on a shelf.
SPECIFIC	The thermometer lay on the topmost shelf of the medicine cabinet.
GENERAL	The place was damaged.
SPECIFIC	The tornado blew out all the windows in the administration building at Midwestern College.

Using general words is much easier, of course, than using specific words; the English language is filled with umbrella terms with broad meanings. These words come readily to mind, whereas the specific words that express exact meaning require thought. To write effectively, you must search until you find the right words to convey to your audience your intended meaning.

Jargon

Jargon is the specialized or technical language of a trade, profession, class, or group. This specialized language is often not understandable by persons outside those fields.

Jargon is appropriate: (1) if it is used in a specialized occupational context and (2) if the intended audience understands the terminology. For example, a computer programmer communicating with persons knowledgeable about computer terminology and discussing a computer subject may use such terms as *input, interface, menu driven,* or *8MB of RAM.* The computer programmer is using jargon appropriately.

If, however, these specialized words are applied to actions or ideas not associated with computers, jargon is used inappropriately. Or if such specialized words are used extensively—even in the specialized occupational context—with an audience who does not understand the terminology, jargon is used inappropriately.

The problem with inappropriate use of jargon is that the speaker or writer is not communicating clearly and effectively with the audience. Study these examples of inappropriate jargon:

Nurse to patient: "A myocardial infarction is contraindicated." (No heart attack)
TV repairperson to customer: "A shorted bypass capacitor removed the forward bias from the base-emitter junction of the audio transistor." (A capacitor shorted out and killed the sound.)

Gobbledygook

Jargon enmeshed in abstract pseudo-technical or pseudo-scientific words is called *gobbledygook* (from *gobble,* to sound like a turkey). Examine these sentences:

The optimum operational capabilities and multiple interrelationships of the facilities are contiguous on the parameters of the support systems.
Integrated output interface is the basis of the quantification.

The message in both sentences is unclear because of jargon and pseudo-technical and pseudo-scientific words—words that *seem* to be technical or scientific but in fact are not. Further, the words are all abstract; that is, they are general words that refer to ideas, qualities, or conditions. They are in contrast with concrete words that refer to specific persons, places, or things that one can see, feel, hear, or otherwise perceive through the senses. Not one word in either sentence above is a concrete word; not one word in either sentence creates an image in the mind of the reader.

Conciseness

In communication, clarity is of primary importance. One way to achieve clarity is to be concise. Conciseness—saying much in a few words—omits nonessential words, uses simple words and direct word patterns, and combines sentence elements.

Remember that you can be concise without being brief and that what is short is not necessarily concise. The essential quality in conciseness is making every word count.

It is indeed misleading to suggest that all ideas can be expressed in simple, brief sentences. Some ideas by the very nature of their difficulty require more complex sentences. And frequently a longer sentence is necessary to show relationships between ideas.

Omitting Nonessential Words

Nonessential words weaken emphasis in a sentence by thoughtlessly repeating an idea or throwing in "deadwood" to fill up space. Note the improved effectiveness in the following sentences when unnecessary words are omitted.

WORDY	The train arrives at 2:30 P.M. in the afternoon.
REVISED	The train arrives at 2:30 P.M.
WORDY	He was inspired by the beautiful character of his surroundings
REVISED	He was inspired by his surroundings.
WORDY	In the event that a rain comes up, close the windows.
REVISED	If it rains, close the windows.

Eliminating unnecessary words makes writing more exact, more easily understood, and more economical. Often, care in revision weeds out the clutter of deadwood and needless repetition.

Using Simple Words and Direct Word Patterns

An often-told story illustrates quite well the value of simple, unpretentious words stated directly to convey a message.

Generally, use simple words instead of polysyllabic words. Avoid giving too many details in needless modifiers. When necessary, show causal relationships or tie closely related ideas together by writing longer sentences.

A Case for Simple Words

A plumber who had found that hydrochloric acid was good for cleaning out pipes wrote a government agency about his discovery. The plumber received this reply: "The efficacy of hydrochloric acid is indisputable, but the corrosive residue is incompatible with metallic permanence." The plumber responded that he was glad his discovery was helpful.

After several more garbled and misunderstood communications from the agency, the plumber finally received this clearly stated response: "Don't use hydrochloric acid. It eats the hell out of pipes." Much effort and time could have been saved if the agency had used this or similar wording in the *first* response.

WORDY AND OBSCURE	Feathered bipeds of similar plumage will live gregariously.
SIMPLE AND DIRECT	Birds of a feather flock together.
WORDY AND OBSCURE	Verbal contact with Mr. Jones regarding the attached notifications of promotion has elicited the attached representations intimating that he prefers to decline the assignment.
SIMPLE AND DIRECT	Mr. Jones does not want the job.
WORDY AND OBSCURE	Believing that the newer model air conditioning unit would be more effective in cooling the study area, I am of the opinion that it would be advisable for the community library to purchase a newer model air conditioning unit.
SIMPLE AND DIRECT	The community library should buy a new air conditioning unit.
IMPLIED RELATIONSHIP	The area was without rain for ten weeks. The corn stalks turned yellow and died.
STATED RELATIONSHIP	Because the area was without rain for ten weeks, the corn stalks turned yellow and died.

Keep sentences within a maximum of four or five units of information. Look at the following two examples.

Unit 1   Unit 2
The dimensions on a blueprint are called scale dimensions.

Unit 1   Unit 2   Unit 3
Because the corners of the nut may be rounded or damaged and because the wrench

Unit 4   Unit 5   Unit 6   Unit 7
may slip off the nut and cause an accident, choose a wrench with size and type suited

to the nut.

Notice that the second example becomes difficult to read and comprehend because of the many units of information.

Write positively worded statements for easier reading and comprehension. To understand negative wording, you must read the negative statement, mentally make it positive, and then change it back to negative, therefore taking a longer time to comprehend the statement.

NEGATIVE	If sales do not increase, the trustees will not vote to build new laboratories.
POSITIVE	The trustees will vote to build new laboratories only if sales increase.
NEGATIVE	The fact that I was not present caused controversy.
POSITIVE	My absence caused controversy.

The positively worded statement is more direct; it can be read and understood more quickly.

Avoid overloading your writing with nominalizations. Business and technical writers sometimes compose wordy, flowery sentences because they express action by using nouns formed from verbs (nominalization). In such sentences the action is buried in the abstract noun; the verb is weak and generalized. Usually such sentences have many prepositional phrases.

ACTION BURIED	All employees have participated in an evaluation of their performance with the goal of improvement of productivity.
ACTION CLEAR	All employees have evaluated their performances to improve productivity.
ACTION BURIED	The product's failure to sell was the result of ineffective advertising.
ACTION CLEAR	The product sold poorly because advertising was ineffective.
ACTION BURIED	There is an education requirement for all entry level personnel.
ACTION CLEAR	Entry level personnel must meet an education requirement.

> Apply the basic rule: Write sentences clearly and exactly to convey specific meaning to the intended audience.

Combining Sentence Elements

Many sentences are complete and unified yet ineffective because they lack conciseness. Parts of sentences, or even entire sentences, often may be reduced or combined.

REDUCING SEVERAL WORDS TO ONE WORD	the registrar *of the college* the *college* registrar
REDUCING A CLAUSE TO A PHRASE OR TO A COMPOUND WORD	a house *that is shaped like a cube* a house *shaped like a cube* a *cube-shaped* house
REDUCING A COMPOUND SENTENCE	Mendel planted peas for experimental purposes, and from the peas he began to work out the universal laws of heredity.
TO A COMPLEX SENTENCE	As Mendel experimented with peas, he began to work out the laws of heredity.
OR TO A SIMPLE SENTENCE	Mendel, experimenting with peas, began to work out the laws of heredity.
COMBINING TWO SHORT SENTENCES	Many headaches are caused by emotional tension. Stress also causes a number of headaches.
INTO ONE SENTENCE	Many headaches are caused by emotional tension and stress.

Coordination and *subordination* are techniques used to combine ideas and to show the relationship between ideas. Showing the relationship between ideas helps the reader comprehend information more quickly and more accurately. The ideas in sentences may be combined to make meaning clearer by adding a word, usually a subordinate conjunction or a coordinate conjunction, to show the relationship between the ideas.

From the two sentences

1. The company did not hire him.
2. He was not qualified.

a single sentence

The company did not hire him because he was not qualified.

makes the relationship of ideas clearer with the addition of the subordinate conjunction "because." The second sentence is changed to an adverb clause, "because he was not qualified," telling why the company did not hire him.

The two sentences

1. John applied for the job.
2. He did not get the job.

might be stated more clearly in

John applied for the job, but he did not get it.

The addition of the coordinate conjunction "but" indicates that the idea following is in contrast to the idea preceding it.

Study the following examples.

1. Magnetic lines of force can pass through any material.
2. They pass more readily through magnetic materials.
3. Some magnetic materials are iron, cobalt, and nickel.

The ideas in these three sentences might be combined into a single sentence.

Magnetic lines of force can pass through any material, but they pass more readily through magnetic materials such as iron, cobalt, and nickel.

The addition of the coordinate conjunction "but" indicates the contrasting relationship between the two main ideas; sentence 3 has been reduced to "such as iron, cobalt, and nickel." The three sentences might also be combined as follows:

Although magnetic lines of force can pass through any material, they pass more readily through magnetic materials: iron, cobalt, and nickel.

Using coordination and subordination, you can eliminate short, choppy sentences. The following five sentences

1. The missions's most important decision came.
2. It was early on December 24.
3. Apollo was approaching the moon.
4. Should the spacecraft simply circle the moon and head back toward Earth?
5. Should it fire the Service Propulsion System engine and place the craft in orbit?

might be combined:

1. As Apollo was approaching the moon early on December 24, the mission's most important decision came.
2. Should the spacecraft simply circle the moon and head back toward Earth, or should it fire the Service Propulsion System engine and place the craft in orbit?

Combining sentences gives a flow of thought as well as clarifies relationships between ideas.

Use coordination and subordination to emphasize details; important details appear in independent clauses and less important details appear in dependent clauses and phrases.

Almost any group of ideas can be combined in several ways. Choose the arrangement of ideas that best fits in with preceding and following sentences. More importantly, arrange ideas to make important ideas stand out.

Consider the following group of sentences:

1. Carmen Diaz is the employee of the year.
2. She has been with the company only one year.
3. She has had no special training for the job she performs.
4. She is highly regarded by her colleagues.

The four sentences might be combined in several ways.

1. Although Carmen Diaz, the employee of the year, has been with the company only one year and has had no special training for the job she performs, she is highly regarded by her colleagues.
2. Carmen Diaz is the employee of the year although she has been with the company only one year and has had no special training for the job she performs; she is highly regarded by her colleagues.
3. Although Carmen Diaz, highly regarded by her colleagues, is the employee of the year, she has been with the company only one year and has had no special training for the job she performs.

Each sentence contains the same information; through coordination and subordination, however, different information is emphasized.

The way you arrange information, as well as what you say, affects meaning.

Active and Passive Voice Verbs

For your audience and purpose, choose whichever voice will most effectively say what you want to say.

- Generally use active voice verbs. Active voice verbs are effectively because the reader knows immediately the subject of the discussion; you mention first *who* or *what* is doing something.

  The machinist values the rule depth gauge.
  Roentgen won the Nobel Prize for his discovery of X rays.

- Use passive voice verbs when the *who* or *what* is not as significant as the action or the result and when the *who* or *what* is unknown, preferably unnamed, or relatively insignificant.

  The lathe *is broken* again. (More emphatic than "Someone has broken the lathe again.")
  The blood sugar test *was made* yesterday.
  The transplant operation *was performed* by an outstanding heart surgeon.
  Light *is provided* for the architects by windows on the south wall.

General Principles in Technical Communication

1. Technical communication is functional. Technical communication serves a need. Typically, that need is to communicate specific information to someone who requests or requires that information.
2. Technical communication is concerned with audience. The technical communicator asks: Who will read or hear this material? Why? What kind of details will the reader or hearer need? What approach to the material and which order in presenting the material are most appropriate for the reader's or hearer's purpose?
3. Technical communication involves choices in format. Purpose and intended audience largely determine whether the communicator uses a memorandum, informal report, computer printout, letter, conference telephone call, formal report, printed form, or some other format for communicating the needed information.
4. In technical communication, the organization is readily apparent. The introductory section includes a statement of purpose or a forecasting statement indicating the major aspects of the topic to be covered. In written communication, headings and subheadings signal movement from one aspect to another.
5. In technical writing, page layout and document design are an integral part of the communication. Layout is concerned with such matters as spacing (paragraphing, ample white space), use of headings, and the placement of material on the page for optimum readability. Design includes such overall considerations as format, sizes and kinds of typefaces, color, and such composition components as balance, unity, and emphasis.
6. In technical communication, visuals are important. Drawings, charts, graphs, tables, photographs, and the like are often as important as the written or spoken material. Visuals clarify verbal explanation, focus attention, and add interest.
7. In technical communication, accurate, precise terminology is essential. Technical communication uses exact, specific, concrete words.
8. In technical communication, conventional standards of grammar, usage, spelling, punctuation, and so on are observed. Since the emphasis in technical communication is on conveying specific information—for a specific purpose, to a specific audience, in an appropriate format—avoid anything that detracts from that emphasis.
9. Technical communication requires the ability to think critically, objectively, thoroughly, and creatively. Effective technical communication is based on logical thinking applied to problem solving. The communicator must solve such problems as audience analysis and adaptation, choice of format, decisions as to organization and page layout and document design, creation of appropriate visuals, and formulating a communication so clear that the recipient knows exactly what the communicator means.

Applications in Technical Communication: Getting Started

Individual and Collaborative Activities

0.1   Complete the following statement:

I want to be a _____ (give your occupational goal, such as nurse, computer programmer, drafter, electronic technician).

Write a brief report on the kinds of communication skills you will need on the job. Consult as resources such books as the *Occupational Outlook Handbook, The Dictionary of Occupational Titles,* or *Encyclopedia of Careers;* or computer software such as *Discovery;* or interview someone currently employed in a similar job.

0.2   In magazines or newspapers, find two different ads for the same general product, ads that are designed to appeal to two different audiences. In a brief written analysis, compare the two ads. Identify the audiences targeted by the ads and discuss the techniques and visuals for appeal to the audiences.

0.3   In small groups of three or four, discuss the following general topics. How could these topics be developed to meet the needs of different audiences? Name at least two possible audiences for each topic. How would the content differ for the two audiences? Why? What visuals would be helpful for each audience? Why?

backpacks	fog	hypodermic needles
hurricanes	handcuffs	thermometers
plastic garbage bags	blueprints	automobiles
eyeglasses	floppy disks	cameras

After your small group discusses each topic, share your ideas with the entire class.

0.4   A. Revise the following material to eliminate wordiness and lack of conciseness.

It has come to my attention that it is time for us to consider replacing some of the computers that we have been using for the past several years. Of course I realize that some of the computers are still quite useful. I wonder if it would be a good idea to move these old computers into the secretarial area. I think most of the existing sofftware can be used with these computers and therefore no new software would need to be purchased. The secretaries should be happy about this idea since they will not have to learn new software, don't you think? Now, first and foremost, how many new computers do you think we should buy? Second of all, what in your opinion would be the best brand of computers to purchase? Should the new computers have added memory and faster response? Third, and finally, how many additional projects do you think we can expect to complete on an annual basis with the additional computers?

B. After revising the material above to eliminate wordiness and lack of conciseness, write the message, using effective page layout and document design.

0.5   Each of the following groups of sentences contains related information stated in simple sentences. Combine this information into one sentence or, if you think necessary, two or more sentences. Be careful to punctuate each sentence in an acceptable way.

Group A Sentences

1. a. All drawings are projections
   b. Two main types of projections are perspective and parallel projection.
2. a. Some lathes are built with two lead screws.
   b. One is used for thread cutting.
   c. The other is used for general lathe-turning operations.
3. a. Btu is the abbreviaitoin for British thermal unit.
   b. It is a unit of heat equal to 252 calories.
   c. It is the quantity of heat required to raise the temperature of one pound of water from 62°F to 63°F.
4. a. Machining is shaping by the removal of excess metal.
   b. It includes many operations.
   c. Some of these operations are grinding, sawing, drilling, and milling.
5. a. Fusion welding involves three processes.
   b. One of these processes is casting.
   c. Another process is heat treating.
   d. A third process is dilution.
6. a. Management is a social process.
   b. It is a process because it comprises a series of actions that lead to the accomplishment of objectives.
   c. It is a social process because these actions are principally concerned with relations between people.
7. a. Drafting students need special equipment.
   b. They need a drawing board.
   c. They need a T square.
   d. They need a set of drawing tools.
8. a. The operating tasks each company must perform are affected by many factors.
   b. One factor is changes in market.
   c. Another factor is new technology.
   d. A third factor is shifts in competition.
   e. A final factor is government regulations.
9. a. One type of mass production is manufacturing parts in large quantities by specially designed machine tools.
   b. Machine tools have helped to change our economy from principally agricultural to industrial.
10. a. The vernier height gauge is a measuring tool.
    b. It is used for measuring heights.
    c. It is used for laying out height dimensions for work resting on a surface plate.
    d. It is used for locating centers for holes that are to be drilled or bored.

Group B Sentences ② Sentence

1. a. In interior decoration straight lines are considered intellectual, not emotional.
   b. They are considered classic, not romantic.
   c. They are sometimes considered severe and masculine.
2. a. The three basic circuits are series, parallel, and series-parallel.
   b. They are distinguished by the way the electrical equipment is connected.
   c. Their names describe their connection.
3. a. Water from newly constructed wells will normally have a very high bacterial count.
   b. This count will not "settle down" for a matter of weeks or even months unless the supply is treated.
   c. This treatment reduces the bacterial population that found its way into the water during construction operations.
4. a. The Edison storage battery is the only nickel-iron-alkaline type on the American market.
   b. It was developed by Thomas A. Edison.
   c. It is built in many sizes and for many uses.
   d. It was put on the market in 1908.
5. a. Tool designers are specialists.
   b. Duties of the tool designer are to design drill jigs, reaming and tapping jigs, etc.
   c. Another duty is to choose correct cutting tool material.
   d. Sometimes she must establish cutting speeds, feeds, and depths of cut.
6. a. The engineer's scale is often referred to by other terms.
   b. It is called the decimal scale because the scale is graduated in the decimal system.
   c. It is called the civil engineer's scale because it was originally used in civil engineering.
   d. It is called the chain scale because it was derived from the surveyor's chain of 100 links.
7. a. Micrometer calipers are measuring instruments.
   b. They are designed to use the decimal divisions of the inch.
   c. They are used to take very precise measurements.
   d. They are made in different styles and sizes.
8. a. One essential part of an air damper box is a vane.
   b. The vane is made of a thin sheet of lightweight alloy.
   c. It is stiffened by ribs stamped into it.
   d. It is stiffened by the bending of its edges.
   e. Its edges conform to the surfaces of the damping chamber.
9. a. Soil may contain unusually high numbers of *Clostridium tetani*.
   b. This is the etiological agent in lockjaw.
   c. Soil that has been fertilized with manure, particularly horse manure, has high numbers of this agent.
10. a. Every time a manager delegates work to a subordinate, three actions are either expressed or implied.
    b. She assigns duties.
    c. She grants authority.
    d. She creates an obligation.

Group C Sentences   ② | Sentence

1. a. Landscape architects employ texture as a valuable tool.
    b. Repetition of dominant plant texture unifies a plan.
    c. Contrast of texture at corners and focal points gives emphasis.
2. a. There are three basic factors in an electrical circuit.
    b. These are pressure, current, and resistance.
    c. To be usable, these factors must be defined in terms of units that can be handled in measurements.
3. a. Any material employed to grow bacteria is called a culture medium.
    b. This medium, to be satisfactory, must have the proper moisture content.
    c. It must contain readily available food materials.
    d. It must have the correct acid-base balance.
4. a. Metallurgy is the science of metals.
    b. Metallurgy has two divisions: extractive and physical.
    c. Extractive metallurgy is the study of extracting metals from their ores.
    d. Physical metallurgy is the study of the properties of metals.
5. a. Of all the microbiologists, none was so accurate, as Leeuwenhoek.
    b. ~~None was so~~ completely honest. And As practical
    c. ~~None had such common sense.~~
    d. ~~He~~ died in 1723.   Who
    e. ~~He was 91.~~   At the age of
    f. Microbiology ~~went~~ into a dormant stage for almost 150 years.   Then sending
6. a. In ammeter shunts, the shunt is usually made up of a manganin strip.
    b. This strip is brazed in fairly heavy copper blocks.
    c. These heavy copper blocks serve two purposes.
    d. They carry heat away from the manganin strip.
    e. They keep all parts of each copper block at nearly the same potential.
7. a. Gaspard Mongé advanced the theory of projection by introducing two planes of projection at right angles.
    b. He was French.
    c. He was a mathematician.
    d. This development provides the basis of descriptive geometry.
8. a. Flat surfaces are measured with common tools.
    b. The steel rule and the depth gauge are common tools.
    c. Round work is measured by feel with nonprecision tools.
    d. The spring caliper and the firm-joint caliper are nonprecision tools.
    e. They have contact points or surfaces.
9. a. Ohm's law states that the current in an electrical circuit varies directly as the voltage and inversely as the resistance.
    b. This law was developed by George Simon Ohm.
    c. He was born in Germany.
    d. He was born in 1787.
    e. He was a physicist.
10. a. Vitruvius wrote a treatise on architecture.
    b. In the treatise he referred to projection drawings for structures.
    c. He wrote the treatise in 30 B.C.
    d. The theory of projections was well developed by Italian architects.
    e. The Italian architects were Brunelleschi, Alberti, and others.
    f. They developed the theory in the early part of the fifteenth century.

Using Part II Selected Readings

0.6   Read the poem "A Technical Writer Considers His Latest Love Object," by Chick Wallace, page 548.

A. Under "Questions on Content," page 549, answer Number 1.
B. Under "Suggestions for Response and Reaction," page 549, complete Number 5.

0.7   Read the excerpt from "Non-Medical Lasers: Everyday Life in a New Light," by Rebecca D. Williams, pages 564–569. Under "Suggestions for Response and Reaction," page 569, do Number 6.

0.8   Read *Journeying Through the Cosmos,* by Carl Sagan, pages 570–575. Describe Sagan's style of writing. In what ways is his style of writing different from stereotypical scientific writing?

0.9   Read "The Active Document: Making Pages Smarter," by David D. Weinberger, pages 579–584.

A. Evaluate the use of headings and visuals in the article.
B. For the boxed information, "How Active Documents Work," add headings to the second, third, and fourth paragraphs.

0.10   Read "Self-Motivation: Ten Techniques for Activating or Freeing the Spirit," by Richard L. Weaver II, pages 589–595.

A. What can you assume about the content of the speech from reading only the title and the subtitle?
B. In what ways is this speech appropriate for Weaver's audience, college students? To what extent is it appropriate for others?

0.11   Examine other selected readings for qualities discussed in "Getting Started," as directed.

Part I
Forms of Communication

Chapter 1

Instructions and Process: Explaining a Procedure

Outcomes

Upon completing this chapter, you should be able to

- Define instructions
- Present instructions for an intended audience
- Use visuals in giving instructions
- Plan the page layout and document design for a set of instructions
- Explain the relationship between planning and giving instructions
- Give instructions in writing and orally
- Define process explanation
- Explain the difference between giving instructions and explaining a process
- Give a process explanation directed to a lay audience
- Give a process explanation directed to a specialized audience
- Use visuals in explaining a process
- Give process explanations in writing and orally

Introduction

You have been a giver and receiver of instructions practically from the beginning of your life. As a child, you were told how to drink from a cup, how to tie your shoes, how to tell time, and so on. As you matured, you became involved with more complex instructions: how to parallel park an automobile, how to throw a block in football, how to tune an electric guitar, how to stock grocery shelves. Since entering college, you have been confronted with even more complex and confusing instructions: how to register as an incoming freshman, how to write an effective report, how to get along with a roommate, how to spend money wisely, how to study.

Since all aspects of life are affected by instructions, every person needs to be able to give and to follow instructions. Clear, accurate, complete instructions frequently save the reader or listener time, help do a job faster and more satisfactorily, or help get better service from a product. Being able to give and to follow instructions is essential for any person on the job. Certainly, in order to advance to supervisory positions, employees must be able to give intelligent, specific, accurate instructions; and they must be able to follow the instructions of others.

Persons on the job, particularly in supervisory positions, often must give instructions to others: how to use a computer system, how to fill out an accident report, how to process an application for a car registration, how to give an antibiotic injection to a patient—many how-to's.

The purpose here is to help you learn how to give how-to's—instructions—that are clear, accurate, and complete.

Intended Audience and Purpose

Giving instructions that can be followed successfully requires thinking and careful planning. Two of the first considerations are (1) who is the intended audience, that is, who will be hearing or reading the instructions, and (2) what is the purpose of the instructions, typically to show the listener or reader how to carry out an operation or procedure. The intended audience determines decisions about such things as level of complexity, word choice and sentence structure, and page layout and document design. An explanation of how to operate the latest model X-ray machine would differ for an experienced X-ray technician and for a student just being introduced to X-ray equipment. Instructions on how to freeze corn would differ for a food specialist at General Mills, for a homemaker who has frozen other vegetables but not corn, and for a seventh-grade student beginning a nutrition course.

Writing or speaking on a level that the intended audience will understand determines the kind and extent of details presented and the manner in which they are presented. Therefore, you must know who will be reading or hearing the instructions and why. You must be fully aware whether a particular background, specialized knowledge, or certain skills are needed in order to understand the instructions.

Consider the following examples. For what readers were these instructions written? In what specific ways has the intended audience determined the kind and extent of details presented and the manner in which they are presented?

The first example is an excerpt from a user's manual for a GE television.* The excerpt shows the user how to erase and add channels in channel memory. The user could be almost anyone; thus the instructions are simply stated. The instructions are arranged in an effective two-column design with the first column explaining how to erase channels and the second explaining how to add channels. The numbered, written instructions include visuals to show which buttons to press and what the screen shows at each stage of the instructions.

*Reprinted by permission from General Electric Company.

Erasing/Adding Channels in Channel Memory

Erasing Channels

1. **To erase a channel,** first press the *CHANNEL* ∧/∨ button until the channel number you want appears on the screen.

2. Then repeatedly press the *SETUP* button until the Channel Memory display appears on the screen.

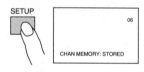

3. Then push the "−" button to erase that channel from memory.

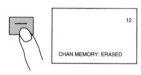

4. Wait until Channel Memory display disappears from the screen. You can then check channels in memory by pressing the *CHANNEL* ∧ or ∨ button. If you want to erase another channel, repeat steps 1 through 3.

Adding Channels

1. Repeatedly press the *SETUP* button until the Channel Memory display appears on the screen.

2. **To add a channel,** press the *CHANNEL* ∧ or ∨ button until the channel number you want to add appears on the screen.

3. Then push the "+" button to add that channel to the memory.

4. If you want to add another channel, repeat steps 2 and 3.

Note: The Channel Memory display will automatically disappear from the screen in a few seconds after you have finished erasing or adding channels.

Compare the two following examples. Both are excerpts from instructions written in simple, easy-to-understand language for a lay audience. The first, a page on maintenance from instructions for using a Sear's shop vacuum cleaner,* gives instructions for filter maintenance. A numbered list of steps and key phrases (TO REMOVE, TO CLEAN, and TO REPLACE) and the visuals guide the user to maintain the vacuum cleaner.

MAINTENANCE

FILTER MAINTENANCE

FILTER SHOULD BE CLEANED OFTEN TO
MAINTAIN PEAK PERFORMANCE

STEP 1. TO REMOVE – Remove wing nut on bottom of filter (See Fig. 3B). Remove filter by lifting off of wire guide (See Fig. 3A).

STEP 2. TO CLEAN –

A. Dry dust applications – shake or tap filter to remove dust.

B. Wet applications – if dirt will not come off with shaking or tapping, run water through filter. Make sure filter is dry before replacing.

NOTE: Don't operate unit with torn, ripped or excessively worn filter. Small tears may be repaired with tape. Dirt will discharge from unit if holes are not repaired.

STEP 3. TO REPLACE – Threaded part of wire guide goes through hole in bottom plate of filter. Make sure filter is seated on lid. Wing nut holds filter in place (See Fig. 3B). Run wing nut down till it is finger tight.

NOTE: If filter is not seated, then a leak is created that will cause dust to discharge from your vac.

DO NOT STORE UNIT WITH A WET FILTER. RUN UNIT WITH FILTER IN PLACE TO AID DRYING.

WARNING: If you use a wet filter on dry applications or store a wet filter, you may damage the filter.

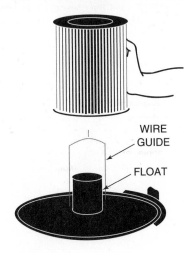

WIRE
GUIDE

FLOAT

Fig. 3A

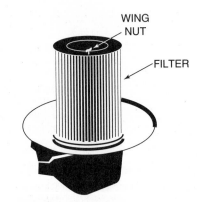

WING
NUT

FILTER

Fig. 3B

Immediately replace a torn, ripped or excessively worn filter.

*Reprinted by permission from Sears, Roebuck and Company.

The second example clearly explains through a sequence of steps how to maintain a cordless power wrench.* Notice this excerpt includes no visuals, for the content does not lend itself to visual presentation. Notice the excerpt includes caution and warning statements; each statement is placed within the sequence of steps where the reader needs the information.

Maintenance

Preventive maintenance of your Cordless Power Wrench is easy. There are just 3 steps to follow:

1. Use a clean cloth to wipe oil, grease and other deteriorating substances from the tool and contacts.

⚠ CAUTION Certain cleaning agents and solvents can damage plastic parts. Some of these are: gasoline, carbon tetrachloride, chlorinated cleaning solvents, ammonia, and household detergents which contain ammonia. Avoiding use of these and other types of cleaning agents will minimize the probability of damage.

2. Store the tool where it is cool and dry. Avoid areas where the temperature will drop to 0° F or exceed +120° F.

3. About once per year (depending on use), return your tool and charger to the nearest Skil Factory Service Center, or Authorized Skil Service Station, or other competent repair service for the following:
- Parts cleaned and inspected
- Relubricated with fresh lubricant
- Electrical system tested
- All repairs

⚠ WARNING All repairs, electrical or mechanical, should be attempted only by trained repairmen. Contact the nearest Skil Factory Service Center, or Authorized Service Station or other competent repair center. Use only identical replacement parts; any other may create a hazard.

⚠ WARNING The use of any other accessories not specified in this manual may create a hazard.

While both examples give instructions for maintenance, the content and the page layout and the document design vary. The writer and the designer made decisions about content, page layout, and document design to meet the needs of the audience and to accomplish the purpose for the instructions. As a writer, remember that you have choices. You make those choices after you answer key questions. Who is my audience? What is my purpose for writing? What do I want the audience to do or be able to do when they finish reading? How can I arrange information on pages to make audience use of the material as easy as possible?

The final example, on handwashing, is designed for a specialized audience. For a lay audience, such would likely include only four or five steps. For a nurse, however, the instructions are more complex. The following set of instructions,* includes 17 steps. Each step includes a rationale to justify the step.

*Reprinted by permission from SKIL Corporation, a subsidiary of Emerson Electric Company, Chicago, IL.

◆ ◆ ◆

PROCEDURE 23-1
Handwashing*

Steps	Rationale
1. Use sink with warm running water, soap or disinfectant, and paper towels.	Running water facilitates removal of organisms. Paper towels are easy to discard.
2. Push wristwatch and long uniform sleeves above wrists. Remove jewelry, except plain band, from fingers and arms.	Provides complete access to fingers, hands, and wrists. Jewelry may harbor microorganisms.
3. Keep fingernails short and filed.	Dirt and secretions that lodge under fingernails contain microorganisms. Long fingernails can scratch skin.
4. Inspect surface of hands and fingers for breaks or cuts in skin and cuticles. Report such lesions when caring for highly susceptible clients.	Open cuts or wounds can harbor high concentrations of microorganisms. Such lesions may serve as portals of exit, increasing client's exposure to infection, or as portals of entry, increasing nurse's risk of acquiring infection.
5. Stand in front of sink, keeping hands and uniform away from sink surface. (If hands touch sink during handwashing, repeat.) Use sink where it is comfortable to reach faucet.	Inside of sink is contaminated area. Reaching over sink increases risk of touching edge, which is contaminated.
6. Turn on water. Press pedals with foot to regulate flow and temperature. Push knee pedals laterally to control flow and temperature. Turn on hand-operated faucets by covering faucet with a paper towel.	When hands contact faucet, they are contaminated. Organisms spread easily from hands to faucet.
7. Avoid splashing against uniform.	Microorganisms travel and grow in moisture.
8. Regulate flow of water so that temperature is warm.	Warm water is more comfortable. Hot water opens pores of the skin, causing irritation.
9. Wet hands and lower arms thoroughly under running water. Keep hands and forearms lower than elbows during washing.	Hands are the most contaminated parts to be washed. Water flows from least to most contaminated area, rinsing microorganisms into sink.
10. Apply 1 ml of regular or 3 ml of antiseptic liquid soap to hands, lathering thoroughly. If bar soap is used, hold it throughout lathering. Soap granules and leaflet preparations may be used.	Bar soap should be rinsed before returning it to soap dish. Soap dish that allows water to drain keeps soap firm. Jellylike soap permits growth of microorganisms
11. Wash hands using plenty of lather and friction for at least 10 to 15 seconds. Interlace fingers and rub palms and back of hands with circular motion at least 5 times each.	Soap cleanses by emulsifying fat and oil and lowering surface tension. Friction and rubbing mechanically loosen and remove dirt and transient bacteria. Interlacing fingers and thumbs ensures that all surfaces are cleansed.
12. If areas underlying fingernails are soiled, clean them with fingernails of other hand and additional soap or clean orangewood stick. Do not tear or cut skin under or around nail.	Mechanical removal of dirt and sediment under nails reduces microorganisms on hands.
13. Rinse hands and wrists thoroughly, keeping hands down and elbows up [illustration omitted].	Rinsing mechanically washes away dirt and microorganisms.
14. Repeat steps 10 through 12 but extend period of washing to 1, 2, and 3 minutes.	Greater the likelihood of hands being contaminated, greater the need for thorough handwashing.
15. Dry hands thoroughly from fingers to wrists and forearms.	Drying from cleanest (fingertips) to least clean (forearms) area avoids contamination. Drying hands prevents chapping and roughened skin.
16. Discard paper towel in proper receptacle.	Prevents transfer of microorganisms.
17. Turn off water with foot and knee pedals. To turn off hand faucet, use clean, dry paper towel.	Wet towel and hands allow transfer of pathogens by capillary action.

*Reprinted by permission from *Basic Nursing: Theory and Practice,* by Patricia A. Potter and Anne G. Perry, Mosby-Year Book, Inc., 1991.

Visuals

To a person reading or listening to a set of instructions, visuals can be very helpful in understanding the procedure. For a reader, visuals such as drawings, photographs, and diagrams can help. For a person listening to instructions, visuals such as real objects, models, demonstrations, slides, and overhead-projected transparencies are particularly helpful.

Visuals allow the audience to *see* what the writer or speaker is explaining. Visuals portray the procedure as a whole as well as particular aspects of the procedure.

Note the use of visuals (line drawings) in the instructions for performing the Heimlich Maneuver below. The drawings are as important as the verbal explanation in conveying what one should do to help the victim.

THE HEIMLICH MANEUVER

FOOD-CHOKING
<u>What to Look For.</u> Victim cannot speak or breathe; turns blue; collapses.

<u>To perform the Heimlich Maneuver when the victim is standing or sitting.</u>

1. Stand behind the victim and wrap your arms around his waist.
2. Place the thumb side of your fist against the victim's abdomen, slightly above the navel and below the rib cage.
3. Grasp your fist with the other hand and press your fist into the victim's abdomen with a quick upward thrust. Repeat as often as necessary.
4. If the victim is sitting, stand behind the victim's chair and perform the maneuver in the same manner.
5. After the food is dislodged, have the victim seen by a doctor.

Position of hands

<u>To perform the Heimlich Maneuver when the victim has collapsed and cannot be lifted.</u>

1. Lay the victim on his back.
2. Face the victim and kneel astride his hips.
3. With one hand on top of the other, place the heel of your bottom hand on the abdomen slightly above the navel and below the rib cage.
4. Press into the victim's abdomen with a quick upward thrust. Repeat as often as necessary.
5. Should the victim vomit, quickly place him on his side and wipe out his mouth to prevent aspiration (drawing of vomit into the throat).
6. After the food is dislodged, have the victim seen by a doctor.

Direction of hand and fist movement

Position of hands

NOTE: If you start to choke when alone and help is not available, an attempt should be made to self-administer this maneuver.

Direction of hand movement

Source: New York City Department of Health

Look back at the visuals for erasing/adding channels for a GE television and for filter maintenance on pages 29–30; then turn forward to pages 40–41 and look at the visuals with the instructions for sharpening a ruling pen.

The visuals in all of these examples of instructions

- Illustrate the verbal explanation
- Clarify the verbal explanation
- Minimize the verbal explanation
- Add interest

For a detailed discussion of visuals, see Chapter 11 Visuals.

Page Layout and Document Design

Especially in instructions, page layout and document design are of prime importance. Arrange material on the page to achieve maximum ease in reading and comprehension. Plan an overall design that provides ready access to needed information, clearly presented in a convenient format.

Review the examples on pages 7, 36, and 40–41. In each example, various techniques such as line drawings, numbered steps, white space, double columns, headings, and boldface contribute to an overall layout and design that are appealing and effective for the intended audience and purpose.

Consider also the opening page of two versions of a set of instructions for determining the resistance of a carbon resistor (see "Example of Negative Page Layout and Design" and "Example of Positive Page Layout and Design," pages 35–36). Both pages give the same information. But which is easier to read and comprehend? Page 36. The page has plenty of *white space;* the material is uncluttered and so placed on the page that the eye easily and quickly sees the relationship of one part of the material to another. The *headings* provide key words for each section. Emphasis is achieved in several ways: *uppercase letters* for the headings, the *box* to set off material, the *closed bullets* for a list in which sequence is not important, the *numbers* for the major steps (where sequence is very important), and the *letters of the alphabet* (for substeps in a sequence).

The main difference in the two presentations of the excerpt from instructions for determining the resistance of a carbon resistor is the use of page layout techniques on page 36 that make reading and comprehension easier:

- Uppercase letters
- Boldface
- White space
- Lists
- Closed bullets
- Headings
- Table
- Box
- Numbered steps

For a detailed discussion, see Page Layout and Document Design, pages 6–11.

Page Layout Makes a Difference: First Draft of Page 1 of a Set of Instructions

35

Chapter 1
Instructions
and Process:
Explaining
a Procedure

HOW TO DETERMINE THE RESISTANCE OF A CARBON RESISTOR

An essential component of most electronic circuits is the carbon resistor. It opposes current flow, thereby providing a usable load into which a circuit may operate. For this load to be effective, the value of the carbon resistor must have the correct resistance. A carbon resistor looks like a firecracker with a fuse on both ends. Resistors normally come in sizes of 1/8 watt, 1/4 watt, 1/2 watt, 1 watt, and 2 watts. The 1/8-watt resistor is about 1/4 inch long and 1/16 inch in diameter, and the 2-watt resistor is about 1 inch long and 5/16 inch in diameter. To facilitate reading the value of a resistor, the electronics technician will need to know the color code or have access to the color code: black = 0, brown = 1, red = 2, orange = 3, yellow = 4, green = 5, blue = 6, violet = 7, gray = 8, white = 9; for tolerances, silver = ±10% and gold = ±5%. The Greek letter Ω (omega) is the universal symbol for *ohms,* the unit in which resistance is measured. Each resistor has four colored bands; the first three indicate the amount of resistance, and the fourth indicates the amount of tolerance.

The first step in determining the resistance is to look at Band 1. Determine its color, note the digit value (0 to 9) of the color from the color code, and record the digit.

Page Layout Makes a Difference: Final Draft of Page 1 of a Set of Instructions

uppercase
letters,
in color

ample
margins

double
spacing

list format

closed bullets

headings,
underlined

table

box

numbers for
major steps
and for
substeps

HOW TO DETERMINE THE RESISTANCE OF A CARBON RESISTOR

An essential component of most electronic circuits is the carbon resistor. It opposes current flow, thereby providing a usable load into which a circuit may operate. For this load to be effective, the value of the carbon resistor must have the correct resistance.

A carbon resistor looks like a firecracker with a fuse on both ends. Resistors normally come in these sizes:

- 1/8 watt (about 1/4 in. long, 1/16 in. diameter)
- 1/4 watt
- 1/2 watt
- 1 watt
- 2 watts (about 1 in. long, 5/16 in. diameter)

Color Code

To facilitate reading the value of a resistor, the electronics technician will need to know the color code or have access to a code chart, as shown in Table 1.

TABLE 1   Color Code for Carbon Resistors[a]

Color	Value	Color	Value
black	0 Ω	green	5 Ω
brown	1	blue	6
red	2	violet	7
orange	3	gray	8
yellow	4	white	9

[a]Tolerances: silver = ± 10%
        gold = ±5%

> **NOTE:** The Greek letter Ω (omega) is the universal symbol for *ohms*, the unit in which resistance is measured.

Each resistor has four colored bands; the first three indicate the amount of resistance, and the fourth indicates the amount of tolerance.

Steps in Determining the Resistance

1. Look at Band 1.
   1.1 Determine its color.
   1.2 Note the digit value (0 to 9) of the color from the color code chart.
   1.3 Record the digit.

Planning and Giving Instructions

Planning instructions requires several steps. As with any planning, you need first to answer two questions:

1. Why am I giving the presentation? (purpose)
2. For whom is the presentation intended? (audience)

Although these two items do not necessarily appear in the actual presentation, purpose and audience directly affect the way you select and present details in a set of instructions, as illustrated by the examples of instructions on pages 29–33, and 40–41.

Review the General Principles in Giving Instructions on page 44. These principles summarize the major points of the chapter. Now review the Procedure for Giving Instructions on pages 42–43 on the form and the content for instructions. In the content outline, note that the suggestions under roman numeral I list kinds of information you may include in the introductory material; suggestions under II identify content for the body; and suggestions under III give choices for the closing material. Then decide if including visuals will make the instructions clearer to the intended audience. Remember within the text of the instructions to refer to visuals and label and title visuals (see pages 463–465).

The Plan Sheet (see the illustration on pages 38–39) includes parts of the content outline that are of major importance in planning the instructions. It serves as a guide to help you plan what you will write or tell; it helps you to clarify your thinking on a topic and helps you to select and organize details.

You may or may not use every bit of information on the Plan Sheet. Nevertheless, fill it in completely and fully so that once you begin to prepare a preliminary draft you will have thought through all major details needed in the presentation.

As you work, refer frequently to the Procedure for Giving Instructions and to the Plan Sheet. These two guides help you plan and develop an acceptable presentation.

Following is a sample presentation of instructions. Included are the filled-in Plan Sheet and the final paper. Marginal notes to outline the development of the instructions have been added to the final paper. The preliminary draft showing peer editing suggestions and subsequent revisions is shown in Revising and Editing, page 601.

PLAN SHEET FOR GIVING INSTRUCTIONS

Analysis of Situation Requiring Instructions

What is the procedure that I will explain?
how to sharpen a ruling pen

Who is my intended audience?
a classmate in drafting

How will the intended audience use the instructions? For what purpose?
to sharpen a ruling pen

Will the final presentation of the instructions be written or oral?
written

Importance or Usefulness of the Instructions

Used regularly, a pen becomes dull. Anyone using one needs to know how to sharpen it. A sharp pen is required for neat, clean lines.

Organizing Content

Equipment and materials necessary:
3- or 4-inch sharpening stone (hard Arkansas knife piece soaked in oil several days), crocus cloth (if desired)

Terms to be defined or explained:
ruling pen, nibs

Overview of the procedure (list as command verbs):
1. *Close the ribs.*
2. *Hold the sharpening stone correctly in the left hand.*
3. *Hold the ruling pen correctly in the right hand.*
4. *Round the nibs.*
5. *Sharpen the nibs.*
6. *Polish the nibs, if desired.*
7. *Test the pen.*

Major steps with identifying information:

What to do	How to do	Why, if applicable
1. *Close the ribs.*	*Turn the adjustment screw (located above nibs) to the right until the nibs touch.*	*So the nibs can be sharpened to exactly the same shape and length*

Major steps, with identifying information, cont.:

What to do	How to do	Why, if applicable
2. Hold the sharpening stone correctly in the left hand.	Lay the stone across the left palm; grasp the stone with the thumb and fingers.	To give the best control of the stone
3. Hold the ruling pen correctly in the right hand.	Pick up the pen with the thumb and index finger. Pick the pen up like a drawing crayon. Rest the other three fingers on the pen handle.	To avoid an injured hand; to avoid a ruined pen
4. Round the nibs.	Stroke the pen back and forth on the stone. Start with the pen at a 30-degree angle and follow through to past a 90-degree angle. Usually 4 to 6 strokes.	Both the nibs must be the same length and shape so the tips of both will touch the paper.
5. Sharpen the nibs.	Open the blades slightly; turn the adjustment screw left. Sharpen the outside of each blade. Hold the stone and the pen as in steps 2 and 3. Hold the pen at a slight angle. Rub the pen back and forth, 6 to 8 times.	Rounding the nibs will leave them dull.
6. Polish the nibs. (optional)	Rub the crocus cloth lightly over the nibs.	To remove any rough places
7. Test the pen.	Add a small drop of ink between the nibs. Draw a line with a T square.	To determine if the pen is properly sharpened

Precautions to emphasize (crucial steps, possible difficulties, dangers, places where errors are likely to occur):
Failure to hold the pen and the stone correctly may cause an injured hand or a damaged pen.

Types and Subject Matter of Visuals I Will Include

drawings to show:
1. hand holding pen for sharpening
2. nibs to show correct shape

The Title I May Use

How a Person Can Sharpen a Ruling Pen

Sources of Information

class lecture in drafting, textbook in drafting, personal experience

Clearly focused title

Identification of
ruling pen

Definition of nib

Reason for
sharpening pen

Equipment and
materials needed

Listing of steps

Each step numbered
and explained in
detail: what to do,
how to do it, and
why (if needed)

Comparison with a
familiar action

Reference to visual

Visual to illustrate
step 3

Visual given a
number and caption

HOW TO SHARPEN A RULING PEN

A major instrument in making a mechanical drawing is the ruling pen, which is used to ink in straight lines and noncircular curves with a T square, a triangle, a curve, or a straightedge as a guide. The shape of the nibs (blades resembling tweezers) is the most important aspect of the pen. The nibs must be rounded (elliptical) to create an ink space between the nibs. To assure neat, clean lines, the ruling pen must be kept sharp and in good condition. It must be sharpened from time to time after extensive use because the nibs wear down.

Equipment and Materials

- 3- or 4-inch sharpening stone (preferably a hard Arkansas knife piece that has been soaked in oil for several days)
- crocus cloth (optional)

Overview of Procedure

Sharpening a ruling pen involves seven steps: close the nibs, hold the sharpening stone in the left hand, hold the ruling pen in the right hand, round the nibs, sharpen the nibs, polish the nibs (optional), and test the pen.

Steps

1. Close the nibs. Turn the adjustment screw, located above the nibs, to the right until the nibs touch. The nibs can then be sharpened to exactly the same shape and the same length
2. Hold the sharpening stone in the left hand in a usable position. With the stone lying across the palm of the left hand, grasp it with the thumb and fingers to give the best control of the stone.
3. Hold the ruling pen in the right hand. Pick up the ruling pen with the thumb and index finger of the right hand as if it were a drawing crayon. The other three fingers should rest lightly on the pen handle. See Figure 1.

CAUTION! Failure to hold the sharpening stone and ruling pen correctly may result in an injured hand and a ruined ruling pen.

Figure 1. Holding the pen for sharpening.

4. Round the nibs. To round (actually to make elliptical) the nibs, stroke the pen back and forth on the stone, starting with the pen at a 30-degree angle to the stone and following through to past a 90-degree angle as the line across the stone moves forward. Usually four to six strokes are needed.

   Be sure that both nibs are the same length and shape so that the tips of both nibs touch the paper as the pen is used. See Figure 2.

Reference to visual

Visual to illustrate important aspect of step 4

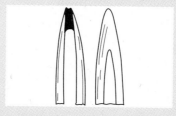

Visual given a number and caption

Figure 2. Correct shape of pen nibs.

5. When the nibs are satisfactorily rounded, they will be left dull; sharpen the nibs. Open the blades slightly by turning the adjustment screw gently to the left. Sharpen only the outside of each blade one at a time.

   To sharpen, hold the stone and the ruling pen in the hands as in rounding the nibs (see step 3); the pen should be at a slight angle with the stone. Rub the pen back and forth, six to eight times, with a rocking, pendulum motion to restore the original shape.

Optional step

6. If desired, polish the nibs with a crocus cloth by rubbing the cloth lightly over the nibs to remove any rough places.

7. Test the ruling pen after sharpening it. Add a small drop of ink between the nibs. Draw a line along the edge of a T square placed on a piece of paper. If the pen is properly sharpened, the pen will draw sharp, clean lines.

Oral Presentation

The content of oral instructions is very similar to that of written instructions; however, the delivery is different. In speaking, you don't have to be concerned with such things as spelling and punctuation, and you have the advantage of a visible audience with whom you can interact.

Your delivery of oral instructions will be improved if you follow these suggestions, whether addressing one or two persons or a large group.

1. *Look at your audience.* Use eye contact with your audience. Avoid continuously looking at your notes, the floor, the ceiling, or a particular individual.

2. *Speak, don't read, to the audience.* Avoid memorizing or reading your presentation. Rather, have it carefully outlined on a note card.

3. *Repeat particularly significant points.* Remember that the audience is *listening*. Repeat main points and summarize frequently. The hearer cannot reread material as the reader can; therefore repetition is essential.

4. *Speak clearly, distinctly, and understandably.* Follow the natural pitches and stresses of the spoken language, and use acceptable pronunciation and grammar. Speak on a language level appropriate for the audience and the subject matter.

5. *Use bodily movements and gestures naturally.* Put some zest in your expression; be alive; show enthusiasm for your subject. Stand in an easy, natural position, with your weight distributed evenly on both feet. Let your movements be natural and well timed.

6. *Involve the audience, if practical.* Invite questions or ask someone to carry out the procedure or some part of it.

7. *Use visuals where needed.* Whenever visuals make the explanation clearer, use them. The visuals should be large enough for everyone in the audience to see clearly.

For a detailed discussion of oral presentations, see Chapter 10 Oral Communication.

Procedure for Giving Instructions

Form

The steps in a set of instructions are usually presented as a list, with numbers indicating the major steps. For nonsequential lists (such as a list of needed materials or equipment) and for indicating emphasis, symbols such as these may be used: the open or closed bullet ○ ●, the open or closed square or box □ ■, the dash —, and the asterisk *. See page 8.

In giving instructions, use the second person pronoun *you* (usually understood) and imperative or action verbs.

Example
First, (*you* understood) *unplug* (imperative verb) the appliance. Then with a Phillips screwdriver (*you* understood), carefully *remove* (imperative verb) the back plate.

Content

In giving instructions, be sure to tell your audience not only *what* to do but also *how* to do each activity, if there is any doubt that the audience might not know how. You may also add *why* the activity is necessary.

Example
Remove the cap (tells *what*). Use the fingers to turn it counterclockwise until it can be removed (tells *how*). You can then insert the spout of the gasoline can to fill the tank with gasoline (tells *why*).

Use good judgment in selecting details. For example, if you are explaining how to pump gas, it is not necessary to say you must have a vehicle and a gas tank.

The following suggestions may also be helpful. Generally include *a, an,* and *the,* unless space is limited. Notice the two examples above which include *a, an,* and *the.* The second example, for instance, is much easier to read than if it were written: "Remove cap. Use fingers to turn counterclockwise until it can be removed. You can then insert spout of gasoline can to fill tank with gasoline." Notice the example of the instructions for the maintenance of a Sears shop vacuum cleaner, page 30. The *a, an,* and *the* have been omitted in most of the sentences because the instructions were printed in a limited space.

When organizing material for giving instructions, whether in writing or orally, divide the presentation into two or three parts, as follows.

I. The identification of the subject is usually brief, depending on the complexity of the operation.
   A. State the operation to be explained.
   B. If applicable, give the purpose and significance of the instructions and indicate who uses them, when, where, and why.
   C. State any needed preparations, skills, equipment, or materials.

II. The development of the steps is the main part of the presentation and thus will be the lengthiest section.
   A. Clearly list each step and develop each step fully with sufficient detail to tell what to do and how to do it. Also tell why a step is necessary, if applicable.
   B. In a more complex operation, subdivide each major step, if necessary.
   C. Explain in clear detail exactly what is to be done to complete the operation.
   D. Sometimes it is helpful to emphasize particularly important points and to caution the reader where mistakes are most likely to be made.
   E. Plan the page layout and document design of the instructions, using headings, visuals, and other elements as needed.

III. The closing may be
   A. The completion of the discussion of the last step.
   B. A summary of the main steps, especially in complex or lengthy instructions.
   C. A comment on the significance of the operation.
   D. Mention of other methods by which the operation is performed.

Length of Presentation

The length of a presentation is determined by the complexity of the operation, the degree of knowledge of the audience, and the purpose of the presentation. In explaining how to do a simple operation, such as taking a temperature or changing a printer ribbon, the entire presentation may be very brief.

General Principles in Giving Instructions

1. *Knowledge of the subject matter is essential.* To give instructions that can be followed, you must be knowledgeable about the subject. If need be, consult sources—knowledgeable people, textbooks, reference works in the library—to gain further understanding and information about the subject.
2. *The intended audience influences what information is to be presented and how it is to be presented.* Consider the audience's degree of knowledge and understanding of the subject. Avoid talking down to the audience as well as overestimating the audience's knowledge or skills.
3. *Effective instructions are accurate and complete.* The information must be correct. Instructions should cover the subject adequately, with no step or essential information omitted.
4. *Visuals help to clarify instructions.* Whenever instructions can be made clearer by the use of such visuals as maps, diagrams, graphs, pictures, drawings, slides, demonstrations, and real objects, use them.
5. *Instructions require careful page layout and document design.* An integral part of effective instructions is consideration of how material is placed on the page and of how the document as a whole is presented.
6. *Conciseness and directness contribute to effective communication.* An explanation that is stated in the simplest language with the fewest words is usually the clearest. If the instructions call for terms unfamiliar to the audience or for familiar terms with specialized meanings, explain them.
7. *Instructions that can be followed have no unexplained gaps in the procedure or vagueness about what to do next.* Well stated instructions do not require the audience to make inferences, to make decisions, or to ask, "What does this mean?" or "What do I do next?"

Applications in Giving Instructions

Individual and Collaborative Activities

1.1 Make a list of three persons to interview about their jobs. After interviewing each of the three, make a list of examples showing how they use instructions on the job. From this experience, what can you speculate about the importance of instructions in your own future work?

1.2 Find and attach to your paper a set of instructions that a manufacturer included with a product.

  a. Evaluate in a paragraph the page layout and document design of the instructions (see pages 6–11 and 34–36).
  b. Evaluate in a paragraph the clarity and completeness of the instructions by applying the General Principles in Giving Instructions above.

In completing applications 1.3 and 1.4,

   a. Fill in the Plan Sheet on pages 47–48, or one like it.
   b. Write a preliminary draft.
   c. Go over the draft in a peer editing group.
   d. Revise. See Part III Revising and Editing and the checklists for revising, front and back endpapers.
   e. Write the instructions.

1.3   Assume that you are a foreman or a supervisor with a new employee on the job. Explain in writing to the employee how to carry out some simple operation.

1.4   From the following list, choose a topic for an assignment on giving instructions or choose a topic from your own experience. Consult whatever sources necessary for information. Review A Note About Plagiarism, page 403.

How to:
1. Copy a disk on a personal computer
2. Take a patient's blood pressure
3. Water ski
4. Install a room air conditioner
5. Sharpen a drill bit
6. Set up a partnership
7. Start an airplane engine
8. Produce a business letter—individualized to several people—on a word processor
9. Cure an animal hide
10. Cut a mat for a picture
11. Operate a piece of heavy-duty equipment
12. Administer an intramuscular injection
13. Open a checking account
14. Change a tire on a hill
15. Plant a garden
16. Change oil in a car
17. Read a micrometer or a dial caliper
18. Hang wallpaper
19. Set out a shrub
20. Replace a capacitor in a television set
21. Sterilize an instrument
22. Use a compass, architect's scale, divider, or French curve
23. Make a tack weld
24. Operate an office machine
25. Prepare a laboratory specimen for shipment
26. Develop black and white film
27. Fingerprint a suspect

28. Reposition copy in a computer-generated document
29. Repair (or rebind) a book
30. Topic of your choosing

Oral Instructions

1.5   As directed by your instructor, adapt for oral presentations the instructions you prepared in the applications above. Ask your classmates to evaluate your speech by filling in the Evaluation of Oral Presentations on page 459, or one like it.

Using Part II Selected Readings

1.6   Read the article "Clear Only If Known," by Edgar Dale, pages 496–498.

a. List Dale's reasons why people have difficulty in giving and receiving instructions.
b. Under "Suggestions for Response and Reaction," page 499, complete Number 1.
c. Under "Suggestions for Response and Reaction," page 499, complete Number 4.

1.7   Read the transcript of the speech "Self-Motivation: Ten Techniques for Activating or Freeing the Spirit," by Richard L. Weaver II, pages 589–595. List Weaver's ten suggestions for dealing with stress.

PLAN SHEET FOR GIVING INSTRUCTIONS

Analysis of Situation Requiring Instructions

What is the procedure that I will explain?

Who is my intended audience?

How will the intended audience use the instructions? For what purpose?

Will the final presentation of the instructions be written or oral?

Importance or Usefulness of the Instructions

Organizing Content

Equipment and materials necessary:

Terms to be defined or explained:

Overview of the procedure (list as command verbs):

Major steps with identifying information:

What to do	How to do	Why, if applicable

Major steps, with identifying information, cont.:

What to do   **How to do**   **Why, if applicable**

Precautions to emphasize (crucial steps, possible difficulties, dangers, places
where errors are likely to occur):

Types and Subject Matter of Visuals I Will Include

The Title I May Use

Sources of Information

Explanation of a Process

Explaining a process—giving the specifics of how something is done—is similar to giving instructions. There are, however, two basic differences: a difference in the purpose and a difference in the procedure of presentation. The purpose in giving instructions is to enable an individual to perform a particular operation. The giver of the instructions expects the reader or hearer to *act*. In explaining a process, however, the purpose is to explain a method, operation, or sequence of events so that the intended audience will comprehend the concept of what is done or what happens. The presenter expects the reader or hearer to *understand* what is done or what happens.

Processes are carried out by people, by machines, or by nature.

Processes carried out by people	**Processes carried out by machines**	**Processes carried out by nature**
how steel is made from iron	how a mainspring clock works	how sound waves are transmitted
how glass is made	how a copier works	how rust is formed
how a computer programmer designs a new program	how a gasoline engine operates	how food is digested
		how mastitis is spread

With instructions, you use commands (for example, *unplug, remove, insert*) so that the audience can act. You present and explain each step the reader or listener must carry out to perform the operation. In the explanation of a process, you emphasize the sequence of actions that is the procedure for an operation. The audience is unlikely to perform the operation. Notice below a comparison of possible procedures for giving instructions and for explaining a process.

Giving instructions	**Explaining a process**
(*you*) Make the cut as flush to the tree trunk as possible.	When oil is found, the raw material is channeled from the well.
1. Imperative mood (orders or commands)	1. Indicative moon (statements of fact)
2. Active voice (subject does the action)—"Make"	2. Passive (subject is acted upon)—"is found," "is channeled"
3. Second person (person spoken to is subject). Subject is understood *you*.	3. Third person (thing spoken about is subject)—"oil," "raw material"

Through reading or hearing an explanation of a process, the audience develops an *understanding* of the operation or sequence of events. It would, in fact, be impossible to perform some processes, for example those carried out by nature. It is possible, however, to understand what happens as these natural processes occur; for instance, you can understand how sound is transmitted or how a tornado develops. Further, you can understand how bricks are made, but you would probably never make a brick.

Intended Audience and Purpose

Just as in giving instructions, in giving an explanation of a process you must aim your presentation at a particular audience for a particular purpose. You then write your explanation so that the intended audience will clearly understand. The intended audience and purpose determine the kind and extent of details you include in the explanation and the manner in which you present the details. Audiences may be grouped into two broad categories.

- lay audiences (cross sections of people)
- specialized audiences (people with assumed shared knowledge in a particular profession)

Process Explanation for a Lay Audience

The lay audience requires a fairly inclusive description. The writer should assume that this audience has little, if any, of the particular background, knowledge, or skill necessary to understand a description of a technical process. Therefore, you need to explain the process as clearly and simply as possible, defining terms (see Chapter 3 Definition) that might have special meaning.

The following description of how the heart works is directed to a lay audience. Note the use of drawings as well as simplified language in describing the process.

How the Heart Works

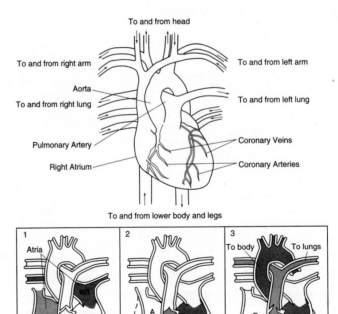

1. During the heart's relaxed stage (diastole), oxygen-depleted blood from body flows into right atrium, oxygenated blood from lungs into left. 2. Natural pacemaker, or sinoatrial node (A), fires electrical impulse and atria contract. Valves open and blood fills ventricles. 3. In pumping stage (systole), the electrical signal, relayed through atrioventricular node (B), causes ventricles to contract, forcing oxygen-poor blood to lungs, oxygen-rich blood to body.

An article in *PC Magazine** includes the following graphic to make clear to the lay audience how the Internet delivers mail.

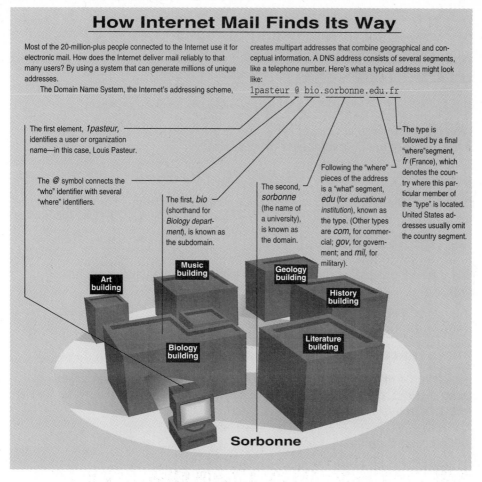

How Internet Mail Finds Its Way

Most of the 20-million-plus people connected to the Internet use it for electronic mail. How does the Internet deliver mail reliably to that many users? By using a system that can generate millions of unique addresses.

The Domain Name System, the Internet's addressing scheme, creates multipart addresses that combine geographical and conceptual information. A DNS address consists of several segments, like a telephone number. Here's what a typical address might look like:

`1pasteur @ bio.sorbonne.edu.fr`

The first element, *1pasteur*, identifies a user or organization name—in this case, Louis Pasteur.

The @ symbol connects the "who" identifier with several "where" identifiers.

The first, *bio* (shorthand for *Biology department*), is known as the subdomain.

The second, *sorbonne* (the name of a university), is known as the domain.

Following the "where" pieces of the address is a "what" segment, *edu* (for *educational institution*), known as the type. (Other types are *com*, for commercial; *gov*, for government; and *mil*, for military).

The type is followed by a final "where" segment, *fr* (France), which denotes the country where this particular member of the "type" is located. United States addresses usually omit the country segment.

Art building

Music building

Geology building

History building

Biology building

Literature building

Sorbonne

The following explanation of how Alexander Fleming discovered penicillin illustrates an oral presentation for a lay audience. A filled-in Plan Sheet and the explanation with directions for oral presentation are included.

In the explanation of how penicillin was discovered, the speaker's purpose is to give the lay listener a general view of the process. The speaker, sufficiently knowledgeable about the subject to give a general explanation of it, is successful in accomplishing that purpose. The speaker remembers the audience throughout the presentation by using a vocabulary and level of speaking that a lay audience can easily understand.

The speaker is accurate and complete in the explanation, and the information is correct. All the necessary stages in the development of penicillin are given. Certainly, there is much more information about the discovery of penicillin that the speaker *could* have included, such as related research or experimental subjects. Or the speaker could have elaborated further about each step in the growth process. The important thing, however, is that the speaker included all *essential* information.

*Used by permission from *PC Magazine,* 11 Oct. 1994, page 121.

PLAN SHEET FOR EXPLAINING A PROCESS

Analysis of Situation Requiring Explanation of a Process

What is the process that I will explain?
how Alexander Fleming discovered penicillin

Who is my intended audience?
a lay audience of adult volunteers at a hospital

How will the intended audience use the process explanation? For what purpose?
for general understanding of how a major antibiotic was accidentally discovered

Will the final presentation of the process explanation be written or oral?
oral

Importance or Significance of the Process

Penicillin is a miracle drug. For 50 years it has eased pain, facilitated healing, cured diseases, even saved lives—for millions of people around the world. Very effective in treating syphilis.

Major Steps and Explanation of Each Step

1. *Experimentation with antibacterial agents*
   —*study of septic wounds in WWI*
   —*experiments with Staphylococcus aureus*
   —*exposed culture plate contaminated with mold*

2. *Analysis of the mold*
   —*failure of bacteria to grow in the area of the mold*
   —*tiny drops of liquid on the mold*

3. *Identification of the mold*
   —*genus Penicillium*
   —*liquid on the mold = penicillin = chemical that killed the bacteria*

4. *Execution of the essential compound*
   —*Flory and Chain research*
   —*treatment of infections in humans (early 1941)*
   —*mass production of penicillin in United States (1941)*

Major Steps and Explanation of Each Step, cont.:

Important Points to Receive Special Emphasis

Although Fleming's discovery of penicillin was accidental, he had spent many years in research and experimentation in trying to develop antibacterial agents from nature.

Types and Subject Matter of Visuals I Will Include

- *a flowchart to outline major stages*
- *real objects*
    - *–petri dish*
    - *–moldy block of cheese and moldy slice of bread*
    - *–prescription vial of penicillin tablets and a prescription tube of penicillin*
    - *–ointment*

The Title I May Use

How Alexander Fleming Discovered Penicillin

Sources of Information

Encyclopaedia Britannica, Great Discoverers in Medicine

ORAL PRESENTATION
How Alexander Fleming Discovered Penicillin

Many fortunate discoveries have resulted from accidents. We call such discoveries serendipitous discoveries. One such serendipitous discovery happened in September 1928 resulting in a miracle antibiotic drug. The drug was first widely used in World War II among the Allied military personnel in the treatment of infections, wounds, and disease. This miracle drug proved also to be very effective against the organism that causes the venereal disease syphilis. Today the drug is commonly available in tablets, capsules, ointments, and drips as well as by injection. The miracle drug is penicillin. For over fifty years this drug has helped to ease pain, facilitate healing, even save lives—for millions of people around the world.

How did an accidental discovery by Sir Alexander Fleming lead to the development of penicillin? These events occurred in four stages: experimentation with antibacterial agents, analysis of the mold, identification of the mold, and extraction of the essential compound.

The speaker's use of visuals:

Hold up a prescription vial of penicillin tablets and a prescription tube of penicillin ointment.

Stage 1   Experimentation with Antibacterial Agents

The young Alexander Fleming, M.D., from Scotland, specializing in bacteriology at the University of London, was interested in developing antibacterial agents from nature. Having made a special study of septic wounds while serving as a doctor in World War I, he continued his research and experimentation. While Fleming was working on a series of experiments with *Staphylococcus aureus,* a fortunate accident occurred. Fleming had left exposed to the open air some culture plates (also called petri dishes) of staphylococci.

A few days later when he returned to the laboratory, he discovered that one of the bacteriological plates had become contaminated with a mold from the air, possibly from an open window.

As each major stage is listed, write on the chalkboard:

Experimentation with antibacterial agents
↓
Analysis of the mold
↓
Identification of the mold
↓
Extraction of the essential compound

Fleming probably thought to himself, "Fiddlesticks! I must learn to be more careful." He started to discard the plate. But the trained eye of the scientist noticed something different about the mold.

Stage 2   Analysis of the Mold

Fleming carefully analyzed the conditions surrounding the mold. He noticed that the bacteria had failed to grow in the area of the mold. Thus, he reasoned, some unknown substance in the mold had killed the bacteria. Too, he noticed tiny drops of liquid on the surface of the mold.

Though Fleming had tainted his staphylococci experiment, he had stumbled upon a discovery that would revolutionize the treatment of wounds and illnesses in human beings.

Stage 3   Identification of the Mold

Hold up a petri dish.

Fleming continued to study and to analyze mold. This mold, by the way, was green and was very similar to the ordinary fungus growth we see on cheese and bread when they get a little old.

Hold up a piece of moldy cheese and a slice of moldy bread.

Eventually Fleming identified the mold as belonging to the genus *Penicillium,* and he called the unknown substance penicillin. The tiny drops of liquid on the surface of the mold—the substance that Fleming called penicillin—was the chemical that had destroyed the neighboring bacteria.

Stage 4   Extraction of the Essential Compound

Some ten years after Fleming had made his accidental discovery of penicillin, two fellow scientists—Howard Walter Flory and Ernst Boris Chain—took up Fleming's work. They extracted the essential compound from the liquid in which penicillin grows.

Knowledge of the drug and experimentation beyond the laboratory quickly grew. By early 1941, amazing results were being obtained in the treatment of infections in human beings. And later in 1941, penicillin began to be mass-produced in the United States.

Conclusion

Alexander Fleming was given widespread recognition and honor for his discovery of penicillin. He was knighted—hence, *Sir* Alexander Fleming. And Fleming, with his two co-workers, was awarded in 1945 the highest award in his field—the Nobel Prize in medicine.

Process Explanation for a Specialized Audience

A specialized audience, as the term implies, has at least an interest in a particular subject and probably has the background, either from reading or from actual experience, to understand an explanation of a process related to that subject. For example, a person whose hobby is working on cars and reading about them would understand a relatively technical description of a fuel injection system. The specialized reader may even have a high degree of knowledge and skill. For example, sanitation engineers, fire chiefs, inhalation therapists, research analysts, programmers, or machinists would understand technical explanations related to their fields of specialization.

The paragraph below explains to nursing students the process of ear irrigation.* This specialized audience would understand the meaning of "cleanse a client's external auditory canal of excess cerumen or exudate from a lesion or inflamed area," "not pulling the auricle excessively," and "introducing fluid under pressure could rupture the tympanic membrane."

> **Ear Irrigations.** Using a small syringe and solution at body temperature, the nurse can cleanse a client's external auditory canal of excess cerumen or exudate from a lesion or inflamed area. The use of solution at body temperature prevents discomfort and severe dizziness. The nurse should avoid irritating the canal's sensitive skin lining by not pulling the auricle excessively, nor introducing the tip of the irrigating syringe into the canal. The auditory canal should never be occluded with the syringe tip because introducing fluid under pressure could rupture the tympanic membrane. A slow, gentle irrigation works best.

---

*Reprinted by permission from *Basic Nursing: Theory and Practice,* by Patricia A. Potter and Anne G. Perry, Mosby-Year Book, Inc., 1991.

If you have any doubt about the audience's level of knowledge, clarify any part of the explanation that you think the reader might have difficulty understanding. (See also Accurate Terminology, page 76.)

Process Explanation Combined with Other Forms

Process explanation may combine with other kinds of information. Look at the following description of a process written for a lay audience. The description explains what happens to a check from the time it appears at a bank to the time it is returned to the check writer as a canceled check in the monthly bank statement. Note the use of visuals: a flowchart and a copy of a check with identifying information. The explanation includes definition ("Helping speed the process is a nine-digit line of a machine-readable code, officially known as magnetic ink character recognition, located in the bottom left of each check.") and cause/effect ("If it's deposited in the morning, the processing begins that afternoon.")

> **Time Is Money In Checking**
>
> *If you've ever wondered what happens to a check after you write it and give it to a bank, here are the facts.*
>
> *By Lee Ragland*
> *Clarion-Ledger* Business Writer*
>
> Checking seems so easy. You write a check, the amount is deducted from your account and the check is returned in the monthly statement.
>
> What the customer doesn't see is what happens to the check behind the scenes. It is balanced and encoded with magnetic ink; sorted by a giant computer and sprayed with a sequence number; microfilmed, recorded on a deposit slip, and in some cases placed on a chartered flight out of town.
>
> All by nightfall.
>
> "I don't think the general public has any concept of what happens to a check or how it is processed," said Paul Carrubba, a Deposit Guaranty National Bank senior vice president and manager of bank operations in Jackson.
>
> Helping speed the process is a nine-digit line of a machine-readable code, officially known as magnetic ink character recognition, located in the bottom left of each check. The characters allow the documents to be read directly by high-speed equipment.
>
> Each number signifies something—Federal Reserve district, Federal Reserve bank or branch number, an institution's identification number and even a "check" number to show the document has been properly coded.
>
> "They are not just there for decoration on the check," Carrubba said. "Those numbers on the bottom have meaning."
>
> Here's a quick, basic look at the life of a check going through the Deposit Guaranty operations center:
>
> • You receive your paycheck, and deposit it in your account. If it's deposited in the morning, the processing begins that afternoon. That includes checks from all 100-plus DGNB branches statewide. Those deposited at faraway branches in Corinth, Southaven, Starkville and Tupelo are flown to Jackson daily.
>
>   A proof-machine operator, working on a 10-key board, balances the deposits and adds a magnetic-ink coding that shows the check's amount.

*Reprinted by permission from the 17 Dec. 1989 issue of *The Clarion-Ledger.*

The check has plenty of company. At Deposit Guaranty, each proofer averages 1,800 checks per hour—that's one every two seconds. There are 32 machines humming in the room from about 2 p.m. until 9:30 p.m.

- The check then is sent to a sorter. Deposit Guaranty uses three giant IBM 3890 document processors. Each can sort about 2,000 checks per minute. The check shoots through the 20-foot-long piece of equipment at 20 miles per hour. The machine sorts the check based on where the check is going, and then drops it into one of 24 slots.

During this if-you-blink-you-miss-it procedure, the machine reads the magnetic ink character recognition, sprays a sequence number on the check, places the bank's endorsement on the back and microfilms the document.

When the run is completed, the bank creates a cash letter—a deposit ticket—with all the checks listed.

"I think you can equate it to a product," Carrubba said. "The raw material is the check."

"The check comes to us and we are adding to that check and capturing information off of it and storing the information. So it is a process."

The bank processes an average of about 450,000 checks a day.

The check then heads to one of three destinations.

- In-House (On Us)—The name says it all. If your employer and you use the same bank, the check stays at that operations center.

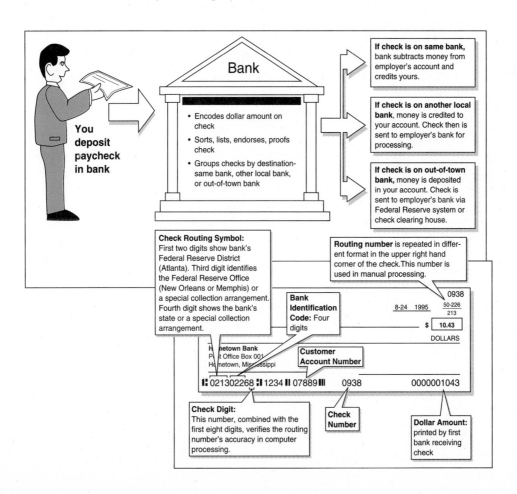

The checks are sent to the bookkeeping department, where a number of functions are performed. The checks are held until the customer's statement is mailed out.

- Local—If your employer uses a different financial institution in the same city, typically your bank would send the check to the employer's bank. There the employer's bank would capture all the information from the bottom of the check—customer's account number, check number and check amount—before placing it in the monthly statement.
- Transit Items—If the employer uses an out-of-town bank, the check is sent by various methods. Popular delivery methods are Federal Reserve Banks and clearinghouses—voluntary associations that banks join to help clear their checks.

Banks send their checks to such facilities in a hurry.

"Time is money in the banking business," Carrubba said. "We need to clear those items quickly so we can get use of those funds."

Deposit Guaranty puts checks on three charter flights a day—flying the checks to facilities in New Orleans, Memphis, and to Birmingham and on to Atlanta. The flights often are shared with other financial institutions.

The last flight departs for New Orleans at 9:30 p.m., to make the Federal Reserve's midnight deadline.

That is why the operation center's busiest times are from 2 p.m. to 9:30 p.m., as workers hurry to make sure the day's work makes the flight.

At the Federal Reserve branch in New Orleans, the checks will join others from parts of Mississippi, Alabama and Louisiana. On the average, 1.8 million go through the clearing center daily.

"Most are processed at night and sent out by 6 in the morning," said Federal Reserve spokesman Charlene Shucker.

Nationally, the Federal Reserve processed 18.1 billion checks last year.

The Fed and clearinghouses will present checks to the employer's banks the next day, then pay your bank the money.

If the checks doesn't clear because of insufficient funds, then the employer's bank contacts your bank, which deletes the amount from your account.

"What is so mind-boggling about this is not so much the volume of checks and how they have to be cleared, but the movement of that data," Carrubba said. "If you think of the millions of bits of information that have to be moved on any given night, that to me is mind-boggling."

The following example explains what happens when blood pressure is measured and why, is designed for a lay audience. To accomplish this purpose, the article combines words and graphics and such forms as definition and cause/effect, along with process explanation.

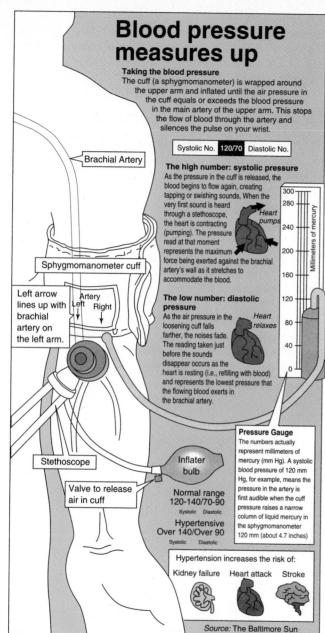

Blood pressure measures up

Taking the blood pressure
The cuff (a sphygmomanometer) is wrapped around the upper arm and inflated until the air pressure in the cuff equals or exceeds the blood pressure in the main artery of the upper arm. This stops the flow of blood through the artery and silences the pulse on your wrist.

Systolic No. **120/70** Diastolic No.

The high number: systolic pressure
As the pressure in the cuff is released, the blood begins to flow again, creating tapping or swishing sounds, When the very first sound is heard through a stethoscope, the heart is contracting (pumping). The pressure read at that moment represents the maximum force being exerted against the brachial artery's wall as it stretches to accommodate the blood.

Heart pumps

The low number: diastolic pressure
As the air pressure in the loosening cuff falls farther, the noises fade. The reading taken just before the sounds disappear occurs as the heart is resting (i.e., refilling with blood) and represents the lowest pressure that the flowing blood exerts in the brachial artery.

Heart relaxes

Millimeters of mercury — 300, 280, 240, 200, 160, 120, 80, 40, 0

Brachial Artery

Sphygmomanometer cuff

Left arrow lines up with brachial artery on the left arm.

Artery
Left Right

Stethoscope

Valve to release air in cuff

Inflater bulb

Normal range
120-140/70-90
Systolic Diastolic

Hypertensive
Over 140/Over 90
Systolic Diastolic

Pressure Gauge
The numbers actually represent millimeters of mercury (mm Hg). A systolic blood pressure of 120 mm Hg, for example, means the pressure in the artery is first audible when the cuff pressure raises a narrow column of liquid mercury in the sphygmomanometer 120 mm (about 4.7 inches)

Hypertension increases the risk of:

Kidney failure   Heart attack   Stroke

Source: The Baltimore Sun

What those numbers mean*

The Baltimore Sun

The cuff is tightened around your arm. The nurse or doctor presses a stethoscope against the inside of your elbow, listens intently, and then tells you your blood pressure is, say 120 over 70.

So what does that mean?

As the heart pumps blood through the body, the force of the blood against the walls of the arteries is your "blood pressure." Its measurement is written as two numbers, representing arterial pressure during two phases of heart activity. When the heart squeezes, blood squirts into the arteries, stretching their elastic walls. As it relaxes, the arterial walls spring back and the heart refills.

The squeezing action creates a high pressure, called the systolic pressure (denoted by the higher, first number). The lesser pressure generated when the heart relaxes is known as diastolic pressure (the lower, second number).

Occasional increases in blood pressure are not unusual. Strong emotion can cause pressure to go up, giving some people artificially high readings.

For example, a person's pressure could rise because he's nervous about being in a doctor's office.

Older people generally have higher blood pressures than younger people, probably because their blood vessels have become less elastic. If the blood pressure increases above normal and remains elevated, the result is called hypertension. Kidney disease, thyroid and adrenal gland disorders, obesity, smoking, excess sodium consumption, and lack of certain minerals in the diet can contribute to hypertension. In most cases, the exact cause of hypertension is unknown.

High blood pressure is a complex disorder that requires a doctor's care. If you have it, your physician may prescribe pressure-lowering drugs, or may advise you to lose weight, exercise, stop smoking, reduce alcohol consumption, cut down on sodium, increase your calcium intake or learn to meditate or do other relaxation or stress-reducing activities.

Harold L. Gater/The Clarion-Ledger

As you look carefully at the article on measuring blood pressure, you will notice the article also includes cause/effect analysis ("As the heart pumps blood through the body, the force of the blood against the walls of the arteries is your 'blood pressure.'") (see Chapter 5) and definition ("High blood pressure is a complex disorder that requires a doctor's care.") (see pages 109–118). Thus to accomplish the main purpose of the article, explaining the process of measuring blood pressure, the writer uses both cause-effect analysis and definition.

Look at the excerpt "Non-Medical Lasers: Everyday Life in a New Light," pages 564–569. The second major heading in the excerpt is "How a Laser Works," pages 565, 567. The writer then gives a simple explanation of the process as a part of an article that also defines *laser* and details the use of lasers in such areas as communication and information, industry and the military, arts and entertainment, and the future. Boxed information on page 566 explains how the Federal Drug Administration (FDA) monitors laser safety. Notice also the simplified diagram of a common laser on page 565. The writer of the article identifies the parts of the laser and includes a miniprocess explanation to show how a laser works.

The article "The Active Document: Making Pages Smarter," pages 579–584, explains how future documents will think and act for themselves (adding information, changing graphics, and even determining what readers may or may not see), and includes boxed data headed "How Active Documents Work," page 581.

Technical communication often combines various forms of communication. As you read the following student-written explanation of how a black and white ad to promote Hinds Community College is produced, notice the use of definition (see pages 109–118), partition (see pages 158–160), and cause/effect (see pages 177–183). The Plan Sheet used in preparing the process explanation is included.

PLAN SHEET FOR EXPLAINING A PROCESS

Analysis of Situation Requiring Explanation of a Process

What is the process that I will explain?
how a black and white ad for a college newspaper is produced

Who is my intended audience?
a lay reader of the newspaper

How will the intended audience use the process explanation? For what purpose?
to understand the basic stages of producing a black and white advertisement

Will the final presentation of the process explanation be written or oral?
a written explanation

Importance or Significance of the Process

Black and white advertisements appear in all kinds of newspapers and magazines.

Major Steps and Explanation of Each Step

1. Discussing ideas
  Answering questions
    What should the advertisement convey?
    What effect should the ad have?
    What information should the ad have?
  Making a decision about the type of ad and text to be used
    Simple ad
    Theme: "Your future's on the line."
    Other text: "Call 1-800 HINDS CC"
          College logo: H superimposed over a C
          College motto: "The College for All People"

Major Steps and Explanation of Each Step, cont.

2. *Producing camera-ready copy*
   Experimenting with design (fonts, sizes of text, and placement)
   Deciding on a visual (rising sun with telephone in the center)
   Perfecting the placement of the elements
   Getting approval from college personnel
   Sending the camera-ready copy to the printer
3. *Preparing the copy for printing*
   Shooting a line negative
   Masking the negative to a masking sheet
   Making a plate
4. *Printing the ad*
   Positioning the plate on a round plate cylinder
   Printing the page with ad and other pages for the newspaper edition

Important Points to Receive Special Emphasis

omit

Types and Subject Matter of Visuals I Will Include

- *camera-ready copy sample*
- *line negative sample*
- *simulation of a masking sheet*

The Title I May Use

How a Black and White Ad to Promote Hinds Community College Is Produced

Sources of Information

Interviews with layout artist and with printer.

HOW A BLACK AND WHITE AD TO PROMOTE
HINDS COMMUNITY COLLEGE IS PRODUCED

Hinds Community College uses a variety of advertisements to promote interest in the college. Recently when the college put into service an 800 number to call for information about the college, the Institutional Advancement Office and the Public Relations Office wanted to advertise the availability of this service. One plan was to announce the service through a black and white ad in *The Hindsonian,* the Hinds Community College newspaper. Once this decision was made, college personnel produced the ad.

Discussing Ideas

First, Institutional Advancement and Public Relations personnel met to discuss ideas for an ad. What should the ad convey? What effect should the ad have? What information should the ad include? After many ideas were discussed, the decision was made to use a simple advertisement with the message that at Hinds Community College a student can get the desired education for the future and can learn about the options available by calling the 800 number. The theme chosen was "Your future's on the line" (a play on the word "line" since the ad would also advertise the 1–800 telephone line). Other text selected was "Call 1-800 HINDS CC." The ad would also include the college logo (an H superimposed over a C) and the college motto, "The College for All People."

Producing Camera-Ready Copy

This information then went to the layout artist. Using a computer, she experimented with different fonts, sizes of text, and the placement of text on the page to achieve the best balance and effect. As she experimented, she decided to use a visual of a rising sun with a telephone in the center to suggest that persons need to make this wake-up call now. She printed a copy to show what the final ad might look like. Using this copy, she worked further with the placement of the different elements on the page, text and graphics, to achieve the best possible effect. She proofread the text and used a computer spellcheck to be sure all words were spelled correctly.

Satisfied with the placement of the elements on the page, the layout artist printed a copy of the planned advertisement and sent it to Institutional Advancement and Public Relations personnel. After their approval, the layout artist printed a final copy of the advertisement for the printer. This copy is called a mechanical or camera-ready copy. See Figure 1.

Figure 1. A mechanical of camera-ready copy for the ad

The markings in the margins are used by the printer to align the ad.

Preparing the Copy for Printing

The printer used the mechanical and a process camera to shoot a line negative showing all details of the advertisement. See Figure 2.

Figure 2. A line negative of the ad

The printer then masked (stripped) the negative to a masking sheet or goldenrod flat (named for the color of the paper) to position the copy so that when the plate was made, the copy would appear in the desired location on the printed page. The masking sheet, available in many different sizes, has markings to guide the printer in placing the material. Also masked to this sheet is all other copy that will appear on the same page with the ad. Figure 3 is an example of a masking sheet without the goldenrod color.

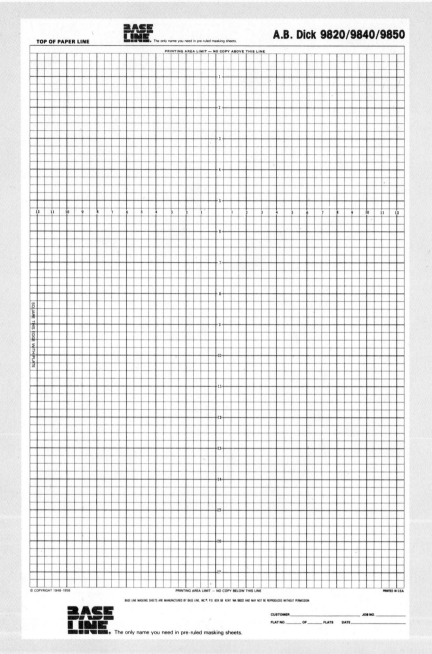

Figure 3. An example of a masking sheet without the goldenrod color

Using the masking sheet, the printer made a plate—a thin, flexible metal sheet from which impressions are transferred to paper during a printing operation.

Printing the Ad

The printer placed the plate on a round plate cylinder. This cylinder transfers the image to a rubber blanket cylinder which prints the image on paper as the paper passes between the blank cylinder and the impression cylinder. This process is called offset printing. Once the ad page and other pages of *The Hindsonian* were printed and collated, the campus newspaper was distributed to various campus locations.

Oral Presentation and Visuals

For detailed discussion of oral presentation and visuals, see Chapter 10 Oral Communication and Chapter 11 Visuals.

General Principles in Explaining a Process

1. *The purpose of an explanation of a process is for audience understanding, not audience action.*
2. *The intended audience influences the kind and extent of details included and the manner in which they are presented.* Consider the audience's degree of knowledge and understanding of the subject, and use an appropriate language level for the particular audience.
3. *Audiences may be grouped into two broad categories: lay audiences and specialized audiences.* Lay audiences are a mixture, a cross section of people. Specialized audiences have mutually shared professional knowledge.
4. *Accuracy and completeness are essential.* The process explanation should be accurate in its information, and it should adequately cover all necessary aspects of the process.
5. *Visuals can enhance a process explanation.* Visuals—such as flowcharts, diagrams, drawings, real objects, demonstrations—can clarify and emphasize a verbal explanation.

Procedure for Explaining a Process

FORM

In explaining a process you may arrange the steps or stages of the process in a numbered list or in paragraphs. Remember, the purpose of the explanation is to make the reader understand *what happens* during a process. The usual way to explain what happens is to use the third person, present tense, active or passive voice.

EXAMPLES

- An *officer* (third person) *fingerprints* (present tense, active voice) a suspect by . . .
- *Glass* (third person) *is made* (present tense, passive voice) by . . .

Or, the explanation might use *-ing* verb forms.

EXAMPLES

- *Peeling* (-*ing* verb form) the thin layers . . .
- The final stage is *removing* (-*ing* verb form) the seeds . . .

Remember to be consistent with person, tense, and voice in a single presentation. For example, if you select third person, present tense, active voice for the major steps or stages, use the same for all major steps or stages throughout the presentation. Do not needlessly shift to some other person, tense, or voice. (See pages 628, 630–631).

Content

When organizing an explanation of a process, divide the explanation into two or three parts, as shown below.

I. The identification of the subject may be brief, perhaps only two or three sentences, depending on the complexity of the process.
   A. State the process to be explained and identify or define it.
   B. If applicable, give the purpose and significance of the process.
   C. Briefly list the main steps or stages of the process, preferably in one sentence.
II. The development of the steps (or stages) is the main part of the presentation and thus will be the lengthiest section.
   A. The guiding statement is a list of the main steps given in the introduction.
   B. Take up each step in turn, developing it fully with sufficient detail.
   C. Subdivide major steps, as needed.
   D. Insert headings for at least the main steps.
   E. Use visuals whenever they will help to clarify, to explain, or to emphasize.
III. The closing is determined largely by the purpose of the presentation.
   A. If the purpose is simply to inform the audience of the specific procedure, the closing may be the completion of the last step, a summary, a comment on the significance of the process, or a mention of other methods for performing the process.
   B. If the presentation serves a specific purpose, such as evaluation of economy or practicality, the closing may be a recommendation.

Length of Presentation

The length of a presentation explaining a process is determined by the complexity of the process, the degree of knowledge of the audience, and the purpose of the presentation. In a simple process, such as how a cat laps milk or how a stapler works, the steps may be developed adequately in a single paragraph, with perhaps a minimum of two or three sentences for each step. In a more complex process, the steps may be listed and numbered, and the explanation of each step may require a paragraph or more. The closing is usually brief.

Applications in Explaining a Process

Individual and Collaborative Activities

1.8 Analyze the use of process explanation in the excerpt from "Non-Medical Lasers: Everyday Life in a New Light," the section headed "How a Laser Works," pages 565–567, and "The Active Document: Making Pages Smarter," the boxed data headed "How Active Documents Work," page 581. Explain how each of the two sections follows the "Procedure for Explaining a Process," pages 67–68.

Writing Explanations of Processes

In completing applications 1.9–1.12,

a. Fill in the Plan Sheet on pages 72–73, or one like it.
b. Write a preliminary draft.
c. Go over the draft in a peer editing group.
d. Revise. See Part III Revising and Editing and the checklists for revising, front and back endpapers.
e. Write the final explanation of a process.

1.9 Identify some event or activity that takes place at your college (students' notification of grades, library's getting a book ready to place on the shelf, disciplinary process), at your place of employment (how a new employee is hired, how payroll is prepared and distributed), or on your job. Write an explanation of the process to show a lay audience the stages from beginning to completion of the event or activity. Write the explanation for printing in a newspaper.

1.10 Choose one of the following processes carried out by people or another, similar topic of interest to you for explaining a process for a lay audience. Consult any sources needed. Review A Note About Plagiarism, page 403.

Processes carried out by people. How:
1. A synthetic heart is transplanted
2. A diamond is cut and polished
3. Ceramic tile, bricks, tires, sugar (or any other material) is (are) made
4. A person becomes a "star"
5. Community colleges and technical institutes are helping to alleviate the skilled labor shortage
6. A site is chosen for a business or industry
7. Flood prevention helps to eliminate soil erosion
8. Gold, silver, or coal is mined
9. An industrial plant works with community leaders
10. Contact lenses were developed
11. An alcoholic beverage is produced
12. A buyer selects merchandise for a retail outlet.
13. A television repairperson troubleshoots a television set
14. A college cafeteria dietician plans meals
15. A topic of your choosing

1.11 Choose one of the following processes carried out by machines or anther, similar topic for writing a process for a specialized audience. Consult any sources needed. Review A Note About Plagiarism, page 403.

Processes carried out by machines. How:
1. A telephone answering service works
2. A space satellite works
3. A tape recorder works
4. A photocopier (or similar machine) reproduces copies
5. A computer copies a disk
6. An automatic icemaker works
7. A jet (diesel or gas turbine) engine works
8. Air brakes work
9. A mechanical cotton picker (or any other piece of mechanical farm or industrial machinery) operates
10. A lathe (or any other motorized piece of equipment in a shop) works
11. An autoclave works
12. A printing press prints material
13. An electronic calculator displays numbers
14. A dental unit works
15. A topic of your choosing

1.12   Choose one of the following processes carried out by nature or another, similar topic for writing a process explanation for either a lay audience or a specialized audience. Consult any sources needed. Review A Note About Plagiarism, page 403.

Processes carried out by nature. How:
1. The human eye works
2. Oxidation occurs
3. Digestion occurs
4. Foods spoil
5. Aging affects one of the senses (such as seeing or hearing)
6. Gold, silver, granite, oil, peat (or any other substance) is formed
7. An acorn becomes an oak tree
8. Infants change during the first year of life
9. Microorganisms produce disease
10. Wood becomes petrified
11. A tadpole becomes a frog
12. Freshwater fish spawn
13. Sound is transmitted
14. An amoeba reproduces
15. A topic of your choosing

Oral Process Explanation

1.13   As directed by your instructor, adapt for oral presentations the process explanations you prepared in the applications above. Ask your classmates to evaluate your speech by filling in the Evaluation of Oral Presentations on page 459, or one like it.

Using Part II Selected Readings

1.14   Read the excerpt from "Non-Medical Lasers: Everyday Life in a New Light," by Rebecca D. Williams, pages 564–569.

a. Evaluate the effectiveness of the process explanations under "How Laser Works," pages 565, 567, and under "How FDA Monitors Laser Safety," page 566.

b. Under "Suggestions for Response and Reaction," page 569, complete Number 1 or Number 3.

1.15   Read the "The Active Document," by David D. Weinberger, pages 579–584. Under "Suggestions for Response and Reaction," page 584, complete Number 1.

PLAN SHEET FOR EXPLAINING A PROCESS

Analysis of Situation Requiring Explanation of a Process

What is the process that I will explain?

Who is my intended audience?

How will the intended audience use the process explanation? For what purpose?

Will the final presentation of the process explanation be written or oral?

Importance or Significance of the Process

Major Steps and Explanation of Each Step

Major Steps and Explanation of Each Step, cont.

Important Points to Receive Special Emphasis

Types and Subject Matter of Visuals I Will Include

The Title I May Use

Sources of Information

Chapter 2

Description of a Mechanism: Explaining How Something Works

Outcomes

75

Chapter 2
Description of a
Mechanism:
Explaining How
Something
Works

Upon completing this chapter, you should be able to

- Define mechanism
- Explain the difference between a description of a representative model and a description of a specific model
- List and explain the three frames of reference (points of view) used to describe a representative mechanism
- Use accurate and precise terminology in describing a mechanism
- Use visuals in describing a mechanism
- Describe a representative model of a mechanism
- Describe a specific model of a mechanism

Introduction

A mechanism is defined broadly as any object or system that has a functional part or parts. Thus, a mechanism is any item that performs a particular function. Items regarded as mechanisms are as diverse as a fingernail file, a diesel engine, a computer, a ballpoint pen, or a lawn mower. Another example of a mechanism is a heart, which receives and distributes blood through dilation and contraction. The human body can also be defined as a mechanism. And systems like the universe or a city, which are composed of parts that work together, can be described as mechanisms. Most often, perhaps, the term *mechanism* suggests tools, instruments, and machines.

We constantly work with mechanisms and need to understand them: what they do, what they look like, what parts they have, how these parts work together, and how to use them. At times an employee—when writing bid specifications for product specialists, a memorandum for a repairperson, or purchase requests for a supervisor; when demonstrating a new piece of equipment to a potential customer; when learning how to perform heart catheterization or kidney dialysis; or when planning changes in a city department's activities—may need to describe a mechanism or a part of it, using a written or an oral presentation.

You too come in contact daily with mechanisms—hair dryers, computers, traffic lights, coffee makers, compact disc players. On the job, you may use cash register terminals, fork lifts, shopping carts, stethoscopes, two-way pagers, milling machines, fax machines, autoclaves, or any number of other mechanisms. You may find that describing a mechanism for a repairperson or a new employee or a budget committee is a part of your daily life on your job. This chapter gives basic guidelines for these types of descriptions.

Purpose and Audience in Description

As always in planning a written or an oral presentation, you must identify the audience and the purpose of the description. The audience and the purpose affect decisions on everything related to the description—including content, language, complexity, organization, type of presentation, length, format, page layout, and document design. For instance, descriptions of a stereo speaker differ in emphasis and in detail, depending on whether the descriptions are given so a music buff can construct a similar speaker, so a prospective buyer can compare the speaker to a similar one, or so the general public can simply understand what a stereo speaker is. A description of a notebook computer for a lay audience who wants to know what a notebook computer is, what it is used for, and how it works differs in emphasis and in detail from a description in a catalog offering notebook computer accessories for sale. The primary readers of accessories descriptions are likely to be professionals and experts who know what a notebook computer is, what it is used for, and how it works. Such persons read the catalog descriptions to find a notebook computer with certain features or characteristics to meet their needs.

See pages 2–4 for a further discussion of audience and purpose.

Accurate Terminology

In describing a mechanism, use accurate and precise terminology. It is easier, of course, to use terms like *thing, good, large, narrow, tall,* and so forth. But your writing will be more effective if you use precise, specific words (see pages 11–14). Take time to think of or look up and use precise words.

Use "The chart shows manufactured goods representing 44 percent of the goods produced in Manitoba," not "The chart shows manufactured goods representing a large percentage of goods produced in Manitoba." Or "Breaker points in use more than 10,000 miles need cleaning or replacing," not "Breaker points in use more than a few thousand miles need attention."

Be careful to show variances. For example, rather than "Dividers may vary in size," use "Dividers may vary in size from 2 to 12 inches." Also be careful to refer to sizes by the standard method of measurement. The size of an electric drill, for instance, is referred to by the maximum diameter of the chuck (such as 1/4-inch or 1/2-inch), not by the length of the barrel or the dimensions of the handle.

If you are not certain about precise words, consult available sources. Such sources include dictionaries; encyclopedias; textbooks; specialized dictionaries, handbooks, and encyclopedias; knowledgeable people; advertisements; mail order catalogs; and instruction manuals.

Visuals

Visuals are especially valuable in describing a mechanism. Generally in describing a mechanism, the goal is to enable the audience to see what the mechanism looks like. Other goals may be to enable the audience to see what the mechanism's parts are and to understand how these parts interact to allow the mechanism to function.

Pictorial illustrations, such as photographs, drawings, and diagrams, that are included with a written or an oral description accomplish the goal more readily. Photographs can provide a clear external or internal view of a subject; they can show the size of a mechanism or part; and they can show color and texture. Drawings, ranging from the simple, freehand sketches in pencil, pen, crayon, or brush, to engineers' or architects' minutely detailed computer designs can do the same. Diagrams can outline an object or a system, showing such aspects as its parts, its operation, and its assembly. These outlines may be picture, schematic, or block diagrams. Drawings and diagrams can show the entire exterior or the relationships between parts, and the functions of parts as they work together; they can also show cross sections or cutaway views, pages 82, 83, 92, and 95. They can be exploded views (pages 81, 82, 95) that show the parts disassembled but arranged in sequence of assembly. Obviously visuals such as these are of great value in describing mechanisms.

Look at the photographs on pages 81, 85, and 86. Look at the additional drawings on pages 84, 92, and 95. Note the use of a poster and real object, page 89, and a transparency, page 90. You can readily see that the visuals enhance the written and oral explanations.

Visuals do not have to be complicated or sophisticated. Visuals can be prepared with pencil or pen and paper. Even a rough sketch may be helpful. Almost everyone, however, has access to computers that offer a wide array of graphic capabilities; from simple software programs to highly complex programs, the means to create effective visuals are available. For a discussion of commonly used visuals and ways to prepare them, see Chapter 11 Visuals.

77

Chapter 2
Description of a
Mechanism:
Explaining How
Something
Works

Two Ways to Describe Mechanisms

Two familiar ways to describe mechanisms are describing a **representative** model and describing a **specific** model. Describing a representative model focuses on characteristics of a group or class of mechanisms. Examples are a word processor, a camera, a ballpoint pen, a power supply, a calculator, a soccer team, or a rose. Describing a specific model, conversely, focuses on particular characteristics of an identified model, style, or brand of mechanism. Examples are a Brother WP2450DS Writer Series Daisy Wheel Word Processor, a Polaroid 600 Business Edition Instamatic Camera, or a Parker Classic Black Matte Ballpoint Pen No. 67839.

Describing a Representative Model of a Mechanism

The representative description identifies and explains aspects *usually* or *typically* associated with the mechanism. These aspects may include what the mechanism can do or can be used for; what it looks like; what its parts (components) are, what each part looks like, what each part does; and how these parts interact to perform the function of the mechanism.

Frames of Reference in Describing a Representative Mechanism

Logically there are three frames of reference (points of view or organizing principles) from which to describe a representative mechanism: its function, its physical characteristics, and its parts.

Function.   A mechanism is created to perform a particular function or task. The wedging board, for example, eliminates bubbles of air in clay. The fire hydrant provides a conduit of water for fire fighting. The hypodermic needle and syringe are used to inject medications under the skin. The microscope makes a very small object, such as a microorganism, appear larger so that it is clearly visible to the human eye. An automobile serves as a means of transportation. The kidney separates water and the waste products of metabolism from the blood. An elevator transports people, equipment, and goods vertically between floors.

Thus, the key element in a description of a representative mechanism is an explanation of its function, that is, the answer to What is it used for? This question automatically raises other questions: When is this mechanism used? By whom? How? These questions require answers if the description is to be adequate.

Physical Characteristics. A description of a representative model points out the physical characteristics of the mechanism. The purpose is to help the audience "see," or visualize, the object, to give an overall impression of the appearance of the mechanism. Physical charactertistics to consider include size, shape, weight, material, finish, color, and texture.

Often, comparison with a familiar object is helpful too. For example, the technical pen, described on pages 81–83, is compared to a fountain pen.

Sometimes a drawing of the parts of the mechanism is even more helpful. Notice the use of drawings with the descriptions, pages 84, 92, and 95; notice the use of cutaway diagrams on pages 82, 83, 92, and 95, and the exploded views on pages 81, 82, and 95.

Parts. A third frame of reference in describing a representative mechanism is its parts, or construction. The mechanism is divided into its parts, the purpose of each part is given, and the way that the parts fit together is explained. A drum, for example, as a percussion musical instrument is described as having two major parts: a hollow shell or cylinder and a drumhead stretched over one or both ends. When the drumhead is beaten with the hands or with an implement such as a stick or wire brush, sound is produced. In the part-by-part description, the hollow shell or cylinder and the drumhead are each described in turn. In essence, each of these parts becomes a new mechanism and is subsequently described according to its function, physical characteristics, and parts.

Identifying the parts is similar to partitioning (see pages 158–160), and explaining how the parts work together is similar to explaining a process (see pages 49–60).

Sample Description of a Representative Model

The following student-written description of a technical pen illustrates one way to write a representative description; the audience is a beginning commercial artist. The purpose of the description is to familiarize this beginning commercial artist with a drawing tool that such artists commonly use.

The description includes the Plan Sheet the writer filled in before writing the actual description. Marginal notes are added to help you analyze the sample description.

PLAN SHEET FOR DESCRIBING A REPRESENTATIVE MODEL OF A MECHANISM

Analysis of Situation Requiring a Description of a Representative Model of a Mechanism

What is the general mechanism that I will describe?
a technical pen

Who is my intended audience?
a beginning commercial artist

How will the intended audience use the description? For what purpose?
to understand the technical pen as a mechanism used by commercial artists

Will the final presentation of the description be written or oral?
written

Analysis of the Mechanism

Definition, identification, and/or special features:
a drawing tool used for marking camera copy to indicate directions as to trim size, centers, and folds; drafting; and lettering and compass work
Function, use, or purpose:
included in definition above
Who uses:
graphic artists and drafters
When:
included in definition
Where:
included in definition
Physical characteristics:
   Size:
    cap 1 1/4 inches long; body 4 5/8 inches
   Shape:
    resembles an ordinary fountain pen but with shorter cap
   Weight:
    2 ounces
   Material:
    sturdy plastic with some metal components
Other:
interchangeble air-ink exchange unit to make possible the use of different types of points for drawing on such surfaces as vellum, drafting film, or a combination

Major parts and a description of each:

Parts	Function	Physical characteristics
cap	protects the nib to keep ink from drying attaches to shirt pocket	lightweight metal
nib	produces marks on drawing surface	stainless steel, tungsten carbide, or jewel tip
needle	allows ink to flow smoothly breaks up dried ink	designed like a plunger kept in place by a safety screw
air-ink exchange unit	allows air to enter the cartridge as ink flows out	metal and plastic threaded into pen body
pen body	connects air-ink exchange unit to the pen body	plastic cylinder threaded to fit over the air-ink exchange unit
ink cartridge	holds the ink	plastic cylinder
clamp ring	seals the cartridge	plastic cylinder with threads
barrel	is handle and protection for cartridge	plastic cylinder threaded

Operation Showing Parts Working Together

not applicable

Variations or Special Features

included in physical characteristics and parts

Other Materials

not applicable

Types and Subject Matter of Visuals I Will Include

picture of a technical pen; drawings of external and internal parts

The Title I May Use

A Description of a Technical Pen

Sources of Information

Staedtler catalog, textbook

80

81

Chapter 2
Description of a
Mechanism:
Explaining How
Something
Works

A DESCRIPTION OF A TECHNICAL PEN

Focused title

Identification and Function

Technical pen defined; including function and uses

Control sentence naming topics to be explained

A technical pen is a drawing tool used for marking camera copy to indicate directions as to trim size, centers, and folds; in drafting; and in lettering and compass work. It is especially useful to graphic artists and drafters. The technical pen can be described generally by looking at physical characteristics and parts of the pen.

Physical Characteristics

Physical characteristics: material made from, shape, size

Made of sturdy plastic with some metal components, the technical pen resembles an ordinary fountain pen. The technical pen has two principal outer parts, the cap and the barrel. The cap is $1^1/_4$ inches long and the barrel is $4^5/_8$ inches long, typical measurements for a technical pen. The cap has a clip attached. A typical technical pen is shown in Figure 1.

Figure 1. The Marsmatic 700 technical pen made by J. S. Staedtler, Inc. (Source: Staedtler Mars Catalog)

weight

variations

The pen weighs 2 ounces.

One end of the barrel holds the nib or point, which can be interchanged to attach the type of point needed to draw a particular line. Different types of points are available for the different types of material the pen is used on; for example different types of points are used on vellum, on drafting film, or on both vellum and drafting film. The point selection is also based on the line thickness desired. Most manufacturers of technical pens recommend the ink to be used and the kind of material on which the ink is to be used.

Sentence renaming a topic

All technical pens are similar in design and have the same basic parts.

Parts

Parts listed

The eight basic parts of a technical pen are the cap, nib, needle (not visible in the figure), air-ink exchange unit, pen body, ink cartridge, clamp ring, and barrel. See Figure 2.

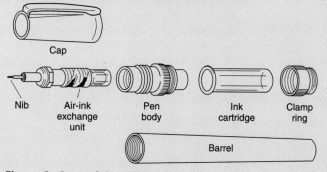

Cap

Nib Air-ink Pen Ink Clamp
exchange body cartridge ring
unit

Barrel

Figure 2. Parts of the technical pen

Parts described by physical characteristics and function of each

Cap

The cap fits over the nib to protect the nib and keeps the ink around the writing tip or nib from drying when the pen is not being used. Inside the cap, threads form an air trap when the cap attaches over the nib. Some technical pens have an internal moisture reservoir inside the cap; the reservoir has a small sponge kept saturated by a drop of water placed in the cap periodically. When the cap attaches over the nib, the sponge contacts the nib and keeps the ink from drying. See Figure 3.

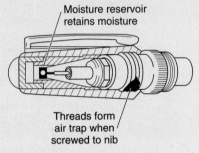

Moisture reservoir
retains moisture

Threads form
air trap when
screwed to nib

Figure 3. Inside the cap

The cap may have a metal clip to attach the pen to a shirt pocket.

Nib

The nib is the writing point. The end of the nib discharges ink to a drawing surface. The pen point is made out of different metals. For example, the point may be made out of stainless steel to be used in drawing on vellum or out of tungsten carbide for drawing on drafting film. Some points may have jewel tips that can be used to draw on either vellum or drafting film.

Needle

Inside the nib is a weighted needle that projects almost to the end of the writing tip. The needle, designed like a plunger, allows ink to flow smoothly. When the pen is in use, placed against a drawing surface or shaken vertically, the needle breaks up any dried ink in the tip of the nib; when not in use, the needle repositions itself and stops the ink flow. A safety screw keeps the needle in position. See Figure 4.

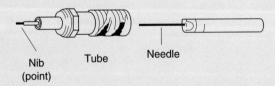

Nib
(point)

Tube

Needle

Figure 4. Inside the air-ink exchange unit

83

Chapter 2
Description of a
Mechanism:
Explaining How
Something
Works

Both the nib and needle attach to a plastic casing that threads into the air-ink exchange unit.

Air-Ink Exchange Unit and Pen Body

The air-ink exchange unit is a valve that allows air to enter the cartridge of the pen as the ink flows out. At one end of the unit is the nib and at the other end is the ink cartridge. Ink flows from the ink cartridge, through the air-ink exchange unit, into the nib, and out through the tip onto a drawing surface. See Figure 5.

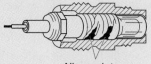

Allows air to
enter pen as
ink is discharged

Figure 5. Air entry in the air-ink exchange unit

The exchange unit is made of plastic. It threads into a cylindrical section of the pen body that allows the exchange unit to connect to the ink cartridge. Staedtler manufactures color coded air-ink exchange units, the color indicating the line thickness that the point will produce.

Ink Cartridge and Clamp Ring

The ink cartridge holds the ink. A clamp ring seals the cartridge to keep the ink from drying. Fitted into one end of the air-ink exchange unit, the cartridge is available either empty or prefilled. Prefilled cartridges are more convenient but are also more expensive.

Ink used in technical pens is a specific ink; it is available in a variety of colors: green, yellow, blue, red, and so on.

Barrel

The barrel is a plastic cylinder used as a handle and as a protection for the ink cartridge. Threaded for easy removal, the barrel may contain a hex wrench in the end; the wrench can be used to loosen the nib for cleaning. Also the end of the barrel that contains the air-ink exchange unit may have a $1^{1}/_{4}$-inch finger grip to allow the user to grip the pen firmly and steadily.

Describing a Specific Model of a Mechanism

In contrast with the description of a representative model, the description of a specific model of a mechanism emphasizes particular characteristics, aspects, qualities, or features. The specific description focuses on what sets this specific model of a mechanism apart from all other models.

For example, a description of the Winsome (brand) Professional Quiet 1800 (model) blow dryer for a potential buyer might include facts about special features, including the dryer's unusual quietness. See the sample description, pages 89–90.

Uses of Description of a Specific Mechanism

Description of a specific mechanism is used in bid specifications, in advertisements for products, or in any description for an audience that wants to know particular features of a specific model of a mechanism.

Bid specifications name a product or service and describe the unique characteristics, qualities, or features the product or service must have to meet a need. A representative of a company or business that sells the product looks at the specifications, determines if his/her company can offer the product and then offers to sell it at a stated price. The audience for bid specifications is a manufacturer, business, or person who can supply a product or service; the purpose of the bid specifications is to locate and buy the product or service.

While there is no standard form for recording or presenting bid specifications, they always identify the required qualities clearly and precisely. The company or individual who will offer the product or service at a stated price must know *exactly* what they are offering.

Bid specifications for a microscope appear on page 94.

Descriptions of specific mechanisms also appear in all types of merchandise catalogs. You might be more familiar with such descriptions in general merchandise catalogs such as those from Office Depot, J.C. Penney, or Spiegel. In addition, many specialized catalogs offer computer hardware and software, automotive parts and accessories, electronic equipment and components, parts for appliances—the list goes on and on. Each item in a catalog is described as a specific model. For example, look at the description of the 6545-H10 and H30 microscopes for teaching, offered in a merchandise catalog; the description appears on page 93.

Specific descriptions are also found in magazine and newspaper sections in which products are advertised. The following description of a water cooler appears in a section on new products in an issue of *The Office.**

Water Cooler

Elkay Manufacturing Company, 2222 Camden Court, Oak Brook, Ill. 60521, offers the Hands-Free water cooler that automatically turns on when an individual steps in front of the cooler and interrupts an infrared light beam sensor. It has a maximum running time of 30 seconds, eliminating waste or damage should sensor be blocked. After blockage is removed, system automatically resets. Built-in time delay prevents activation by individuals walking past the cooler. Two models are available.

* THE OFFICE, August 1991, Stamford, Conn.

In the Modern Equipment Review section of an issue of *Modern Machine Shop,* this description of a milling machine appears.*

85

Chapter 2
Description of a
Mechanism:
Explaining How
Something
Works

Direct Drive Hydraulic Milling Machine Features Only Two Gear Contacts

The Kent-Owens 2-20 hydraulic milling machine features balanced design and simple direct drive. Wide spindle speed range is provided by wide-face, quiet running pick-off gears which are spline mounted and are easily removed for spindle speed changes.

The 2-20 hydraulic milling machine implements a simple direct drive with only two-gear contacts—motor to cutter. This design allows anti-friction bearings throughout, and quiet, smooth, steady cutting action without vibration.

A 42" x 12" table, a 20" table travel, and a fully automatic cycle make this model extremely versatile. Practically any desired cycle can be obtained. The table can be fed or rapid traversed automatically in either direction, and reversed at both ends of a stroke. The minimum feed distance permits taking very short cuts without loss of time.

"Chatter-proof" operation is another advantage of the Kent-Owens 2-20 hydraulic milling machine. Two cylindrical ground steel posts provide unusual rigidity and accuracy in either direction of table travel.

The Kent-Owens Model No. 2-20 hydraulic milling machine is used by manufacturers of screw machine and plastic products, medical equipment precision and electric hand tools, and by manufacturers of parts for the aircraft, automotive and computer industries.

LeLand-Gifford, 1497 Exeter Road, Akron, OH 44306.

Advertisements incorporate description to highlight a particular product or model. In the following NordicTrack** advertisement, two sets of characteristics are listed: overall qualities of the exerciser and particular features.

Audiences for these three descriptions—water cooler, milling machine, exerciser—would vary. Architects designing and furnishing a new building might be interested in the Elkay water cooler description. A machinist or manufacturer might use the specific description of the Kent-Owens 2-20 hydraulic milling machine to determine whether purchase of the machine would be beneficial. Anyone, lay person or professional, who is interested in keeping physically fit might consider the exerciser.

---

*Reprinted by permission of *Modern Machine Shop,* Oct. 1994, page 226.
**Reprinted by permission of NordicTrack, Inc. A CML Company, 1994.

Sample Description of a Specific Model

Now follows a description of a specific mechanism prepared by a student for an oral presentation. The description includes the Plan Sheet the speaker filled in. Such planning, for written or oral presentations, enables a writer or speaker to prepare carefully.

As you read the student-prepared description, notice that the description emphasizes the features which make the Winsome Professional Quiet 1800 blow dryer desirable to use. The function, physical characteristics, and parts of this specific blow dryer are very similar to those of any other blow dryer; therefore, treatment of these usual aspects is kept to a minimum. Unusual aspects are emphasized. Marginal notes are added to outline the development of the description.

You might also review Chapter 10 Oral Communication; then try reading the description aloud to approximate an oral presentation.

PLAN SHEET FOR DESCRIBING A SPECIFIC MODEL OF A MECHANISM

Analysis of Situation Requiring a Description of a Specific Mechanism

What is the specific mechanism that I will describe?
Winsome Professional Quiet 1800 blow dryer

Who is my intended audience?
professional hairdressers; individuals who use a hair dryer a lot, particularly around wet clothes or wet surfaces

How will the intended audience use the description? For what purpose?
to learn about a model of blow dryer that has many safety features

Will the final presentation of the description be written or oral?
oral

Analysis of the Mechanism

Who uses:
professional hairdressers; individuals with special safety needs

When:
to dry the hair

Where:
bathroom, bedroom

Physical characteristics:
  Size:
1800 watts

  Shape:
omit

  Weight:
20 ounces

  Material:
heat resistant plastic

  Other:
cost—$59.95

Special features to be emphasized:
1. *Built-in GFCI (ground fault circuit interrupter)*
    GFCI automatically stops electricity flow if a current leak occurs.
    A typical hair dryer could fatally shock you, if it comes in contact with water.
    This feature increases the cost.
2. *Prevention of overheating: two built-in safety devices*
    electronic chip for power surge
    electronic chip for cycling off High Heat and then to OFF
3. *Extra quiet 1800-watt motor with dual ventilation*
    high power fan
    air vents on both sides of dryer
4. *Heavy duty switches*
    four switches
    two speeds: Low, High
    four heat settings: Low Warm, Low Hot, High Warm, High Hot
    two cool settings: Low Cool, High Cool
    On, Off

Variations available:
omit

Types and Subject Matter of Visuals I Will Include

—poster listing the four safety features
—my own Winsome 1800 dryer
—transparency showing switches on the dryer

The Title I May Use

A Description of the Winsome Professional Quiet 1800 Hair Dryer

Sources of Information

"Use and Care" pamphlet that came with the hair dryer; telephone interview with Shunya Fenton, a Winsome Products distributor

ORAL PRESENTATION

Description: The Winsome Professional Quiet 1800 Blow Dryer

Designed for use by hair care professionals, the Winsome Professional Quiet 1800 is an especially safe, hand-held blow dryer. It may seem a bit pricey at $59.95 for home use.

I opted to purchase it rather than one of the $9 to $15 models because of the safety features. Here is the Winsome 1800 I bought at a local department store.

I am an avid swimmer and also a scuba diver. As you can see, I am a woman with long hair. Like professionals in hair-care salons, I use a hair dryer quite often—sometimes three or four times a day. I need a hair dryer with variable settings. If I'm off to class, I need a quick, thorough dry—HIGH, HEAT. For personal comfort at home, I like the LOW COOL or LOW HEAT setting, depending on outside temperatures when I come in from swimming or diving.

The safety features, though, are what make the Winsome 1800 my choice as a hair dryer. When I come in dripping wet and take a quick shower, I must be presentable for class in a few minutes. I need a hair dryer that is safe to use when I drop wet clothes or stand on wet floors.

Four Distinguishing Safety Features

The four distinguishing safety features of the Winsome 1800 blow dryer are listed on the poster.

[Listing might be generated by computer or manual lettering and displayed in hard copy, on computer monitor, or by screen projection.]

Winsome Professional Quiet 1800 Blow Dryer Safety Features
- Built-in GFCI (ground fault circuit interrupter) for safe use near water
- Two built-in safety devices to prevent overheating
- 1800-watt extra quiet motor with dual ventilation
- Heavy duty switches (two speeds, four heat settings, two cool settings)

First Feature

The most outstanding safety feature of the Winsome Professional Quiet 1800 dryer is the GFCI. GFCI means ground fault circuit interrupter. GFCI is a sensing device that immediately stops electricity flow when a small electric current leak occurs.

This GFCI could save your life. Why? Electricity and water are a deadly mix. If a typical hair dryer falls in water while it's plugged in, the electric shock could kill—even if the switch is "off." The Winsome 1800 has a built-in GFCI. This feature runs up the price of the dryer, but I'm willing to pay the extra dollars for the protection.

Second Feature

The second distinguishing safety feature is the prevention of overheating. The Winsome 1800 has two built-in safety devices to prevent overheating. The wiring in the dryer contains two electronic chips for overheating protection. One electronic chip is activated if there is a surge in electric current. Such a surge can occur if the electricity supplier is experiencing difficulty. For instance, the lights may blink on and off, suddenly become dim, or go off for a couple of minutes and then come back on. In such cases, the power surge chip in the hair dryer prevents a small internal "explosion" that could overheat the dryer and cause it to burn out. In this situation, typically there is no smoke or unusual odor—the hair dryer just doesn't work any more.

Identification of mechanism

Emphasis: safety features

[Hold up the dryer so that all can see.] Personal requirements for a hair dryer

Control sentence for the description

Key phrase: four distinguishing safety features

[Show the poster listing each safety feature. As each feature is explained, point to its listing on poster; hold up dryer and point to each feature.]

two electronic chips

Chip for power
surge
  Examples of
  power surge
Chip for
extended HI H
  Examples

[*Hold up the dryer
and point out the
dual air vents.*]

Another electronic chip in the wiring of the hair dryer activates when overheating could occur. If the High Heat (HI H) setting runs for more than 60 seconds, the dryer automatically reverts to Low Heat (LO H). After another 60 seconds, the dryer cycles to OFF.

Let's say I'm trying to quick dry my hair and the phone rings. I go to the bedside phone and forget to turn off the hair dryer. Ten minutes later, I'm dressed and skip off to class. When I return two hours later, I remember the hair dryer. I left it on. Thank goodness it automatically switched itself off after two minutes.

Third Feature

The third safety feature of the Winsome 1800 is the 1800-watt extra quiet, heavy-duty motor with dual ventilation. An 1800-watt hair dryer puts out a lot of wattage. That's the same as eighteen 100-watt bulbs a few inches from your head. Quick, intense heat!

That quick heat could burn out an ordinary hair dryer. The Winsome 1800, however, with its unusually rugged construction has a high power fan for maximum velocity. In addition, the dryer has dual ventilation, that is, air vents on both sides of the dryer.

The dual ventilation system is needed to accommodate the high power fan and the 1800 watts of power.

Fourth Feature

[*Point out the
switches on the
dryer.*]

[*Point out the dryer
switches, using the
transparency.*]

The last of the four safety features I would like to emphasize is the heavy-duty switches. The dryer has two speeds, four heat settings, and two cool settings.

This transparency provides a close-up view of the four switches.

The Four Switches	
Warm	② Low
Hot	③ High
Cool	
①	④ On Off

The fact that the switches are heavy duty helps to insure many years of trouble-free use in changing rapidly from, say, High Hot to Low Warm or Low Cool.

By the way, the manufacturer recommends in the "Use and Care" pamphlet that the dryer be switched to Low Cool for a few seconds before switching the dryer to Off. This action permits the dryer motor to cool at a slower rate, prolonging the life of the dryer.

Summary of Four Main Safety Features

[*As each feature is
mentioned point to
the feature on the
poster*]

Thus we have the Winsome Professional Quiet 1800 blow dryer. In summary, the characteristics that distinguish this hair dryer are its safety features:

- GFCI
- Overheating prevention
- Motor with dual ventilation
- Heavy duty switches

Closing Warning

Re-emphasis of
personal need for
safety features

In closing, let me warn you that most other hair dryers do not have these safety features, particularly the ground fault circuit interrupter (GFCI). Be careful about using a hair dryer around water.

I use a hair dryer a lot—sometimes several times a day. The extra cost of this model is a good investment; for me, the safety features are a must.

Comparing Descriptions of Representative and of Specific Mechanisms

91

Chapter 2
Description of a
Mechanism:
Explaining How
Something
Works

The description of a representative mechanism emphasizes what the mechanism does or can be used for and what it looks like. It includes a description of the parts, what each part looks like, and what each does. The representative description describes the parts as they relate to one another—that is, the mechanism in operation.

The representative description concerns a mechanism typical of a class. It describes what all types of a stated mechanism or all mechanisms in a class have in common, even though variations may exist for different brands and different models.

A representative description of a mechanism is the kind of description given in a reference book such as a dictionary or encyclopedia. The following description of a microscope appears in *The World Book Encyclopedia.**

> **MICROSCOPE** is an instrument that magnifies extremely small objects so they can be seen easily. It produces an image much larger than the original object. Biologists and other scientists use the term *specimen* for any object studied with a microscope.
>
> The microscope ranks as one of the most important tools of science. With it, researchers first saw the tiny germs that cause disease. The microscope reveals an entire world of organisms too small to be seen by the unaided eye. Physicians and other scientists use microscopes to examine such specimens as bacteria and blood cells. Biology students use microscopes to learn about algae, protozoa, and other one-celled organisms. The details of nonliving things, such as crystals in metals, can also be seen with a microscope.
>
> There are three basic kinds of microscopes: (1) *optical,* or *light;* (2) *electron;* and (3) *ion.* This article discusses optical microscopes. For information on the other types, see the WORLD BOOK articles on ELECTRON MICROSCOPE and ION MICROSCOPE.
>
> **How a microscope works.** An optical microscope has one or more lenses that bend the light rays shining through the specimen (see Lens). The bent light rays join and form an enlarged image of the specimen.
>
> The simplest optical microscope is a magnifying glass (see Magnifying Glass). The best magnifying glasses can magnify an object by 10 to 20 times. A magnifying glass cannot be used to magnify an object any further because the image becomes fuzzy. Scientists use a number and the abbreviation X to indicate (1) the image of an object magnified by a certain number of times or (2) a lens that magnifies by that number of times. For example, a 10X lens magnifies an object 10 times. The magnification of a microscope may be expressed in units called *diameters.* A 10X magnification enlarges the image by 10 times the diameter of the object.
>
> Greater magnification can be achieved by using a *compound* microscope. Such an instrument has two lenses, an *objective lens* and an *ocular,* or *eyepiece, lens.* The objective lens, often called simply the *objective,* produces a magnified image of the specimen, just as an ordinary magnifying glass does. The ocular lens, also called the *ocular,* then magnifies this image, producing an even larger image. Many microscopes have three standard objective lenses that magnify by 4X, 10X, and 40X. When these objective lenses are used with a 10X ocular lens, the compound microscope magnifies a specimen by 40X, 100X, or 400X. Some microscopes have *zoom* objective lenses that can smoothly increase the magnification of the specimen from 100X to 500X.

* Excerpted by permission from *The World Book Encyclopedia.* © 1992 World Book, Inc. By permission of the publisher.

In addition to magnifying a specimen, a microscope must produce a clear image of the structure of the object. This capability is called the *resolving power* of the microscope. Optical microscopes can resolve only objects that are larger than the wavelength of light. Even the best optical microscopes cannot resolve parts of objects that are closer together than about 0.000008 of an inch (0.0002 of a millimeter). For this reason, such tiny structures as atoms, molecules, and viruses cannot be seen with an optical microscope.

Parts of a microscope. Most microscopes used for teaching have three parts (1) the *foot,* (2) the *tube,* and (3) the *body.* The foot is the base on which the instrument stands. The tube contains the lenses, and the body is the upright support that holds the tube.

The body, which is hinged to the foot so that it may be tilted, has a mirror at the lower end. The object lies on the *stage,* a platform above the mirror. The mirror reflects light through an opening in the stage to illuminate the object. The operator can move the tube within the body by a course-adjustment knob. This focuses the microscope. Most microscopes also have a *fine-adjustment* knob, which moves the tube a small distance for final focusing of a high-power lens.

The lower part of the tube holds the objective lens. In most cases, this lens is mounted on a revolving *nosepiece* that can be rotated to bring the desired lens into place. The upper end of the tube holds the ocular lens.

Using a microscope. A microscope is an expensive instrument and can be damaged easily. When moving one, be sure to handle it carefully.

To prepare a microscope for use, turn the nosepiece so that the objective with the lowest power is in viewing position. Lower the tube and lens by turning the coarse-adjustment knob until the lens is just above the opening in the stage. Next, look through the eyepiece and adjust the mirror so a bright circle of light appears in the eyepiece. The microscope is now ready for use. Most people keep both eyes open when looking into the eyepiece. They concentrate on what they see through the microscope and ignore anything seen with the other eye.

Parts of a Microscope

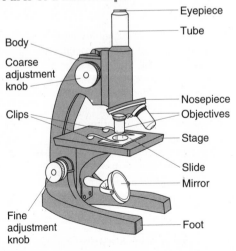

Body
Coarse adjustment knob
Clips
Fine adjustment knob

Eyepiece
Tube
Nosepiece
Objectives
Stage
Slide
Mirror
Foot

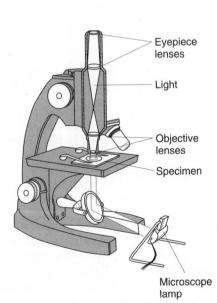

Eyepiece lenses
Light
Objective lenses
Specimen
Microscope lamp

The diagram at the left shows the external parts of a microscope. A person adjusts these parts to view a specimen. The cutaway diagram at the right shows the path that light follows when passing through the specimen and then through the lenses and tubes of the microscope.

93

Chapter 2
Description of a
Mechanism:
Explaining How
Something
Works

Most specimens viewed through a microscope are transparent, or have been made transparent, so that light can shine through them. Objects to be viewed are mounted on glass slides that measure 3 inches long and 1 inch wide (76 by 25 millimeters). The technique of preparing specimens is called *microtomy* (see MICROTOMY). See also the *Science project* with this article.

To view a slide, place it on the stage with the specimen directly over the opening. Hold the slide in place with the clips on the stage. Look through the eyepiece and turn the coarse-adjustment knob to raise the lens up from the slide until the specimen comes into focus. In order to avoid breaking the slide or lens, never lower the lens when a slide is on the stage.

When the specimen is in focus, turn the nosepiece to an objective lens with higher power. This lens will reveal more details of the specimen. If necessary, focus the objective with the fine-adjustment knob. A zoom microscope is changed to a higher power by turning a part of the zoom lens. Different parts of the specimen can be brought into view by moving the slide on the stage.

The description of a specific mechanism, on the other hand, describes the outstanding or unique features or characteristics of a stated brand or model. It emphasizes the aspects of the stated brand or model that set it apart from other brands or models of the same mechanism. The specific description below appears in a merchandise catalog from Arthur H. Thomas Company (note the omission of noncontent words to save space).*

6545-H10 MICROSCOPES, Teaching
6545-H30 American Optical Series Sixty

Designed to withstand classroom handling. Advanced features include unique, spring-loaded focusing mechanism which protects slide and objective, avoids conventional rack and pinion. Equipped with in-base illuminator.

Revolving ball-bearing nosepiece is raised or lowered to effect focus, by means of coaxial coarse and fine adjustments, low-positioned for convenience. Adjustable stop (Autofocus) limits objective travel, permitting rapid focus while safeguarding objective and slide.

Stage, with clips, is 125 × 135 mm. Integral in-stage condenser with adjustable diaphragm has reference ring for centering specimen.

Inclined body is reversible for viewing from front or back of stage. Huygenian eyepiece, 10×, with pointer and measuring scale, is locked in.

Optics on both models have Americote magnesium fluoride coating on air-glass surfaces to reduce internal reflection and improve image contrast.

6545-H10 Microscope has 5-aperture disc diaphragm, double revolving nosepiece, 10× and 43× achromatic objectives, providing 100× and 430× magnification.

6545-H30 Microscope has iris diaphragm; triple revolving nosepiece; 4×, 10×, and 43× achromatic objectives, providing 40×, 100×, and 430× magnification.

With switch, cord, and plug, for 120 volts. For replacement bulb, see 6625-D30.

Still another type of specific description is illustrated in the following bid specifications for a microscope.

Note that the audiences for the three descriptions of the microscope are different, and the reason for reading each of the three descriptions is different. A student in a laboratory class or a person interested in the microscope might read the *World*

---

*Reprinted by permission.

Bid No. 215
Opening Date: May 10, 1995
Name: Microscope, AO Series 10

Item No.	Quantity	Description	Unit Price	Total Price
	50	MICROSCOPE, AO Series 10 Microscope to permit simultaneous viewing of the same image by two persons in side-by-side position. To have dual viewing adapter binocular body, 10×, widefield eyepieces. To have illuminated green arrow, to be powered by microscope transformer, to be controlled by lever at base of adapter, and positioned anywhere in the field of view. To be equipped with ungraduated mechanical stage, Aspheric, N.A. 1.25, condenser with auxiliary swing-in lens, 2 pair 10× widefield eyepieces. SPECIFICATIONS: Viewing position—side by side Planachromatic 4×, 40×, 100×		

Book Encyclopedia description to gain basic knowledge of what a microscope is, how it works, and how to use it. A lab instructor might read the catalog entry to locate and purchase a microscope to meet specific needs. A salesperson might read the bid specifications to determine if his/her company has such a microscope and to determine the price for such a microscope with these features when 50 are ordered. In each description the details about the microscope were selected with the audience and the audience's needs in mind.

Mechanism Description Combined with Other Forms

The sample description of a technical pen (pages 81–83) might have included instructions for using or caring for the pen. The description might have included manufacturing specifications or other kinds of information about the pen, depending on the audience and the purpose for the description. Most technical communications, whether written or oral, are made up of several forms of communication. A description of a mechanism might include a set of instructions for using the mechanism. A shop manual likely would describe shop machines, use visuals to identify parts of the machines, and give instructions for using the machines correctly and safely. A booklet describing health benefits might classify and define health problems and related benefits, explain the process for receiving benefits, and give instructions for filing for benefits. Many mechanisms have an accompanying user's manual that explains how the mechanism works, shows the parts, gives instructions for use and care, and explains how to replace parts that frequently need replacement.

Look at the example on page 95 showing excerpts from a flier advertising Sanford mechanical pencils.* The entire flier is one sheet of paper 12 inches by 4 3/4 inches and is folded three times to a 4 3/4-inch-long and 3-inch-wide size that can be displayed easily in a rack. A person interested in the pencils can open the accordion-folded flier to find eight panels of information, four front-printed panels,

*Reprinted by permission of the Sanford Corporation.

95

Chapter 2
Description of a
Mechanism:
Explaining How
Something
Works

LOGO/LOGO®II

Change your mind? Make a mistake? Then you need a Logo or Logo II with a jumbo exposed eraser (6 1/2 times BIGGER than other mechanical pencil erasers) that advances with a twist! No need to fumble with a tiny end cap and eraser assembly.

Logo pencils also feature a unique sliding lead sleeve. As you write, this sleeve actually moves back as lead is used to reduce lead breakage. It also fully retracts to protect pockets and briefcases when not in use.

Logo features clean, contemporary design, and Logo II adds a dash of style with a mirror-finish clip and tip!

Logo

Item Number	Lead Size	Barrel Color	List/Each
64001	0.5mm	Black	2.98
64002	0.5mm	Burgundy	2.98
64003	0.5mm	Gray	2.98
64008	0.7mm	Black	2.98
64009	0.7mm	Burgundy	2.98
64010	0.7mm	Gray	2.98

Logo

Item Number	Lead Size	Barrel Color	List/Each
64051	0.5mm	Black	3.98
64052	0.5mm	Burgundy	3.98
64053	0.5mm	Blue	3.98
64058	0.7mm	Black	3.98
64059	0.7mm	Burgundy	3.98
64060	0.7mm	Blue	3.98

Panel 1

PENCIL OPERATION

To Advance Lead While Using Pencil

Depress end cap (or eraser for Logo and Logo II). Lead will advance with each click.

To Feed a New Lead

When lead starts to slip back into tip, this indicates the lead is used up and a new lead is needed. Depress the end cap 10–15 times to push old lead out and advance a new piece from the lead tube. If tube is empty, see Refilling Instructions. Never refill through the tip of the pencil, as this could cause damage.

To Retract Lead and Sleeve

Depress end cap (or eraser for Logo and Logo II) and hold. On a solid surface, push lead straight down and at the same time relieve pressure on end cap. The lead and sleeve will retract into the pencil. Extend will automatically retract by depressing the end cap fully.

Sanford Quality Leads and Erasers

Item Number	Description	List/Each
64751	HB 0.5mm 25 Leads per Cartridge	1.59
64752	B 0.5mm 25 Leads per Cartridge	1.59
64761	HB 0.5/12 Leads per Tube	.79
64762	B 0.5/12 Leads per Tube	.79
64767	HB 0.7/12 Leads per Tube	.79
64768	B 0.7/12 Leads per Tube	.79
64771	HB 0.7mm 25 Leads per Cartridge	1.59
64772	B 0.7mm 25 Leads per Cartridge	1.59
64881	ZeZe Eraser/5 per tube	.89
64891	Regular Eraser/5 per Tube	.89
64892	Logo Twist Eraser/2 per Tube	.89

Panel 2

REFILLING PENCIL

To Refill the Lead Tube

Remove end cap and eraser assembly. Insert up to three pieces of Sanford lead in the lead tube. Replace eraser assembly and end cap. To refill Logo and Logo II, pull apart pencil at mid-section and insert lead into the lead tube.

IMPORTANT: When refilling lead, make sure to match the lead size to the number indicated on the pencil. Never refill through the tip of the pencil. Always refill into the lead tube.

Refill or Adjust Eraser

Remove end cap and eraser assembly. Gently pull apart the sides of the metal clip holding the eraser. Pull eraser up to desired height, or replace. Squeeze the metal clip together and insert assembly back into pencil. The Logo and Logo II eraser twists up and can be replaced by pulling eraser out and inserting a new eraser.

Diagram labels: ERASER — LEAD TUBE — BARREL — FERRULE — LEAD SLEEVE — ERASER CAP — ERASER — ERASER ASSEMBLY — LEAD TUBE — BARREL — FERRULE — LEAD SLEEVE — LOGO II

MPB (592) SANFORD CORPORATION, BELLWOOD, IL 60104

Panel 3

and four back-printed panels. These eight panels present a wealth of information about Sanford mechanical pencils in an attractively presented design. Basically the flier gives specific descriptions of five models of pencils, but it also includes instructions for operating and refilling the pencil.

Shown on page 95 are three panels from the flier. Panel 1 shows one of the five models advertised. Notice the specific description of the Logo/Logo II model included with a picture of the pencil and the chart listing the different lead sizes, barrel colors, and cost of the Logo/Logo II model. In the actual flier, the pen is shown in a maroon color on a blue background.

Panels 2 and 3 give instructions on how to operate the pencil and how to refill the pencil. Notice that *a, an,* and *the* have generally been left out because of space limitations. Under the heading "To Refill the Lead Tube," notice the use of the word IMPORTANT to call attention to the use of correct lead size and to the correct refilling procedure. Included is a visual which labels the parts of the pencil for easy reference from the instructions.

The flier is effectively presented for the intended audience and purpose— anyone interested in selecting and using a Sanford mechanical pencil for writing, drawing, drafting, or lettering.

Oral Description

As a worker employed in such areas as sales, supervision, or training, you may find occasions when you must describe a mechanism orally. This description might be given to a large group of potential buyers, to several newly hired workers, to students in a training program, or to a single individual. Review the sample oral description on pages 89–90.

The content of an oral description might be organized according to the procedures outlined within this chapter. Presenting the material, however, requires other considerations. For a detailed discussion of oral presentations, see Chapter 10 Oral Communication.

General Principles in Describing a Mechanism

1. *Audience and purpose control all decisions you make about mechanism description.* The audience and the purpose must be clear. These two basic considerations determine the extent and the kind of details given and the manner in which they are given.
2. *Use accurate and precise terminology.* Sources for such terminology are dictionaries, encyclopedias, textbooks, knowledgeable people, instruction manuals, and merchandise catalogs.
3. *Visuals can enhance mechanism description effectively.* Typical visuals include pictorial illustrations, such as photographs, drawings, and diagrams.
4. *The description of a mechanism may be a description of a representative model.* In a representative description, emphasis is on giving an overall view of what the mechanism can do or can be used for, what it looks like, and what components it has.

97

Chapter 2
Description of a
Mechanism:
Explaining How
Something
Works

5. *The three frames of reference in a description of a representative model are function, physical characteristics, and parts.* Although these three frames of reference are closely related, ordinarily they should be kept separate and in logical order.
6. *The description of a mechanism may be of a specific model.* In a specific-model description, emphasis is on particular characteristics or aspects of a mechanism identifiable by brand name, model number, and the like.
7. *The description may be of a mechanism at rest or in operation.* In an at-rest description, emphasis is on parts of the mechanism and their location in relationship one to another. In an in-operation description, emphasis is on the parts and the way these parts work together to perform the intended function.
8. *The description of a mechanism may be combined with other forms.* For example, descriptions of mechanisms may be combined with instructions, specifications for manufacture, classification, or definition.

Procedure for Describing a Representative Model of a Mechanism

I. The identification of the mechanism is usually simple and requires only a few sentences.
   A. Define or identify the mechanism and indicate why this description is important, if appropriate.
   B. In a sentence, list the points (frames of reference) about the mechanism to be described.
II. The explanation of the points (frames of reference) is the lengthiest section of the presentation.
   A. Give the function, use, or purpose of the mechanism.
      1. If the mechanism is part of a larger whole, show the relationship between the part and the whole.
      2. If applicable, state who uses the mechanism, when, where, and why.
   B. Give the physical characteristics of the mechanism.
      1. Try to make the reader "see" the mechanism.
      2. Describe, as applicable, such physical characteristics as size, shape, weight, material, color, texture, and so on.
   C. Give the parts of the mechanism.
      1. List the major parts of the mechanism in the order in which they will be described.
      2. Identify each part.
      3. State what the part is used for—its function.
      4. Tell what each part looks like—its physical characteristics.
      5. Give the relationship of each part to the other parts.
      6. If necessary, divide the part into its parts and give their functions, physical characteristics, and parts.
   D. Use headings and visuals, when appropriate, to make meaning clear.
III. The closing, usually brief, emphasizes particular aspects of the mechanism.
   A. Show how the individual parts work together.

B. If applicable, mention variations of the mechanism, such as optional features, other types, and other sizes; or comment on the importance or significance of the mechanism.

Procedure for Describing a Specific Model of a Mechanism

I. The identificaiton of the mechanism is usually simple and requires only a few sentences.
  A. Identify the mechanism by giving the brand name or model number.
  B. Tell the function or purpose of the mechanism, who uses the mechanism, when it is used, and where.
  C. Introduce the features or characteristics of the mechanism.

II. The features or characteristics of the particular brand or model make up the main section of the description.
  A. Identify each feature or characteristic.
    1. Select information about the mechanism to set it apart from other similar mechanisms.
    2. Consider such features as size, shape, weight, material, overall appearance, available colors, cost, warranty.
  B. Describe each feature or characteristic in detail.
  C. Use headings and visuals, when appropriate, to make meaning clear.

III. The closing, if included, is usually brief.
  A. The description may end after the discussion of the last feature or characteristic.
  B. The features or characteristics discussed may be summarized; or any variations of the mechanism—such as other models, different colors, other prices, or different designs—may be mentioned.

Applications in Describing a Mechanism

Individual and Collaborative Activities

2.1   Sort the sentences below into three general groups under the following three headings: Function, Physical Characteristics, and Parts. Write the numbers of the sentences under the appropriate headings.

  a. The jaws are made of cast iron and have removable faces of hardened tool steel.
  b. The machinist uses a small stationary holding device called the machinist's bench vise to grip the work securely when performing bench operations.
  c. For a firm grip on heavy work, serrated faces are usually inserted.
  d. In addition to the typical bench vise described here, there are many other varieties and sizes.

99

Chapter 2
Description of a
Mechanism:
Explaining How
Something
Works

e. It is essential for holding work pieces when filing, sawing, and clipping.

f. A vise consists of a fixed jaw, a movable jaw, a screw, a nut fastened in the fixed jaw, and the handle by which the screw is turned to position the movable jaw.

g. To protect soft metal or finished surfaces from dents and scratches, false lining jaws are often set over the regular jaws.

h. This holding device is about the size of a small grinding wheel and is fastened to the workbench in a similar manner.

i. These lining jaws can be made from paper, leather, wood, brass, copper, or lead.

j. A smooth face is inserted to prevent marring the surface of certain work pieces.

After sorting the sentences into three groups, arrange the sentences in a logical order to form a paragraph. The reorganized paragraph thus would read:

Sentence __b__ , ____ , ____ , __f__ , ____ , ____ , ____ , ____ , ____ , __d__

Now write out the reorganized paragraph.

2.2   Select a mechanism that you use in one of your lab courses.

a. Describe the mechanism so that an employee can find it and put an inventory number on it.

b. Describe this mechanism for a new lab student who must use the mechanism.

c. Describe this mechanism for a technician who will repair or replace a broken part.

Writing Descriptions of Mechanisms

In completing applications 2.3–2.5,

a. Fill in the appropriate Plan Sheet on pages 101–104, or one like it.
b. Write a preliminary draft.
c. Go over the draft in a peer editing group.
d. Revise. See Part III Revising and Editing and the checklists for revising, front and back endpapers.
e. Write the description.

Describing a Representative Model

2.3   Write a description of one of the following mechanisms or of a mechanism in your major field.

a. A disk drive, monitor, keyboard, or other computer component
b. A human heart, hand, big toe, or other body part
c. A drill press, power saw, sander, or other shop tool

d. A washing machine, toaster, microwave oven, room air conditioner, or other appliance

e. A lie detector, pair of handcuffs, service revolver, or other law enforcement articles

f. A vaccinating needle, autoclave, dental unit, stethoscope, incubator, or other health-related mechanism

g. A fax machine, telephone answering machine, photocopier, or other office machine

h. A compact disc player, modem, bag phone, or other electronic mechanism

i. A flashlight, wristwatch, camera, calculator, stapler, or other common mechanism

j. Topic of your choosing

Describing a Specific Model

2.4   From the list in 2.3 above, choose a mechanism and write a description of a specific model. Examples: Swingline 90 stapler, a 20-gauge Remington automatic shotgun, a Zenith portable radio—model number 461, or a MacIntosh Performa 550 computer.

2.5   Research a mechanism to find out a variety of possible options. Then create your own mechanism for a specific audience and purpose by adding your choice of options to the basic mechanism. After studying a sampling of brochures and fliers advertising mechanisms, plan and present a sales brochure to advertise your mechanism. Use appropriate visuals (see Chapter 11).

Oral Descriptions of Mechanisms

2.6   Give an oral description of a mechanism in 2.3 above. Ask your classmates to evaluate your speech by filling in the Evaluation of Oral Presentations on page 459, or one like it.

2.7   Give orally the specific description of a mechanism in 2.4 above. Use a visual of the mechanism, either a sample mechanism, a picture, or a line drawing, or arrange for the class to visit a shop or laboratory and view the mechanism. Ask your classmates to evaluate your speech by filling in the Evaluation of Oral Presentations on page 459, or one like it.

Using Part II Selected Readings

2.8   Read the excerpt from *Zen and the Art of Motorcycle Maintenance,* by Robert M. Pirsig, pages 526–530. Identify a mechanism that you have had to cope with recently. Write a description of the mechanism.

2.9   Read "The Science of Deduction," by Sir Arthur Conan Doyle, pages 532–535. Under "Suggestions for Response and Reaction," page 536, write Number 5.

2.10   Read the excerpt from "Non-Medical Lasers: Everyday Life in a New Light," by Rebecca D. Williams, pages 564–569. Under "Suggestions for Response and Reaction," page 569, respond to Number 2.

PLAN SHEET FOR DESCRIBING A REPRESENTATIVE MODEL OF A MECHANISM

Analysis of Situation Requiring a Representative Description of a Mechanism

What is the general mechanism that I will describe?

Who is my intended audience?

How will the intended audience use the description? For what purpose?

Will the final presentation of the description be written or oral?

Analysis of the Mechanism

Definition, identification, and/or special features:

Function, use, or purpose:

Who uses:

When:

Where:

Physical characteristics:
    Size:

    Shape:

    Weight:

    Material:

    Other:

Major parts and a description of each:

Parts	Function	Physical characteristics

How the Parts Work Together

Variations or Special Features

Other Materials

Types and Subject Matter of Visuals I Will Include

The Title I May Use

Sources of Information

PLAN SHEET FOR DESCRIBING A SPECIFIC MODEL OF A MECHANISM

Analysis of Situation Requiring a Description of a Specific Mechanism

What is the specific mechanism that I will describe?

Who is my intended audience?

How will the intended audience use the description? For what purpose?

Will the final presentation of the description be written or oral?

Analysis of the Mechanism

Who uses:

When:

Where:

Physical characteristics:
   Size:

   Shape:

   Weight:

   Material:

   Other:

Special features to be emphasized:

Variations available:

Types and Subject Matter of Visuals I Will Include

The Title I May Use

Sources of Information

Chapter 3

Definition: Explaining What Something Is

Outcomes

Upon completing this chapter, you should be able to

- List the conditions under which a term should be defined
- Demonstrate a variety of ways to define terms
- Demonstrate the three steps in arriving at a formal definition
- Give a formal definition
- Give an extended definition
- State definitions according to their purpose and the knowledge level of the reader

Introduction

You might sometimes dismiss the idea of including definitions in a written or oral presentation, thinking perhaps, "Why bother with definitions? Aren't there dictionaries around for people who run into words they don't know?" But giving very specific definitions is often indispensable in your writing and speaking.

Consider a common, frequently used word in the English language: *pitch*. Certainly, *pitch* is not a strange word to most people. A drafter writing about *pitch* for other drafters can be sure that they know the term refers to the slope of a roof, expressed by the ratio of its height to its span. An aeronautical technician, however, would automatically associate *pitch* with the distance advanced by a propeller in one revolution. A geologist would think of *pitch* as being the dip of a stratum or vein. A machinist thinks of *pitch* as the distance between corresponding points on two adjacent gear teeth or as the distance between corresponding points on two adjacent threads of a screw, measuring along the axis. To a musician, *pitch* is that quality of a tone or sound determined by the frequency or vibration of the sound waves reaching the ear. To a construction worker, *pitch* is a black, sticky substance formed in the distillation of coal tar, wood tar, petroleum, and such, and used for such purposes as waterproofing, roofing, and paving. To the average layperson, *pitch* is an action word meaning "toss" or "throw," as in "*Pitch* the ball." Such expressions as "It's *pitch* dark" or "I see the neighborhood's chief con man has a new *pitch*" illustrate other uses of *pitch*.

This one word illustrates quite well the importance of knowing when to define words or concepts for different readers, and it implies that the writer or speaker must know *how* to define.

Definition Adapted to Audience and to Purpose

In business and industry the need for defining a term frequently arises because many words have precise, specific meanings. You must know how and when to define a term, and you must word the definition for a particular purpose and according to the knowledge level of the audience. You must give a clear, accurate definition that is appropriate for the situation. See also Jargon, page 13.

Consider the following definitions of the Internet.

Definition 1

The Internet is an on-line service allowing users to retrieve information, send electronic mail, establish private links between sites, and connect sellers and customers.

Definition 2

The Internet is one of several bibliographic networks that give easy access to the catalogs of other libraries. A telecommunications network, the Internet is an informal organization made up of some 40 types of libraries—research, government, and academic. The catalog of each library is the main data. Once a user gets on-line, that user can search the catalog of any of the libraries. Quickly and easily the user can locate a needed book or periodical. Other resources available on the Internet are on-line databases such as full-text periodicals, books, and government documents; statistics; business and travel information; and indexes. Software, pictures, and sounds can be located and downloaded from the Internet.

Definition 3

The Internet is the nation's largest electronic information exchange. It offers a potential user three paths to connect to the Internet.

1. Computer connection to a LAN (Local Area Network) whose server is an Internet host
2. Dial access to an Internet host using SLIP (Serial Line Internet Protocol) or PPP (Point-to-Point Protocol)
3. Dial access into an on-line service

The network connection gives the user access to everything the Internet has to offer: mail, news, Gopher servers, Web servers, and more. The SLIP/PPP connection also gives access to everything the Internet has to offer but at slower speeds. The on-line service connection gives you access to the Internet services your on-line service offers. Some services offer only e-mail; others offer e-mail, news groups, Gopher services, and more.

Each of the three definitions above is designed for a different audience and a different purpose. There are occasions when each of the definitions is appropriate. The first definition is very general, designed to give a lay audience a basic understanding of what the Internet is. Such a definition might appear in a general desk dictionary or a listing of various computer networks. (Remember that a general desk dictionary does not include many of the specialized terms and meanings that technicians often need.) The second definition, which gives more detailed information, is for a reader interested in using the Internet for library research or for a librarian who is considering connection to the Internet services. The third definition is for a reader who wants to connect to the Internet and understand the services each connection makes available to the user. Such a definition might appear in a computer magazine article.

When to Define a Term

A writer or a speaker should define a term under specific circumstances.

- The term is unfamiliar to the audience.
  A term is defined when the audience might not know its meaning but should. An audience needs such a definition simply because they are unfamiliar with the term. A metallurgist, communicating with a lay adult audience, would probably have to explain the use of *anneal* or *sherardize;* or an electronics

technician, writing a repair memorandum for a customer's information, would have to explain the meaning of *zener diode* or *signal-to-noise ratio*.

- The term has multiple meanings.

    A term that has multiple meanings is defined. A term may not be clear to an audience because it has taken on a meaning different from that which the audience associates with it. For instance, if the company nurse instructs the carpenter to put a new 2" by 4" and 4" by 8" on the cuts on his leg each morning, the carpenter may leave in disbelief—and certainly in misunderstanding. To the nurse, of course, 2" by 4" and 4" by 8" are common terms for *gauze squares* in those inch dimensions used in dressing wounds. But to the carpenter 2" by 4" and 4" by 8" mean *boards* in those inch dimensions. If there is any possibility that the intended meaning of a term may be misunderstood, define it.

- The term is used in a special way in a presentation.

    Occasionally a writer or speaker gives a term or concept a special meaning within a presentation. The writer or speaker then certainly lets the audience know exactly how the term is being used.

Ways to Define a Term

Terms are defined in a variety of ways. The extent to which a term is defined, thus the length of a definition, depends on the writer's or speaker's purpose and the knowledge level of the audience.

Sometimes merely a word or a phrase is sufficient explanation. Notice the various ways in which definitions are given in the following sentences.

Three Examples
- The optimum (most favorable) cutting speed of cast iron is 100 feet per minute.
- Consider, for instance, lexicographers—those who write dictionaries.
- The city acquired the property by eminent domain, that is, the legal right of a government to take private property for public use.

The miniglossary of cooking terms, shown below, is from a nutrition textbook.* In each example, a brief, yet satisfactory, definition is supplied.

Miniglossary of Cooking Terms

Bake to cook in an oven surrounded by heat.

Braise to cook by browning in fat and then simmering in a covered container with a little liquid.

Broil to cook quickly over or under a direct source of intense heat, allowing fats to drip away.

Steam to cook foods suspended over boiling water.

Poach to cook foods (fish, an egg without its shell, etc.) in water near the boiling point.

Sauté (saw-TAY, the French word for stir-fry) to cook in a pan using little fat; foods are stirred frequently to prevent sticking.

*Reprinted with permission from Mary A. Boyle and Gail Zyla, *Personal Nutrition,* 2nd ed., West Publishing Company, 1992.

At other times a definition may require a sentence, several sentences, or an entire paragraph. Occasionally a definition requires several paragraphs for a clear explanation, as when explaining an unfamiliar idea or item that is the central focus of a presentation. The following paragraph defines "adolescence."

> Adolescence is the period of physical and emotional transition when a person discards childish ways and prepares for the duties and responsibilities of adulthood. During this period the adolescent is neither child nor adult, although having the characteristics of both. "To grow to maturity," the meaning of the derivative Latin word *adolescere,* describes the state of the adolescent. This period of transition is customarily divided into three phases: preadolescence, or puberty, the approximately two-year period of sexual maturation; early adolescence, extending from the time of sexual maturity to the age of 16 1/2; and late adolescence, extending from the age of 16 1/2 to 21.

Further, the extent to which a term is defined depends on the complexity of the term. In a presentation for the general public, a definition of Ohm's law, for instance, undoubtedly requires more details than a definition of a Phillips screwdriver because few readers or listeners are familiar with the principles and terms of a specialized field like electronics and thus would have difficulty understanding the unfamiliar concept of Ohm's law. Most people, on the other hand, are familiar with a screwdriver; thus a definition of a Phillips screwdriver would be concerned only with how the Phillips is different from other screwdrivers. In addition, an abstract term like *Ohm's law* is usually more difficult to explain than a concrete item like *Phillips screwdriver*. The general knowledge and interest of the audience are certainly important in determining just how far to go in giving a definition. Most important of all, however, is the *purpose* for which the definition is given. In defining a term you must keep clearly in mind your purpose and a way to accomplish it. It may be that a definition is merely parenthetical. Or it may be that a definition is of major significance, as when an audience's understanding of a key idea hinges on the comprehension of one term or concept. And occasionally, though not often, an entire presentation is devoted to an extended definition.

Formal Definition

A well-established, three-step method for giving a formal definition includes stating

1. The term (species)—the word to be defined
2. The class (genus)—the group or category of similar items
3. The distinguishing characteristics (differences)—essential qualities that set the term apart from all other terms of the same class.

Class

The class identifies a group or category of similar terms. For instance, a *chair* is a piece of furniture; a *stethoscope* is a medical listening instrument. When determining a class, be as precise and specific as possible. Placing a screwdriver in the class of small hand tools is more specific than placing it in a class such as objects, instruments, or pieces of hardware. The more specific a class is, the simpler it is to give the distinguishing characteristics, that is, the qualities that separate the term being defined from all other terms in the class.

For instance, if, in a definition of *trowel,* the class is given simply as a device, many distinguishing characteristics will have to be given to set a trowel apart from innumerable other devices (staple guns, rockets, tractors, phonographs, keys, pencil sharpeners, washing machines, etc.). If however the class for *trowel* is given as "a hand-held, flat-bladed implement," then the distinguishing characteristics might be narrowed to "having an offset handle and used to smooth plaster and mortar."

Distinguishing Characteristics

The distinguishing characteristics make the definition accurate and complete. They identify the essential qualities that set the term apart from all other terms of the same class. A *chair* (term) is a piece of furniture (class). But there are other pieces of furniture that are not chairs; thus the chair as a piece of furniture must be distinguished from all other pieces of furniture. The characteristics "that has a frame, usually made of wood or metal, forming a seat, legs, and backrest and is used for one person to sit in" differentiate the chair from other items in the class. The formal definition would be: "A chair is a piece of furniture that has a frame, usually made of wood or metal, forming a seat, legs, and backrest and that is used for one person to sit in."

To avoid confusing peripheral information with the essential distinguishing characteristics, you must have an understanding of the *essence* of the term being defined. Essential to the nature of brick, for instance, is that it is made out of baked clay. The color, methods of firing, size, shape, cost, and so on are peripheral information. Essential to the nature, the essence, of a refrigerator is that it preserves food by keeping it at a constant, cold temperature. Information that this large home appliance may be self-defrosting or may come in combination with a freezer has no place, ordinarily, in a formal definition.

Study the following examples.

Term	Class	Distinguishing Characteristics
Program	set of organized instructions	composed of words and symbols and used to direct the performance of a computer
Rivet	permanent metal fastener	shaped like a cylinder with a head on one end; when placed in position, the opposite head formed by impact
Photojournalist	photographer	works with newsworthy events, people, and places; works for newspapers, magazines, television; must possess skill in using cameras and in composing pictures
Scrolling	movement on a computer screen	vertically from one line to another, up or down; continuously through the text, or horizontally one-quarter inch at a time or continuously. Allows user to move from page to page within an open document

Now look at the details restated in complete sentences.

- A program (term) is a set of organized instructions (class) composed of words and symbols and used to direct the performance of a computer (distinguishing characteristics).

- A rivet (term) is a permanent metal fastener (class) that is shaped like a cylinder and has a head on one end. When placed in position, the opposite head is formed by impact (distinguishing characteristics).
- A photojournalist (term) is a photographer (class) who works with newsworthy events, people, and places. This person works for newspapers, magazines, and television and must possess skill in using cameras and in composing pictures (distinguishing characteristics).
- Scrolling (term) is a movement on a computer screen (class) vertically from one line to another, up or down; continuously through the text; or horizontally one-quarter inch at a time or continuously. Scrolling also allows the user to move from page to page within an open document (distinguishing characteristics).

Study the definitions below from a TransAmerica Life Companies annuity newsletter to investors.*

An Annuity Primer

When you bought your annuity contract, you received a multi-paged document that defined all the terms, conditions and contingencies of your annuity. To help you better understand your annuity, here is a review of some of the most frequently used annuity terms.

Accumulation Period
The time period in which you pay money into your annuity contract and allow interest to compound, allowing funds to build up. These funds will be the source of annuity payments some time in the future.

Deferred Annuity
An annuity that will provide payments sometime in the future, generally after the annuitant has retired.

Fixed Annuity
An annuity that earns a fixed rate of return. This rate is usually locked in for a year or more. When the time comes to receive income from the annuity, payments will be in fixed-dollar amounts.

Immediate Annuity
An annuity contract that provides payments either immediately or within a relatively short period of time, such as one year.

Principal
The amount of money placed in the annuity contract, separate from the interest earned on those funds. Also referred to as the "premium" amount.

The left column headnote clearly identifies the purpose of the definitions, "to help you better understand your annuity." The audience is anyone who is an investor; thus the definitions are stated simply and concisely.

As the examples above show, a formal definition covers in a prescribed order (term, class, distinguishing characteristics) only one meaning of a word.

Inadequate Formal Definitions

The formal definition, if it is to be adequate, must give the term, state the class as specifically as possible, and give characteristics that are really distinguishing. Such definitions as "Scissors are things that cut" or "Slicing is something you ought not do in golf" or "Oxidization is the process of oxidizing" or "A stadium is where games are played" are inadequate if not completely useless. These definitions give little if any insight into what the term means.

*Reprinted by permission from TransAmerica Life Companies.

The definitions "Scissors are things that cut" or "Scissors are used for cutting" equate scissors with razors, saw, drills, cookie cutters, plows, knives, sharp tongues, and everything else that cuts in any way. The definitions are not specific in class ("things" and no class named in the second definition), and the characteristics ("that cut" or "used for cutting") can hardly be called distinguishing. A more adequate definition might be "Scissors are a two-bladed cutting implement held in one hand. The pivoted blades are pressed against opposing edges to perform the cutting operation."

"Slicing is something you ought not do in golf" gives *no* indication of what slicing is. The so-called definition is completely negative. The class ("something") is about as nonspecific as it can be. Furthermore, there are no distinguishing characteristics ("you ought not do in golf" is meaningless—a golfer ought not slow down the players coming up behind, ought not move other players' balls, ought not damage the green, ought not start at any hole except the first, ought not cheat on the score, etc.). A better definition is this: "Slicing is the stroke in golf that causes the ball to veer to the right."

"Oxidization is the process of oxidizing" as a definition is completely useless because the basic word *oxidize* is still unexplained. If a person knows what oxidizing means, he or she has a pretty good idea of what oxidization means; however, if someone has knowledge of neither word, the stated "definition" is only confusing. This kind of definition, which uses a form of the term as either the class or the distinguishing characteristic, is called a circular definition because it sends the reader in circles—the reader never reaches the point of learning what the word means. Oxidization might be defined more satisfactorily as the following: "Oxidization is a process of combining oxygen with a substance that reduces its strength."

"A stadium is where games are played" is inadequate as a definition because no class is given and the distinguishing characteristic is not specific. The "is where" or "is when" construction does not denote a class. Furthermore, only certain types of games—sports events—are typically associated with a stadium. An adequate definition is this: "A stadium is a large, usually unroofed building where spectator events, primarily sports events, are held."

Extended Definition

The extended definition gives information beyond stating the essence, or primary characteristic, of a term. The extended definition is concerned with giving enough information so the audience can gain a thorough understanding of the term. This definition in depth may contain such information as synonyms; origin of the term or item; data concerning its discovery and development; analysis of its parts; physical description; mention of necessary conditions, materials, or equipment; description of how it functions; explanation of its uses; instructions for operating or using it; examples and illustrations; comparisons and contrasts; different styles, sizes, and methods; and data concerning its manufacture and sale. The central focus in an extended definition is on stating what something is by giving a full, detailed explanation of it.

The extended definition is closely related to other forms of explanation, particularly instructions and process explanations (see Chapter 1) and descriptions of mechanisms (see Chapter 2), and often includes them.

Following is an example of a student-written extended definition (including the Plan Sheet that was filled in before the paper was written). Marginal notes have been added to help in analyzing the definition.

PLAN SHEET
FOR GIVING AN EXTENDED DEFINITION

Analysis of Situation Requiring Definition

What is the term that I will define?
tornado

Who is my intended audience?
students involved in a tornado preparedness week

How will the intended audience use the extended definition? For what purpose?
for their knowledge and understanding

Will the final presentation of the extended definition be written or oral?
written

Formal Definition

Term:	Class:	Distinguishing characteristics:
tornado	*destructive windstorm*	*whirling winds (up to 400 miles an hour) accompanied by a funnel-shaped cloud that progresses in a narrow path over the land*

As Applicable

Synonyms:
omit

Origin of the term or item:
from Spanish <u>tronada</u>, "thunderstorm," from Latin <u>tonare</u>, "to thunder"

Information concerning its discovery and development:
omit

Analysis of its parts:
omit

Physical description:
appears like a rotating funnel cloud; extends from thundercloud toward ground; gray or black

Necessary conditions, materials, equipment:
(1) cold front; thunderstorm; (2) layers of air with contrasting temperature, moisture, density, and wind flow; cool, dry air from west or northwest; (3) warm, moist air near earth's surface; narrow bands of spiraling wind

113

Description of how it functions:
omit

Explanation of its uses:
omit

Instructions for operating or using it:
omit

Examples and illustrations:
Destructive power: combination of strong rotary winds and partial vacuum in center of vortex. Example: building has outside twisted and torn by winds; abrupt reduction of pressure in center of tornado causes building to "explode" outward; walls collapse or fall outward, windows explode, debris flies through air; train cars and airplanes moved

Comparisons and contrasts:
appearance of funnel cloud and thunderstorm like huge mushroom; sounds like roar of hundreds of airplanes or several trains racing through the house

Different styles, sizes, methods:
omit

Data concerning its manufacture and sale:
omit

When and where used or occurring:
anywhere in the world; central part of U.S. most common; in Southeast, most likely in March; in Midwest most likely in May, June; occur any time of day; most likely 3 to 7 P.M.

History:
March 18, 1925, 740 persons killed in MO, IN, KY, TN, and IL by a series of 8 tornadoes; 2,000 injured; millions of dollars of property damage. March 21, 1932, 268 killed in AL by tornado series; 2,000 injured. April 11, 1965, 257 killed in Midwest by series of 27 tornadoes; over 5,000 injured

Other:
omit

Types and Subject Matter of Visuals I Will Include

photograph of tornado; line drawings of stages of how a tornado forms

The Title I May Use

Extended Definition of Tornado

Sources of Information

pamphlets from the National Weather Service; <u>Weather Almanac</u>; <u>Encyclopedia Americana</u>

EXTENDED DEFINITION OF A DESTRUCTIVE KILLER: A TORNADO

Introduction
Anecdote to arouse
interest

"It sounded like a freight train. The wind had been blowing, it was raining hard, and it was humid. Then all of a sudden things got real still. The next thing I knew, windows blew out, the chimney crashed in, and a tree smashed through the roof." So reported a victim to her insurance adjustor. That kind of experience is repeated many, many times each year, often with the loss of hundreds of lives and millions of dollars of property destruction. The cause of such destructions in lives and property? A tornado.

Term to be defined

Questions focusing
on definition

Formal definition

Kinds of infor-
mation for further
definition

What is a tornado? How does a tornado form? What is a tornado like?

A tornado is a severe storm with winds whirling at up to 400 miles per hour, occurring primarily in the central area of the United States, usually in the spring. To explain what a tornado is—to define a tornado—requires explaining how a tornado forms and giving descriptive characteristics of a tornado.

First defining aspect:
formation

Formation of a Tornado

A tornado forms in several stages within a few hours—or even within minutes. A tornado usually forms along a cold front, and it begins to form several thousand feet above the earth. The formation requires layers of air with contrasting temperatures, moisture, density, and wind flow. There is warm, moist air near the earth's surface with spiraling cooler air above.

Conditions for
formation

Three stages in
formation with
visuals illustrating
the three stages

In the first stage, warm air in violent updrafts causes a thunderstorm, as shown in Figure 1. These warm updrafts in a thunderstorm are the basis for a tornado. In the second stage, shown in Figure 2, the updrafts of warm air begin to spiral around a column of downdrafts of cool air. Then in the third stage, shown in Figure 3, the spiraling winds gain speed as they are pulled toward the bottom of the funnel (the axis of rotation). The highest winds occur just above the ground where air pressure is the lowest. Objects are sucked upward and dropped as spiraling wind speed—higher in the vortex of the tornado—slows.

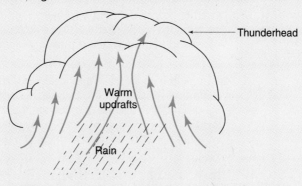

Visuals with a
number and a
caption

Figure 1. Updrafts of warm air during a thunderstorm.

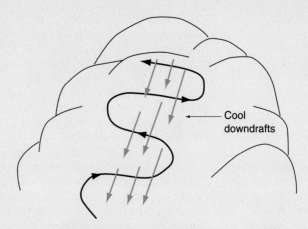

Figure 2. Warm updrafts spiraling around cool downdrafts.

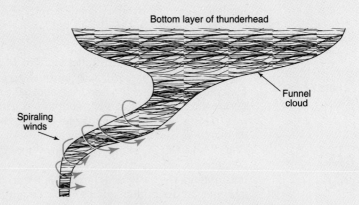

Figure 3. Spiraling winds at their
most destructive near the ground.

Labels for
components of a
tornado

Arrows to show wind
direction

Second defining
aspect:
characteristics

Listing of five
descriptive
characteristics

First characteristic

Comparison with
familiar objects

Second characteristic

Characteristics of a Tornado

Another way of defining a tornado is giving characteristics that
describe it. These characteristics include appearance, sound,
speed and path, destructive power, and occurrence.

Appearance

A tornado is shaped like a funnel. A tornado, a violently rotating
column of air descending from a thundercloud system toward
the ground, can also be described as looking like a mushroom. A
tornado looks foreboding—dark, gray, black. In the photograph
in Figure 4, notice the typical shape of a tornado.

Sound

The sound of a tornado has been described as being like the
roaring of hundreds of airplanes or like the sound of several

Photograph

Figure 4. Photograph of a tornado. (Courtesy of Bob Hodges)

Source of visual in parentheses

Comparison with familiar experience

freight trains racing through one's house. According to people who have heard the sound of a tornado, "It is a sound you never forget."

Third characteristic

Speed and Path
The speed and path of tornadoes may vary widely. Most tornadoes move about 30 miles an hour. However, some move very slowly and others may move as fast as 60 or more miles an hour. Winds may exceed 200 miles an hour and even reach 400 miles an hour.

Varying speeds
Example

The average tornado is about 300 to 400 yards wide and about 4 miles long. However, tornadoes have been a mile or more wide and 300 miles long. On May 26, 1917, a tornado traveled 293 miles across Illinois and Indiana over a period of 7 hours and 23 minutes.

The direction of movement is usually southwest to northeast.

Fourth characteristic

Destructive Power

Cause

The destruction caused by a tornado occurs by the combination of the strong rotary winds and the partial vacuum in the center of its vortex. For example, as a tornado passes over a building, the winds twist and tear the outside, and the abrupt reduction of pressure in the tornado's center causes the building to "explode" outward. Walls may collapse or fall outward, windows explode, and the debris from the building may be driven through the air.

Example

The destructive power of a tornado is immeasurable. Tornadoes have picked up train cars and airplanes and deposited them some distance away.

Fifth characteristic	**Occurrence**
Time	Tornadoes may occur at any time of the day, but they are most likely to occur during the afternoon between 3 and 7 P.M.
Location	While tornadoes may occur anywhere in the world, the central part of the United States is the most likely place. For the Southeast, tornadoes occur most likely in March. Midwestern states, especially Oklahoma, Kansas, Iowa, and Nebraska, have a peak tornado season during May and June.
Closing	**Closing**
Examples of major destruction	A tornado—within minutes, either singly or in a series—can wield untold destruction. On March 18, 1925, a series of eight tornadoes killed 740 persons in Missouri, Indiana, Kentucky, Tennessee, and Illinois; almost 2,000 individuals were injured, and millions of dollars of property damage occurred. On March 21, 1932, in Alabama, a series of tornadoes killed 268 and injured almost 2,000. On April 11, 1965, a series of 27 tornadoes in the Midwest killed 257 and injured over 5,000. This destructive killer strikes time and time again every year.
Restatement of title focus	

Intended Audience and Purpose

Whether a definition involves a word, a phrase, one or more sentences, a paragraph, or several paragraphs, the writer or speaker must be aware of the audience to whom the communication is directed and the purpose of the communication.

Definitions in General Reference Works

Professional writers are keenly aware of who will be reading their material and why. Consider, for instance, lexicographers—those who write dictionaries. Lexicographers know that people of all ages and with varying backgrounds turn to the dictionary to discover or ascertain the meanings that most people attach to words. Lexicographers are aware that dictionary definitions must be concise, accurate, and understandable to the general, or lay, reader. The following definition of *measles,* for instance, from a standard desk dictionary,* is adequate for its purpose.

> **mea • sles** / 'mē-zelz / *n pl but sing or pl in constr* [ME *meseles,* pl. of *mesel* measles, spot characteristic of measles; akin to MD *masel* spot characteristic of measles] (14c) **1a:** an acute contagious viral disease marked by an eruption of distinct red circular spots **b:** any of various eruptive diseases (as German measles) **2** [ME *mesel* infested with tapeworms, lit., leprous, fr. OF, fr. ML *misellus* leper, fr. L, wretch, fr. *misellus,* dim. of *miser* miserable]: infestation with or disease caused by larval tapeworms in the muscles and tissues

---

*By permission. From *Webster's Tenth New Collegiate Dictionary* © 1994, by Merriam-Webster, Inc., publisher of the Merriam-Webster® dictionaries.

A specialized dictionary or encyclopedia is not designed to give a concise definition for a lay audience (a cross section of people). The specialized reference book aims its information at a well-defined, select audience. The following entry, "Measles," from the *McGraw-Hill Encyclopedia of Science and Technology*,* illustrates such a definition. Note the number of cross-references suggested for the reader.

Measles

An acute, highly infectious viral disease, with cough, fever, and maculopapular rash. It is of worldwide endemicity *See Animal Virus*

The infective particle is a ribonucleic acid (RNA) virus about 100–150 nanometers in diameter, measured by ultrafiltration, but the active core is only 65 nm as measured by inactivation after electron irradiation. Negative staining in the electron microscope shows the virus to have the helical structure of a paramyxovirus with the helix being 18 nm in diameter. Measles virus will infect monkeys easily and chick embryos with difficulty; in tissue cultures the measles virus may produce giant multinucleated cells and nuclear acidophilic inclusion bodies. Measles, canine distemper, and bovine rinderpest viruses are antigenically related. *See Embryonated egg culture; Paramyxovirus; Tissue culture; Viral inclusion bodies.*

The virus enters the body via the respiratory system, multiplies there, and circulates in the blood. Prodromal cough, sneezing, conjunctivitis, photophobia, and fever occur, with Koplik's spots in the mouth. A rash appears after 14 days' incubation and persists 5–10 days. Serious complications may occur in 1 out of 15 persons; these are mostly respiratory (bronchitis, pneumonia), but neurological complications are also found. Encephalomyelitis occurs rarely. Permanent disabilities may ensue for a significant number of persons. Laboratory diagnosis (which is seldom needed since 95% of cases have the pathognomonic Koplik's spots) is by virus isolation in tissue culture from acute-phase blood or nasopharyngeal secretions, or by specific neutralizing, hemagglutination-inhibiting, or complement-fixing antibody responses.

In unvaccinated populations, immunizing infections occur in early childhood during epidemics which recur after 2–3 years' accumulation of susceptible children. Transmission is by coughing or sneezing. Measles is spread chiefly by children during the catarrhal prodromal period; it is infectious from the onset of symptoms until a few days after the rash has appeared. By the age of 20 over 80% of persons have had measles. Second attacks are very rare. Treatment is symptomatic.

At one time, prevention was limited to use of gamma globulin, which protects for about 4 weeks, and can modify or prevent the disease. *See Immunoglobulin.*

Killed virus vaccine was once available, but it should not be used, as certain vaccinees become sensitized and develop local reactions when revaccinated with live attenuated virus, or a severe illness upon contracting natural measles. Live attenuated virus vaccine effectively prevents measles; vaccine-induced antibodies persist for years. Prior to the introduction of the vaccine, over 500,000 cases of measles occurred annually in the United States. However, with emphasis on immunization, the number of cases declined to 3,000 annually. *See Biologicals; Hypersensitivity; Skin test.*

Measles antibodies cross the placenta and protect the infant during the first 6 months of life. Vaccination with the live virus fails to take during this period. Timely use of vaccine should lead to the eradication of measles. That goal has been thwarted, however, by repeated outbreaks among unvaccinated inner-city preschool-age children. Lowering

---

*Reprinted by permission from *McGraw-Hill Encyclopedia of Science and Technology,* Seventh edition, Vol. 10 (New York: McGraw-Hill, 1992).

the minimum age for routine vaccination from 15 months to 9 months of age could prevent measles outbreaks, but children who receive single-antigen vaccine before their first birthday should be revaccinated with trivalent measles-mumps-rubella (MMR) vaccine at 15 months of age. Vaccination for measles is not recommended in persons with febrile illnesses, with allergies to eggs or other products that are used in production of the vaccine, and with congenital or acquired immune defects.

Because early lots of measles vaccine proved to be unstable, people vaccinated before 1980 may not be immune. This may explain the measles outbreaks that occurred in high schools and colleges during the late 1980s. When a measles case occurs among school children, therefore, any student at the school who was initially vaccinated before 1980 should be vaccinated again.

Measles virus is responsible for subacute sclerosing panencephalitis (SSPE, Dawson's inclusion body encephalitis), a rare chronic degenerative brain disorder. The disease develops a number of years after the initial measles infection. Virus is not localized only in brain tissues, since isolations have been made from lymph nodes. The presence of a latent intracellular measles virus in lymph nodes suggests a tolerant infection with defective cellular immunity. With the widespread use of measles vaccine, SSPE has almost disappeared as a clinical entity. SEE VIRUS INFECTIONS, LATENT, PERSISTENT, SLOW.

A laboratory-produced, defective (temperature-sensitive) mutant of measles virus has caused hydrocephalus when inoculated intracranially into newborn hamsters; this finding shows the need for caution in use of experimentally induced virus variants. SEE VIRUS, DEFECTIVE.

Joseph L. Melnick

Bibliography. F. L. Black, Measles active and passive immunity in a worldwide perspective. *Prog. Med. Virol.,* 36:1–33, 1989; A. R. Hinman et al., Elimination of indigenous measles from the United States, *Rev. Infect. Dis.,* 5:538–545, 1983; S. Krugman, S. L. Katz, and T. C. Quinn (eds.), Measles: Current impact, vaccines and control, *Rev. Infect. Dis.,* 5:389–626, 1983; L. E. Markowitz et al., Patterns of transmission in measles outbreaks in the United States, 1985–1986, *N. Engl. J. Med.,* 320:75–81, 1989.

Definition as Part of a Longer Communication

Giving a definition may be the main purpose of a communication; perhaps more often, however, a definition is an integral part of a longer communication. Consider, for instance, the paragraph definition of the term *report* on page 305 in Chapter 8 Reports. While the chapter is concerned with much more than definition, the definition of the focal term *report* serves as a framework for the entire chapter.

The article "Battle for the Soul of the Internet," pages 500–507, includes as a part of the introduction a paragraph definition of the Internet. A definition of the Internet is also included as an answer to a frequently asked question, What is the Internet?, page 502.

The excerpt from "Non-Medical Lasers: Everyday Life In a New Light," pages 564–569, includes a section entitled "What Is a Laser?," page 564.

Visuals

Visuals are very helpful in defining. For example, the definition of a tornado on pages 115–118 is much clearer because of the visuals showing the formation of a tornado and a photograph of an oncoming tornado. Look on page 565 at the simplified diagram of a common laser. This diagram showing the major parts of the laser and explaining basically how it works helps the reader to understand what a laser is as well as what it does.

In the definition of concepts, visuals can be especially helpful. Consider, for instance, the following definition of *horsepower*. Note how each of the three distinguishing characteristics (raising 33,000 pounds, distance of 1 foot, in 1 minute) is visually illustrated below.

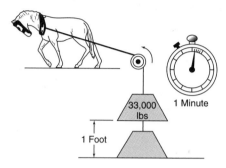

One horsepower is the rate of doing work equivalent to raising 33,000 pounds a distance of 1 foot in 1 minute.

Any time that you can illustrate a term you are defining and thus help your audience understand it more easily and clearly, do so. For a discussion of types of visuals and guidance in preparing them, see Chapter 11 Visuals.

Oral Definition

In giving a definition orally, you would generally follow the procedures suggested for a formal or extended definition. Suggestions for effective oral presentation of information appear in Chapter 10 Oral Communication.

General Principles in Giving a Definition

1. *Knowing when to define a term is essential.* Define a term if the audience does not know the meaning of a word but should, if you use a word in a meaning different from that which the audience ordinarily associates with it, or if you give a term a special meaning within a presentation.
2. *The extent to which a term should be defined depends on several factors.* The complexity of the term, the general knowledge and interest of the audience, and, primarily, the purpose for the definition determine the extent for defining a term.
3. *A formal definition has three parts: the term, the class, and the distinguishing characteristics.* The term is simply the word to be defined. The class is the group or category of similar items in which the term can be placed. The distinguishing characteristics are the essential qualities that set the term apart from other terms in the same class.
4. *An extended definition gives information beyond stating the essence, or the primary characteristic, of a term.* The extended definition includes such

information as origin, development, analysis of parts, physical description, function, and so on.

5. *The crucial factor in giving a definition is understanding the essence of the term being defined.* You must understand the nature of the object or concept to define it accurately.

Procedure for Giving an Extended Definition

An extended definition, whether one paragraph or several or an independent presentation or part of a longer whole, usually has two distinct parts: identification of the term and additional information. A formal closing is optional, although frequently the presentation ends with a comment or summarizing statement.

I. The identification of the term is usually brief.
   A. State the term to be defined.
   B. Give a formal definition.
   C. Indicate the reason for giving a more detailed definition.
   D. State the kinds of additional information to be given.
II. The additional information forms the longest part of the presentation.
   A. Select additional information (as applicable): synonyms; origin of the term or item; information concerning its discovery and development; analysis of its parts; necessary conditions, materials, equipment; description of how it functions; explanation of its uses; instructions for operating or using it; examples and illustrations; comparisons and contrasts; different styles, sizes, methods; and data concerning its manufacture and sale.
   B. Organize the selected additional information.
   C. Give the additional information, including whatever details are needed to give the audience an adequate understanding of the term.
   D. Use connecting words and phrases so that each sentence flows smoothly into the next and so that all the sentences in a paragraph and all paragraphs fit together as a unit.
   E. Include visuals to enhance understanding of the term.
III. Formal closing is optional although a comment or summarizing statement is often included.

Applications in Giving Definitions

Individual and Collaborative Activities

3.1 By giving the distinguishing characteristics, complete the following to make general sentence definitions.

1. An orange is a citrus fruit . . . .
2. A speedometer is a gauge . . . .
3. A hammer is a hand tool . . . .

4. Measles is a disease . . . .
5. A printer is a data output device . . . .
6. An ambulance is a vehicle . . . .
7. A library is a building . . . .
8. A molar is a tooth . . . .
9. An anesthetic is a drug . . . .
10. Handcuffs are a restraining device . . . .

3.2 Analyze the following definitions, noting their degree of accuracy and usefulness. Make whatever revisions are needed for adequate formal definitions for a lay audience.

1. A compass is for drawing circles.
2. A fire lane is where you should not park.
3. A T square, helpful in drawing lines, is a device used by drafters.
4. Immunity means to be immune to disease.
5. A crime is a violation of the law.
6. An anesthetic is a drug.
7. A board foot is a piece of material 1 inch thick, 12 inches long, and 12 inches wide.
8. Sterilization is the process of sterilizing.
9. A pictorial drawing is a drawing that is drawn from a drawing that was first drawn flat.
10. Airplanes without engines are called gliders.

Written Definitions

3.3 Write a formal definition of five of the following terms.

1. Architects' scale	16. Network	31. On-line service
2. Technology	17. Database	32. Cable
3. Computer language	18. Dividers	33. Lineup
4. Leader	19. Sawhorse	34. Disinfectant
5. Calipers	20. Depreciation	35. Arrest
6. Fax	21. Terminal	36. Upgrade
7. E-mail	22. Thermometer	37. Graphics
8. Antenna	23. Data	38. Multimedia
9. Feedback	24. Serum	39. Module
10. Market	25. Coagulation	40. Advertising
11. Thermostat	26. Antibiotic	41. Disk
12. Cellular phone	27. Memorandum	42. Airbrush
13. Consumer	28. Software	43. Flowchart
14. Herbicide	29. Enzyme	44. Management
15. Hydraulic lift	30. CD-ROM	45. Fee

3.4 Identify your major field. Then make a list of five words from your major (do not list words from application 3.3 above) that you should be thoroughly familiar with. Write an adequate formal definition of each of the words.

3.5 Choose from your major field a term that can be found in each reference work specified below. Note how the term is defined in each reference work. In presenting each definition, include the title of each reference work. You may prefer photocopying a lengthy definition.

1. Give the meaning of the term as stated in a standard desk dictionary.
2. Give the meaning as stated in a technical handbook or dictionary.
3. Give the meaning as stated in the *McGraw-Hill Encyclopedia of Science and Technology* and its yearbooks or a similar encyclopedia pertinent to your field. (Find the term in the index and then turn to the pages referred to.)
4. Write a paragraph explaining the differences between the definitions in the reference works, and the reasons for the differences.

3.6 Read the following excerpt "Clinical Laboratory Technologists and Technicians" from the *Occupational Outlook Handbook,* 1994–1995 edition.

1. Write a formal definition of (a) medical technologist and (b) medical laboratory technician. Use your own wording; do not copy phrases from the article.
2. Write a paragraph definition of a medical laboratory worker. Use your own wording; do not copy phrases and sentences from the article.

Clinical Laboratory Technologists and Technicians
Nature of the Work

Clinical laboratory testing plays a crucial role in the detection, diagnosis, and treatment of disease. Clinical laboratory technologists and technicians, also known as medical technologists and technicians, perform most of these tests.

Clinical laboratory personnel examine and analyze body fluids, tissues, and cells. They look for bacteria, parasites, or other microorganisms; analyze the chemical content of fluids; match blood for transfusions, and test for drug levels in the blood to show how a patient is responding to treatment. They also prepare specimens for examination, count cells, and look for abnormal cells. They use automated equipment and instruments that perform a number of tests simultaneously, as well as microscopes, cell counters, and other kinds of sophisticated laboratory equipment to perform tests. Then they analyze the results and relay them to physicians.

The complexity of tests performed, the level of judgment needed, and the amount of responsibility workers assume depend largely on the amount of education and experience they have.

Medical technologists generally have a bachelor's degree in medical technology or in one of the life sciences, or have a combination of formal training and work experience. They perform complex chemical, biological, hematological, immunologic, microscopic, and bacteriological tests. Technologists microscopically examine blood, tissue, and other body substances; make cultures of body fluid or tissue samples to determine the presence of bacteria, fungi, parasites, or other micro-organisms; analyze samples for chemical content or reaction; and determine blood glucose or cholesterol levels. They also type and cross-match blood samples for transfusions.

They may evaluate the effects a patient's condition has on test results, develop and modify procedures, and establish and monitor programs to insure the accuracy of tests. Some medical technologists supervise medical laboratory technicians.

Technologists in small laboratories perform many types of tests, while those in specialty laboratories or large laboratories generally specialize. Technologists who prepare specimens and analyze the chemical and hormonal contents of body fluids are *clinical chemistry technologists.* Those who examine and identify bacteria and other micro-organisms are *microbiology technologists. Blood bank technologists* collect, type, and prepare blood and its components for transfusions; *immunology technologists* examine elements and responses of the human immune system to foreign bodies. *Cytotechnologists,* who have specialized training, prepare slides of body cells and microscopically examine these cells for abnormalities which may signal the beginning of a cancerous growth.

Medical laboratory technicians generally have an associate degree from a community or junior college, or a diploma or certificate from a vocational or technical school. They perform routine tests and laboratory procedures. Technicians may prepare specimens and operate automatic analyzers, for example, or they may perform manual tests following detailed instructions. Like technologists, they may work in several areas of the clinical laboratory or specialize in just one. *Histology technicians* cut and stain tissue specimens for microscopic examination by pathologists and *phlebotomists* draw and test blood. They usually work under the supervision of medical technologists or laboratory managers.

In completing applications 3.7 and 3.8,

a. Fill in the appropriate Plan Sheet on pages 128–129, or one like it.
b. Write a preliminary draft.
c. Go over the draft in a peer editing group.
d. Revise. See Part III Revising and Editing and the checklists for revising, front and back endpapers.
e. Write the definitions.

3.7 Choose a term in your major field for writing extended definitions that serve different purposes. Review page 403, A Note About Plagiarism.

a. Define the term for a fifth-grade *Weekly Reader.*
b. Define the term for your English teacher, who wants to understand the term well enough to judge the accuracy and completeness of students' formal definitions of the term.
c. Define the term as if you were having an examination in your technical field and you were asked to write a paragraph definition of it.

3.8 Choose six terms that an audience new to your technical field needs to understand. Review the glossary on page 108. Note that this glossary appeared in a newsletter. Assume that you are producing a newsletter to give basic information about your technical field. Design a glossary to include in the newsletter to give basic definitions of the six terms.

Oral Definitions

3.9 As directed, adapt for oral presentations the definitions you prepared in the applications above. Ask your classmates to evaluate your speech by filling in the Evaluation of Oral Presentations on page 459, or one like it.

3.10   Read "Battle for the Soul of the Internet," by Philip Elmer-Dewitt, pages 500–507. Under "Suggestions for Response and Reaction," page 508,

a. Complete Number 1.
b. Write Number 3.

3.11   Read "The Powershift Era," by Alvin Toffler, pages 509–514. Under "Suggestions for Response and Reaction," page 515, answer Number 1.

3.12   Read the excerpt from *Zen and the Art of Motorcycle Maintenance,* by Robert M. Pirsig, pages 526–530.

a. In what ways is definition used in this selection?
b. Under "Suggestions for Response and Reaction," page 531, write Number 4.

3.13   Read "Clear Writing Means Clear Thinking . . . ," by Marvin H. Swift, pages 542–547. Under "Suggestion for Response and Reaction," page 547, write Number 4.

3.14   Read "Blue Heron," by Chick Wallace, page 550. Under "Suggestions for Response and Reaction," page 551, write Number 5.

3.15   Read "The Effects of Perceived Justice on Complainants' Negative Word-of-Mouth Behavior and Repatronage Intentions," by Jeffrey B. Blodgett, Donald H. Granbois, and Rockney G. Walters, pages 552–562. Under "Suggestions for Response and Reaction," page 563, answer Number 1.

3.16   Read "Non-Medical Lasers: Everyday Life in a New Light," by Rebecca D. Williams, pages 564–569.

a. In what ways is definition used in the selection?
b. Under "Suggestions for Response and Reaction," page 569, write Number 4.

3.17   Read "Journeying Through the Cosmos," by Carl Sagan, pages 570–575. Under "Suggestions for Response and Reaction," page 576, answer Number 1.

3.18   Read the commentary on "Work," by William Bennett, pages 577–578. Under "Suggestions for Response and Reaction," page 578, complete Number 1.

3.19 Read "The Active Document," by David D. Weinberger, pages 579–584. Under "Suggestions for Response and Reaction," page 584, write Number 2.

3.20 Read "The Power of Mistakes," by Stephen Nachmanovitch, pages 585–587. Under "Suggestions for Response and Reaction," page 588,

a. Answer Number 1.
b. Write Number 2.

PLAN SHEET
FOR GIVING AN EXTENDED DEFINITION

Analysis of Situation Requiring Definition

What is the term that I will define?

Who is my intended audience?

How will the intended audience use the extended definition? For what purpose?

Will the final presentation of the extended definition be written or oral?

Sentence Definition

Term:	Class:	Distinguishing characteristics:

As Applicable

Synonyms:

Origin of the term or item:

Information concerning its discovery and development:

Analysis of its parts:

Physical description:

Necessary conditions, materials, equipment:

Description of how it functions:

Explanation of its uses:

Instructions for operating or using it:

Examples and illustrations:

Comparisons and contrasts:

Different styles, sizes, methods:

Data concerning its manufacture and sale:

When and where used or occurring:

History:

Other:

Types and Subject Matter of Visuals I Will Include

The Title I May Use

Sources of Information

Chapter 4

Analysis Through Classification and Partition: Putting Things in Order

Outcomes

131

Chapter 4
Analysis Through
Classification and
Partition: Putting
Things in Order

Upon completing this chapter, you should be able to

- Define classification
- State the basis of division into categories in a classification system
- Select for a classification system a basis of division that is useful and purposeful
- Set up a classification system whose categories are coordinate, mutually exclusive, nonoverlapping, and complete
- Present classification data in outlines, in verbal explanations, and in visuals
- Select an appropriate order for presentation of classification categories
- Give an analysis through classification
- Define partition
- State the basis of division in a partition system
- Select for a partition system a basis of division that is useful and purposeful
- Set up a partition system whose divisions are coordinate, mutually exclusive, nonoverlapping, and complete
- Present partition data in outlines, in verbal explanations, and in visuals
- Select an appropriate order for presentation of partition divisions
- Give an analysis through partition
- Give an analysis using both classification and partition
- Describe situations in which an employee would need to know how to give an analysis using classification or partition, or both

Introduction

We as human beings try to make sense out of the world in which we live. We try to see how certain things are related to other things. We try to impose some kind of order on our environment. More specifically, the technical student tries to see how a skill in building trades or in engineering technology or health occupations is related to getting and keeping a good job and being able to support a family. Or perhaps the student is trying to devise a practical plan for getting enough money for college expenses.

Persons on the job try to make sense out of the industrial and business world in which they work. There are situations in which they need to put things in some kind of organized relationship. The situation may be a request from a supervisor for a list of parts that must be replaced in a machine. The situation may be a weekly report of the number of units produced in a particular department. The situation may be a memorandum to the division head on the problems encountered in a new manufacturing process. The situation may be a report to the doctor on the temperature changes of a patient.

All of these kinds of situations, whatever their nature and wherever they are experienced, call for analysis—looking at a subject closely so that it can be put into a useful, meaningful order. Establishing an order, or relationship, is the basic step in solving a problem, whether the problem is how to reduce air pollution, how to operate a blood bank more efficiently, or how to improve a variety of cotton.

The purpose of this chapter is to help you give order to information—to give order by classifying subjects into related groups or by partitioning a subject into its components.

Definition of Classification

Classification is a basic technique in organization, and thus in writing and speaking. It recognizes that different items have similar characteristics and sorts these items into related groups. In a letter of application, for instance, details about places of former employment, dates of employment, or names of supervisors all have similar characteristics in that they all have to do with work experience. In organizing the letter of application, items could be sorted into one group on the basis of work experience, into another group on the basis of education, and so on. Classification, then, is the grouping together, according to a specified basis, of items having similar characteristics.

Note the following examples of classification and the format used to present the information.

In preparing to explain how batteries generate power, batteries might first be broadly classified as primary and secondary types. The explanation might be given in simple paragraph form.

> Batteries are broadly classified as primary and secondary types. Primary batteries generate power by irreversible chemical reaction and require replacement of parts that are consumed during discharge (or, more commonly, the batteries are discarded when discharged). Secondary batteries, on the other hand, involve reversible chemical reactions in which their reacting material will be restored to its original "charged" state by applying a reverse, or charging, current. In general, primary batteries provide higher energy density, specific energy, and specific power than do secondary batteries.

A more extensive classification may be given using a chart format with illustrations, as shown in a Black & Decker owner's manual for a power saw that uses circular blades. The chart classifies the blades according to type. Each type of blade has a different tooth shape that determines what the blade can cut most effectively and efficiently. The chart showing the types of blade and the tooth shapes appears on page 133.*

Notice the effective combination of the verbal and the visual material. To describe each saw blade verbally would be difficult; the visual allows the reader (audience) to picture immediately what each blade looks like and then choose the appropriate blade for a particular cutting job (purpose).

A chapter on safety in a manual teaching instrumentation classifies fire extinguishers according to type of extinguishers, types of fires, and method of operation. A single chart (see page 134) shows the multiple classifications systems quickly and effectively.

Almost any literate person could use this chart to select an appropriate fire extinguisher for a specific type of fire and know the basic method for operating the extinguisher.

---

*Reprinted with the permission of the Black & Decker Corporation.

133

Chapter 4
Analysis Through
Classification and
Partition: Putting
Things in Order

TYPE OF BLADE	TOOTH SHAPE
COMBINATION Chisel tooth configuration means this blade is the fastest cutting blade in our line. Specifically designed for general-purpose ripping and cross-cutting where the finish of the cut is not critical.	
MASTER COMBINATION General-purpose blade designed for ripping, cutting off and mitering wood where a fine, smooth cut is needed. No sanding is necessary.	
FRAMING/RIP An all-purpose blade for smooth, fast cutting in any direction. Rips, crosscuts, miters, etc. Gives especially fast, smooth finishes when cutting with the grain of both soft and hard woods.	
HOLLOW GROUND PLANER Specially ground for satin-smooth finish cuts (cross-cuts, rips and miters) in all solid woods. A professional quality blade for use in cabinet work, furniture, etc. Specifically designed to make extremely smooth cuts in wood.	
FLOORING For use where nails or other metal objects may be encountered, such as cutting reclaimed lumber, flooring, opening crates. Allows crosscuts as well as miters.	
HOLLOW GROUND PLYWOOD Special taper grinding on the sides of this thin-rim blade gives an absolutely smooth cut in plywood, veneers and laminates, etc. Can be used in crosscutting and mitering for a professional finish on all types of cabinet work.	
CROSS-CUT Specifically designed for smooth, fast cutting cross the grain of both hard and soft woods where finish is an important factor. May also be used for rips and crosscuts on extremely hard woods.	

Classification is used in all aspects of life. Because of classification, we can find books in a library, apples in a supermarket, a pair of shoes in a department store, or a drinking glass in the kitchen. Military personnel can review the different types of airplane camouflage paint to decide the kind of camouflage design best suited for planes in desert warfare, a gardener can choose the proper spade to plant a tree, a journalist can choose a computer program appropriate for his or her needs, and a football coach can select players according to position on the team. Classification gives order and organization, and it can be shown in a variety of ways, depending on audience and purpose.

The Process of Classification

Before you can classify information, you need to know and understand some basic guidelines. These guidelines are discussed under the following topics: items that

KNOW YOUR FIRE EXTINGUISHER

TYPE EXTINGUISHER	WATER TYPE				FOAM	CARBON DIOXIDE	DRY CHEMICAL	
	STORED PRESSURE	CARTRIDGE OPERATED	WATER PUMP TANK	SODA ACID	FOAM	CO2	CARTRIDGE OPERATED	STORED PRESSURE
TYPES OF FIRES								
CLASS A: WOOD, PAPER, TRASH HAVING GLOWING EMBERS	YES	YES	YES	YES	YES	NO	NO	NO
CLASS B: FLAMMABLE LIQUIDS, GASOLINE, OIL, PAINT, GREASE, ETC.	NO	NO	NO	NO	YES	YES	YES	YES
CLASS C: ELECTRICAL EQUIPMENT	NO	NO	NO	NO	NO	YES	YES	YES
CLASS D: COMBUSTIBLE METALS	*	*	*	*	*	*	*	*
METHOD OF OPERATION	SQUEEZE HANDLE OR TURN VALVE	TURN UPSIDE DOWN AND BUMP	PUMP HANDLE	TURN UPSIDE DOWN	TURN UPSIDE DOWN	PULL PIN, SQUEEZE LEVER	RUPTURE CARTRIDGE, SQUEEZE LEVER	PULL PIN, SQUEEZE LEVER

*DO NOT USE FIRE EXTINGUISHER, SMOTHER FIRE WITH DRY SAND, GRAPHITE, DIRT, OR SODA ASH.

SOURCE: Davis Instrumentation Training Program, *Instrumentation Installer Course Outline,* Davis Electrical Constructors, Inc., Greenville, SC.

can be classified, bases for classification, and characteristics of a classification system. An example of the classification process is also included.

Items That Can Be Classified

Any group of items plural in meaning can be classified. These items can be objects, concepts, or processes; the items may be classified into a variety of categories. Objects such as men's beach robes might be classified by size, style, fabric, color, and so on. Concepts or ideas, such as the cause of the decline of unions in the United States, may make use of the principles of classification; the causes, for instance, may be classified according to significance, origin, historical influences, and so on. Process explanations frequently employ classification—methods of desalinizing water may be classified according to cost, required time, materials needed, and so on, or particular steps may be classified as one phase of a desalinization method.

Bases for Classification

Usefulness

The bases on which classifications are made should be useful. Classification of trees according to bark texture or according to leaf structure would be of significance to one interested in botany; classification of trees according to the climate and geographical area where they will grow best would be important information for a wholesale nursery. It would take a stretch of the imagination, however, to under-stand the usefulness of classifying trees according to the number of leaves pro-duced and shed over a fifty-year span.

135

Chapter 4
Analysis Through
Classification and
Partition: Putting
Things in Order

Audience and Purpose

The usefulness of a classification system depends on its audience and purpose. And depending on the audience and purpose, the writer or speaker as classifier will em-phasize certain aspects of the subject. Suppose, for instance, that a salesperson in a garden center has three pieces of mail that must be answered regarding lawn mow-ers. First, there is a letter from a man who wishes to purchase a lawn mower that he can start easily. He wants suggestions for such models. Second, there is a letter from a meticulous gardener who gives certain motor, cutting blade, and attachment specifications. She would like suggestions for several models that most nearly meet her requirements. Third, there is a memorandum from the store manager requesting a list of the best-selling lawn mowers. Obviously each of the three persons is seek-ing quite different information for different purposes—but the information concerns the same group of lawn mowers, the lawn mowers that the garden center stocks. In order to meet each request, the salesperson must classify the lawn mowers accord-ing to at least three different bases: ease in starting (for the man); motor, cutting blade, and attachment specifications (for the gardener); and popularity of models (for the store manager).

Characteristics of a Classification System

The categories in a classification system, as illustrated below in organizing the in-formation about a student body, must be *coordinate, mutually exclusive, nonover-lapping,* and *complete.* A classification system must have all of these qualities if it is to be adequate.

Coordinate

The categories in a classification system must be coordinate, or parallel. The groups, or categories, that items are sorted into must be on the same level in grammatical form and content. For instance, classification of the refrigerators in Mr. Lee's appli-ance store, according to source of power, into the following categories is inade-quate: natural gas, butane gas, and electric. The error is in the grammatical form of the word *electric,* which is an adjective; *natural gas* and *butane gas* here are nouns. These categories, however, are adequate: natural gas, butane gas, and electricity.

Categories must be coordinate in content as well as in grammatical form. The classification of automobile tires as whitewall, blackwall, red stripe, and tubeless is not coordinate because *tubeless* does not refer to the same content, or substance, that *whitewall, blackwall,* and *red stripe* do. These three terms have to do with dec-orative coloring; *tubeless* does not.

See also Parallelism in Sentences, pages 621–622, and Conjunctions, pages 615–616.

Mutually Exclusive

The categories in a classification system must be mutually exclusive; that is, each cat-egory must be independent of the other categories in existence. A category must be composed of a clearly defined group that would still be valid even if any or all of the other categories were unnecessary or nonexistent. For instance, if a shipment of coats is classified according to the amount of reprocessed wool each coat contains, and the categories are designated "Coats in Group A," "Coats with more reprocessed wool than Group A," and "Coats with less reprocessed wool than Group A," the categories are inadequate. Two of the categories depend upon "Group A" for their meaning, or existence. A more logical classification is "Coats containing 40 to 60 percent reprocessed wool," "Coats containing more than 60 percent reprocessed wool," and "Coats containing less than 40 percent reprocessed wool." In this grouping, each category is independent, that is, mutually exclusive, of the other categories.

Nonoverlapping

The categories in a classification system must be nonoverlapping. It should be possible to place an item into only *one* category. If an item, however, can reasonably be placed under more than one category, the categories should be renamed and perhaps narrowed. For instance, if the fabrics in a sewing shop were being classified as natural fibers, synthetics, or blends, how would a fabric that is part cotton and part wool be classified? The fabric is of natural fibers but it is also a blend. Thus the given categories are inadequate. More satisfactory would be categories such as these: fabrics of a pure natural fiber, fabrics of blended natural fibers, synthetic fabrics, and fabrics that have a blend of synthetic and natural fibers.

Complete

The categories in a classification system must be complete. Every item to be classified must have a category into which it logically fits, with no item left out. For instance, the categories in a classification of American made automobiles according to the number of cylinders would not be sufficient unless the categories were these: four cylinders, six cylinders, and eight cylinders. The omission of any one of the categories would make the classification system incomplete, for a number of automobiles then would have no category into which they would logically fit.

An Example of the Classification Process

Suppose that a newspaper reporter visits a campus to gather data from students for an informative article about the student body. During interviews with students she jots down some of their comments. The notes include the following:

> A lot of the students live at home and commute to college.
> Some students have real hangups, particularly when it comes to sex.
> There are more freshmen than sophomores.
> I'm a technical student in document processing. When I finish the two-year program here, I'll be able to get a good job in my specialized field.
> I'm not taking a full number of courses this semester because I work eight hours a day to support my family.
> Many students have a real money problem. Like my roommate. He's completely paying his own way.
> I chose this college because it's close to home.
> They say you gotta go to college to get anywhere. OK, here I am.
> Somehow I didn't do too well on my ACT score, so I can't take all the courses I want to take.

Some students are lucky enough to have their own apartment.
The biggest thing I've had to cope with is finding enough time to study and to play—all in the same weekend.
This college is really growing. This year we have the biggest freshman class in the history of the college.
At spring ceremonies, 1200 seniors will graduate.

137

Chapter 4
Analysis Through
Classification and
Partition: Putting
Things in Order

The newspaper reporter must now put these items in order so that she can write an article her readers can follow.

Analysis of the reporter's notes shows possible groupings of information. Two statements deal with the residence of students, several deal with problems of students, and several deal with reasons for attending college. Three deal with the number of students enrolled, and two deal with the status of the student according to the number of courses currently enrolled in. The notes are organized below:

Residence of Students While Attending College

A lot of the students live at home and commute to college.
Some students are lucky enough to have their own apartment.

Problems of Students

Some students have real hangups, particularly when it comes to sex.
Many students have a money problem. Like my roommate. He's completely paying his own way.
The biggest thing I've had to cope with is finding enough time to study and to play—all in the same weekend.

Reasons for Students Attending College

I'm a technical student in document processing. When I finish the two-year program here, I'll be able to get a good job in my specialized field.
I chose this college because it's close to home.
They say you gotta go to college to get anywhere. OK, here I am.

Status of Students According to Enrollment

There are more freshmen than sophomores.
This college is really growing. This year we have the biggest freshman class in the history of the college.
At spring ceremonies, 1200 seniors will graduate.

Status of Student According to Number of Courses Currently Enrolled In

I'm not taking a full number of courses this semester because I work eight hours a day to support my family.
Somehow I didn't do too well on my ACT score, so I can't take all the courses I want to take.

Organizing the newspaper reporter's notes into groups, or categories, is analysis through classification. This analysis is carried out logically, not haphazardly. Clearly, the subject being classified is the reporter's notes (plural in meaning), not students or courses or factors affecting student achievement. Moreover, the newspaper reporter, in order to sort the notes into useful categories, has to be knowledgeable about the various aspects of a student body.

Five different bases help the reporter to sort the information:

- Residence of students while attending college
- Problems of students
- Reasons for students attending college
- Status of students according to enrollment
- Status of students according to number of courses currently enrolled in

These bases are clear and logical. If, for instance, the last two bases had been considered as one basis, such as "Status of student," there would be confusion. The term *status of student* has two different meanings here, depending on whether reference is made to the student as a freshman, sophomore, junior, or senior ("Status of students according to enrollment") or whether reference is made to the student as a full-time student, part-time student, probationary student, or whatever ("Status of students according to number of courses currently enrolled in").

In the example, *the categories are coordinate;* that is, they are all on the same level, with no confusion of major categories with subcategories. If, however, along with the five major categories there had been included "Students who have money problems," the major categories, or classes, would not be coordinate. They would no longer be on the same level because "Students who have money problems" is a subcategory, or subclass, or "Problems of students."

Each of the categories is mutually exclusive; that is, each of the five categories is composed of a clearly defined group that would still exist without the other categories. "Residence of students while attending college," for example, would be a valid category even if some or all of the other categories were unnecessary.

In the example, *the categories do not overlap;* an item can be placed in only one category. For instance, the item "I chose this college because it's close to home" can be placed in only one of the groups: "Reasons for students attending college." The speaker of the "close to home" comment may live at home and commute to school, but the comment itself could not possibly be grouped under "Residence of students while attending college" or any of the groups except "Reasons for students attending college."

Finally, in the classification of the reporter's notes into five categories, each item of information fits into a category; no item is left out. *The categories are complete.*

Order of Data Presentation

The categories, or classes, in a classification system should be presented in the order that best accomplishes your purpose. In classifying steps in a process, probably a *chronological,* or *time, order* would be used. In classifying items such as breeds of cattle in the United States, it might be wise to use an *order of familiarity,* that is, to start with the best known or most familiar and go to the least known or least familiar. In classifying the qualities a nurse should have, the qualities might be listed in *order of importance.* At times, *order according to complexity,* that is, movement from the simple to the more difficult, might be best to use, as in classifying swimming strokes or in classifying spreadsheet programs. A *spatial order,* movement from one physical point to another (from top to bottom, inside to outside, left to right, etc.), may be the most practical arrangement, as in classifying the parts of an automobile engine or the furnishings in a house. Order according to maximum effectiveness is appropriate if beginning with the most crucial point or building to a climax. If the categories are such that order is not important, an *alphabetical* or a *random presentation* may be used. The important thing to remember is this: Use whatever order is appropriate for your audience and your purpose.

For additional discussion of order, see page 334 and Part III Revising and Editing, pages 608–611.

139

Chapter 4
Analysis Through
Classification and
Partition: Putting
Things in Order

Forms of Data Presentation

Classification may be the major organizing principle in a presentation or in a section of a presentation, or it may be the major organizing principle within a paragraph or in a part of a paragraph. Whichever way classification is used, data may be presented in several forms: in outlines, in verbal explanations, and in visuals.

Outline

Outlines are particularly helpful for grouping items in an orderly, systematic arrangement. The outline may be purely for personal use in preparing information for presentation in a verbal explanation or in a visual and thus may be very informal. Or the outline may be the form selected for the actual presentation of the information. The outline that others will read or see should be clear, consistent, and logical. If the outline meets these requirements, it follows certain accepted standards.

1. Choose either the traditional number-letter outline form or the decimal outline form. (Illustrations of both outline forms follow.)
2. Write the outline, using a topic outline (illustrations follow) or a sentence outline (illustrated on page 149 and in the other procedure sections of chapters 1–5). In a topic outline each heading is a word, phrase, or clause. In a sentence outline each heading is a complete sentence. *Do not combine both the topic style and the sentence style in one outline.* Such a combination usually confuses the reader.
3. Label headings with an appropriate number or letter. In the traditional number-letter outline form, first-level, or major, headings are designated with roman numerals (I, II, III, etc.); second-level headings with uppercase letters (A, B, C, etc.); third-level headings with Arabic numerals (1, 2, 3, etc.); and four-level headings with lowercase letters (a, b, c, etc.). In the decimal outline form, use a system of decimal points to designate the various levels of headings.
4. Indent headings to indicate the degree of classification.
5. Use the same grammatical structure for headings on a given level.
6. Include at least two headings on each level. Compose each level as divisions of the preceding heading. If, however, a heading calls for a list, it is possible to have only one item in the list.

Topic Outlines: Classification of Saws
Two Bases: Uses and Types

Traditional Number-Letter Outline Form	*Decimal Outline Form*
I. Basic uses	1. Basic uses
A. Crosscutting	1.1 Crosscutting
B. Ripping	1.2 Ripping
II. Types	2. Types
A. Hand-operated saws	2.1 Hand-operated saws
1. Handsaw	2.1.1 Handsaw
2. Backsaw	2.1.2 Backsaw
3. Keyhole saw	2.1.3 Keyhole saw
4. Coping saw	2.1.4 Coping saw
5. Hacksaw	2.1.5 Hacksaw
B. Power saws	2.2 Power saws
1. Portable power saws	2.2.1 Portable power saws
a. Circular saw	2.2.1.1 Circular saw
b. Saber saw	2.2.1.2 Saber saw
c. Reciprocating saw	2.2.1.3 Reciprocating saw
2. Stationary power saws	2.2.2 Stationary power saws
a. Radial arm saw	2.2.2.1 Radial arm saw
b. Table saw	2.2.2.2 Table saw
c. Motorized miter saw	2.2.2.3 Motorized miter saw

A well-organized classification presentation can be easily outlined. For example, the classification of rotary tillers on pages 000–000 might be outlined as follows:

Rotary Tillers

I. Two basic types of rotary tillers
  A. Front-tine tillers
  B. Rear-tine tillers

II. Types of tines
  A. Bolo tines
  B. Pick and chisel tines
  C. Slasher tines

Verbal Explanation

The outline, as a presentation form of a classification system, may be used alone. More often, perhaps, a classification system is presented as a verbal explanation and frequently includes visuals, as in the following classification system of rotary tillers and tines.

The classification of rotary tillers and tines was written for an audience of beginning gardeners. Such an audience might vary widely in educational background but as beginning gardeners all are novices. The purpose of the classification is to provide information to help beginning gardeners select a tiller with appropriate tines, depending on each gardener's individual needs.

Tillers are divided into two categories on the basis of where the tines are located; tines are divided into three categories on the basis of their shape. Notice also the interrelationship of classification and definition.

The Plan Sheet that the student filled in before writing the classification is included. To help you analyze the sample description, marginal notes are added.

PLAN SHEET
FOR GIVING AN ANALYSIS THROUGH
CLASSIFICATION

Analysis of Situation Requiring Classification

What is the plural group that I will classify?
rotary tillers and tines

Who is my intended audience?
beginning gardeners

How will the intended audience use the classification? For what purpose?
to decide on a tiller and the type of tines to purchase

Will the presentation of the classification be written or oral?
written

Setting Up the Classification System

Definition or identification of subject:
machines used to cultivate the soil in areas larger than a quarter acre. To turn under a crop to enrich the soil, to get a bed ready for planting, to turn the soil, to mulch plant stems after plants have quit producing

Basis (or bases) for classification:
basic types of tillers according to location of tines; where used most effectively; and advantages and disadvantages of basic types of tines according to type of blade and type of ground where best used

Significance or purpose of basis (or bases):
to save time and do work more efficiently

Categories of the subject, with identification of each category:
Types of tillers
 Front-tine tillers
  —tines located beneath the engine and in front of back wheels
  —best used in flower beds and vegetable gardens where the soil is well prepared, soft and smooth; tines will not pierce hard ground
  —advantage: low cost ($350–$500)
  —disadvantage: useful only for light or medium tilling

Categories of the subject, with identification of each category, cont.:

> *Rear-tine tillers*
>> —*tines located behind the wheels and beneath the handles*
>> —*useful for breaking ground, even rocky or compacted soil*
>> —*advantage: heavy and easily steered*
>> —*disadvantage: expensive ($850–$2,000)*
>
> *Basic types of tines*
>> *Bolo tines*
>>> —*broad, heavy blades*
>>> —*used to dig and mulch and will not clog easily*
>>
>> *Pick and chisel tines*
>>> —*medium-length, slightly curved blades*
>>> —*used for cultivating hard, rocky ground but may clog easily, especially in vegetation*
>>
>> *Slasher tines*
>>> —*short, sharp points*
>>> —*used to easily cut thick vegetation and into soft ground*

Presentation of the System

Most logical order for presenting the categories:
random listing

Types and Subject Matter of Visuals I Will Include

tillers showing front and rear locations of tines, blades of each of the types of tines

The Title I May Use

Classification of Rotary Tillers According to Location and Type of Tines, Area Where Best Used, Advantages, and Disadvantages

Sources of Information

lecture in landscape management class; literature from a garden center

143

Chapter 4
Analysis Through
Classification and
Partition: Putting
Things in Order

Focused title

CLASSIFICATION OF ROTARY TILLERS ACCORDING TO LOCATION AND TYPE OF TINES

Subject identified

Importance
of subject

Rotary tillers are machines used to cultivate the soil. They are especially useful if the area to be cultivated is more than a quarter of an acre. The tillers can be used for a variety of activities: to turn under a crop such as rye grass to enrich the soil, to get a bed ready for garden seed, to turn the soil over, and to mulch plant stems after the plants have quit producing. The tiller can save time and do work more efficiently than a shovel or hand cultivator. Rotary tillers are classified according to the location of the tines; the tines are classified according to the type of blade.

First major topic

Two Basic Types of Tillers

First topic divided
into two broad
categories

Depending on the placement of the tines, there are two basic types of rotary tillers: front-tine tillers and rear-tine tillers.

Front-Tine Tillers

First broad category
described by
location of tines

Front-tine tillers have the tines located on the front of the tiller, beneath the engine and in front of the back wheels. See Figure 1.

Drawing showing a
front-tine rotary tiller

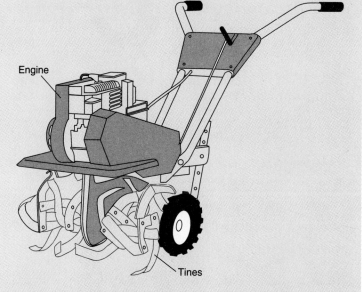

Engine

Tines

Figure 1. Front-tine rotary tiller.
SOURCE: How to Select, Use and Maintain Garden Equipment, Ortho Books, Chevron Chemical Company.

Area of effective use

Chains, drive belts, or gears link the tines to the engine; the action of the tines pulls the tiller forward. These tillers, therefore, are most effectively used in flower beds and vegetable gardens where the soil is well prepared, soft and smooth. If used on hard, rocky, or weedy soil, the tiller can go out of control because the tines will not pierce the hard ground.

Advantage

Disadvantage

A major advantage of the front-tine tiller is low cost. These tillers can be purchased for $350–$500. The disadvantage is that the tiller can be used only for light or medium tilling.

Rear-Tine Tillers

Second broad category described by location of tines

Rear-tine tillers have tines located behind the wheels and beneath the handles. See Figure 2.

Drawing showing a rear-tine tiller

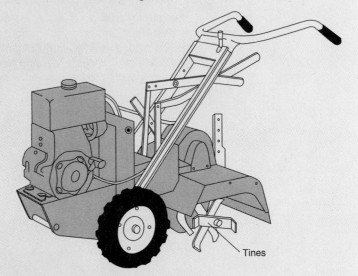

Tines

Figure 2. Rear-tine rotary tiller.
SOURCE: How to Select, Use and Maintain Garden Equipment, Ortho Books, Chevron Chemical Company.

Area of effective use

The tines are also linked to the engine by chains, drive belts, or gears. The rear tines, however, dig down, pulling themselves into the soil. The rear-tine tiller, therefore, can be used for breaking ground, even rocky or compacted soil.

Disadvantage

Advantages

A disadvantage of the rear-tine tiller is high cost, which may range from $800 to $2,500, depending on engine and manufacture; advantages are the tiller is heavy and easily steered.

Basic Types of Tines

Second major topic

Second topic divided into three broad categories

Either the front-tine tiller or the rear-tine tiller is available with one of three types of tines: bolo tines, pick and chisel tines, or slasher tines.

Bolo Tines

First category described by blade characteristics, area of effective use

Bolo tines have broad, heavy blades. They can be used to dig and mulch. A major advantage is these tines do not clog easily. See Figure 3.

145

Chapter 4
Analysis Through
Classification and
Partition: Putting
Things in Order

Bolo tines illustrated

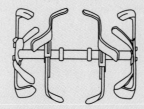

Figure 3. Bolo tines.
SOURCE: How to Select, Use and Maintain Garden Equipment, Ortho Books,
Chevron Chemical Company.

Second category
described by blade
characteristics, area
of effective use

Disadvantage

Pick and Chisel Tines

Pick and chisel tines have medium-length, slightly curved blades.
These blades, designed for cultivating hard, rocky ground, may
clog easily, especially when used where there is vegetation. See
Figure 4.

Pick and chisel tines
illustrated

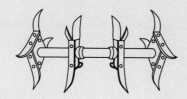

Figure 4. Pick and chisel tines.
SOURCE: How to Select, Use and Maintain Garden Equipment, Ortho Books,
Chevron Chemical Company.

Slasher Tines

Third cateogry
described by blade
characteristics, area
of effective use

Slasher tines have short, sharp points, and they are most efficient
when kept sharp. These tines will easily cut through thick vegetation and into soft ground. See Figure 5.

Slasher tines
illustrated

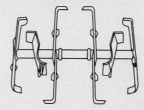

Figure 5. Slasher tines.
SOURCE: How to Select, Use and Maintain Garden Equipment, Ortho Books,
Chevron Chemical Company.

Closing

Suggestions for
selecting a tiller.

Selecting a Tiller

Consider the following factors before selecting a tiller: the size of
the area to be tilled, the type of soil to be tilled, the reason for
tilling the soil, and the amount of money you can spend.

Visuals

In addition to presenting data in a classification system as an outline or as a verbal explanation, you may use visuals. Such visuals as charts, diagrams, maps, photographs, drawings, graphs, and tables frequently make a mass of information understandable. Notice how much information is shown clearly on the chart on page 133. Also look again at the visuals on pages 143–145, and 148. Each visual makes meaning clearer for the reader.

Study the bar chart below. Determine the data being organized; then determine the bases on which the data are organized. Why are the data presented in this form?

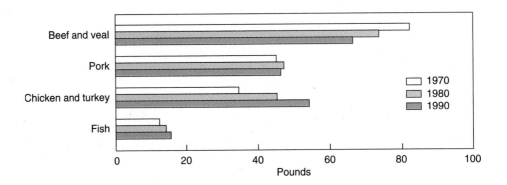

Per Capita Meat Consumption, by Selected Products: 1970 to 1990
SOURCE OF DATA: U.S. Dept. of Agriculture.

Of course, the information in the chart might have been presented in outline form, as follows:

I. Beef and veal consumption
   A. 1970—83 pounds
   B. 1980—73 pounds
   C. 1990—65 pounds

II. Pork consumption
   A. 1970—45 pounds
   B. 1980—48 pounds
   C. 1990—46 pounds

III. Chicken and turkey consumption
   A. 1970—35 pounds
   B. 1980—46 pounds
   C. 1990—55 pounds

IV. Fish consumption
   A. 1970—12 pounds
   B. 1980—14 pounds
   C. 1990—15 pounds

Or the information might have been presented as a verbal explanation, as follows:

An interesting pattern has evolved over the years from 1970–1990 as the American consumer has altered the individual consumption of beef and veal, pork, chicken and turkey, and fish. A look at per capita consumption during these years shows a decrease in the pounds of beef and veal eaten and an increase in the pounds of pork, chicken and turkey, and fish eaten. In 1970 each person in the United States ate 83 pounds of beef, in 1980 ate 73 pounds, and in 1990 ate 65 pounds. Per capita consumption of pork was 45 pounds in 1970, 48 pounds in 1980, and 46 pounds in 1990. Chicken and turkey eaten was 35 pounds in 1970, 46 pounds in 1980, but 55 pounds in 1990. Fish consumed was 12 pounds in 1970, 14 pounds in 1980, and 15 pounds in 1990. The greatest increase is in the consumption of chicken and turkey; during the 20 years surveyed, the per capita consumption of chicken and turkey has increased by 20 pounds. The pounds of fish consumed has also increased but by a much smaller percentage. In spite of the increased consumption of chicken and turkey and fish, people ate more beef and veal than any of the other three meats.

147

Chapter 4
Analysis Through
Classification and
Partition: Putting
Things in Order

Of the three forms—chart, outline, and verbal explanation—used for presenting the data concerning per capita meat (beef, pork, chicken, fish) consumption between 1970–1990, the chart obviously is the clearest and most easily understood. The reader can easily comprehend the data because each of the years charted (1970, 1980, and 1990) is shown in a clearly identifiable shaded bar. To the left of the bars, a list of the four common meats appears. Across the bottom of the chart, pounds are marked from 0 to 100 in increments of 20. Almost with a glance the reader can comprehend various facts related to the increase and decrease in consumption of each type of meat.

Combination

Frequently a combination of forms is used to present data. Architects and builders, for example, certainly understand the classification of roofs according to shape. A client who hires the architect and builder to design and construct a house, however, might not understand the available roof options or suggested roof shapes. Even upon hearing the shapes named, the client may still have difficulty understanding exactly how each shape looks. In such a circumstance, the varieties (classification) could best be explained in both verbal and visual form. Given both verbal and visual design, a client has a better idea of the appearance of a particular roof design.

COMMON ROOF SHAPES

| Lean-to roof | Gable roof | Hip roof |

Gable and valley roof   Hip and valley roof

<u>Shed or Lean-to Roof.</u> A single slope roof.
<u>Gable or Pitch Roof.</u> Two slopes meet at the center or ridge to form a gable.
<u>Hip Roof.</u> A four-sided roof. All sides slope toward the center of the building. Rafters run diagonally to meet the ridge into which other rafters are framed.
<u>Gable and Valley, or Hip and Valley Roof.</u> Combination of two gable or two hip roofs intersecting each other. The valley is the place where two slopes of the roof, running in different directions, meet. Intersections are usually at right angles.

General Principles in Giving an Analysis Through Classification

1. *Classification is a basic approach in analysis.* It places related items into categories, or groups.
2. *Only a plural subject or a subject whose meaning is plural can be classified.* If a subject is singular, it can be partitioned but not classified.
3. *The basis on which classification is made should be clear, useful, and purposeful.*
4. *The categories in classification must be coordinate, or parallel.* All categories on the same level must be of the same rank in grammatical form and in content.
5. *The categories must be mutually exclusive.* Each category should be composed of a clearly defined group that would still exist without the other categories on the same level.
6. *The categories must not overlap.* An item can have a place in only one category.
7. *The categories must be complete.* There should be a category for every item, with no item left out.
8. *The order of presentation of categories depends on their audience and their purpose.* Among the possible orders are time, familiarity, importance, complexity, space, maximum effectiveness, and alphabetical and random listing.

9. *The data in a classification analysis may be presented in outlines, in verbal explanation, and in visuals.* The form or combination of forms that presents the data most clearly for the audience and the purpose should be used.

149

Chapter 4
Analysis Through
Classification and
Partition: Putting
Things in Order

Procedure for Giving an Analysis Through Classification

Analysis through classification is generally a part of a larger whole, as have been the other forms of communication in the preceding chapters. Regardless of whether the classification analysis is a dependent or an independent communication, however, the structure of such an analysis is as follows:

I. The presentation of the subject and of the bases of division into categories is usually brief.
 A. State the subject that is to be divided into categories and identify or define the subject.
 B. If applicable, list various bases by which the subject can be divided into categories.
 C. State explicitly the bases of the categories and point out
 1. the reasons the categories are significant
 2. the purpose they serve
II. The listing and the discussion of the categories are the longest parts of the presentation.
 A. List the categories and, if any, the subcategories.
 B. Give sufficient explanation to clarify and differentiate among the given categories.
 C. Present the categories in whatever order best serves the purpose of the analysis.
 D. Use outlines and visuals whenever they will help clarify the explanation.
III. The closing (usually brief) depends on the purpose of the presentation.
 A. The closing may be the completion of the last point in the analysis.
 B. The closing may be a comment on the analysis or a summary of the main points.

Applications in Giving an Analysis Through Classification

Individual and Collaborative Activities

4.1 Some of the following classifications are satisfactory and some are not. Indicate the sentences containing unsatisfactory classifications. Rewrite them to make them satisfactory.

a. The students on our campus can be classified either as blonds or as brunets.
b. Automobiles serve two distinct functions: They are used either for pleasure or for business.
c. Cattle may be classified as purebred, crossbred, or unbred.
d. The courses a technical student takes at this college can be divided into technical courses, related courses, and general education courses.
e. Structural material may be wood, brick, aluminum, steel, concrete blocks, or metal.
f. Personal computers can be classified according to the size of disk they require: $5 1/4$- or $3 1/2$-inch disk.

4.2   Each of the following classification systems is lacking in *coordination, mutual exclusiveness, nonoverlapping,* and/or *completeness.* Working in small groups, point out the specific errors. Then make whatever changes are necessary for adequate classification systems.

I. Classification of Watches According to Shape
   A. Round
   B. Oblong
   C. Thin
   D. Oval

II. Classification of Phonograph Records According to Type of Music
   A. Classical
   B. Jazz
   C. Michael Jackson
   D. Country and Western
   E. Instrumental

III. Classification of Mail According to Collection Time
   A. Day
      1. Weekdays
      2. Saturdays
      3. Holidays
   B. Hour
      1. A.M.
      2. Afternoons

IV. Classification of Television Shows
   A. Soap operas
   B. Situation comedies
   C. Today Show
   D. NBC

V. Classification of College Students
   A. Academic classification
      1. Freshmen
      2. Sophomores
      3. Third year
      4. Seniors and graduate students

B. Residence
    1. Dormitory
    2. Local
    3. In state
    4. Out of state

151

Chapter 4
Analysis Through
Classification and
Partition: Putting
Things in Order

4.3   A. For each of the following groups of items, suggest at least three bases of classification.

Example

Group of items:	Books
Bases of classification:	Subject matter, nationality of author, cost, date of writing, publisher, alphabetical listing by author

a. Metals   f. Birds
b. Office machines   g. Tools
c. Diseases   h. Computers
d. Buildings   i. Televisions
e. Clothing   j. Sports

B. From the list above, choose one group of items and one basis of classification. For that basis of classification write an explanatory paragraph.

4.4   Joe received the following injuries in an automobile accident: black eyes, broken nose, fractured ribs, crushed pelvis, broken thumb, twisted ankle, cuts on the forehead, teeth knocked out, dislocated knee, gash on the right leg, bruised shoulder, and tip of little finger cut off. Organize Joe's injuries into related groups. Use a formal outline.

4.5   With three or four students working together, from the following list of words pick out as many groups of related items as possible; identify the relationship within the group of words. Use your dictionary to look up any unfamiliar words. As a class, share the word groups and the relationship between the words.

Example

List of Words	*Word Group*	*Relationship*
Electrical pressure	Electrical pressure	Basic electric quantities
Magnets	Ohm	
Ohm	Ampere	
Ampere	Coulomb	
Copper wire		
Coulomb		

a. Gothic
b. Propeller
c. Airbrush
d. Dry cells
e. WordPerfect
f. Blood pressure
g. Horn
h. Non-photographic pencil
i. Drawing board
j. Drill press
k. Keyboard
l. Tractor
m. Dissecting microscope
n. Storing
o. Micrometer
p. Macwrite
q. X-acto knife
r. Dial indicator

s. Plaintiff
t. Breathing
u. Ruling pen
v. Planer
w. Text
x. Wing
y. Stirrup
z. Primary cells
aa. CD-RAM
bb. Windows
cc. Preserved specimen
dd. Monitor
ee. Pulse
ff. Arrest
gg. Lathe
hh. T square
ii. Slide
jj. Hard drive

kk. Dial caliper
ll. Roman
mm. Lead-acid storage cells
nn. Reflexes
oo. Flexible disk
pp. Printer
qq. Witness
rr. Strap
ss. Shaper
tt. Hay baler
uu. Compass
vv. Triangles
ww. Seat
xx. Italic
yy. Drawing set
zz. Combine

4.6   Assume that an insurance company will give you an especially good rate if you will insure all your possessions with it. Prepare an *organized* list of your belongings so that the insurance company can suggest the amount of coverage you need.

4.7   Working in groups, discuss the different types of visuals that could be produced using the statistics shown in the table on page 153. Chapter 11 discusses visuals in more detail. What conclusions could be drawn from the data in each visual? Who would be interested in the data? Then select appropriate data and show it in another visual form. Add a sentence explaining why you selected the particular visual form and identify who would likely want the information in the visual.

4.8   Select a subject suitable for classification. Present the classification system in a visual.

153

Chapter 4
Analysis Through
Classification and
Partition: Putting
Things in Order

Employed Persons by Occupation, Sex, and Age
(in thousands)

Occupation	Total 16 years and over		Men 16 years and over		Women 16 years and over	
	1991	1992	1991	1992	1991	1992
Total	116,877	117,598	63,593	63,805	53,284	53,793
Managerial and professional specialty	31,012	31,153	16,656	16,416	14,356	14,736
Executive, administrative, and managerial	14,954	14,767	8,890	8,641	6,064	6,126
Officials and administrators, public adminstration	591	619	336	361	255	258
Other executive, administrative, and managerial	10,412	10,187	6,611	6,384	3,801	3,802
Management-related occupations	3,951	3,961	1,943	1,896	2,008	2,065
Professional specialty	16,058	16,386	7,767	7,775	8,292	8,611
Engineers	1,846	1,751	1,694	1,603	152	148
Mathematical and computer scientists	923	935	583	622	339	313
Natural scientists	438	459	324	334	114	125
Health diagnosing occupations	849	914	696	747	153	167
Health assessment and treating occupations	2,376	2,517	328	332	2,048	2,184
Teachers, college and university	773	737	457	435	316	302
Teachers, except college and university	4,029	4,216	1,038	1,062	2,992	3,154
Lawyers and judges	772	788	626	620	146	167
Other professional specialty occupations	4,051	4,068	2,020	2,018	2,031	2,050
Technical, sales, and administrative support	36,086	36,808	12,734	13,269	23,352	23,539
Technicians and related support	3,794	4,253	1,921	2,169	1,873	2,084
Sales occupations	13,958	13,919	7,142	7,252	6,816	6,667
Administrative support, including clerical	18,334	18,636	3,671	3,848	14,663	14,788
Service occupations	15,986	16,096	6,429	6,494	9,557	9,602
Precision production, craft, and repair	13,162	13,128	12,030	12,000	1,132	1,128
Mechanics and repairers	4,427	4,441	4,264	4,293	163	147
Construction trades	4,808	4,790	4,721	4,702	88	89
Other precision production, craft, and repair	3,927	3,897	3,045	3,005	881	892
Operators, fabricators, and laborers	17,172	16,957	12,842	12,720	4,330	4,237
Machine operators, assemblers, and inspectors	7,696	7,524	4,610	4,535	3,086	2,989
Transportation and material moving occupations	4,878	4,878	4,441	4,451	437	427
Handlers, equipment cleaners, helpers, and laborers	4,597	4,556	3,791	3,734	806	821
Farming, forestry, and fishing	3,459	3,456	2,903	2,905	557	551

Note: Data for 1992 are not fully comparable with data for prior years because of the introduction of the occupational classification system used in the 1990 census.

Source: Bureau of Labor Statistics, U.S. Dept. of Labor

Written Classification Analyses

In completing applications 4.9 and 4.10,

a. Fill in the Plan Sheet on pages 156–157, or one like it.
b. Write a preliminary draft.
c. Go over the draft in a peer editing group.
d. Revise. See Part III Revising and Editing and the checklists for revising, front and back endpapers.
e. Write the classification analysis.

4.9　A. Select one of the following subjects. Divide the subject into classes in accordance with at least four different bases; subdivide wherever necessary. Use outline form to show the relationship between divisions.

Example

Automobiles

I. Body style
   A. Sedan
   B. Coupe
   C. Convertible
   D. Hard top

II. Body size
   A. Regular
   B. Intermediate
   C. Compact
   D. Subcompact

III. Kind of fuel
   A. Gasoline
   B. Diesel
   C. Gasohol
   D. Butane
   E. Propane

IV. Cost
   A. Inexpensive—under $12,000
   B. Average—$12,000–$20,000
   C. Expensive—over $20,000

a. Courts	i. Transistors	r. Tires
b. Farm machinery	j. Muscles	s. College courses
c. Cattle	k. Home air conditioners	t. Meters
d. Hand tools	l. Commercial aircraft	u. Furniture
e. Drawing instruments	m. Computer games	v. Wrenches
f. Scales	n. Clocks	w. Books
g. Precision measuring instruments	o. Clothing	x. Radios
h. Building materials	p. Food	y. Paint
	q. Transportation	z. Disks

B. For the subject you chose in A above (or another of the listed subjects), present the data in a written explanation.

4.10   Modern automobiles can be bought with numerous options. List these options, and then group them into categories. Use outline form to show the relationship between divisions. Then give a written classification analysis.

Oral Classification Analyses

4.11   As directed by your instructor, adapt for oral presentations the classification analyses you prepared in the applications above. Ask your classmates to evaluate your speech by filling in the Evaluation of Oral Presentations on page 459, or one like it.

4.12   Read "Hardest Lessons for First-Time Managers," by Linda A. Hill, pages 516–519. Under "Suggestions for Response and Reaction," page 520, complete Number 1.

4.13   Read "Are You Alive?," by Stuart Chase, pages 537–540.

   a. Outline the article to show its use of classification.
   b. Under "Suggestions for Response and Reaction," page 541, write Number 1.

4.14   Read the poem "Blue Heron," by Chick Wallace, page 550.
   a. List types of environmental pollution.
   b. Under "Suggestions for Response and Reaction," page 551, complete Number 2.

4.15   Read the excerpt from "Non-Medical Laser: Everyday Life in a New Light," by Rebecca D. Williams, pages 564–569. Under Suggestions for Response and Reaction, page 569, write Number 2.

PLAN SHEET
FOR GIVING AN ANALYSIS
THROUGH CLASSIFICATION

Analysis of Situation Requiring Classification

What is the plural group that I will classify?

Who is my intended audience?

How will the intended audience use the classification? For what purpose?

Will the presentation of the classification be written or oral?

Setting Up the Classification System

Definition or identification of subject:

Basis (or bases) for classification:

Significance or purpose of basis (or bases):

Categories of the subject, with identification of each category:

Categories of the subject, with identification of each category, cont.:

Presentation of the System

Most logical order for presenting the categories:

Types and Subject Matter of Visuals I Will Include

The Title I May Use

Sources of Information

Analysis Through Partition

Definition of Partition

Partition is analysis that divides a singular item into parts, steps, or aspects. Only *singular* subjects can be partitioned; plural subjects are classified. Partition breaks down into its components a concrete subject, such as a tree (parts: roots, trunk, branches, and leaves), or an abstract subject, such as how to build a herd of cattle (steps: select good stocker cows, select a good herd bull, breed the cows at the right time, keep the heifers, sell the bull calves, and sell the old cows), or such as inflation (aspects: causes, effects on consumers, etc.).

Basis for Partition

Partition, like classification, requires a useful, purposeful basis. A carpenter's partitioning, or dividing, the tasks in the construction of a proposed porch on the basis of needed materials is useful and purposeful. Partitioning the activities carried out in an assembly line process to streamline the process could result in money-saving decisions. Reviewing a business or organization to show divisions and responsibilities may lead to a stronger, more effective business or organization. Partitioning a building on the basis of use may lead to greater efficiency of use.

Characteristics of a Partition System

If a partition system is to be adequate, that is, if it is to fulfill its purpose, the divisions must have certain characteristics (same as in a classification system):

1. The divisions must be coordinate.
2. The divisions must be mutually exclusive.
3. The divisions must not overlap.
4. The divisions must be complete.

Consider the partition of a concrete subject, a lamp socket (on page 159), on the basis of its construction.

The partition of the lamp socket is adequate. The divisions are **coordinate:** they are of equal rank in grammatical form and in content, as shown below:

Lamp Socket Partition

Outer parts } *First level of partition*
    Outer metal shell
    Socket cap   } *Second level of partition*
    Bushing with screw

Inner parts } *First level of partition*
    Fiber insulating shell
    Socket base   } *Second level of partition*

On each level, the divisions are **mutually exclusive;** that is, each division could exist without the other divisions. For instance, the outer metal shell could still exist even if there were no socket cap. The divisions do **not overlap;** a part has a place in only *one* division. The divisions are **complete;** every part of the lamp socket is accounted for; no part is left out.

159

Chapter 4
Analysis Through
Classification and
Partition: Putting
Things in Order

PARTITION OF A LAMP SOCKET

For the do-it-yourself person, replacing the base in a lamp socket can be simple. The parts of a lamp socket are few and uncomplicated as shown in the following drawing.

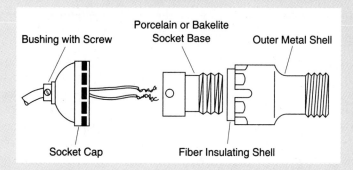

A lamp socket.

The lamp socket has an outer metal covering with an interlocking cap topped with a bushing and screw. Lodged inside in a fiber insulating shell is the socket base, usually made of porcelain or Bakelite. To replace the socket base, the repairer has only to disconnect the wires from the old base, remove that base, insert a new socket base, reconnect the wires, and pull the socket cap over the outer metal shell so that the lamp socket is one piece again.

Or, consider the claw hammer. It can be partitioned into two main parts: the handle and the head, as illustrated in the outline and the visual following.

I. Handle
II. Head
    A. Neck
    B. Poll
    C. Face
    D. Cheek
    E. Claw
    F. Adze eye

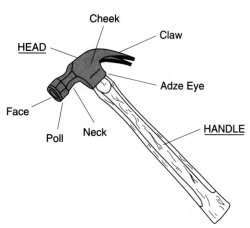

Figure 1. A claw hammer.

A carpenter would be concerned about the main parts and the subparts as they function in various ways when the hammer is in use.

Purpose, Audience, and Order

Purpose and audience determine the order of data presentation in a partition system. As in classification, among the possible orders are time, familiarity, importance, complexity, space, maximum effectiveness, and alphabetical and random listing. For example, the work of a general duty nurse can be partitioned, or divided, into what he does the first hour, second hour, and so on (time order), if the purpose of partition is to show how he spends each working hour. The audience for such a partition might be a supervisor who is trying to decide if more nurses need to be hired for certain hours. The nurse's work can be partitioned according to the activities a shift might require (order of importance or complexity), if the purpose of the partition is to show what the nurse does on the job. The audience for this partition could be a group of student nurses who want to know what they will do on a shift. The nurse's work on a shift can be partitioned to show where he works (the work space). The audience for such a partition could be a consultant who is judging the efficiency or inefficiency of the work space.

Forms of Data Presentation

The forms of presentation of analyses through partition are the same as those for classification. Logical forms are outlines, visuals, and verbal explanations.

Look at the following floor plan of the United States Capitol (Figure 1). It partitions the building to show the design and layout and to show space utilization. Several audiences might find this partitioning useful. Tourists or guests visiting the capitol could easily locate the different sections of the building from the floor plan. A congressional committee could use the floor plan to reassign space. An architect could use the floor plan as a basis for renovation ideas.

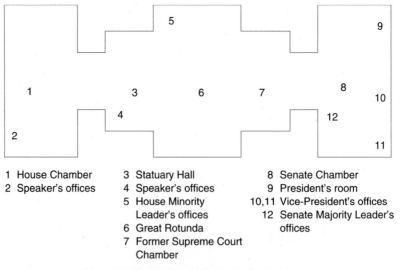

1 House Chamber
2 Speaker's offices
3 Statuary Hall
4 Speaker's offices
5 House Minority Leader's offices
6 Great Rotunda
7 Former Supreme Court Chamber
8 Senate Chamber
9 President's room
10,11 Vice-President's offices
12 Senate Majority Leader's offices

Figure 1. Floor plan of the U.S. Capitol.

The same information could be shown in outline form:

161

Chapter 4
Analysis Through
Classification and
Partition: Putting
Things in Order

Floor Plan of the U.S. Capitol

1. House of Representatives wing }   *(First level of partition)*
  1.1 House Chamber   *(Second level of partition)*
  1.2 Speaker's offices }
2. Great Rotunda and center wings }   *(First level of partition)*
  2.1 Great Rotunda
  2.2 Statuary Hall
  2.3 Speaker's offices   *(Second level of partition)*
  2.4 House Minority Leader's offices
  2.5 Former Supreme Court Chamber }
3. Senate wing }   *(First level of partition)*
  3.1 President's room
  3.2 Senate Chamber   *(Second level of partition)*
  3.3 Vice-President's offices
  3.4 Senate Majority Leader's offices }

A report from the U.S. Government's Office of Technology Assessment titled "Facing America's Trash: What Next for Municipal Solid Waste?" uses partition in different ways. One way is an informal outline to show the system that produces municipal solid waste (MSW). Intended readers are environmentalists and representatives of city governments who make decisions about safely and efficiently handling solid wastes.

Box 1-D—The MSW "System"

The "system" that produces MSW is complex and dynamic, and different parts of it are linked together in ways that are not always clear. Illustrating the elements of the system can help identify leverage points at which strategies can be developed and options applied most effectively.

1. Materials/products lifecycle:
   - Products: design, production, distribution, purchase, use, discard
   - Non-product materials (yard and food waste): generation, discard
   - Management: collection, processing, treatment/disposal, etc.
2. Actors that touch materials and products:
   - Designers, manufacturers, distributors, retailers, etc.
   - Waste managers (haulers, landfills, incinerators, recyclers)
   - Citizens (purchasing, generation, siting decisions)
3. Private infrastructure:
   - Collection system, reclamation/other processing (e.g., scrap industry)
   - Landfills, incinerators, recycling facilities
   - Vertical integration of waste management industry
   - Structure and dynamics of materials industries:
     —dynamics of prices and disposal costs
     —international aspects
   - Financial sector
4. Public institutional structure:
   - Local decision making, collection programs
   - Government programs:
     —dynamics of Federal and State roles and plans
     —subsidies and incentives (PURPA, tax credits, etc.)
     —effects on private sector

5. Social attitudes:
  • Value judgments, perspectives affect how potential options are viewed
  • Resource policies:
    —extent to which federal government is involved
    —how the nation deals with materials and energy policies
  • Siting of facilities, degree of acceptable risks

Another use of partition is shown in a diagram of the configuration of selected engineering features at municipal solid waste landfills. City engineers can study the diagram to determine the feasibility of constructing such a system at their city's landfill.

Diagram of configuration of selected engineering features at MSW landfills

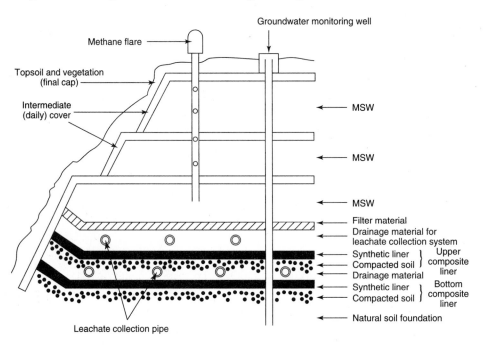

SOURCE: 52 *Federal Register* 20226, May 29, 1987.

Visuals frequently used in oral or written partition presentations include diagrams (such as the diagram above showing engineering features at municipal solid waste landfills), flowcharts, and drawings. Another often-used visual is the organization chart, illustrated in the explanation of the criminal justice system on pages 166–170. See Chapter 11 Visuals for a fuller discussion of visuals.

The following student-written report is an analysis through partition of an abstract subject, the criminal justice system. Written for a beginning class at a police academy, the subject is given careful examination by separating it into its constituent parts. Three main divisions or categories are identified: law enforcement; prosecution and courts; and sentencing and corrections. Law enforcement is subdivided into two divisions: (1) reported crimes and (2) arrest and booking. Prosecution and courts is subdivided into four divisions: (1) initial appearance,

163

Chapter 4
Analysis Through
Classification and
Partition: Putting
Things in Order

(2) preliminary hearing, (3) arraignment, and (4) trial. Through individual consideration of each part, the reader gains a better understanding of the whole. The writer thus accomplishes the intended purpose of the report: to provide a basic understanding of the criminal justice system.

The Plan Sheet that the student filled in before writing is included. The marginal notes are added as an aid to help you study the organization of the material.

PLAN SHEET
FOR GIVING AN ANALYSIS THROUGH PARTITION

Analysis of Situation Requiring Partition

What is the subject that I will partition?
the criminal justice system

Who is my intended audience?
a beginning class at a police academy

How will the intended audience use the partition? For what purpose?
for a basic understanding of the overall system

Will the final presentation of the partition be written or oral?
written

Setting Up the Partition

Definition or identification of subject:
an elaborate system set up to enforce criminal laws; partition will present only basic information about the three major divisions

Basis (or bases) for partition:
major divisions and their functions

Significance or purpose of basis (or bases):
The divisions of the system are intertwined. Decisions made in one division influence decisions made in another. Each division has a clearly defined role in administering the system.

Divisions of the subject, with identification of each division:
Law enforcement
   Reported crimes
      —Initiates the investigation of a crime and apprehension of a suspect
   Arrest and booking
      —Identifies the suspect. May initiate several activities: jailing of the suspect, booking
Prosecution and courts
   Initial appearance
      —Appearance before a judge for advising of rights and allowing or denying bail

Divisions of the subject, with identification of each division, cont.:
Preliminary hearing
   —Determination of continuing a case and taking a suspect to trial
Arraignment
   —Formal charges. Suspect enters plea to charges.
   Possibility of plea bargaining
Trial
   —Examination of crime-related facts by judge or jury
   Bench trials and jury trials. Judgment of guilty or not guilty
   determined
Sentencing and corrections
   —Sentence based upon penalties already established
   —Judge decides on degree of penalty. May consider circumstances: mitigating
   or aggravating

Presentation of the System

Most logical order for presenting the divisions:
sequential order

Types and Subject Matter of Visuals I Will Include

organization chart to depict visually the basic system

The Title I May Use

A Simplified Partition of the Criminal Justice System: Major Divisions and Their Functions

Sources of Information

textbook; interview with an attorney

Focused title

Partition of criminal
justice system
on basis of divisions
and functions.

A SIMPLIFIED PARTITION OF THE CRIMINAL JUSTICE SYSTEM: MAJOR DIVISIONS AND THEIR FUNCTIONS

The criminal justice system in the United States is an elaborate system set up to enforce criminal laws. Simply put, the system can be divided into three major divisions: law enforcement, prosecution and courts, and sentencing and corrections. These three divisions and their components are delineated in the accompanying organization chart. The divisions of the criminal justice system are intertwined in various ways; decisions made in one division influence decisions made in another. Each division, nevertheless, has a clearly defined role in administering the system.

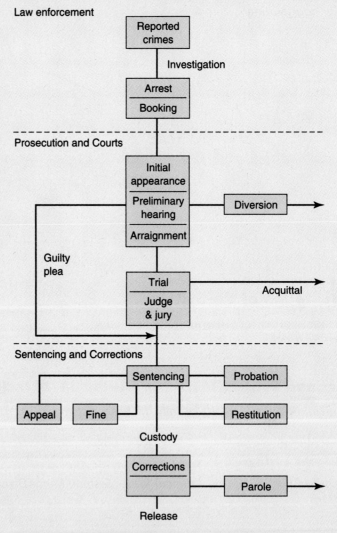

Figure 1. Organization chart showing a simplified version of the criminal justice system.

167

Chapter 4
Analysis Through
Classification and
Partition: Putting
Things in Order

First major division
described including
function

Law Enforcement

Law enforcement is the most familiar and the largest division of the criminal justice system. Police and highway patrol officers are highly visible as they investigate crimes, write traffic tickets and citations, apprehend criminals, walk a neighborhood beat, control traffic flow, and settle domestic disturbances. According to a 1988 U.S. Department of Justice report, the number of full-time law enforcement employees ranged from 1.8 to 3.9 per 1,000 inhabitants, depending on the section of the United States surveyed. The law enforcement division of the system investigates reported crimes and handles arrest and booking.

Its function is represented in the first third of the organization chart of the criminal justice system.

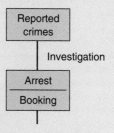

First subdivision
described including
function

Reported Crimes

Law enforcement officers are sworn to uphold the law within their area of authority or jurisdiction. Once a crime is reported, a law enforcement officer may investigate the crime and apprehend the individual accused of the crime.

Second subdivision
described including
function

Arrest and Booking

Law enforcement officers arrest criminal suspects. These suspects may be taken to jail or brought before a judge and booked. The booking process involves several activities. The suspect is photographed, fingerprints are taken, and background data about the suspect is obtained. A formal record is completed about the arrest: the time, place, and the circumstances.

Second major
division described
including function

Prosecution and Courts

After a suspect is arrested, prosecution representatives review the facts of the case and determine what action to take. Various factors influence the decision: the evidence, the circumstances under which the crime took place, the seriousness of the crime, and the criminal history of the suspect. The prosecution and courts division of the criminal justice system covers the initial appearance of the suspect, a preliminary hearing, and arraignment; a trial by a judge and jury may follow.

The midsection of the organization chart given earlier reflects these functions.

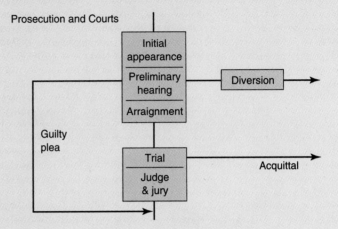

Prosecution and Courts

Initial Appearance

Anyone charged with a crime is brought before a judge or magistrate, who informs the individual of the charges filed. The judge or magistrate advises the suspect of legal right and gives the suspect an opportunity to make bail or denies bail. Bail is a fixed amount of money or a guarantee of payment of a fixed amount by responsible persons or organizations such as a bonding company. If bail is allowed, the suspect is released but must return to court for trial. The bail is designed to guarantee that the suspect will appear for the trial.

Preliminary Hearing

Preliminary hearings are held for serious crimes to determine if there is a basis for continuing a case and taking the suspect to trial. At a preliminary hearing, probable cause must be shown that a crime was committed and that the suspect committed the crime. Some suspects waive the preliminary hearing.

In some instances, a case may be temporarily removed from the criminal justice system; during this time a defendant may be required to attend counseling sessions, attend school (such as a school for drunk drivers), perform community service, or carry out some other specified activity. The term describing this procedure is diversion.

Arraignment

Arraignment is a formal proceeding where formal charges against the defendant are stated, the suspect enters a plea to the charges, and a trial date is set unless the suspect pleads guilty. Often plea bargaining occurs at this stage, and a suspect agrees to plead guilty for some concession from the state or government; the suspect, for example, might get a lesser charge, might be placed on probation rather than jailed or imprisoned, or might receive some leniency from the judge hearing the case.

169

Chapter 4
Analysis Through
Classification and
Partition: Putting
Things in Order

Fourth subdivision
described including
function

Trial

A trial is a proceeding where facts relating to the crime are examined by a judge or jury. Bench trials are trials where guilt is determined by a judge; jury trials are trials where guilt is determined by a jury, a group of persons selected through a screening process to hear the case. Both prosecution and defense attorneys try to persuade the judge or the jury. Evidence is presented; witnesses offer testimony. Many questions are asked. The goal is to discover the truth about the crime and the suspect's involvement in the crime. Finally the judge or the jury must decide, based upon information heard and viewed during the trial, whether the suspect is not guilty or guilty. If found not guilty, the suspect goes free (acquittal). If found guilty, the suspect is sentenced.

Third major division
described including
function

Sentencing and Corrections

The processes of sentencing and serving time are shown in the last section of the organization chart.

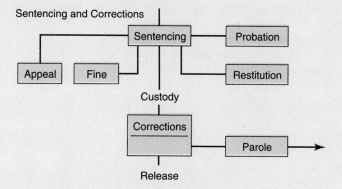

According to the crime committed, a suspect is sentenced based upon maximum penalties already established. The sentencing may be to pay a fine or to serve time in jail or prison. Sentencing may include placing the accused on probation, which is a conditional release from jail or prison; or the accused may have a sentence suspended and be placed under the supervision of a probation officer. Sentencing may also include requiring the accused to pay restitution—that is, to reimburse the victim for physical or property damage.

The judge may impose the maximum sentence possible under law for the crime committed, or the judge may impose some lesser sentence, based upon mitigating or aggravating circumstances. Mitigating circumstances considered by a judge might include such things as age, mental health, cooperation with police, or the degree of physical harm to a victim. Aggravating circumstances considered might include such things as the violence of the crime; the outcome of the crime, the seriousness of injury; or the status of the convicted suspect, if on probation or parole.

Following sentencing, a convicted criminal may be held in custody until placed in a jail or prison (a corrections facility—

local, state, or federal) to serve whatever time the sentencing dictates. After a period of time, some criminals may be considered for parole, which is a conditional early release with supervision by a parole officer. In some cases the criminal may be released through pardoning or by completing the time required by sentencing.

Contrast of Classification and Partition

Both classification and partition are approaches to analyzing a subject; classification and partition differ, however, in the number of the subjects being analyzed, in the relationship of divisions, and in the overall purpose.

Classification divides plural subjects into classes. The subject may be singular in form, but if the meaning is plural (such as mail, furniture, food) the subject is classifiable. Partition, on the other hand, divides a singular subject, or item, into its component parts.

In partition, the parts do not necessarily have anything in common, other than being parts of the same item. For instance, the handle, shank, and blade of a screwdriver share no relationship beyond their being parts of the same item. In classification, however, all the items in a division have a significant characteristic in common. For example, in classifying pencil sharpeners according to their power source, all manual pencil sharpeners use a person as the source of energy, and all electric pencil sharpeners use electricity as the source of energy.

Classification and partition differ also in their end results. Since their purposes are different—the purpose of classification being to sort related items into groups and the purpose of partition being to divide an item into its parts—their outcomes are different. For instance, regardless of the category into which screwdrivers are classified (common, Phillips, spiral ratchet, powered, etc.), they are still screwdrivers. In the partition of a screwdriver, however, a part—whether the handle, blade, or shank—is still only a part of the whole.

General Principles in Giving an Analysis Through Partition

1. *Partition is a basic approach in analysis.* It divides a subject into parts, steps, or aspects so that through individual consideration of these, a better understanding of the whole can be achieved.
2. *Only a singular subject can be partitioned.* If a subject is plural, it can be classified but not partitioned.
3. *The basis on which partition is made should be clear, useful, and purposeful.*
4. *The divisions in partition must be coordinate, or parallel.* All divisions on the same level must be of the same rank in grammatical form and in content.

5. *The divisions must be mutually exclusive.* Each division should be composed of a clearly defined group that would still exist without the other divisions on the same level.
6. *The divisions must not overlap.* A part can have a place in only one division.
7. *The divisions must be complete.* Every part must be accounted for, with no part left out.
8. *The order of presentation of divisions depends on their purpose and the audience.* Among the possible orders are time, familiarity, importance, complexity, space, maximum effectiveness, and alphabetical and random listing.
9. *The data in a partition analysis may be presented in outlines, in visuals, and in verbal explanations.* Use the form or combination of forms that presents the data most clearly for the stated audience and purpose.
10. *Classification and partition may be used together or separately.*

171

Chapter 4
Analysis Through
Classification and
Partition: Putting
Things in Order

Procedure for Giving an Analysis Through Partition

Analysis through partition is generally a part of a larger whole, like the other forms of communication discussed in the preceding chapters. Regardless of whether the partition analysis is a dependent or an independent communication, however, the structure of such an analysis is as follows:

I. The presentation of the subject and of the bases of partition is usually brief.
  A. State the subject to be partitioned and identify or define the subject.
  B. If applicable, list various bases by which the subject can be partitioned; state explicitly the bases of the divisions and list the divisions.
II. The listing and discussion of the divisions and subdivisions are the longest parts of the analysis.
  A. Point out the reasons why the divisions are significant and what purpose they serve.
  B. Give sufficient explanation to clarify and differentiate among the divisions.
  C. Present the divisions in whatever order best serves the purpose of the analysis.
  D. Use outlines and visuals to clarify the analysis.
III. The closing (usually brief) depends on the purpose of the presentation.
  A. The closing may be the completion of the last point in the analysis.
  B. The closing may be a comment on the analysis, or a summary of the main points.

Applications in Giving an Analysis Through Partition

Individual and Collaborative Activities

4.16   Select a subject suitable for partition. Present the partition system in a visual.

4.17   A. Working in small groups, choose one of the following for partition.

a. Key
b. Shoe
c. Restaurant menu
d. Book
e. Rocking chair

f. Egg
g. Golf club
h. Pencil
i. College year
j. Worship service

B. Give the partition in outline form.
C. Give the partition as a visual.

Written Partition Analysis

In completing applications 4.18 and 4.19,

a. Fill in the Plan Sheet on pages 174–175, or one like it.
b. Write a preliminary draft.
c. Go over the draft in a peer editing group.
d. Revise. See Part III Revising and Editing and the checklists for revising, front and back endpapers.
e. Write the partition analysis.

4.18   Select one of the following for a partition analysis.

a. Part of the body (ear, eye, heart, etc.)
b. Musical instrument (piano, trumpet, guitar, etc.
c. Electronic item (calculator, beeper, microwave oven, microprocessor, etc.)
d. Sports team (soccer, basketball, football, softball, etc.)

e. Family genealogy
f. Medical or dental clinic
g. Computer system
h. College education
i. Training of a technician or technologist
j. Topic of your own choosing from your major field

4.19   Select a business or organization (store, industry, club, farm, etc.) that you are familiar with (or can become familiar with). Show how this organization functions by dividing it into departments or areas. Include an organization chart (review pages 473–474 in Chapter 11 Visuals). Your presentation will be an explanation of a process; within this framework you will be relying heavily on analysis through classification and through partition. (One way to get information is interviewing knowledgeable people. For suggestions for informational interviews, see pages 455–456 in Chapter 10 Oral Communication.)

Oral Partition Analyses

4.20   As directed by your instructor, adapt for oral presentations the partition analyses you prepared in the applications above. Ask your classmates to evaluate your speech by filling in the Evaluation of Oral Presentations on page 459, or one like it.

4.21   Read "Are You Alive?," by Stuart Chase, pages 537–540.

    a. Show the structure of the essay by a partition analysis.
    b. Under "Suggestions for Response and Reaction," page 541, write Number 3.

4.22   Read the poem "A Technical Writer Considers His Latest Love Object," by Chick Wallace, page 548. Under "Suggestions for Response and Reaction," page 549, answer Number 2.

4.23   Read the excerpt from "Non-Medical Lasers: Everyday Life in a New Light," by Rebecca D. Williams, pages 564–569. Study the laser diagram on page 565; show the partition system of the laser in an outline (see pages 139–140).

PLAN SHEET
FOR GIVING AN ANALYSIS THROUGH PARTITION

Analysis of Situation Requiring Partition

What is the subject that I will partition?

Who is my intended audience?

How will the intended audience use the partition? For what purpose?

Will the final presentation of the partition be written or oral?

Setting Up the Partition

Definition or identification of subject:

Basis (or bases) for partition:

Significance or purpose of basis (or bases):

Divisions of the subject, with identification of each division:

Divisions of the subject, with identification of each division, cont.:

Presentation of the System

Most logical order for presenting the divisions:

Types and Subject Matter of Visuals I Will Include

The Title I May Use

Sources of Information

Chapter 5

Analysis Through Effect-Cause and Comparison-Contrast: Looking at Details

Outcomes

Upon completing this chapter, you should be able to

- Establish the cause of an effect
- Differentiate between an actual cause and a probable cause
- Differentiate between logical and illogical or insufficient causes of an effect
- List and explain the steps in problem solving
- Solve a problem
- Give an analysis through effect and cause
- List and explain three organizational patterns for a comparison and contrast analysis
- Give an analysis through comparison and contrast

177

Chapter 5
Analysis Through
Effect-Cause and
Comparison-
Contrast:
Looking at
Details

Introduction

Analysis through effect and cause and through comparison and contrast involves an intense examination of details. This examination may be as extraordinary as Sherlock Holmes's solving a case or as ordinary as a technician troubleshooting an air conditioning unit. Both Mr. Holmes and the technician have the same problem: discovering the cause of a particular effect.

Similarly, a comparison-contrast analysis may be as complex as a scientist's comparison of life in Hiroshima before and after the dropping of the atom bomb. Or a comparison-contrast analysis may be as pointed as a technician's explaining that a chip in a computer is like a note in a piece of music: Remove one chip or one note and the whole is disrupted but not destroyed.

The purpose of this chapter is to help you develop skill in analyzing details through effect-cause and comparison-contrast and in reporting the results of these analyses.

Analysis Through Effect-Cause

In an effect-cause analysis you try to find a relationship or link between two or more specific bits of information. You examine why something happens or happened by looking at the cause or causes; or you look at a condition or circumstance, or a set of conditions or circumstances, and explain what effects result or may result from the conditions or circumstances.

Causes lead to effects; effects result from causes. As you reason through effect-cause, you can follow at least three basic strategies. You can reason from effect to cause, from cause to effect, and effect to effect.

Reasoning from Effect to Cause

Trying to find the cause of a situation is a common problem. Not at all unusual are such questions as these: Why was production down last week? Why does this piece of machinery give off an unusual humming sound after it has been in operation a few minutes? Why didn't this design sell? Why does this hairspray make some customers' hair brittle? Why does this automobile engine go dead at a red light when the air conditioner is on? Why does this patient have recurring headaches?

When persons begin to answer these questions, they are using an investigative, analytical approach; they are seeking a logical reason for a situation, event, or condition. Explaining the cause of an effect is a kind of process; in this analysis, however, the concern is with answering *why* rather than *how*. In explaining the process of making a print by the diazo method, for example, the emphasis is on *how* the operation is done. In analyzing the effect-cause relationship in the diazo process, the concern might be *why* this particular process is used in reproducing a print and not another method, or the concern might be *why* exposed diazo film must be subjected to ammonia gas for development.

In the following report, the writer is analyzing the reasons for the loss of 1,500 lives aboard the *Titanic*. The report, written for a lay audience, addresses the question of *why* so many people perished in the tragedy. (Marginal notes have been added.)

WHEN THAT GREAT SHIP WENT DOWN

Effect investigated

On the evening of 14 April 1912, on its maiden trip, the "unsinkable" luxury liner the *Titanic* struck an iceberg in the freezing waters of the North Atlantic. Within a few hours the ship sank, taking with it some 1,500 lives. Why did so many people perish, particularly when there was another ship within ten miles? Subsequent investigation by national and international agencies showed that the *Titanic* crew was small and insufficiently trained, that the ship did not have sufficient lifesaving equipment, and that international radio service was inadequate.

Actual causes
(conclusions)
established

Causes discussed in order of least to most important

First cause explained in detail

Small and Insufficiently Trained Crew

One reason so many people perished in the *Titanic* tragedy was that the crew was small and insufficiently trained for such an emergency. In the face of grave danger, evidently the crew of the *Titanic* was simply not able to meet the situation. Many of the lifeboats left the ship only half full and others could have taken on several more people. The passengers were not informed of their imminent danger upon impact with the iceberg, investigation suggests, and the officers deliberately withheld their knowledge of the certain sinking of the ocean liner. Therefore, when lifesaving maneuvers were finally begun, many passengers were unprepared for the seriousness of the moment. Further indication of poor crew leadership is that when the *Titanic* disappeared into the sea, only one lifeboat went back to pick up survivors.

Insufficient Lifesaving Equipment

Second cause
explained in detail

Another reason so many people lost their lives was that the ship was not properly equipped with lifeboats. Although millions of

179

Chapter 5
Analysis Through
Effect-Cause and
Comparison-
Contrast:
Looking at
Details

dollars had been spent in decorating the ship with palm gardens, Turkish baths, and even squash courts, this most luxurious liner afloat was lacking in the vitally important essentials of lifesaving equipment. Records show that the *Titanic* carried only 16 wooden lifeboats, capable of carrying fewer than 1,200 persons. With a passenger and crew list exceeding 2,200, there were at least 1,000 individuals unprovided for in a possible sea disaster—even if every lifeboat were filled to capacity.

Inadequate International Radio Service

Third cause explained in detail

Undoubtedly, however, the primary reason that so many of the persons aboard the *Titanic* were lost is that international radio service was inadequate. Within ten miles of the *Titanic* was the *Californian,* which could have reached the *Titanic* before she sank and could have taken on all of her passengers. When the *Titanic* sent out emergency calls, near midnight, the radio operator of the *Californian* has already gone to bed. A crew member, however, was still in the radio room and picked up the signals. Not realizing that they were distress signals from a ship in the immediate vicinity, the crew member decided not to wake up the radio operator to receive the message. At that time there were no international maritime regulations requiring a radio operator to be on duty around the clock.

Comment on significance of causes

Had there been a more adequate international radio system, regardless of the *Titanic*'s small and insufficiently trained crew and its insufficient lifesaving equipment, probably those 1,500 lives could have been saved when that great ship went down.

In "When That Great Ship Went Down," the writer begins by identifying the topic and stating the circumstances. Next, the question is posed, "Why did so many people perish," with the crucial modification, "particularly when there was another ship within ten miles?" Then the writer summarizes in one sentence the three main causes. In the following three paragraphs, each cause is explained in turn, with sufficient supporting details. The writer discusses the causes in the order of their significance, leading up to the most critical cause. Then the writer closes the analysis with a brief comment that includes a relisting of the three main causes and emphasis on the most critical one.

You may present ideas in an effect-cause arrangement to gain suspense or to make a point dramatically, as the following paragraph illustrates.

Then rose a frightful cry—the hoarse, hideous, indescribable cry of hopeless fear—the despairing animal cry man utters when suddenly brought face to face with Nothingness, without preparation, without consolation, without possibility of respite. . . . *Sauve qui peut!* Some wrenched down the doors; some clung to the heavy banquet tables, to the sofas, to the billiard tables: during one terrible instant, against fruitless heroisms, against futile generosities, raged all the frenzy of selfishness, all the brutalities of panic. And then, then came, thundering through the blackness, the giant swells, boom on boom! . . . One crash! The huge frame building rocks like a cradle, seesaws, crackles. What are human shrieks now? The tornado is shrieking! Another! Chandeliers splinter; lights are dashed out; a sweeping cataract hurls in: the immense hall rises, oscillates, twirls as upon a pivot, crepitates, crumbles into ruin. Crash again! The swirling wreck dissolves into the wallowing of another monster billow; and a hundred cottages overturn, spin in sudden eddies, quiver, disjoint, and melt into the seething. . . . So, the hurricane passed.*

Effect-cause analysis is used in trouble-shooting, that is, trying to find out the causes of observed effects. A doctor tries to determine what causes a patient's dizziness. A mechanic must find out why a haybaler does not cut the cord when the bale is ejected. An office worker determines why faxes are not received. A consumer wonders why an appliance is not functioning properly. The example below shows a page from an owner's manual for a GE color television. For common problems (effects), suggestions (possible causes) guide the owner in solving the problem.

Reasoning from Cause to Effect

Often persons are confronted with a cause and want to know what the effects are, or will be. For instance, if a piece of stock is not perfectly centered in a lathe (cause), what will happen (effect)? If a television antenna is not properly installed (cause), what will be the consequences (effect)? If the swelling of an injured foot (cause) is not attended to, what will happen to the patient (effect)? If assembly line procedures are modernized (cause), how will the final product be changed (effect)?

The following sample paragraph explains that aspirin (cause) has many advantages as well as some disadvantages (effects).

Since its introduction in Germany in 1853, aspirin (the acetyl derivative of salicylic acid) has proved to be a miracle drug. Aspirin banishes headache, reduces fever, and eases pain. It is not without its dangers, however. Aspirin taken in excess will cause stomach upset, bleeding ulcer, and even death. Nevertheless, aspirin is responsible for more good than harm. It may even prevent heart attacks. According to current research, it is possible that aspirin contains anticoagulant properties that prevent thrombi, the blood clots that clog the coronary arteries, causing heart attacks.

The excerpt below from a pamphlet on quitting smoking shows that giving up smoking (cause) brings health benefits (effect). Notice also the conversational tone of the excerpt; realizing the reader of the pamphlet could be anyone, the writer puts the information in simple terms.

WHAT HAPPENS AFTER YOU QUIT SMOKING**

Within 12 hours after you have your last cigarette, your body will begin to heal itself. The levels of carbon monoxide and nicotine in your system will decline rapidly, and your heart and lungs will begin to repair the damage caused by cigarette smoke.

Within a few days, you will begin to notice some remarkable changes in your body. Your sense of smell and taste will return. Your smoker's hack will disappear. Your digestive system will return to normal.

Most important of all, you feel really alive—clear-headed, full of energy and strength. You're breathing easier. You can climb a hill or a flight of stairs without becoming winded or dizzy. And you will be free from the mess, smell, inconvenience, expense and dependence of cigarette smoking. *(Continued on page 182)*

*From *Chita,* by Lafcadio Hearn (New York: Harper & Brothers, 1889).
**From a U.S. Department of Health and Human Services pamphlet *A Guide to Quitting Smoking.*

Excerpt from an Owner's Manual for a GE Color Television*

181

Chapter 5
Analysis Through
Effect-Cause and
Comparison-
Contrast:
Looking at
Details

Problems You Can Solve

- Problems sometimes are caused by simple "faults" that you can easily correct without the help of a service technician by checking a few basic remedies.

- Before you call or take your TV to an Authorized GE TV Servicenter, look below in the left-hand column for the type of problem you are experienc-

ing. Then perform the simple checks and adjustments listed for that problem.

- If your TV is still in warranty, these checks and adjustments could save you time and the cost of unneeded service. They also could save you the cost of a diagnosis not covered by your warranty.

- If your TV is out of warranty, these checks and adjustments could save you the cost of an unneeded diagnosis.

- If service should be necessary, the warranty information on the back cover page tells you what your warranty covers and how to get service.

Trouble	Checks and Adjustments
TV will not turn on	• Check to make sure it is plugged in. • Check the wall receptacle (or extension cord) to make sure it is "live" by plugging in something else. • Maybe batteries in remote control are "dead." • Maybe remote control was not aimed at remote sensor.
Turns off while playing	• Electronic projection circuit may have been activated because of a power surge. Wait 30 seconds and then turn on again. If this happens frequently, the voltage in your house may be abnormally high (especially if you burn out a lot of light bulbs).
Blank screen	• Try another channel.
No sound, picture okay	• Maybe sound is adjusted very low. Try pressing *VOLUME* ∧ button to restore sound.
Can't select certain channel	• Channel may not be in Channel Memory. • If using VCR, check to make sure *TV/VCR* button on VCR is in correct position.
No picture, no sound	• Maybe someone changed *CABLE/AIR* function to wrong position. • Maybe a vacant channel is tuned. • If watching VCR (connected only through antenna input), make sure TV is tuned to channel 3 or 4–same as *CH3/CH4* switch on VCR. Also check to make sure *TV/VCR* button on VCR is in correct position
Sound okay, picture poor	• Check antenna connections. • Try adjusting *SHARPNESS* function to improve weak signals.
Picure okay, sound poor	• Try another channel.
Controls don't work	• Try unplugging set for 30 seconds and then turning it on again.
Intermittent or no remote	• Maybe something was between the remote and the remote sensor. • Maybe the remote was not aimed directly at the TV. • Maybe batteries in remote are weak or dead. Try replacing batteries.

*Reprinted by permission from General Electric.

WHAT HAPPENS AFTER YOU QUIT SMOKING *(continued)*

Long Range Benefits

Now that you've quit, you've added a number of healthy productive days to each year of your life. Most important, you've greatly improved your chances for a longer life. You've significantly reduced your risk of death from heart disease, stroke, chronic bronchitis, emphysema, and cancer.

Recovery Symptoms

As your body begins to repair itself, instead of feeling better, you may feel worse. These "withdrawal pangs" are really symptoms of recovery. Immediately after quitting, many ex-smokers experience "symptoms of recovery" such as temporary weight gain caused by fluid retention, irregularity and sore gums or tongue. You may also feel edgy and more short-tempered than usual.

It is important to understand that the unpleasant after effects of quitting are only temporary and signal the beginning of a healthier life.

This same topic, what happens after you quit smoking, is treated in another format in the *HOPE HEALTH LETTER,* November 1994. The health letter indicates the effect-cause relationship between quitting smoking and improved health through a graphic (see page 183) showing a hand snubbing out a cigarette in an ashtray; the cigarette then becomes a time line showing improved health from 20 minutes after the last cigarette is smoked to 15 years after.*

Reasoning from Effect to Effect

Analysis through effect and cause may involve a chain of reasoning in which the cause of an effect is the effect of another cause. In the following paragraph (from the *Occupational Outlook Handbook*) dealing with technological progress and labor, for instance, industrial applications of scientific knowledge and invention (cause) result in increased automation (effect). Increased automation (cause) calls for skilled machine operators and service people (effect). This (cause), in turn, results in occupational changes in labor (effect).

> Technological progress is causing major changes in the occupational makeup of the nation's labor force. Rapid advances in the industrial applications of scientific knowledge and invention are making possible increasing use of automatic devices that operate the machinery and equipment used in manufacturing. Nonetheless, the number of skilled and semiskilled workers is expected to continue to increase through the 1990s, despite this rapid mechanization and automation of production processes. It is expected that our increasingly complex technology generally will require higher levels of skill to operate and service this machinery and related equipment.

*Reprinted from the *Hope Health Letter,* November 1994, p.6, by permission The Hope Heart Institute, Seattle, WA.

183

Chapter 5
Analysis Through
Effect-Cause and
Comparison-
Contrast:
Looking at
Details

What happens after you quit smoking?

Within 20 minutes of smokng your last cigarette, your body begins a series of changes that continues for years. All benefits are lost by smoking just *one* cigarette a day, according to the American Cancer Society.

20 MINUTES
• Blood pressure drops to normal
• Pulse rate drops to normal
• Body temperature of hands and feet increases to normal

8 HOURS
• Carbon monoxide level in blood drops to normal
• Oxygen level in blood increases to normal

48 HOURS
• Nerve endings start regrowing
• Ability to smell and taste is enhanced

1 TO 9 MONTHS
• Coughing, sinus congestion, fatigue, and shortness of breath decrease
• Cilia regrows in lungs, increasing their ability to handle mucus, clean the lungs, and reduce infection
• Body's overall energy increases

10 YEARS
• Lung-cancer death rate is similar to that of a nonsmoker
• Precancerous cells are replaced
• Risk of cancer of the mouth, throat, esophagus, bladder, kidney, cervix, and pancreas decreases

24 HOURS
• Chance of heart attack decreases

2 WEEKS TO 3 MONTHS
• Circulation improves
• Walking becomes easier
• Lung function increases up to 30%

1 YEAR
• Excess risk of coronary heart disease is half that of a smoker

5 YEARS
• Lung-cancer death rate for average former smoker (one pack a day) decreases by almost half
• Stroke risk is reduced to that of a nonsmoker five to 15 years after quitting
• Risk of cancer of the mouth, throat, and esophagus is half that of a smoker's

15 YEARS
• Risk of coronary heart disease is that of a nonsmoker □

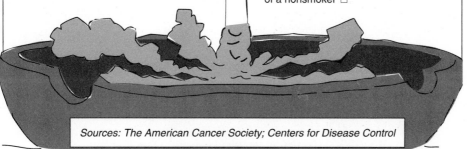

Sources: The American Cancer Society; Centers for Disease Control

© HHI

Intended Audience and Purpose

You must be aware of who will read an effect-cause analysis and why, for these two basic considerations determine the extent of the investigation and the manner in which the investigation will be reported. For instance, the analysis may be presented for a lay audience who wants a general understanding of the topic, or the analysis may be for a division manager who will base decisions and actions on the analysis.

Relationship to Other Forms of Communication

Analysis through effect and cause is a method of thinking and of organizing and presenting material. The effect-cause analysis frequently includes other forms of communication, such as definition, description, process explanation, classification, and partition, as well as oral presentation and the use of visuals. (For discussion of oral communication, see Chapter 10; for discussion of visuals, see Chapter 11.) Several of these forms are found in the following student-written effect-cause analysis. Given first is the Plan Sheet that the student filled in before writing the analysis.

PLAN SHEET FOR GIVING AN ANALYSIS THROUGH EFFECT AND CAUSE

Analysis of Situation Requiring Effect-Cause Analysis

What is the subject that I will analyze?
why people get hungry

Who is my intended audience?
people who want to lose weight

How will the intended audience use the analysis? For what purpose?
to understand how to deal with hunger pangs; to change eating habits in order to lose weight

Will the final presentation of the analysis be written or oral?
written

Effect to Be Investigated

the feeling of being hungry, particularly the sometimes overpowering feeling in overweight people

Significance of the Effect

When overweight people feel the need to satisfy hunger, they often eat junk food that just adds to the problem. Understanding why people get hungry can lead to healthy eating habits.

Causes (as applicable)

Possible causes:
omit

Probable cause (or causes):
omit

Actual cause (or causes):
- *Daily eating routine*
  The body programs itself on a daily routine.
- *Daily nutrient/vitamin intake*
  The body needs vitamins and healthy meals.

Actual cause (or causes), cont.

- *Blood sugar level*
  The body reacts to a too-high or too-low blood sugar level.
- *Metabolism rate*
  Persons with a low metabolism don't feel full as quickly as those with a high metabolism rate.

Supporting Evidence or Information for the Probable or Actual Cause (or Causes)

see above

Organization of the Analysis

by the four major factors that affect feeling hungry, in the order of what overweight people can control

Types and Subject Matter of Visuals I Will Include

a humorous picture illustrating insulin regulating sugar in the bloodstream

The Title I May Use

Why People Get Hungry

Sources of Information

nutrition textbook; Diet Centers, Inc.; Cindy Bracken and Steve Fasano

187

Chapter 5
Analysis Through
Effect-Cause and
Comparison-
Contrast:
Looking at
Details

WHY PEOPLE GET HUNGRY

The human body is a very complicated system. Appetites differ from one individual to the next, but the cause for feeling hungry can be attributed to four major factors: daily eating routine, daily nutrient/vitamin intake, blood sugar level, and individual metabolism rate.

Daily Eating Routine

When a person eats on a schedule, the body gets programmed to that schedule and therefore gets ready to begin the fat-burning process before the person actually begins to eat. For instance, someone who has meals at the same time each day finds herself hungry at that time each day because her body has programmed itself for food. On the other hand, someone else who eats sporadically and skips meals will find himself having a bigger appetite and getting hungry more often than the person with regular meals on a daily routine.

Daily Nutrient/Vitamin Intake

The nutrients and vitamins that people daily consume affect their appetite immensely. People who take vitamins regularly and eat healthy meals consisting of the basic food groups will not get hungry as often as those who do not take vitamins and eat junk food regularly. The reason for this is that eating sweets and fats triggers the release of insulin into the bloodstream, and elevated levels of insulin can make a person hungry. Also, those who do not consume high-fiber foods get hungry more often because these foods are bulky, causing the person to fill up faster.

Blood Sugar Level

The blood sugar level in the body is being used for fuel even when a person is sleeping. After eight or nine hours of sleep, the blood sugar level has dropped drastically and the body is ready for food. When the blood sugar level is below the normal range, people get hungry. A simple example illustrates this point. Joe feels hungry; he eats a candy bar, and he feels full for twenty minutes or so. Then Joe feels hungry again and wants another candy bar. All this happened because Joe's blood sugar level was low and he got hungry. When Joe ate the candy bar with the high concentration of sugar, his body produced too much insulin (a hormone). Insulin is the agent that "opens the valve" in the bloodstream to overreact and remove the excess sugar, as shown in Figure 1, leaving Joe weak and craving more food.

Figure 1. Insulin regulates the amount of sugar in the bloodstream.
SOURCE: Cindy Bracken and Steve Fasano

The excess sugar is stored as body fat. If he had eaten an apple or other piece of fruit, his blood sugar level would have risen just enough to curb the appetite.

<u>Metabolism Rate</u>
Metabolism is the rate at which the body burns calories. The higher the metabolism of a person, the less hungry that person gets. This occurs because if the body burns the calories faster, the energy and bulk of the food are distributed to the body faster, therefore making the person feel full and not hungry. In contrast, the lower the metabolism, the hungrier a person gets and the longer it takes to feel full. The reason for this is the body takes so long to distribute the energy and bulk that the person is fooled into thinking he hasn't eaten enough when in reality if he would just wait twenty minutes or so, the body would be finished with the process of assimilating the food and he would feel full.

<u>Closing</u>
People are constantly going on crash diets or starving themselves. One of the major ways people try to curb their appetite is taking diet pills. This is a BIG mistake because these pills can be dangerous, and they also do no good. Yes, the pills curb the appetite as long as the person is taking them; but if she quits, her body goes crazy and her hunger pangs are multiplied.

   The best way people can curb their appetite—and thus lose weight—is to gain a full understanding of why they get hungry and to make their eating habits healthy and regular.

Establishing the Cause of an Effect

As we have seen, a common problem is to discover the cause, or causes, of a given effect. Through reading, observation, consultation with knowledgeable people, and so on, an answer is sought. Why was there a high rate of absenteeism last week? Investigation shows that 97 percent of the absenteeism was due to a highly contagious virus. Why has the new paint already begun to peel on a car that was repainted only two months ago? Investigation shows that the cause is inferior paint, not an improper method of paint application or improper conditioning of the old finish, as at first suspected. In these two situations, it was possible to establish a definite or *actual cause*—one that evidence proves beyond question is the true cause.

   Often, however, a definite cause cannot be established. A possible or *probable cause* must suffice. Why was the Edsel car a flop? Why is this patient antagonistic toward the nurses? Why is red hair associated with a quick temper? Why have sales this year more than tripled those of last year? Why is my roommate so popular? Although such questions as these cannot be answered with complete certainty, causes that answer them, nevertheless, need to be established.

   In the process of establishing a probable cause, consider all possible causes. After examining each possible cause, eliminate it entirely, keep it as merely a possibility, or decide that—all things considered—it is a probable cause. In the investigation of possible causes, the reasoning process must be logical and relevant. Likewise, each possible cause must be logical and relevant. Each possible cause must be examined on the basis of reason, not on the basis of emotions or preconceived ideas; and each possible cause must *really* be a possible cause.

189

Chapter 5
Analysis Through
Effect-Cause and
Comparison-
Contrast:
Looking at
Details

Establishing the Cause: Why Sam Didn't Get 150 Bucks

Consider the plight of Sad Sam, who asks, "Why didn't I get the 150 bucks I asked my parents for to get new front tires on my car?" Perhaps—not on paper but at least mentally—Sam thinks of these possible reasons:

1. My parents don't love me.
2. They don't have the money.
3. They didn't receive my letter.
4. They mailed me a check but something happened to the letter.
5. They just don't realize what bad shape those front tires are in.
6. They think I should pay my car expenses out of the money I earn from my part-time job.

Sam turns over each possibility in his mind. He immediately dismisses possibility 1 (My parent's don't love me). He knows this was only a childish reaction and that love cannot be equated with money. Sometimes parents who don't love their children send them money; and sometimes, for very good reasons, parents who love their children don't send them money.

There is a real possibility Sam's parents didn't have 150 dollars to spare (possibility 2). But Sam reasons if that is the cause, they would have asked him to wait a while longer or at least made some response.

Perhaps his parents did not receive his request (possibility 3). Maybe they mailed him a check but something happened to the letter (possibility 4). Either of the letters could have gotten lost in the mail. To his knowledge, however, none of his other letters has ever gotten lost.

Possibility 5 (They just don't realize what bad shape those front tires are in) could certainly be a true statement. But Sam realizes that his parents could be aware of the condition of the tires and still have reason not to send the money.

The more Sam considers possibility 6 (They think I should pay my car expenses out of the money I earn from my part-time job), the more this seems the likely reason. After all, Sam had promised his parents that if they would help him buy a car, he would keep it up. Furthermore, in his excitement of owning a car at long last, he had cautioned his parents to ignore any requests for car money, no matter how desperate they sounded.

So, as Sam carefully considers each possibility, he comes to the conclusion that the probable reason he did not receive the money is that his parents expect him to pay for his car expenses.

Whether the conclusion that is reached is the definite cause or the most probable cause of a situation, it must be arrived at through *logical* investigation, as in Sam's case. A conclusion, to be logical, must be based on reliable evidence, on relevant evidence, on sufficient evidence, and on an intelligent analysis of the evidence. Reliable evidence is the *proof* that can be gathered from trustworthy sources: personal experience and knowledge, knowledgeable individuals, textbooks, encyclopedias, and the like. Relevant evidence is information that directly influences the situation. Sufficient evidence means enough or adequate information with no omission of significant facts that would alter the situation. Evidence that is reliable, relevant, and sufficient must be analyzed intelligently. The meaning and significance of each individual piece of information and of all the pieces of information together must be considered if the most plausible conclusion is to be reached.

Illogical and Insufficient Causes

In an investigation of the "why" of a situation, especially guard against four pitfalls in reasoning, lest you arrive at an illogical or insufficient cause.

Pitfall 1. A Following Event Caused by a Preceding Event

One pitfall in reasoning assumes that a preceding event causes a following event, simply because of the time sequence. This fallacy in logic is called *post hoc, ergo propter hoc,* which literally means "after this, therefore because of this." Many superstitions are based on this false assumption. For instance, if a black cat runs across the street in front of Tom and later his automobile has a flat tire, blaming the cat is illogical. The two events (black cat running across street and flat tire) are not related in any way, except that they both concern the same person and they happened one after the other.

Pitfall 2. Hasty Conclusion

A common pitfall in effect-cause reasoning is jumping to a conclusion before all the facts are in. This results from not thoroughly investigating all aspects of a situation before announcing or indicating the cause. Consider this telephone call from a teenager to her father:

> "Hello. Dad? I'm calling from City Drug Store. I've been in a slight automobile accident."
> "Automobile accident? Just how fast were you driving?"
> "I had stopped at a traffic light and this guy just rammed me from the rear."

Note the hasty conclusion that the daughter had caused the accident by driving too fast. The father had jumped to a conclusion before all the facts were made known.

Pitfall 3. Oversimplification

Avoid the pitfall of oversimplifying a situation by thinking the situation has only one cause when it has several causes. If May says the reason she did not get a job she wanted is that she was nervous during the interview, she is oversimplifying the situation. She is not taking into consideration such possibilities as her training was inadequate, she lacked experience in the type of work, another applicant was better qualified, or she was not recommended highly enough.

Pitfall 4. Sweeping Generalization

All-encompassing general statements that are not supportable reflect illogical cause-effect thinking. Consider these statements:

> Labor unions are never concerned with consumer safety.
> Our products always completely satisfy the customer.

The two preceding statements are sweeping generalizations. They are broad statements that are difficult if not impossible to support with sufficient evidence. Be wary of using all-inclusive terms such as *never, always, completely, everyone,* and the like.

191

Chapter 5
Analysis Through
Effect-Cause and
Comparison-
Contrast:
Looking at
Details

Problem Solving

Closely related to answering the "why"—the effect–cause—of a situation is answering the "what should I do" in a situation. What courses should I take next semester? Should I drop out of school for a while until I can make some money? Should I report a fellow worker who does not follow posted safety rules? Should I take another job that has been offered to me? Which tires should I buy? What action should I take to stop a supervisor from asking personal questions?

All these questions indicate problems that people are trying to solve. Everyone has problems—some of them insignificant, some very significant—that must be faced and dealt with. Such problems as which shoes to wear today or which vegetables to choose at lunch or which movie to go to tonight require only a few moments consideration and, viewed a few days later, are quite insignificant. But other problems, other decisions, require thoughtful consideration and will still be significant a few days or a few years from now. How does one go about solving such problems?

Problem solving is not easy; several suggestions, however, may help in reaching a decision:

1. Recognize the problem for what it is.
2. Realize that there are *various* possibilities for solution.
3. Consider *each* possibility on its own merits.
4. Decide which possibility is *really* best.
5. Determine to make the chosen course of action successful.

1. Recognition of the Real Problem

The first step in solving any problem is recognizing the problem for what it is. The student who wrote only three pages on a test and who thinks a classmate received an A because the classmate filled up ten pages is avoiding the *real* problem: what the pages contain, not how many pages there are. In looking at a problem for what it is, reason and logical thinking must prevail. Glorifying oneself as a martyr or exaggerating the circumstances is avoiding the real issue. Sometimes emotions and pressures from others cloud reason and thus camouflage the real problem.

2. Various Possible Solutions

Various choices, or possibilities, or answers are available for solving problems. Sometimes, in a hastily posed problem, there seems to be only two alternatives, thus an either-or situation. An example is the problem of the student who does not have money for college and says, "I'll either have to go so far in debt it will take me ten years to get a clean start in life, or I'll have to postpone college until I have enough money in the bank to see me through two years." This student is certainly not considering *all* the alternatives available. The student has not considered that many colleges have an agreement with various industries whereby the student goes to college for a semester and works for a semester. Or the possibility of evening school, through which the student can attend courses one or more evenings per week after work and earn college credit. Or the possibility of taking a part-time job to help defray most or all of the college expenses. Or the possibility of taking a reduced college load and working full-time. Thus a problem that seemed to have only two possible solutions really has a number of possible solutions.

3. Merits of Each Possible Solution

The third step in solving a problem is considering each possible choice on its own merits. Each possible solution must be analyzed and weighed carefully from every angle. This may involve investigation: consulting with knowledgeable people, drawing upon the experiences of family and friends, turning to reference materials, and so forth.

4. Choosing the Best Solution

Once each of the possible choices has been given an honest, thoughtful consideration, it is time to decide which *one* choice is best. This is the most difficult step in problem solving, but if it is preceded by a mature analysis of the problem and all of its possible solutions, the final decision is more likely to be satisfactory even in a few weeks or in several years. Sometimes it may seem easier to let someone else make the decision; in that way, if things do not turn out as wished, someone else gets the blame. But shunning responsibility by avoiding making a decision goes back to the first step in problem solving: recognizing the problem for what it is.

5. Determination to Succeed

Finally, in problem solving there must be a determination to make the chosen course of action successful. This may involve acquainting others with the decision and the reasons for it and perhaps even persuading them to accept the decision. A positive, open-minded attitude toward a well-thought-out decision is the best assurance that the decision, after all, was the best one.

General Principles in Giving an Analysis Through Effect and Cause

1. *The precise effect (situation, event, or condition) being investigated must be made clear.* The focus is on gathering information that relates directly to that one effect.
2. *Awareness of who will read the analysis, and why, is essential to the writer.* These two considerations—the who and the why—determine the extent of the investigation and the manner of reporting it.
3. *Additional forms of communication may be needed.* Definition, description, process explanation, classification, and partition—plus visuals—may help clarify the effect-cause analysis.

4. *If the actual cause can be established, sufficient supporting evidence should be given.* The evidence must be reliable, relevant, and sufficient.
5. *If the actual cause cannot be established, adequate support for the probable cause should be given.* The support must be reliable, relevant, and as sufficient as possible.
6. *A preceding event may or may not cause a particular following event.* You must determine whether there is a causal relationship between two events.
7. *A hasty conclusion indicates lack of thorough analysis.* Investigate and weigh carefully all aspects of a situation before reaching a conclusion.
8. *A situation may be oversimplified by attributing to it only one cause when it may have several causes.* Significant situations are usually more complex than they at first seem.
9. *Sweeping generalizations weaken a presentation because they are misleading and are often untrue.* Caution should guide the use of such all-inclusive terms as *never, always,* and *every person.*
10. *Problem solving is a logical process.* The steps in problem solving are recognizing the problem for what it is, realizing that there are various possible solutions, considering each possible solution on its own merits, deciding which possibility is really best, and determining to make the chosen course of action successful.

193

Chapter 5
Analysis Through
Effect-Cause and
Comparison-
Contrast:
Looking at
Details

Procedure for Giving an Analysis Through Effect and Cause

The analysis may be an independent presentation, or it may be a paragraph or a part of a presentation. Whichever the case, the procedure is the same and presents no unusual problems. (Although the procedure suggested here is for analyzing from effect to cause, the particular communication situation may call for analyzing from cause to effect.) Generally, the analysis should be divided into three sections:

I. Stating the problem usually requires only a few sentences.
   A. State the effect (situation, event, or condition) that is being analyzed.
   B. If applicable, give the scope and limitations of the investigation (as in reporting the causes of traffic accidents involving fatalities at the Highway 76-Justice Avenue intersection from January 1 to July 1).
   C. Give the assumptions, if any, on which the analysis or interpretation of facts is based (as in assuming that the new machine that is not working properly today was correctly installed and serviced last month).
   D. If applicable, give the methods of investigation used (reading, observation, consultation with knowledgeable people, etc.).
   E. Unless the order of the presentation requires otherwise, state the conclusions that have been reached.

II. Reporting the investigation of the problem is the longest section of the presentation.
   A. Consider in sufficient detail possible causes; eliminate improbable causes.
   B. If the actual cause can be established, give the cause and sufficient supporting evidence.

C. If the actual cause cannot be established, give the probable cause and sufficient supporting evidence.

D. Interpret facts and other information when necessary.

E. If necessary, especially in a longer presentation, divide the subject into parts and analyze each part individually.

F. Organize the information around a logical pattern or order.

    1. The topic may suggest moving from the less important to the more important, or vice versa.

    2. The topic may suggest using a time order of what happened first, second, and so on.

    3. The topic may suggest moving from the more obvious to the less obvious.

    4. The topic may suggest going from the less probable to the more probable.

    5. The nature of the presentation may suggest some other order.

III. The conclusion (usually brief) reflects the purpose of the analysis.

A. If the purpose of the analysis is to give the reader a general knowledge of an effect-cause situation, summarize the main points or comment on the significance of the situation.

B. If the analysis is a basis for decisions and actions, summarize the main points, comment on the significance of the analysis, *and* make recommendations.

Applications In Analysis Through Effect-Cause

Individual and Collaborative Activities

5.1 Each of the following statements contains an illogical or insufficient cause for an effect. Point out why the stated or implied cause is inadequate.

1. The reason my cow died is that I didn't go to church last Sunday.
2. Strikes occur because people are selfish.
3. Mr. Smith's television appearance on election eve won him the election.
4. Automobile insurance rates are higher for teenagers than for adults because teenagers are poorer drivers.
5. Susan isn't dependable. She was supposed to turn in a report today but she hasn't shown up.
6. The atom bomb won World War II.
7. If I were you, I wouldn't switch kinds of drinks at a party. I did so last night, and this morning I have a terrible headache.
8. I can't stand pizza. I ate some once and it was as soggy as a wet dishrag.
9. Most people who are successful in business have large vocabularies. If I develop a large vocabulary, I'll be successful in business.
10. "I joined the Confederacy for two weeks. Then I deserted. The Confederacy fell."—Mark Twain

195

Chapter 5
Analysis Through
Effect-Cause and
Comparison-
Contrast:
Looking at
Details

5.2   In each of the following situations, indicate whether a probable cause or an actual cause is the more likely to be established. Point out why.

1. I thought that I had a B for sure in history, but the grade sheet shows a D. Why did I get a D?
2. This is the fourth time that the same jaw tooth has been filled. Why won't the filling stay in?
3. For the third night in a row I have asked my neighbors to please be a little quieter. Tonight they have the stereo turned up even higher. Why won't they be quieter?
4. In the 1984 presidential election the Republican candidate won. Why did he receive more votes than the Democratic candidate?
5. Every time my mother eats a lot of tomatoes, she breaks out in a rash. Why?

Written Effect-Cause Analysis

5.3   Make a list of five problems regarding your schoolwork, job, or vocation that require thoughtful consideration. Choose one of these problems and make a plan for solving it by listing:

1. The real problem
2. Various possibilities for solution
3. Consideration of each possible solution
4. The solution
5. Ways to assure that the chosen course of action will be successful

---

In completing applications 5.4 and 5.5,

a. Fill in the Plan Sheet on pages 198–199, or one like it.
b. Write a preliminary draft.
c. Go over the draft in a peer editing group.
d. Revise. See Part III Revising and Editing and the checklists for revising, front and back endpapers.
e. Write the effect-cause analysis.

5.4   Choose one of the general subjects below (or one that is similar). Restrict the subject to a specific topic. Then write an effect-cause analysis.

Example of Subject Restriction
General subject: Decrease or increase of students in certain majors
Specific topic: Increase in students majoring in building trades

1. Improper functioning of a mechanism
2. A mishap of national importance
3. A decrease or increase in jobs in your city, county, or state

4. Variations in fringe benefits with different companies
5. A decrease or increase in sales during a particular period of time
6. Factors in a company's consideration of a plant site, a new piece of equipment, decrease or increase of employees
7. A change in attitude toward a decision
8. Increased cost of an item or service

5.5   Write an explanation of the cause (or causes) for one of the topics below. Review page 403 A Note About Plagiarism.

1. Why an automobile engine starts when the ignition is turned on
2. Why a light comes on when the switch is turned on
3. Why water and oil do not mix
4. Why computer literacy is important
5. Why two houses on adjacent lots and with the same floor space may require different amounts of electricity for air conditioning or for heating
6. Why an earthquake occurs
7. Why artificial breeding has become a popular practice among ranchers
8. Why extreme care should be taken in moving an accident victim
9. Why different fabrics (such as silk, cotton, wool, linen) supposedly dyed the same color may turn out to be different shades
10. Why proper insulation decreases heating and cooling costs
11. Why an appliance wired for alternating current will not operate on direct current
12. Why a dull saw blade should be sharpened
13. Why a stain should not be applied over a varnish
14. Why the Rh factor is important in blood transfusions and in pregnancy
15. Why the blood bank insists on having complete identification of a patient's sample of blood
16. Why a new book in the library does not appear on the shelf immediately after it is received
17. Why computer documentation is essential
18. Why vitamins are essential in human growth
19. A topic of your own choosing

Oral Effect-Cause Analyses

5.6   As directed by your instructor, adapt for oral presentations the analyses through effect-cause you prepared in the applications above. Ask your classmates to evaluate your speech by filling in the Evaluation of Oral Presentations on page 459, or one like it.

Using Part II Selected Readings

5.7   Read the short story "Quality," by John Galsworthy, pages 521–525. In what ways are effect-cause relationships used in the story? Under "Suggestions for Response and Reaction," page 525, complete Number 1 and Number 5.

197

Chapter 5
Analysis Through
Effect-Cause and
Comparison-
Contrast:
Looking at
Details

5.8   Read "The Science of Deduction," by Sir Arthur Conan Doyle, pages 532–535. In what ways are effect and cause used in the article? Under "Suggestions for Response and Reaction," pages 535–536, complete Number 2, Number 3, and Number 4.

5.9   Read "Blue Heron," by Chick Wallace, page 550. Under "Suggestions for Response and Reaction," page 551,

a. Complete Number 3.
b. Write Number 6.

5.10   Read "The Effects of Perceived Justice on Complainants' Negative Word-of-Mouth Behavior and Repatronage Intentions," by Jeffrey B. Blodgett, Donald H. Granbois, and Rockney G. Walters, pages 552–562. Under "Suggestions for Response and Reaction," page 563, complete Number 4.

5.11   Read the commentary on "Work," by William Bennett, pages 577–578, and the short story "Quality," by John Galsworthy, pages 521–525. Under "Suggestions for Response and Reaction," page 578, write Number 2, and page 525, complete Number 3.

5.12   Read "The Active Document," by David D. Weinberger, pages 579–584. Under "Suggestions for Response and Reaction," page 584, write Number 3.

5.13   Read "The Power of Mistakes," by Stephen Nachmanovitch, pages 585–587. Under "Suggestions for Response and Reaction," page 588, complete Number 3.

PLAN SHEET FOR GIVING AN ANALYSIS THROUGH EFFECT AND CAUSE

Analysis of Situation Requiring Effect-Cause Analysis

What is the subject that I will analyze?

Who is my intended audience?

How will the intended audience use the analysis? For what purpose?

Will the final presentation of the analysis be written or oral?

Effect to Be Investigated

Significance of the Effect

Causes (as applicable)

Possible causes:

Probable cause (or causes):

Actual cause (or causes):

Actual cause (or causes), cont.:

Supporting Evidence or Information for the Probable or Actual Cause (or Causes)

Organization of the Analysis

Types and Subject Matter of Visuals I Will Include

The Title I May Use

Sources of Information

Analysis Through Comparison-Contrast

A basic method of looking at things closely is comparison and contrast, that is, showing how two or more ideas, people, or objects are alike and how they are different. Comparison-contrast may be used singly, of course, but most often they are used together; further, the term *comparison* is commonly used to encompass both likenesses and differences.

Frequently, comparison-contrast is used to explain the unfamiliar. For instance, an airbrush looks like a pencil and is held like one; a computer terminal resembles a typewriter keyboard.

Often, too, comparison-contrast is used to highlight specific details. Llamas, for instance, are related to camels but are smaller and have no hump. *Approve* and *endorse,* though synonyms, have slightly different meanings: approve means simply to express a favorable opinion of, and *endorse* means to express assent publicly and definitely.

Analysis through comparison-contrast is a part of the life of every consumer: Should I buy Brand A or Brand B tires; which is the better buy in a color TV set—Brand X or Brand Y; which of these hospital insurance plans is better? Comparison of products is such an important aspect of the economy that some periodicals, such as *Consumers' Research, Consumer Reports,* and *Changing Times,* are devoted to test reports and in-depth comparisons.

See also the feasibility report, pages 344–349.

Organizational Patterns
of Comparison-Contrast Analysis

An analysis developed by comparison or contrast must be carefully organized. The writer must decide in what order the details will be presented. Although various organizational patterns are possible, three are particularly useful.

Point By Point

In the point-by-point pattern, sometimes called "comparison of the parts," a point (characteristic, quality, part) of each subject is analyzed, then another point of each subject is analyzed, and so on. The following sample paragraph, which contrasts the alligator and the crocodile, uses this arrangement.

TOPIC STATEMENT
POINT 1
POINT 2
POINT 3

The alligator is a close relative of the crocodile. The alligator, however, has a broader head and blunter snout. Alligators are usually found in fresh water; crocodiles prefer salt water. The alligator's lower teeth, which fit inside the edge of the upper jaw, are not visible when the lipless mouth is closed. The crocodile's teeth are always visible.

The report on pages 206–208 on a comparison of two electronics technology programs is organized point by point. The first point of comparison is cost. Cost details are given for each subject—Stevens College and Tinnin Community College. Then the next point, lab equipment, is analyzed for each program; and then the last point, instruction, is analyzed for each program. The point-by-point pattern can be delineated in this way:

201

Chapter 5
Analysis Through
Effect-Cause and
Comparison-
Contrast:
Looking at
Details

Point A (cost) of Subject I (Stevens College)
Point A (cost) of Subject II (Tinnin Community College)

Point B (lab equipment) of Subject I (Stevens College)
Point B (lab equipment) of Subject II (Tinnin Community College)

Point C (instruction) of Subject I (Stevens College)
Point C (instruction) of Subject II (Tinnin Community College)

The point-by-point organizational pattern is usually preferred if the entire analysis is complex, that is, if the comparison is longer than a paragraph or two or if the analysis treats in depth more than two or three points.

Subject By Subject

The subject-by-subject organizational pattern, sometimes referred to as "comparison of the whole," gives all the details concerning the first subject, then all the details concerning the next subject, and so on. The following paragraph, which discusses similarities in the care of the caliper and the micrometer, uses the subject-by-subject arrangement; the first part discusses the care of the caliper and the second part discusses the care of the micrometer. Notice the transition from the first subject to the second subject with the word *similarly*.

TOPIC STATEMENT   Although the caliper is a semiprecision measuring tool and the micrometer is a precision measuring tool, both must receive
SUBJECT 1   proper care to ensure accuracy of measurement when they are used. The caliper has slender legs that can be sprung easily. When not in use, it should be hung in the tool cabinet, never placed on a bench where heavy tools can be laid on it. When stored, the caliper should be protected from rusting by a light
SUBJECT 2   coat of oil. *Similarly,* the micrometer should be kept in a protected place away from emery dust or chips and away from heavier tools. It should be checked for accuracy often and necessary adjustments should be made. When not in use, the spindle and anvil should not be in contact. Further, the micrometer should not be completely closed and allowed to remain closed because rust may form on the ends of the spindle and on the anvil. Finally, the micrometer should be oiled occasionally with a light-grade oil to prevent rust and corrosion.

The report on pages 206–208 on a comparison of two electronics technology programs could be reorganized around the subject-by-subject pattern: All the details concerning the first subject, Stevens College (cost, lab equipment, and instruction), could be given first. Then would follow all the details concerning the second subject, Tinnin Community College (cost, lab equipment, and instruction). The subject-by-subject pattern can be delineated in this way:

Subject I (Stevens College)
    Point A (cost)
    Point B (lab equipment)
    Point C (instruction)

Subject II (Tinnin Community College)
    Point A (cost)
    Point B (lab equipment)
    Point C (instruction)

Similarities/Differences

In the similarities/differences pattern, a comparison of the similarities of the subjects is given first, followed by a contrast of differences between the subjects. Again, the report on a comparison of two electronics technology programs could be reorganized. If the similarities/differences pattern were used, the similarities between Stevens College and Tinnin Community College would be given, and then the differences between the two colleges would be given. The similarities/differences pattern can be delineated in this way:

Similarities of Subjects I and II (Stevens College and Tinnin Community College)
    Point A (cost)
    Point B (lab equipment)
    Point C (instruction)

Differences between Subjects I and II (Stevens College and Tinnin Community College)
    Point A (cost)
    Point B (lab equipment)
    Point C (instruction)

The following paragraph illustrates the similarities/differences pattern.

Social workers involved with case work and group work have similarities and differences. Case workers and group workers both have contact with people. Both groups of workers share a common goal, to make life better for their clients. Moreover, both case workers and group workers must complete an intensive training program. Case workers differ from group workers in that the case workers have direct contact with individuals and families. They help individuals and families solve problems and develop healthy relationships. In contrast, group workers direct programs for varying ages with activities such as arts, crafts, and recreation. The group worker seeks to develop individual capacities and fulfill individual needs. The case worker visits in the home, tries to motivate family members, to teach them good money management, and to guide them to effective use of available resources. The group worker, conversely, may direct activities in churches, in schools, and in various institutions such as those for the physically and mentally disadvantaged, the aged, and the delinquent. The group worker uses involvement in community projects to develop civic responsibility and pride. In preparing to become a case worker, an individual studies interview and motivation techniques, learns how to use community resources, and experiences supervised field training. The group worker studies principles of group dynamics, planning and assessing group activities, and leadership skills.

The similarities/differences (or differences/similarities) organizational pattern is particularly effective when positive and negative emphases are desired. The subject or emphasis presented last makes the greatest impression.

203

Chapter 5
Analysis Through
Effect-Cause and
Comparison-
Contrast:
Looking at
Details

Three Patterns Compared

These three organizational patterns can be delineated as follows (there can be any number of points and any number of subjects; here it is assumed that two subjects are being compared/contrasted on three points):

Point by point	Subject by subject	Similarities/differences
Point A of Subject I	Subject I	Similarities between
Point A of Subject II	Point A	Subjects I and II
	Point B	Point A
Point B of Subject I	Point C	Point B
Point B of Subject II		Point C
	Subject II	
Point C of Subject I	Point A	Differences between
Point C of Subject II	Point B	Subjects I and II
	Point C	Point A
		Point B
		Point C

The following student-written report is an analysis through comparison-contrast. Note that the report is organized around three points of comparison: cost, lab equipment, and instruction. The Plan Sheet that the writer filled in before writing is included. Marginal notes have been added to the final paper to show how the writer developed and organized the paper.

PLAN SHEET FOR GIVING AN ANALYSIS THROUGH COMPARISON AND CONTRAST

Analysis of Situation Requiring Comparison-Contrast Analysis

What are the subjects that I will analyze?
the electronics technology program at Stevens College and the electronics technology program at Tinnin Community College

Who is my intended audience?
prospective electronics students

How will the intended audience use the analysis? For what purpose?
to decide which college the student should attend

Will the final presentation of the analysis be written or oral?
written

Major Points (Characteristics, Qualities, Parts) to be Compared/Contrasted

cost, lab equipment, instruction

Details Concerning the Major Points of Each Subject

Stevens College		Tinnin Community College
	1. Cost	
Tuition $7,100	*Tuition ($500 a semester) . . . $2,000*	
No charge for books	*Books (after reselling for 1/2 price) $300*	
	2. Lab Equipment	
15	*Power supply*	*50*
12	*Digital multimeter*	*40*
4	*Microprocessor trainer*	*12*
6	*Sine-square wave generator*	*25*
0	*Transistor curve tracer*	*1*
0	*Decade boxes*	*40*
4	*Digital trainer*	*20*
2	*Function generator*	*30*
5	*Oscilloscope*	*34*
1	*Personal computer system*	*4*
	Value	
$50,000		*over $150,000*

Details Concerning the Major Points of Each Subject (cont.)

3. Instruction

1 Instructor

A.A.S. from Stevens, taught 3 yrs at Stevens, worked in industry 2 years (IBM)

3 Instructors

Instructor A: B.S., taught at 1 other college, taught 5 yrs, worked in industry 3 yrs

Instructor B: B.S., taught in 1 other college, taught 7 yrs, worked in industry 4 yrs and alternate summers

Instructor C: M.S., taught in 2 other colleges, taught 18 yrs, worked in industry 7 yrs

Classes 8–12, 1–2, M–Th
Electronics lectures 8–10:30, lab 10:30–12

Variable schedules 8–3, M–F
Electronics—usually 2-hr block

A.A.S. degree: Eng 1 & 2, Bus Law 1 & 2, Soc, Math for Electronics, Digital Math, Computer Math, Basic Electronics, Fund of Elec, Digital Circuits, Semiconductors, Transistor Circuits

A.A.S. degree: Tech Writing 1 & 2, Soc Sci, P.E., Ind Psy, Tech Math 1 & 2, Tech Physics 1 & 2, Fund of Drafting, Elec for Electronics, Electron Devices & Circuits, 4 sophomore-level electronics courses, 1 elective

Organization of the Analysis

Point by point. First I'll discuss the cost of attending each college. Then I'll talk about the lab equipment in each, and, finally, the instruction in each.

Types and Subject Matter of Visuals I Will Include

table showing the cost of tuition and books for each college
table showing the major pieces of lab equipment and their value for each college

The Title I May Use

A Comparison of Two Electronics Technology Programs

Sources of Information

catalog from each college; chair of the electronics technology dept. at each college

A COMPARISON OF TWO ELECTRONICS TECHNOLOGY PROGRAMS: STEVENS COLLEGE AND TINNIN COMMUNITY COLLEGE

Subjects compared

General conclusion reached

Sources of information

Specific conclusions and points of comparison

A comparison of the electronics technology program at Stevens College and at Tinnin Community College indicates that the Tinnin program has more to offer the student. This conclusion is based on a thorough examination of the catalog from each college and a visit to each college for a personal interview with the chair of the electronics technology department. This investigation shows that the Tinnin program costs less, has more lab equipment, and possibly has better instruction.

Cost

First point of comparison for each subject

Table visually depicts supporting information

The cost of tuition and books for each college is as follows:

Stevens College

Tuition (18-month program)	$7,100.00
Books—included with tuition	0
	$7,100.00

Tinnin Community College

Tuition (4-semester program) $500.00 per semester		$2,000.00
Books (about $150.00 per semester)		300.00
4 semesters × $150.00 = $600.00		
resale value = 300.00		
$300.00		
		$2,300.00

Interpretation of information

The Stevens program costs $4,800.00 more than the Tinnin program. This large difference in cost is due primarily to Stevens being a private college and Tinnin being a public community college.

Lab Equipment

Each college has these major pieces of lab equipment:

Equipment	Stevens	Tinnin
Power supply	15	50
Digital multimeter	12	40
Microprocessor trainer	4	12
Sine-square wave generator	6	25
Transistor curve tracer	0	1
Decade resistance/capacitance boxes	0	40
Digital trainer	4	20
Function generator	2	30
Dual-trace oscilloscope	5	34
Personal computer system	1	4

207

Chapter 5
Analysis Through
Effect-Cause and
Comparison-
Contrast:
Looking at
Details

Support details

The total value (approximate retail value) of the Stevens equipment is approximately $50,000. The total value of the Tinnin equipment is well over $150,000.

The program at Tinnin Community College has more kinds and more pieces of each kind of lab equipment than does Stevens College.

Third point of comparison for each subject divided into three aspects

Instruction

First aspect of instruction (instructors) compared for both subjects

The programs in both colleges differ in three aspects of instruction: qualifications of the instructors, daily class schedule, and required courses. Stevens College has one electronics instructor. He has an Associate in Applied Science degree from Stevens, has taught three years (all at Stevens), and has worked two years in industry (at IBM). Tinnin Community College has three electronics instructors. Two of them have Bachelor of Science degrees and the other has a Master of Science degree. Each has taught in at least one other college, has taught at least five years, and has worked in industry at least three years. One instructor works in industry every other summer.

Second aspect of instruction (schedule) compared for both subjects

The daily class schedule is set up differently at each school. At Stevens, the students are in class from 8:00 A.M. until 12:00 noon and from 1:00 P.M. to 2:00 P.M. Monday through Thursday (there are no classes on Friday). For the electronics courses, the students have lecture from 8:00 to 10:30 and lab from 10:30 to 12:00. The afternoon hour is for a nonelectronics course. At Tinnin, the students have variable schedules from 8:00 A.M. to 3:00 P.M. five days a week, with electronics courses typically in two-hour blocks. The distribution of lecture and lab time is at the discretion of the instructor.

Third aspect of instruction (courses) compared for both subjects

Both Stevens and Tinnin offer an Associate in Applied Science degree upon successful completion of their programs. Requirements for the associate degree differ slightly at the two schools. At Stevens, the required courses are College English I and II, Business Law I and II, Sociology, Math for Electronics, Digital Mathematics, Computer Mathematics, Basic Electronics, Fundamentals of Electricity, Digital Circuits, Semiconductors, and Transistor Circuits. At Tinnin the required courses are Technical Writing I and II, any course in social science, Physical Education, Industrial Psychology, Technical Mathematics I and II, Technical Physics I and II, Fundamentals of Drafting, Electricity for Electronics, Electron Devices and Circuits, any four additional sophomore-level electronics courses, and one elective. Both programs offer the same amount of classroom instruction—18 months—but the Stevens program does not have a three-month summer break.

Summarizing statement about the third (and lengthiest) point of comparison

Thus, the instructors at Tinnin may be better qualified than the instructors at Stevens, the daily class schedule is more flexible at Tinnin, but the requirements for graduation are similar at both colleges.

Conclusions

Comment

A person planning to specialize in electronics technology should thoroughly explore the programs in prospective colleges. This report shows that two colleges (both in the same city) vary widely in their electronics programs. The electronics technology program at Tinnin Community College costs less, has more lab equipment, and possibly has better instruction than does the electronics technology program at Stevens College.

Relationship to Other Forms of Communication

Comparison-contrast analyses are closely related to other forms of communication, especially to description, definition, classification, and partition, and often involve them. Often, too, the comparison-contrast analysis is given orally. (For a discussion of oral communication, see Chapter 10.)

Various kinds of visuals—drawings, diagrams, tables, graphs, charts—are particularly helpful in showing comparison-contrast. (For a discussion of visuals, see Chapter 11). Table 1, for instance, shows the average annual energy use and annual cost of home appliances in a typical American home. Note how the arrangement of information makes it easy for the reader to see similarities and differences.

Table 1   Home Appliance Annual Energy Use and Cost
Use of electricity is stated in the number of kilowatt-hours (kwh) used per year. Cost is based on an average charge by utilities of 9¢ per kwh.

Electronic Appliance	Annual Energy Use	Annual Cost
Air conditioner	2,000 kwh	$180.00
Can opener	1 kwh	.09
Clock	17 kwh	1.53
Clothes dryer	1,200 kwh	108.00
Coffee maker	100 kwh	9.00
Dishwasher	350 kwh	31.50
Floor furnace	480 kwh	43.20
Freezer (16 cu. ft.)	1,200 kwh	108.00
Frying pan	240 kwh	21.60
Hair dryer	15 kwh	1.35
Hot plate	100 kwh	9.00
Iron	150 kwh	13.50
Lighting	2,000 kwh	180.00
Radio	20 kwh	1.80
Radio-phonograph	40 kwh	3.60
Range	1,500 kwh	135.00
Refrigerator (12 cu. ft.)	750 kwh	67.50
Shaver	1 kwh	.09
Television (black and white)	400 kwh	36.00
Television (color)	550 kwh	49.50
Toaster	40 kwh	3.60
Vacuum cleaner	45 kwh	4.05
Washer (clothes)	100 kwh	9.00
Water heater (all electric)	4,200 kwh	378.00

SOURCE: U.S. Department of Agriculture

The excerpt below* from the Good Housekeeping Institute compares eight types of flashlights. The comparative data provide information for a consumer who wishes to purchase a flashlight. Notice the chart data, which is similar to the description of a specific mechanism (see pages 83–90). Notice the classification (see pages 131–149) of bulbs according to the degree of light each gives.

209

Chapter 5
Analysis Through
Effect-Cause and
Comparison-
Contrast:
Looking at
Details

GH INSTITUTE
BUY & USE GUIDE

FLASHLIGHTS

For our hands-on flashlight review, GH Institute engineers looked at everything from the tiniest of penlights to submergible lights for scuba divers. From these they selected eight types designed to handle the most frequently encountered situations and gave them a rigorous going-over. Waterproof lights were subjected to a blast of running water; then their O-ring seals were completely disassembled to check for the slightest trace of interior moisture. Flashlights with focusing lenses (these can narrow or widen the light beam) had their light patterns analyzed in a dark room. On all models, the brightness of the beam was checked, as were the sturdiness of the case and the ease of operation.

BULBS/BATTERIES Standard *filament* bulbs have the lowest light level. *Fluorescent* bulbs give more light; their shape and energy efficiency make them ideal for lighting large areas for long periods of time. The brightest light is offered by the *halogen* and *krypton* bulbs. *Rechargeable* batteries don't last as long as alkalines (but you may want to consider a rechargeable *flashlight,* always ready for use if you keep it plugged into a socket). Some flashlights that run on car lighters offer instant light when plugged in; others take a few hours to recharge.

FEATURES like shatterproof lenses, water-resistance, and floatability are worth looking for in flashlights that will be used for rough outdoor work or recreation, or on the water. Water resistance is a valuable feature on any flashlight, since it minimizes internal corrosion and helps the flashlight last longer.

THE CHART BELOW shows flashlights chosen from those that survived our scrutiny in the labs.

TYPE	MODEL/ PRICE	BULB/ BATTERIES	LIGHT HOURS	WEIGHT (with batteries)	FEATURES
COMPACT	Mini Maglite/ $16.95	Halogen/2 AA	5 hours	4 ounces	Strong aluminum case; focusing lens; spare lamp; water- and shock-resistant
CAMPING	Eveready Area Light/$15	Fluorescent/ 4 AA	6.5 hours	10 ounces	Shatterproof lens; water-resistant; floats
STANDARD	RAY-O-Vac Roughneck/$7	Krypton/2 D	8 hours	13 ounces	Sturdy; water-resistant; focusing lens
MULTI-PURPOSE	Radio Shack 3-in-1/$15	Filament, krypton, fluorescent/4 AA Or works on car lighter or AC adapter	4.5 hours	13 ounces	Area light; spotlight; amber emergency flasher
TIGHT SPACES	Panasonic Flexible/$6.76	Filament/2 AA	6 hours	3 ounces	Flexible cord for directing light into hard-to-reach areas
CAR EMERGENCY	Coleman Spotlight/ $20-$25	Halogen/Works on car lighter	Not applicable	1 pound, 6 ounces	Strong beam (400,000 candlepower, a measure of brightness and area illuminated) for long-range viewing; lights instantly
HOME EMERGENCY	Eveready Lantern/$19	Halogen/ 6-volt battery	5 hours	2 pounds	Rugged construction; bright light; floats
SPORT/ HIKING	Petzl Headlamp/$28	Halogen/ 2 AA	5 hours	5 ounces	Adjusting head strap for hands-free use; focusing lens; spare lamp; water-resistant

*Reprinted by permission from *Good Housekeeping,* Sept. 1994, page 22.

A page from a do-it-yourself water heater installation guide compares the appearance of the installed electric hot water heater and the gas hot water heater. Other pages in the guide give instruction for installing each of the two types of water heaters.*

YOUR COMPLETED INSTALLATION SHOULD APPEAR SIMILAR TO THIS:

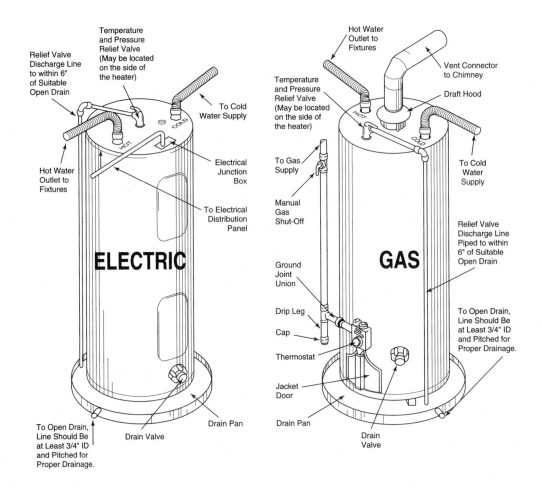

*Reprinted by permission from Richmond Water Heaters, 101 Bell Road, Montgomery, AL 36124.

The chart below compares basic trusts. The chart also clearly illustrates cause-effect relationships between types of trust and the goals and tax consequences.*

211

Chapter 5
Analysis Through
Effect-Cause and
Comparison-
Contrast:
Looking at
Details

Basic Trusts

Type of Trusts	Goals	Tax Consequences
Bypass	• Provide income and assets for beneficiary • Avoid estate taxes at beneficiary's death	• Income taxed to person who receives income • Assets not in taxable estate of beneficiary
Asset Protection	• Keep income and assets for beneficiary who may "lose" assets to creditors	• Income taxed to beneficiaries • Assets may or may not be taxed in beneficiary's estate
Irrevocable Living	• Avoid estate tax on asset appreciation (life insurance) • Give income and appreciation to others	• Pay gift tax when trust is established • Assets not in estate of donor
Revocable Living	• Avoid probate court • Administer estate easily • Provide for management of property in case of incapacity	• No tax costs • No tax benefits
Q-TIP	• Income to spouse • Remainder to beneficiary chosen by donor	• Estate tax deferred until spouse dies
Charitable Remainder	• Increase current income • Make major gift to charity	• Receive income tax deduction when you set the trust up

General Principles in Giving an Analysis Through Comparison-Contrast

1. *Comparison-contrast can be used to explain the unfamiliar or to highlight specific details.*
2. *The comparison-contrast analysis must be carefully organized.* Various organizational patterns are possible.
3. *The point-by-point organizational pattern can be used.* In this pattern, a point of each subject is analyzed, then another point of each subject, and so on.
4. *The subject-by-subject pattern can be used.* In this pattern, all the details concerning the first subject are presented, then all the details about the next subject and so on.
5. *The similarities/differences pattern can be used.* In this pattern, a comparison of similarities of the subjects is given, followed by a contrast of differences between the subjects.

*Reprinted from pamphlet on farm finance, Fall 1994. Reprinted by permission from *Progressive Farmer*.

Procedure for Giving an Analysis Through Comparison-Contrast

The comparison-contrast analysis may be independent or it may be part of a longer presentation. Whichever the case, the analysis is typically divided into three sections.

I. Stating the subjects and points to be compared and contrasted requires only a few sentences.
   A. State the subjects that are being compared and contrasted.
   B. State the points (characteristics, qualities, parts) by which the subjects are being compared and contrasted.
   C. Give any needed background information.
   D. If applicable, give the source of information or methods of investigation used (reading, observation, consultation with knowledgeable people, etc.).
   E. Unless the order of the presentation requires otherwise, state the conclusions that have been reached.

II. Reporting the details is the longest section of the presentation.
   A. Give sufficient supporting details for each point of each subject.
   B. When needed, subdivide the major points.
   C. Interpret facts and other information when necessary.
   D. Use visuals if they will help clarify the analysis.
   E. Organize the information around a logical pattern or order.

III. The conclusion (usually brief) reflects the purpose of the analysis.
   A. If the purpose of the analysis is to give the reader a general knowledge of the subject, summarize the main points or comment on the significance of the comparison.
   B. If the analysis is a basis for decisions and actions, summarize the main points, comment on the significance of the analysis, *and* make recommendations.

Applications in Analysis Through Comparison-Contrast

5.14   Identify two jobs (types of employment) in your field that you might be interested in doing after you complete your training. Compare the two jobs, using three to five points of comparison. Choose one of the organizational patterns discussed in the chapter and present your comparison in a written paragraph(s).

In completing applications 5.15 and 5.16,

a. Fill in the Plan Sheet on pages 215–216, or one like it.
b. Write a preliminary draft.
c. Go over the draft in a peer editing group.
d. Revise. See Part III Revising and Editing and the checklists for revising, front and back endpapers.
e. Write the comparison-contrast analysis.

5.15   Choose one of the topics below for a comparison-contrast analysis.

213

Chapter 5
Analysis Through
Effect-Cause and
Comparison-
Contrast:
Looking at
Details

a. Two brands of a product (examples: an RCA and a Zenith color television set, a John Deere and a Massey-Ferguson tractor, Levi and Dickey blue jeans, a Pontiac Grand AM and a Honda Accord)
b. Two synonyms (examples: job-career, technician-technologist, education-training, friend-acquaintance)
c. Before-and-after situations (examples: a piece of furniture or a building before and after renovation, an engine before and after overhauling, efficiency before and after using a new product)
d. A topic of your own choosing

5.16   Identify a piece of equipment or a tool that is commonly used in your field. Assume that you must recommend for purchase a replacement for the identified piece of equipment or tool. Locate two possible brand names or models. Then write a comparison of the two to convince your employer that the company should buy one brand or model, not the other.

Oral Comparison-Contrast Analyses

5.17   As directed by your instructor, adapt for oral presentations the analysis through comparison-contrast that you prepared in the applications above. Ask your classmates to evaluate your speech by filling in the Evaluation of Oral Presentations on page 459, or one like it.

Using Part II Selected Readings

5.18   Read "Battle for the Soul of the Internet," by Philip Elmer-Dewitt, pages 500–507. Under "Suggestions for Response and Reaction," page 508, complete Number 1.

5.19   Read "The Powershift Era," by Alvin Toffler, pages 509–514. Under "Suggestions for Response and Reaction," page 515, complete Number 2.

5.20   Read "Hardest Lessons for First-Time Managers," by Linda A. Hill, pages 516–519. Under "Suggestions for Response and Reaction," page 520, complete Number 2 and Number 3.

5.21   Read "Zen and the Art of Motorcycle Maintenance," by Robert M. Pirsig, pages 526–530. Under "Suggestions for Response and Reaction," page 531, complete Number 1.

5.22   Read "Are You Alive?," by Stuart Chase, pages 537–540. In what ways are comparison and contrast used in this article? Under "Suggestions for Response and Reaction," page 541, complete Number 1.

5.23 Read "A Technical Writer Considers His Latest Love Object," by Chick Wallace, page 548. Under "Suggestions for Response and Reaction," page 549, complete Number 5.

5.24 Read "Blue Heron," by Chick Wallace, page 550. Under "Suggestions for Response and Reaction," page 551, complete Number 1.

5.25 Read "The Effects of Perceived Justice on Complainants' Negative Word-of-Mouth Behavior and Repatronage Intentions," by Jeffrey G. Blodgett, Donald H. Granbois, and Rockney G. Walters, pages 552–562. Under "Suggestions for Response and Reaction," page 563, complete Number 4.

5.26 Read "Journeying Through the Cosmos" from *The Shores of the Cosmic Ocean,* by Carl Sagan, pages 570–575. Under "Suggestions for Response and Reaction," page 576, write Number 2.

5.27 Read "Work" from *The Book of Virtues,* by William J. Bennett, pages 577–578. Under "Suggestions for Response and Reaction," page 578,

a. Write Number 2.
b. Complete Number 3.

PLAN SHEET FOR GIVING AN ANALYSIS THROUGH COMPARISON-CONTRAST

Analysis of Situation Requiring Comparison-Contrast Analysis

What are the subjects that I will analyze?

Who is my intended audience?

How will the intended audience use the analysis? For what purpose?

Will the final presentation of the analysis be written or oral?

Major Points (Characteristics, Qualities, Parts) to be Compared/Contrasted

Details Concerning the Major Points of Each Subject

Organization of the Analysis

Types and Subject Matter of Visuals I Will Include

The Title I May Use

Sources of Information

Chapter 6

The Summary: Getting to the Heart of the Matter

Outcomes

Upon completing this chapter, you should be able to

- Define the terms *descriptive summary, informative summary,* and *evaluative summary*
- List the specific purposes that each form of summary may serve
- Prepare each form of summary
- Give bibliographic information for a summary
- Describe specific situations in which a student and an employee need to know and use skills in preparing summaries

---

Introduction

Why do you need to know how to give a summary? Of what importance is this to you? First, because you are a *student,* you should be able to give a summary orally and in writing. As a student, for example, you must be able to grasp the main points in a reading assignment. One very effective way of studying textbook material is to write down the significant points, the important facts, in the material. Writing summaries of class lectures and of textbook assignments helps to clarify information and makes it stick in the mind longer.

This practice is quite helpful not only in day-to-day preparations but also in reviewing for tests. Pulling the main points and ideas from pages and pages of textbook material and class notes and then writing them down coherently and understandably is a sound approach to studying for tests. Since tests cannot possibly cover everything you are supposed to learn in a course, test questions usually concentrate on major points and aspects of the subject.

Often a class requirement is to make a report or review (a summary) of reading assignments. Usually the only directions you are given have to do with the sources to be read and the length of the report. Rarely are you told *how* you are to present the report.

Thus, for very practical reasons of immediate importance, you need to learn how to give a summary.

Giving summaries may be a part of your work, such as condensing reports, articles, speeches, or discussions for a supervisor too busy to read or hear the original. Further, you *read* summaries, and thus should recognize a summary for what it is.

Purpose and Intended Audience

In preparing a summary, first consider *why* you are writing, the purpose the summary will serve. The primary purpose may be to *describe* very briefly, perhaps in only a sentence or two, what the article or situation is about. The primary purpose may be to *inform*—to present, in general, the principal facts and conclusions given in the original work. Or the primary purpose may be to *evaluate,* to judge the accuracy, completeness, and usefulness of the work.

Closely related to the purpose is the intended audience. If the summary is a study aid, the audience may be the student writing the summary. Or the audience may be a student trying to determine the sources most helpful for a research paper. If the summary is an assignment, the audience may be an instructor. In the workplace, the audience may be a business executive who wants only the main points and conclusions of a report. Or the audience may be a nuclear scientist looking in *Nuclear Science Abstracts* to identify articles on nuclear waste regulations.

The direction that a summary takes is not haphazard; the writer has a clear purpose and a specific audience.

219

Chapter 6
The Summary:
Getting To the
Heart of the
Matter

Understanding the Material to Be Summarized

Before you can summarize material, you must thoroughly understand it. Since a summary concerns main ideas, you must distinguish main ideas from minor ideas and supporting details.

The first step in understanding the material to be summarized is to scan it. In looking over the material, note the title, introductory matter, opening and closing paragraphs, headings, marginal notes, major topics, organization, and method of presentation. This basic familiarity with the material is preparation for reading it closely.

Now read the material very carefully. Underline the key ideas or jot them down on a separate sheet of paper (particularly if the original text belongs to someone else). Think through the relationship of these key ideas to get the overall significance of the material.

See the study-reading method in the introduction (pages 492–495) to Part II Selected Readings.

Accuracy of Fact and Emphasis

A summary must agree with the original material in two ways: in the factual information and in the emphasis given the information.

Misrepresentation of factual information may occur through careless error, as in omitting a digit or a decimal point in a number. Through the process of omitting words, misleading information may be given. Consider this sentence: "Wood posts can be expected to last at least thirty years *if they are pressure treated and handled properly.*" The last part of the sentence is essential; omission of the *if* qualification changes the meaning of the entire sentence. The sentence might be condensed as follows: "Pressure-treated and properly handled wood posts can last at least thirty years."

Misrepresentation of emphasis may occur because of overemphasizing or underemphasizing information. If a one-page summary of a ten-page report on a new weed killer stresses the possible danger to animals when this was only a minor point in the original report, the summary misrepresents the emphasis. A summary must agree in emphasis with the original; a summary should convey to the intended audience the same impression that the original work conveys to its audience.

Documentation

Material being summarized should be cited, that is, completely identified. Such an identification is referred to as documentation. Documentation is complete bibliographical information on the original material. In addition to author and title, it includes facts about publication (such as place, publisher, and date). See pages 408–409 for examples of bibliographical forms. Depending on the purpose of the summary, documentation may also include such information as price, number of pages, and availability in paper or cloth binding. See pages 404–409 for a discussion of documentation.

Placement of Documentation

The placement of documentation may vary. Although the bibliographical information may be given within the summary, it usually appears as a formal bibliographical entry at the beginning of the summary.

For examples of bibliography given as a separate entry at the beginning (or the end) of a summary, see pages 223, 225, 226, 233, 234, 235.

Note that the arrangement of information and the punctuation in an entry may vary; the information itself, however, usually remains the same: author, title, and publication facts.

Forms of Summary

The three basic types of summary are descriptive, informative, and evaluative, depending on the writer's primary purpose and the intended audience. The descriptive summary and the informative summary can be written as independent summaries. The evaluative summary, however, must include either description or information in addition to evaluation to give the reader a point of reference. Thus, summaries can be written to describe only, to inform only, to describe and evaluate, or to inform and evaluate.

Descriptive Summary

A summary may describe; it may state what a work is about in a very general way. The descriptive summary simply indicates the main idea of a complete work and may include the main topics that are discussed. It is usually very brief, sometimes only one or two sentences.

Following are two sample descriptive summaries.

> The article "Take Charge of Your Life" presents four strategies for turning adversity to advantage.
> The article "Take Charge of Your Life" presents four strategies for turning adversity to advantage: assume responsibility for yourself, make tough choices, seek relationships that enrich your life, and affirm self-worth.

While the second example gives more information about the original article, both of the two samples are acceptable descriptive summaries.

Descriptive summaries frequently occur in the headnotes to chapters in books, in the headnotes to magazine articles, on the table of contents pages in magazines, in publicity for new books and pamphlets, and in annotated bibliographies. Readers (audiences) of such summaries may vary widely; the purpose for reading the summaries, however, is likely to be the same: to determine whether to read the entire work. Readers ask such questions as these: Does the summary suggest material helpful in my work? In my research? Of interest for leisure reading?

The following excerpt from a table of contents page of *The Physician and Sportsmedicine** illustrates a common use of descriptive summary. Each summary includes details about the subject and the focus of the article.

*Modern Machine Shop*** uses a terse descriptive summary in its table of contents, as shown in the following excerpt.

*Reprinted by permission from the July 1994 issue of *The Physician and Sportsmedicine.*
**Reprinted by permission from the June 1994 issue of *Modern Machine Shop.*

221

Chapter 6
The Summary:
Getting To the
Heart of the
Matter

52 Tips For Better Boring

Klaus Lohner offers these step-by-step guidelines to
help you establish optimal feeds and speeds for
boring operations.

62 Lasers Take On Hole-Making
Challenges

Today's multi-axis lasers can take on odd angles and
difficult-to-machine materials. Terry VanderWert
shows how.

72 Hole Making Has Strategic
Importance

Norman Scott explains how a recreational vehicle
manufacturer found that breaking old drilling habits
brought big benefits.

Both excerpts illustrate clear, effective use of the descriptive summary. Readers
of either magazine can use the summaries to make an informed decision on
whether or not to read an article.

Such summaries may also be used in catalogs for certain products. The Bergwall 1992 *Vocational/Technical Video Catalog,** in the example below, uses

#E16
**BASIC ELECTRICITY:
DC CIRCUITS**
Copyright 1990

1 Series Circuits
• Describes various components • Explains how the concept of voltage – current – resistance is
affected by series circuit design • Details procedure for building a DC series circuit on a breadboard
• Explains standard Color Code Identification Chart
#E161–1 VHS cassette, 22:06 min. *$95*

2 Series Circuit Analysis
• Explains how to determine voltage – current – resistance – power at various points • Shows
schematic diagrams • Details use of Ohm's Law
#E162–1 VHS cassette, 19:38 min. *$95*

3 Parallel Circuits
• Details the relationship between voltage – current – resistance – power • Describes reciprocal –
product/sum formulas • Compares schematic for series – parallel circuits using a breadboard
#E163–1 VHS cassette, 23:58 min. *$95*

4 Series/Parallel Circuits
• Details use of schematic diagram • Shows how to use product/sum formula – equivalent resistance
• Describes how to break down a complex schematic
#E164–1 VHS cassette, 17:50 min. *$95*

Set of 4 VHS cassettes
#E16-V12   **$339**

---

*Reprinted with permission from Bergwall Productions, Inc., Chadds Ford, PA 19317.

223

Chapter 6
The Summary:
Getting To the
Heart of the
Matter

descriptive summaries to aid the consumer (the audience) in selecting videos (the purpose).

Another use of the descriptive summary is the *descriptive abstract*. An abstract is a brief presentation of the essential contents of a document. The descriptive abstract frequently indicates the usefulness of the original material.

Descriptive abstracts are very important to persons who want to keep abreast of what is going on in their field. These abstracts are so helpful in determining the relevance of an article to a specific interest that entire books and periodicals—for instance, *Chemical Abstracts, Nuclear Science Abstracts,* and *Biological Abstracts*— are devoted to descriptive abstracts of publications in particular fields. Persons interested in the field can read the abstracts to determine whether they want or need to obtain and read copies of the original works.

Other professional journals include within each issue a section for abstracts. These may be abstracts of papers presented at meetings of professional groups or they may be abstracts of articles appearing in related literature. Again, the abstract serves as an indicator of usefulness, directing persons to articles of concern and interest.

The first example below is an abstract of a report, "What Scientists Funded by the Tobacco Industry Believe About the Hazards of Cigarette Smoking," which appeared in the *American Journal of Public Health.** Every article in this journal is preceded by an abstract.

Abstract

Despite overwhelming evidence documenting the hazards of cigarette smoking, the tobacco industry denies that smoking has been proven to cause disease. The industry professes a desire to clear up the smoking and health "question" and often points to its support of the Council for Tobacco Research (CTR) as evidence of its interest in investigating the health dangers of smoking. This paper presents results of a survey of CTR-funded scientists regarding their beliefs about the health dangers posed by smoking cigarettes. The vast majority of scientists funded by the CTR believe that cigarette smoking is an addiction that causes a wide range of serious, often fatal, diseases. This result suggests that the tobacco industry is unwilling to accept even the opinions of scientists it has deemed worthy of funding. Scientists should consider the ethical implications of accepting funds from the CTR and other tobacco industry-supported institutions. (*Am J Public Health*, 1991; 81:894–896).

The descriptive summary or descriptive abstract is an essential part of a report. It brings together in one section the major points discussed in the report. (For a detailed discussion of the summary or abstract in reports, see Chapter 8 Reports, page 310.) See also the abstract included with an observation report on page 329, the abstract included with a research report on page 414, and the executive summary, abstract, and summary included with the formal report on pages 552–554, 562.

The summary that follows appeared in a student-written report, "Police Work as an Occupation":

The continued need for police officers is obvious. The opportunities are not overwhelming, but they can be found by the qualified applicant. It is an occupation with a long history and tradition and is continuously being upgraded. It provides a secure job even though the salaries are not as high as those in other occupations requiring similar qualifications.

In the classroom, professors frequently mention standard resource materials in particular fields and suggest that students become familiar with them. The student who examines the material and jots down several facts about the kind and manner of presentation of information is writing an annotation, a kind of descriptive summary, as in the following:

> *The McGraw-Hill Encyclopedia of Science and Technology,* 7th edition, 1992, from McGraw-Hill, 20 volumes, including an index. Updated by yearbooks. Specialized encyclopedia written for knowledgeable persons with a high reading level. Includes drawings, diagrams, and schematics. Long, detailed, highly technical articles.

Such a summary makes clear that the material in this work is designed for a very specialized audience.

Below is a more detailed descriptive summary of the magazine *RN,* written by a student. Such a summary interests nurses entering the profession or students preparing to become nurses. The summary clearly explains details about the magazine: frequency of issues, intended audience, contributing writers and submission procedure, subjects of articles, regular columns, cost, publisher, and circulation.

> *RN* is a monthly magazine published by Medical Economics Publishing, based in Montvale, New Jersey. The specific audience targeted is practicing registered nurses and, on a smaller scale, nursing students. Each month the magazine offers approximately six full-length articles on various topics of interest to the practicing nurse. For example, topics of articles in the June 1994 issue are anaphylaxis, latex allergy, specialty certification, a new drug rispiridone, breast cancer, and assessing chest pain. Articles include effective visuals (color photographs and diagrams, charts, and graphs) to support the written word. Departments featured in each issue include Professional Update, New Products, Clinical Highlights, and Management Help Line, as well as letters to the editor. The magazine is full of advertisements for products of interest to nurses. The June 1994 issue contains 80 pages.
>
> The yearly subscription rate is $35.00 ($17.50 for students). The magazine's editorial board members represent many areas of the nursing profession; a peer review group further represents those who teach and practice nursing.
>
> Subscribers, mostly registered nurses, number 302,000. Most of the articles are contributed by registered nurses, and the magazine includes a one-page guide for submitting articles.

Informative Summary

The primary purpose of an informative summary is to inform; that is, it is designed to present the principal facts and conclusions given in the original work. When most people use the word *summary,* they are probably referring to the informative summary. No personal feelings or thoughts are injected; the main points of the material are presented objectively. Unlike the descriptive summary that tells *about* the work in only a few sentences, the informative summary tells what is *in* the work and may run from a paragraph to several pages, depending on the length of the original text.

Two examples of an informative summary appear below. The first is from the *Allied Journal of Health,* * a publication that includes a section of informative sum-

*Reprinted by permission from the Spring 1990 issue of the *Allied Journal of Health,* University of Illinois at Chicago.

maries of articles from other health-related periodicals. Obviously the intended readers of this journal are professionals interested in health-related fields.

225

Chapter 6
The Summary:
Getting To the
Heart of the
Matter

Study Indicates Smoking Cessation Improves Workplace Absenteeism Rate. Susan E. Jackson, David Chenoweth, E.D. Glover, Don Holbert, David White. *Occupational Health and Safety*. December 1989, pp. 13–18.

Employee assistance programs have become very popular and have proven to be cost-effective in dealing with personnel problems in the workplace. One of the many work-related problems these programs help with is smoking cessation. These researchers studied absenteeism among smokers and the financial strain placed on employers because of the excessive absenteeism.

The study used a time-series control group design to identify patterns of absenteeism among smokers and ex-smokers. The company's annual lifestyle appraisal questionnaire provided data on employee smoking. This questionnaire included demographic information as well as a limited family history and information on lifestyle behaviors such as diet and exercise. It also included questions pertaining to the year in which the person began smoking, the number of cigarettes smoked per day, and whether the person stopped smoking, and if so, in what year.

The company in this study is a North Carolina pharmaceutical company with 1,400 employees. There were 734 questionnaires administered and of these, 161 respondents said they were ex-smokers and 188 were current smokers. Out of the 161 ex-smokers, a random sample of 50 was chosen. This sample was matched with 50 current smokers by age, race, gender, and number of cigarettes smoked per day. Each ex-smoker's year of smoking cessation was treated as the midpoint year for the matched smoker. The list of matched pairs was then taken to the personnel department to obtain absenteeism data.

Absenteeism data were adjusted to exclude all absences due to personal leave, death in the family, or jury duty. The time period studied was three years prior to smoking cessation, the year of cessation, and three years after cessation.

The original sample size of 100 subjects was reduced to 70 (35 pairs) due to absence of personnel records, and to some subjects having an excessive number of sick days (more than 90 days for a given year). The mean number of cigarettes smoked per day was 19 for all subjects.

After comparing data on the 35 pairs, a significant difference ($P < 0.01$) was found in absenteeism rates between ex-smokers and current smokers, and in absenteeism rates for the ex-smokers before and after cessation; however, no significant difference was found in absenteeism rates relevant to the number of cigarettes smoked per day.

Smoking has become a widely publicized issue in the last few years. Smokers, ex-smokers, and nonsmokers are all involved with this issue. Not only are the smokers harming themselves, but research has also shown that second-hand smoke has harmful effects on nonsmokers.

Health professionals have several ways in which they can become involved. They can lobby for clear air, provide counseling on the harmful effects of cigarette smoke, and support the facilities that have no-smoking policies. They also need to remember that smoking is an addiction and there are several factors that may cause people to smoke. Support and incentives should be provided to those people who would like to quit smoking and those who are in the process of quitting.

Dianne Martin
College of Health Related Professions
Medical University of South Carolina

The second example is a book review from the *Monthly Labor Review*.*

Pensions For a Mobile Work Force

Private Pensions and Employee Mobility. By Izzet Sahin. New York, NY, Greenwood Press, Inc., 1989. 116 pp.

In *Private Pensions and Employee Mobility,* Izzet Sahin presents a mathematical study of the effect of changing jobs on pension benefits. A worker who changes jobs during a career will usually receive a smaller pension than a worker who remains at one firm during a career, all other things being equal. Inflation can erode pension benefits for workers who terminate their jobs before retirement. In addition, workers may lose some or all of their benefits if they are not vested in a plan. Sahin creates a model to measure lost pension benefits for various types of pensions and levels of mobility.

Because studying the effects of mobility requires measurement over time, the model makes assumptions about economic conditions. Inflation is assumed to stay constant at 5 percent, real growth in wages (above inflation) at 2 percent, and real rate of return on investment at 2 1/2 percent. Sahin arrives at these figures by taking long-term averages of economic data.

Using his model, Sahin demonstrates how changing jobs can reduce pension benefits even if a worker is fully vested in a pension upon termination of employment. Because pension benefits are commonly based on earnings, inflation will lower the value of the pension. A worker who leaves a firm in 1965 and retires in 1990 will receive pension benefits based on a salary that does not reflect current living standards. . . .

Sahin uses the model to show how the three vesting options of the Employee Retirement Income Security Act (ERISA) affect the mobility of workers. Vesting is the nonforfeitable right to future pension benefits. The first option, cliff vesting, requires no vesting until 10 years, and full vesting thereafter. The second option, graded vesting, requires 25 percent vesting after 5 years, with an increase of 5 percentage points a year until 10 years, and then an increase of 10 percentage points a year until full vesting at 15 years. The third option, the rule of 45, requires 50 percent vesting when service plus age equal 45, then increases by 10 percentage points in each of the next 5 years. (The Tax Reform Act of 1986 reduced the years of service requirements for vesting that may be imposed.) . . .

Sahin's model shows that most forms of private pensions tend to discourage mobility. By assuming the economic conditions and levels of mobility, Sahin is able to provide calculations to support his argument. His model allows him to explore a wide variety of issues in the field of pensions, from plan types to vesting rules. *Private Pensions and Employee Mobility* provides a framework for analyzing decisions on policy affecting this timely issue.

—Jason L. Ford
Division of Occupational Pay
and Employee Benefit Levels
Bureau of Labor Statistics

Two Techniques for Writing an Informative Summary

A summary is a shorter, briefer version of an original work. The original work may be summarized by abridgment or by condensation. No matter which technique you use, the summary omits introductory matter, details, examples, and illustrations—everything except the major ideas and most important facts.

Informative Summary by Abridgment.
A summary produced by abridgment retains the essential content of a work in the original wording. Supporting details and supplementary matter are omitted.

*Reprinted by permission from the October 1990 issue of the *Monthly Labor Review*.

Below is a paragraph from "Remote Computing: Telecommunications Made Practical," by Marti Remington, which appeared in *PC Novice*.* In the abridgment process, the supporting matter is crossed out, leaving the essential content of the original paragraph and the abridged paragraph the same.

227

Chapter 6
The Summary:
Getting To the
Heart of the
Matter

Original Paragraph With Supporting Matter Crossed Out

Telephone lines are a fast and inexpensive medium for transferring information long distances, ~~which is what telecommunications is all about~~. And to access phone lines you'll need a modem. ~~A modem is~~ a peripheral device ~~that allows a computer to understand the analog or voice signals that telephone lines are designed to carry. The computer needs a modem~~ to translate the computer's outgoing digital or binary signals into the voice signals that telephone lines can handle. ~~A modem is also necessary~~ to translate the incoming voice signals into the digital signals a computer can understand.

Paragraph Summarized By Abridgment

Telephone lines are a fast and inexpensive medium for transferring information long distances. To access phone lines, you'll need a modem, a peripheral device to translate the computer's outgoing digital or binary signals into the voice signals that telephone lines can handle and to translate the incoming voice signals into the digital signals a computer can understand.

Informative Summary by Condensation. A summary that results from the condensation of the original material rephrases the main content of the original. That is, the writer or speaker carefully reads the original and then restates the details in his or her own words.

The summary may include direct quotations from the original work. The quoted material may be a phrase, a sentence, a paragraph, or perhaps a longer section. Quoted material should be in quotation marks. Sometimes page or chapter numbers for quotations are helpful for reference. Footnotes are unnecessary since only one source is involved and that source is identified.

Below is the same paragraph from *PC Novice*. Following the original paragraph is a summary by condensation.

Original Paragraph

Telephone lines are a fast and inexpensive medium for transferring information long distances, which is what telecommunications is all about. And to access phone lines you'll need a modem. A modem is a peripheral device that allows a computer to understand the analog or voice signals that telephone lines are designed to carry. The computer needs a modem to translate the computer's outgoing digital or binary signals into the voice signals that telephone lines can handle. A modem is also necessary to translate the incoming voice signals into the digital signals a computer can understand.

Paragraph Summarized By Condensation

With the addition of a modem, telephone lines can be used as a quick and inexpensive way to exchange information over long distances. The modem translates the telephone line's voice signals (analog) and the computer's binary signals (digital) so that information can be handled by the telephone lines and understood by the computer.

*Reprinted by permission of the Peed Corporation from the June 1991 issue of *PC Novice*, page 35.

In both examples, abridgement and condensation, the original paragraph is cut approximately 50 percent. The main points are retained and explanatory material is omitted.

The following is a three-part example of a student-written condensation. First, the original article is reprinted with key points and ideas shown in boldface type; marginal notes explain how to determine what information to include and what to omit in the summary. Next is the Plan Sheet for Giving an Informative Summary filled in by the student. And finally there is the informative summary. Note that the length of the original article is approximately 1,250 words; the length of the summary is approximately 225 words.

Computer Buyer Fears*
by Rachel Pred

Large purchases cause anxiety. Buying a new car, house, boat, plane or island can bring on stress symptoms akin to choosing a college major. But at least with a college major you can change your mind six or seven times. OK, eight, but in my case I was trying to find myself.

A computer purchase can cause stress, too. People fear spending too much, too little, buying the wrong equipment. . . . *But read on and you'll find ways to alleviate some common buying fears and things to consider when making your decision.*

Main point of article. Include in a summary

I asked Harold Gruesner, a sales representative for Computer-Land in Edina, Minn. *what people feared most about buying a computer.* His reply: *"Turning on the machine.* Most of the people feel that it's this . . . this black hole."

A fear. Include in a summary

Marty Ackermann, president and founder of the Mr. Micro/MicroAge store in Dallas said he also runs into *people* who *are scared that they won't be able to use the machine.* He said people sometimes fear they won't be able to take full advantage of the technology available to them.

A fear. Include in a summary

However, Ackermann also mentioned that people are less inclined to be scared of computers now than a few years ago. "Customers today are more comfortable with technology," he said. "We're not seeing nearly as many highly intimidated people as we used to five years ago. Most people have seen computers, have seen them work, have probably played with one a little bit."

"Example. Omit in a summary

If you want to receive some training on your system, your local retailer is a good place to start. Some dealers offer free training with the purchase of a system; others charge a small fee. Those retailers who don't provide training will probably be able to suggest places that offer good classes.

A tip for the fearful purchaser. Include in a summary

The continuing trend toward user-friendly systems means you may not even need any training. "A lot of the computers today, because of Windows, because of the Macintosh ease-of-use, don't require formal training at all," Ackermann said. "If you can read and follow a manual or dedicate maybe ten or twenty hours

Examples. Omit in a summary

*Reprinted with permission of the Peed Corporation from the January 1991 issue of *PC Today Magazine,* pages 54–55.

of your time, you can learn to use it [the computer] very well without formal training."

Behind the Times

Are you feeling smug because you have no innate fear of turning on your computer? Even people who don't suffer from technophobia can experience a lot of anxieties when buying a computer.

Purchasing equipment that will become out of date is a major fear that many buyers face. "People think they're buying the latest and greatest and then six months or a year later it's not available anymore," Gruesner said. "That's probably the biggest fear that I run up against." No one wants equipment that may be obsolete in a matter of months.

Anyone stuck with 200 8-track tapes knows what Gruesner is talking about. In the computer industry, where a new technology is announced approximately every 7.8 seconds, it's easy to second-guess yourself when it comes to making a purchase. How can you receive some assurance that the company that made your system won't abandon it (or worse yet, disappear itself) and leave you in the cold?

One way to avoid being left in the technological twilight zone is to purchase products that are well established in the industry. Gruesner said, "Basically I'll tell people either stick with the more well-known DOS machines or the Macintosh line of the Apple series. You know there's a large amount of units already out on the street."

Another way to give yourself some confidence in what you are buying is to *carefully assess your needs before making any purchase.* If you know that in six months or a year you will be using the computer for additional tasks (such as starting a newsletter, keeping track of business accounts, etc.), you can buy a system now that can handle additional power needs later. "When you're buying today you're probably buying for a three or four year period," Ackermann said, adding that product life used to be about five years but that the industry is accelerating and technology changes are being made faster. Ackermann feels the dealer should take time to determine the present and continuing needs of the buyer.

Other Options

If you're not able to anticipate what your future needs will be, don't worry; you can always upgrade your system later. Upgrading a system involves adding higher-end components and peripherals that can speed up your computer and make it capable of accomplishing a wider variety of tasks. Some manufacturers construct systems with easy upgrades in mind, so that users can have a machine that grows with them.

Michael Jossett, a sales representative for Radio Shack in Austin, Texas, said the *most popular upgrades are internal or external hard drives, modems* (for hooking up your system to a telephone line), *and serial ports* (outlets for hooking

229

Chapter 6
The Summary:
Getting To the
Heart of the
Matter

A fear. Include in a summary

Examples. Omit in a summary

Statistic. Might include in a summary

One way to alleviate the fear. Include in a summary

Example. Omit in a summary

A second way to alleviate the fear. Include in a summary

Example. Omit in a summary

A third way to alleviate the fear. Include in a summary

Ways to upgrade. Include in a summary

up other peripherals to your computer). Other upgrades that are often performed are **adding memory** to your system, or adding **a math co-processor,** a special chip that helps your computer perform mathematical functions more efficiently.

Bells and Whistles

One important thing to remember is this: **if you buy a system that handles your needs and a year later some new technological advance is made, don't feel like your current system is suddenly inadequate.** Depending on your uses, the machine you have may be all you will ever need. "Someone who's just using it for occasional word processing, then it [a new technology] doesn't make any difference," Ackermann said. "You can use the same machine for ten years."

A fourth way to alleviate the fear. **Include in a summary**

Gruesner agrees. "You can still use [less powerful] machines. If it's doing the applications that you need it to do, there's no need to jump up."

Example. **Omit in a summary**

Try to avoid getting caught up in all the gee-whiz technology. Some buyers feel that if they don't buy the "Mazerati" of computers their system will forever be inadequate in the computer world. The retailers I spoke with all mentioned that **consumers can be overwhelmed by media hype and advertisements.** If you are dazzled by all the hype associated with high-end computers based on 80386 and 80486 microprocessors, remember that a less powerful 80286 or even an 8088 may be enough for your personal needs. Gruesner said that people see advertisements for "386, 486 machines which are overkill for half the market." If you have money to burn, go ahead, get the best. Buy a machine with the power of a university mainframe. The rest of us, however, can usually have all our needs met by a lower-end machine.

Caution about advertisements. **Include in a summary**

Examples. **Omit in a summary**

Nothing to Fear

There are few surprises when it comes to exploring the fears of buying a computer. They're similar to the fears associated with any major purchase: Will it meet my needs? Will it be quickly outdated? Will it work? Will I need another degree to learn how to use it? Did I pay too much? The best advice is to **know your needs and your budget, and find a retailer you can trust.** Sound easy enough? See you at the store.

Summary of main ideas. **Omit in a summary**

Restatement of main point of article. **Include in a summary.**

PLAN SHEET
FOR GIVING AN INFORMATIVE SUMMARY

Analysis of Situation Requiring Summary

What type of summary will I write?
informative

Who is my intended audience?
a consumer who has anxieties about buying a computer

How will the intended audience use the summary? For what purpose?
to learn ways to get rid of anxieties about buying a computer

Will I summarize the original material by condensation or by abridgment?
condensation

Will the final presentation of the summary be written or oral?
written

Identification of Work (Bibliographical Information), as Applicable

Title:
"Computer Buyer Fears"

Author:
Rachel Pred

Date:
January 1991

Type of material (magazine article, book, report, etc.)
magazine article

Title of publication:
PC Novice

City of publication and publishing company:
omit

Volume number:
omit

Page numbers:
54–55

Other:

Key Facts and Ideas in the Work

Introduction
  —*stress involved in buying computers*
  —*ways to avoid stress*
  —*things to consider when choosing a computer*
Major fears
  —*turning on the computer*
  —*not being able to use the computer*
  —*computer equipment becoming out of date quickly (statistic—"new technology announced approximately every 7.8 seconds")*
Ways to alleviate the fear
  —*buy brand name computers*
  —*consider what you want the computer to do, how you want to use it, now and in the future*
  —*realize that if unanticipated needs arise you can upgrade your computer by adding internal or external hard drives, modems, serial ports, memory, or a math coprocessor*
  —*don't feel that you must purchase computers enhanced with technological advances; keep your computer as long as it meets your needs; don't be swayed by advertisements and media*
Conclusion
  —*good advice: "Know your needs and your budget, and find a retailer you can trust."*

The Title I May Use

An Informative Summary of "Computer Buyer Fears"

233

Chapter 6
The Summary:
Getting To the
Heart of the
Matter

AN INFORMATIVE SUMMARY OF "COMPUTER BUYER FEARS"

Pred, Rachel. "Computer Buyer Fears." *PC Today Magazine* January 1991:54–55.

Buying a computer can be a stressful undertaking. Rachel Pred suggests some ways to avoid the stress involved and some things to consider as you choose a computer.

Major Fears

Individuals fear turning on the computer and not being able to use it. Of course, training is available for those who want or need it, although some user-friendly computers may not require formal training. One major fear a purchaser may have is that the computer equipment will become out of date in a short period of time.

Ways to Alleviate the Fear

A guideline to remember is to buy brand name computers that are established, that many people have been buying and using satisfactorily. Also consider what you want the computer to do and how you want to use it, both now and in the future. If unanticipated needs arise, you can always upgrade your system by adding internal or external hard drives, modems, serial ports, additional memory, or a math coprocessor.

If technological advances improve computers, do not think you need an improved computer. As long as the computer you have meets your needs, continue to use it. Do not be swayed by advertisements and media.

Conclusion

Keep this "best advice" in mind: "Know your needs and your budget, and find a retailer you can trust."

Evaluative Summary

A summary may evaluate; it may be the primary purpose of the summary to analyze the accuracy, completeness, usefulness, appeal, and readability of a piece of writing. The evaluative summary emphasizes the writer's assessment of the original work. In the evaluative summary you include your own comments and reactions—your thoughts and feelings regarding the material and the presentation of the material.

The evaluative summary cannot exist, however, without either descriptive or informative material for a point of reference. In addition to the evaluative documents, the summary may state what the subject is about in a very general way. In effect, the summary becomes a descriptive/evaluative summary. Note the following example.

> The *thought-provoking* article "Take Charge of Your Life" describes in *detail* four strategies for turning adversity to advantage.

The addition of "thought-provoking" and "in detail" gives the writer's assessment of the effectiveness and scope of the original article he or she is summarizing. Without this addition, the summary is descriptive only.

Another technique for adding evaluative commentary to a descriptive summary is to write the evaluation in a separate sentence or sentences.

> The article "Take Charge of Your Life" describes four strategies for turning adversity to advantage. The article is thought-provoking, with each strategy described in detail and with examples.

The first sentence is descriptive only; the second sentence is evaluative. The second sentence alone, however, makes little sense; sentence one gives a point of reference. Together the two sentences are a clear descriptive/evaluative summary.

Either technique—adding evaluative words or phrases to an existing descriptive summary or adding evaluative assessment in a separate sentence—can be used effectively.

The evaluative summary may also include facts from the original work as well as evaluative commentary. In effect, such a summary becomes an informative/evaluative summary.

The following student-written summary begins with a descriptive/evaluative summary sentence and adds the main points of the article as well as additional evaluation. Evaluative commentary is underlined; notice that evaluative commentary is included throughout the summary. This example includes all primary types of summary material: descriptive, informative, and evaluative. The original article appears on pages 496–498.

EVALUATIVE SUMMARY: "CLEAR ONLY IF KNOWN"

The <u>well-written, clearly illustrated</u> article "Clear Only If Known," by Edgar Dale,* answers the question, "Why do people give directions poorly and sometimes follow excellent directions inadequately?"

In answering the first half of the question, "Why do people give directions poorly?," Dale states six specific reasons.

1. People do not always understand the complexity of the directions they attempt to give.
2. People give overly complex directions that include unnecessary elements.
3. People overestimate the experience of the person asking directions.
4. People make explanations more technical than necessary.
5. People are unwilling to say, "I don't know."
6. People use the wrong medium for giving directions.

<u>Each reason is easily identified</u> because Dale uses words like "First of all," "Another frequent reason," and "Another difficulty in communicating directions" as introductory phrases. Further, each reason is <u>explained in detail</u>, as Dale gives <u>at least one excellent illustration for each reason.</u>

The last half of the article <u>briefly</u> answers the second part of the question, "Why do people sometimes follow excellent directions inadequately?" Dale gives two reasons:

1. People don't understand directions but think they do.
2. People often are in too big a hurry when they ask for directions.

In closing, Dale emphasizes the need for clear communication, saying that "clarity in the presentation of ideas is a necessity."

<u>Anyone reading this article will have a clear understanding of why the communication process often breaks down.</u>

*Edgar Dale, "Clear Only If Known," reprinted in Nell Ann Pickett and Ann Laster, *Technical English,* 7th ed. (New York: HarperCollins, 1996), pp. 496–498.

The book review is also a popular form of evaluative summary. The following book review appeared in *The Conservationist*.* Compare the length of the original work, 198 pages, to the length of the evaluative summary, some 225 words. Clearly the purpose of this review is to evaluate the usefulness of the book as well as to describe what it covers. Evaluative commentary appears in boldface type.

235

Chapter 6
The Summary:
Getting To the
Heart of the
Matter

Beavers, Water, Wildlife and History, text and photographs by Earl L. Hilfiker, 198 pages, Earl L. Hilfiker, 284 Somershire Dr., Rochester, NY 14617, (716) 342–8620, $24.95 plus $2 shipping, add $1.75 sales tax NYS.

This is a **well written** and **nicely illustrated** book by a person who has made a career studying and photographing wildlife and, in particular, the beaver. Earl Hilfiker, a former wildlife photographer for our department of conservation, has assembled a **tremendous mass** of information about the beaver from published material and has added a **wealth** of his personal experiences with the animal. He combines the text with a **profusion** of pictures of beavers and other wildlife species that he has photographed over the last 40 years.

The first part of the book deals with the natural history of the beaver as a unique animal in our aquatic-based ecosystem. The relationship of the beaver to the land, water and other animals is **explored in depth.**

The second section tells how the fur trade of beaver pelts helped shape the settling of this country. Due to exploitation, the beaver in northeastern America was nearly exterminated.

The author then recounts the steps in beaver restoration through stocking and control of trapping seasons, which brought them back to abundant numbers.

The fourth section **covers in greater detail** the existing status of the beaver throughout this and other countries. This is followed by a life cycle of a beaver pond and how, in its various stages of growth and decay, the pond attracts a great variety of plants and animals.

Mr. Hilfiker **with good writing** gives the reader an intimate portrait of the dynamics of nature. Not only are the creatures themselves and their interdependence on the pond environment **well defined,** but the pond itself seems to take on a life of its own through the eyes and talents of this **competent naturalist/photographer/writer.**

Mr. Hilfiker, scientist, author and Audubon film tour speaker, has done an **outstanding job** with this book about one of his favorite animals. I **heartily recommend** this book. It is **well balanced** and deserves to be read by those interested in nature in general as well as in the beaver in particular.

—Wayne Trimm

Another form of evaluative summary—closely related to the book review—is the video review. The following video review is from the *Community Health Journal*.**

This summary describes the video as well as evaluates it for audiences interested in rehabilitation and mental health techniques. The title of the magazine in which the summary appears suggests a specialized audience.

Free Dive

10 minutes, color. Producer/Director: Wendy Campbell
Mental Health Video Library, 331 Dundas Street East, Toronto, Ontario M5A 2A2, Canada

This moving tape explores the uses of swimming and scuba diving, both as a rehabilitation technique for persons with spinal cord injury, and as a mental health

*Reprinted by permission from the July/August 1991 issue of *The Conservationist Magazine.*
**Reprinted from the February 1991 issue of *Community Health Journal.*

technique. Mastering scuba for these individuals, many of whom have been devastated following their spinal injury, becomes, in the hands of sensitive and supportive instructors, a way to reintroduce self-confidence, pleasure, and greater independence into their lives. The tape was shown on ABC television several years ago, and is now being used in schools in both the USA and Canada, together with an instruction guide.

General Principles in Giving a Summary

1. *The emphasis of a summary is determined by the purpose and the intended audience.* The primary purpose may be to describe, to inform, or to evaluate.
2. *Summaries require a thorough understanding of the material.* Scanning, reading carefully, and underlining or taking notes are important in comprehending the material and thus in distinguishing which information to include in the summary and which to omit.
3. *A summary should be accurate in fact and in emphasis.* Unintentional distortion of the original may occur through careless error, through omission of essential information, or through unequal treatment of ideas presented as equally significant in the original.
4. *Identify material you are summarizing.* A formal bibliographical entry may appear in the beginning (or end) of the summary, or you may include bibliographical facts within the summary.
5. *A descriptive summary states what the material is about in a very general way.* Frequently you need only a few sentences to indicate the major concern of the material.
6. *An informative summary states objectively what is in the material.* You present the principal facts and conclusions in a paragraph to several pages, depending on the length of the original text.
7. *An evaluative summary emphasizes assessment of the material.* You judge the accuracy, completeness, and usefulness of the information. Include your own reactions, thoughts, and feelings along with the main facts and ideas in the original text.

Procedure for Giving a Summary

1. Decide whether the primary purpose of the summary is to describe what the work is about; to inform by presenting the principal facts and conclusions given in the original work; or to evaluate the accuracy, completeness, and usefulness of the work.
2. Look over the work. Note the opening and closing paragraphs, the major concern of the work, the organization, and the method of presentation.
3. Read the work very carefully. Underline or jot down the key ideas on a separate sheet of paper (particularly if the work does not belong to you). Think

through the relationship of these key ideas; be sure that you understand what the writer is trying to communicate.

4. Identify the original material specifically by author, title, and publication facts. Give this information within the summary or as a formal bibliographical entry at the beginning or end of the summary. (See pages 408–409 for bibliographical forms.)
5. Write a draft of the summary, keeping the intended audience and purpose clearly in mind.
6. Reread the original work; then check the summary for accuracy of fact and emphasis and for completeness.

237

Chapter 6
The Summary:
Getting To the
Heart of the
Matter

Applications in Giving a Summary

Descriptive Summaries

6.1   Bring to class two *descriptive* summaries, as explained on pages 220–224, written by others. State the specific sources. Identify the intended audience. Comment on the effectiveness of the summaries.

6.2   Write a two-to-three sentence *descriptive* summary of "Computer Buyer Fears" on pages 228–230.

6.3   Write *descriptive* summaries of any five books (specialized encyclopedias, dictionaries, etc.) related to your major field of study. Before you can write the summary, you need to become familiar with each book. Examine its physical characteristics (size, type of binding, number of pages), organization (table of contents, index, illustrations), content (chapter or entry headings, kind and extent of information, intended reader), and publishing information (publisher, year of publication, method of updating information).

6.4   Write *descriptive* summaries of two periodicals and two other sources (pamphlets, reports, bulletins, audiovisual materials, etc.) that relate to your major field.

6.5   Write two *descriptive* summaries of a book that you have selected. First, write a descriptive summary for a librarian who will order the book; second, write a descriptive summary for a student who will use the book as a reference for a report.

6.6   Locate a magazine that does not include descriptive summaries in the table of contents. Write appropriate summaries for the table of contents.

Informative Summaries

In completing applications 6.7–6.10, 6.12

a. Fill in the appropriate Plan Sheet on pages 241–244, or one like it.
b. Write a preliminary draft.
c. Go over the draft in a peer editing group.
d. Revise. See Part III Revising and Editing and the checklists for revising, front and back endpapers.
e. Write the final summary.

6.7 Write an *informative* summary of the following excerpt from the *Occupational Outlook Handbook,* 1994–1995 edition.

Engineering Technicians

Nature of the Work

Engineering technicians use the principles and theories of science, engineering, and mathematics to solve problems in research and development, manufacturing, sales, construction, and customer service. Their jobs are more limited in scope and more practically oriented than those of scientists and engineers. Many engineering technicians assist engineers and scientists, especially in research and development. Others work in production or inspection jobs.

Engineering technicians who work in manufacturing follow the general directions of engineers. They may prepare specifications for materials, devise and run tests to ensure product quality, or study ways to improve manufacturing efficiency. They may also supervise production workers to make sure they follow prescribed procedures.

Engineering technicians who work in research and development build or set up equipment, prepare and conduct experiments, calculate or record the results, and help engineers in otherc ways. Some make prototype versions of newly designed equipment. They also assist in routine design work, often using computer-aided design equipment.

Civil engineering technicians help civil engineers plan and build highways, buildings, bridges, dams, wastewater treatment systems, and other structures and do related surveys and studies. Some inspect water and wastewater treatment systems to ensure that pollution control requirements are met. Others estimate construction costs and specify materials to be used. (See statement on cost estimators elsewhere in the *Handbook.*)

Electronics engineering technicians help develop, manufacture, and service electronic equipment such as radios, radar, sonar, television, industrial and medical measuring or control devices, navigational equipment, and computers, often using measuring and diagnostic devices to test, adjust, and repair equipment. Workers who only repair electrical and electronic equipment are discussed in several other statements elsewhere in the *Handbook.* Many of these repairers are often called electronics technicians.

Industrial engineering technicians study the efficient use of personnel, materials, and machines in factories, stores, repair shops, and offices. They prepare layouts of machinery and equipment, plan the flow of work, make statistical studies, and analyze production costs.

Mechanical engineering technicians help engineers design and develop machinery, robotics, and other equipment by making sketches and rough layouts. They also record data, make computations, analyze results, and write reports. When planning production, mechanical engineering technicians prepare layouts and drawings of the assembly process and of parts to be manufactured. They estimate labor costs, equipment life, and plant space. Some test and inspect machines and equipment in manufacturing departments or work with engineers to eliminate production problems.

Chemical engineering technicians are usually employed in industries producing pharmaceuticals, chemicals, and petroleum products, among others. They help design, install, and test or maintain process equipment or computer control instrumentation, monitor quality control in processing plants, and make needed adjustments.

239

Chapter 6
The Summary:
Getting To the
Heart of the
Matter

6.8 Write an *informative* summary of a reading assignment in one of your other courses or of an article in a recent issue of a periodical relating to your major field.

Evaluative Summaries

6.9 In a paragraph of approximately seventy-five words, write a *descriptive/ evaluative* summary of "Computer Buyer Fears" on pages 228–230.

6.10 Write an *informative/evaluative* summary of reading assignment in one of your other courses, or of an article in a recent issue of a periodical relating to your major field.

All Three Types of Summaries

6.11 Find and attach to your paper an example of a *descriptive* summary, of an *informative* summary, and of an *evaluative* summary. Write a one-page report on what is accomplished in the summaries. Be sure to include in your report specific identification of the source of each summary, the form of each summary, the purpose and audience of each, and any other pertinent information.

6.12 Select a piece of writing, such as a magazine article, pamphlet, or chapter in a book, relating to your major field. For the selection, write:

- A *descriptive* summary
- An *informative* summary

Then explain how you could expand the descriptive and informative summaries to include evaluation.

Oral Summaries

6.13 As directed by your instructor, adapt for oral presentations the summaries you prepared in the applications above. Ask your classmates to evaluate your speech by filling in the Evaluation of Oral Presentations on page 459, or one like it.

Using Part II Selected Readings

6.14 Read "The Effects of Perceived Justice on Complainants' Negative Word-of-Mouth Behavior and Repatronage Intentions" by Jeffrey G. Blodgett, Donald H.

Granbois, and Rockney G. Walters, pages 552–562. Under "Suggestions for Response and Reaction," page 563, write Number 4.

6.15 Read "Clear Only If Known," by Edgar Dale, pages 496–498. Under "Suggestions for Response and Reaction," page 499, write Number 3.

6.16 Read "Are You Alive?" by Stuart Chase, pages 537–540. Under "Suggestions for Response and Reaction," page 541, write Number 4.

6.17 Read "The Active Document: Making Pages Smarter," by David D. Weinberger, pages 579–584. Under "Suggestions for Response and Reaction," page 584, write Number 4.

6.18 Give *descriptive, informative,* and *evaluative* summaries of other selections from Part II as directed.

PLAN SHEET
FOR GIVING AN INFORMATIVE SUMMARY

Analysis of Situation Requiring Summary

What type of summary will I write?

Who is my intended audience?

How will the intended audience use the summary? For what purpose?

Will I summarize the original material by condensation or by abridgment?

Will the final presentation of the summary be written or oral?

Identification of Work (Bibliographical Information), as Applicable

Title:

Author:

Date:

Type of material (magazine article, book, report, etc.):

Title of publication:

City of publication and publishing company:

Volume number:

Page numbers:

Other:

Key Facts and Ideas in the Work

The Title I May Use

PLAN SHEET
FOR GIVING AN EVALUATIVE SUMMARY

Analysis of Situation Requiring Summary

What type of summary will I write?

Who is my intended audience?

How will the intended audience use the summary? For what purpose?

Will the final presentation of the summary be written or oral?

Identification of Work (Bibliographical Information), as Applicable

Title:

Author:

Date:

Type of material (magazine article, book, report, etc.)

Title of publication:

City of publication and publishing company:

Volume number:

Page numbers:

Other:

Key Facts and Ideas in the Work/My Evaluative Comments

The Title I May Use

Chapter 7

Memorandums and Letters: Sending Messages

Outcomes

Upon completing this chapter, you should be able to

- List and define three electronic methods for business correspondence
- List and define the two regular parts of a memorandum
- List and define three electronic methods for business correspondence
- Set up a heading for a memorandum
- Write a memorandum
- Analyze the form and content of a letter
- Show and identify by label the three most often used layout forms
- Head a second page properly
- List, define, label, and write an example of the six regular parts of a letter
- List, define, label, and write an example of seven special parts of a letter
- List, define, label, and write an example of the two regular parts on the envelope
- Write a letter of inquiry or request
- Write an order letter
- Write a complaint letter and an adjustment letter
- Write a collection series
- Write a job application letter
- Write a resumé
- Fill out a job application form
- Write application follow-up letters

Introduction

Writing effective memorandums and letters, according to business people and industrialists, is one of the major writing skills that an employee needs. These employers say that the employee must be able to handle the aspects of work that involve correspondence, such as communicating with fellow employees, making inquiries about processes and equipment, requesting specifications, making purchases, answering complaints, and promoting products. Over half of all business is conducted in part or wholly by correspondence. Thus it is impossible to overemphasize the importance of effective business correspondence.

Too, memorandums and letters are business records; copies show to whom messages have been written and why. Since more than one message frequently is required in settling a matter, copies become a necessity for maintaining continuity in the correspondence. Even now you should begin to develop the habit of keeping a copy of every memorandum and letter you send. Use whatever storage form is available or preferred. You may, for example, keep a copy in print form, on a magnetic media form (a floppy disk, either a 3.5 or 5.25 inch disk, or a hard drive, either internal or external), or tape.

Whether or not you get a job may be determined by your application materials (letter of application, resumé, application form). Employers, particularly personnel managers, are concerned that a potentially good worker often does not get the job because the application materials do not make a good impression.

Methods for Transmitting Business Correspondence

247

Chapter 7
Memorandums
and Letters:
Sending
Messages

Methods for transmitting business correspondence are changing rapidly. Until a few years ago, the United States Postal Service handled the delivery of correspondence. Then other mail handling businesses evolved such as United Parcel Service (UPS) and Federal Express (FedEx). Improved sorting methods and delivery systems decreased the time for delivering correspondence. Although mail delivery had taken anywhere from one to several days, now same-day, next-day, or two-day delivery is guaranteed, depending on the mail service and the geographical area.

Electronic transmission of business correspondence has made delivery even faster through the use of facsimile (fax) machines, voice mail, and electronic mail (e-mail). A recent survey revealed that a large percentage of midsize companies (companies with 100 to 1000 employees) use various electronic delivery systems. Ninety-seven percent of these midsize companies use fax transmission, 57 percent use voice mail, and 49 percent use e-mail. (Of these same companies, 75 percent use an 800 number and 20 percent use video conference for oral communication.) No doubt a higher percentage of larger companies use electronic delivery systems, and small businesses and individuals also use fax machines, voice mail, and e-mail.

Facsimile (Fax) Machines, Voice Mail, and E-Mail

When both the sender and the receiver have fax machine capabilities (a fax machine/computer), a letter, a memorandum, or any printed material can be sent through the machine (via telephone lines) in a matter of seconds; within a few more seconds, a reply can be returned. Typically faxed messages include a fax cover sheet and the message page(s).

Using a computer, a typewriter or even handwriting, you can design and print cover sheets as you need them. The example below was keyboarded and printed from a computer.

FAX COVER SHEET

TO: KAREN NEWMAN, SCETC JOURNAL EDITOR
FROM: ANN LASTER
DATE: 7 SEPTEMBER 1995
SUBJECT: ARTICLE ON 1995 INSTITUTE IN TECH COMM

COVER SHEET PLUS 1 PAGE

LASTER FAX NO.   601-857-3591
NEWMAN FAX NO.   205-937-5140

Or you may have a cover sheet designed and printed in large amounts if you use a fax machine often. Various kinds of information are included on fax cover sheets: the name of the person to whom the message is sent, the name of the person sending the message, the date, and the subject of the message. Other data often included are the name and address of the sender, the fax numbers of both the sender and the receiver of the message, the number of pages faxed, a number and a person to call if the fax is received at a wrong location, and space for comments or a message. Fax cover sheets have no standard design or content.

An example of a more detailed cover sheet is shown below.

OCT 10 '95 02:18PM MS. LIBRARY COMM.   P.1/3

**MISSISSIPPI
LIBRARY COMMISSION**
Serving the Information Generation

1221 Ellis Avenue
Post Office Box 10700
Jackson, Mississippi 39289-0700
(601) 359-1036
Tele-Fax (601) 354-4181

Mary Ellen Pellington
Executive Director

FACSIMILE TRANSACTION

DATE: _Oct. 10, 1995_

TO: _Nancy Tenhet_
(Name)
Hinds C.C. Library
(Organization)
_(601) 857-3585_
(Fax Number)

FROM: _Jolinda Ralston_

TOTAL NUMBER OF PAGES: _3_
(including this page)

MESSAGE: _____
_____
_____
_____

Please call _____Jolinda_____ at, (601) 359-1036 Ext. _153_ if this transmission is incomplete or has been sent to the wrong location.

Notice at the top of the received cover sheet are printed such facts as the time and date of the transmission, the sender, the sender's fax number, and the page number.

With voice mail, callers can leave messages in an electronic mail box. It is an automatic call-answering service that takes messages for you in the caller's own voice. Persons who have an electronic mail box can retrieve messages from any touchtone telephone at any time. With voice mail, you can leave messages for yourself (such as reminders of meetings and appointments). Also, unlike some answering machines, from which anyone can retrieve messages, your voice mail box is confidential; you typically access messages through a password or code.

E-mail requires both the sender and the receiver to have computers and a connection through a local area network (LAN) or an on-line service. With a LAN, individuals can communicate, for example, between computers within a building. With an on-line service, individuals can communicate between computers anywhere in the world. Communication may be to one person or to any number of persons through options such as forums and bulletin boards. All of this communication can take place within minutes.

Below are examples of e-mail transmissions.

Sample E-Mail Request for Information

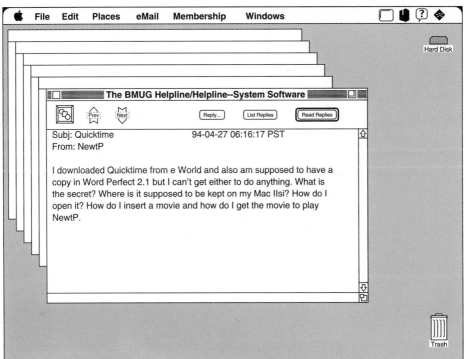

249

Chapter 7
Memorandums
and Letters:
Sending
Messages

In addition to increasing the speed of delivery of messages, fax transmissions and e-mail are changing the format and the content of business communication. As illustrated by the examples above, messages tend to be less formal, less structured, and briefer.

Sample E-Mail Response to a Request

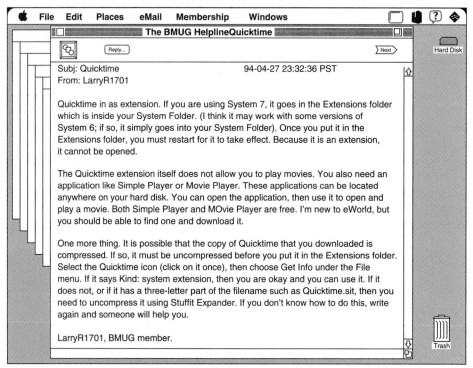

Much e-mail is transmitted by special systems that create multipart addresses combining geographical and conceptual information. One such system is the Domain Name System (DNS) used by the Internet, the nation's largest electronic information exchange. The DNS includes a six-segment address.

The first four and the last items of information tell who and where.

1. An identification of the user or organization (who)
2. The @ symbol (connects the who with the where identification)
3. A subdomain (department, building)
4. A domain (name of school, business)
6. The country (United States addresses usually omit the country segment)

The fifth item of information tells what.

5. Type of organization (edu – education, com – commercial, gov – government, mil – military)

A sample address in domain name system: alaster@eng.hindscc.edu.usa

Memorandums

251

Chapter 7
Memorandums
and Letters:
Sending
Messages

Memorandums are communications typically between persons within the same company. The correspondents may be in the same building or in different branch offices of the company. Memorandums are used to convey or confirm information. They serve as records for transmittal of documents, policy statements, instructions, meetings, and the like. Although the term *memorandum* formerly was associated with a temporary communication, usage of the term has changed. Now a memorandum is regarded as a communication that makes needed information immediately available or that clarifies information.

The memorandum, unlike the business letter, has only two regular parts: the heading and the body. The formalities of an inside address, salutation, complimentary close, and signature usually are omitted. (Some companies, however, prefer the practice of including a signature, as in the memorandum on page 252.) The memorandum may be initialed in handwriting, as illustrated in the memorandum on page 253. This initialing (or signature at the end) indicates official verification of the sender. As in business letters, an identification line (see page 261), a copy line (see page 261), and an enclosure line (see page 261) may be used, if appropriate.

See also the memorandums in Chapter 8 Reports pages 306, 348–349, 352, and 353.

Heading

The heading in a memorandum typically is a concise listing of:

- *To* whom the message is sent
- *From* whom the message comes
- The *subject* of the message
- The *date*

For ease in reading, the guide words To, From, Subject, and Date usually appear on the memorandum. (These guide words are not standardized as to capitalization, order, or placement at the top of the page.) Most companies use a memorandum form, printed with the guide words and the name of the company or the department or both. It is perfectly permissible, however, to make your own memorandum form by simply keyboarding the guide words.

Body

The body of the memorandum is the message. It is composed in the same manner as any other business communication. The message should be clear, concise, complete, and courteous. If internal headings will make the memo easier to read, insert them (see Headings, pages 9–11 and 311–312).

In the following two examples of memorandums, note how the messages serve different purposes. The first memorandum urges cooperation in company strategy for contractors' viewing of prototype testing. The second memorandum announces information for fall classes given by a professional organization within the insurance industry.

TILLMAN MARSHALL

Marine Company

OFFICE MEMORANDUM

Date: July 5, 1995

To: Earlene Beckwith

From: Don C. Eckermann   *Don C. Eckermann*

Subject: Elevating Unit Tests, Customer Visits, and
             Regulatory Agency Involvement

The new Tillman Marshall elevating system and the new
mobile offshore drilling rig designs continue to be a major
undertaking for our company.

I believe that marketplace acceptance can be accelerated by
arranging for selected drilling contractors to view the
prototype testing at Longview in August 1996. Similarly,
the necessary regulatory agency approvals can be expedited
by inviting selected agency representatives to the
prototype testing.

These visitors will be important people on an important
mission, particularly to us. These visitors will convey
their experiences and observations to the marketplace.

I seek your help in planning the proposed visits for
targeted drilling contractors and for regulatory agency
representatives.

Within the next two weeks would you send me

1. a list of selected contractors you think we should
   invite
2. a list of all the regulatory agencies that will be
   involved in the production and marketing of the
   elevating systems

ljb

c: Longview: New Orleans: Houston:

       Rudy Harrison   Will Trimble   Pharr Ingahorn
       Kristi Baich

Women in Insurance

P.O. Box 31580
PORTLAND, OREGON 97233-0294
(503) 932-5990 FAX (503) 932-5981

DATE: July 9, 1996

  TO: All Education Coordinators and IIA Students

FROM: Raja Rojillio, CPIW Chair   *RR*
          Katina Brewer, Co-Chair   *KB*

SUBJECT: Fall 1996 Registration

We are pleased to enclose registration materials for our
fall classes.

Additional information:

<u>STUDY MATERIALS</u> Students will be responsible for obtaining
study materials from the Insurance Institute, other
publishers, or their employer. For your convenience,
enclosed are excerpts from the Institute's Key Information
Booklet. Please order the materials as soon as possible so
that they will arrive in time for the first class.

<u>TUITION</u> The class schedule lists the tuition fees
applicable to each course. Please note the reduced fees for
WIN members.

<u>CLASSES</u> Classes will begin the week of August 19, 1996, at
the time and location shown on the schedule. As always, our
classes are subject to enrollment minimums. If the number
of registrants for a class falls short of the minimum, we
will provide a list of the other applicants to assist in
forming a study group.

We encourage your participation in our education program
and welcome any suggestions that you may have.

Enclosures

Business Letters

The basic principles of composition apply to letters as they do to any other form of writing. The writer should have the purpose of the letter and its intended reader clearly in mind and should carefully organize and write the sentences and paragraphs so that they say what the writer wants them to say.

Business letters are usually not very long, often not more than one page; and sentences and paragraphs tend to be short and direct. Often the opening and closing paragraphs contain only a single sentence and seldom more than two sentences. In letters that are longer than average, paragraphs are likely to be longer. The longer paragraphs should stay on a single topic, and the sentences should be clearly related one to another. Long letters or letters containing a great deal of information on several subtopics often have headings to mark movement from one subtopic to another. (See the memorandum on page 253.)

Study the sample letters throughout this chapter, noting sentence and paragraph length and development.

One of the main differences between the business letter and other forms of writing discussed in previous chapters is that in the business letter there is a specifically named reader to whom the communication is directly addressed.

The "You" Emphasis

Since the business letter is directed to a named reader, it becomes more personal; courtesy becomes important. To achieve a personal, courteous tone, stress the "you."

Focus on qualities of the addressee and minimize references to "I," "me," "my," "we," and "us." For example, note the emphasis on "we."

> We were pleased to receive your order for ten microscopes. We have forwarded it to the warehouse for shipment.

The message can easily be rephrased to

> Thank you for your order for ten microscopes. You should receive the shipment from our warehouse within two weeks.

The first sentence is writer centered; the rephrased sentence is reader centered. Keep your letters reader centered.

The Positive Approach

The "you" emphasis helps the writer attain a positive approach. Study this negative statement.

> We regret that we cannot fill the order for ten microscopes by December 1. It is impossible to get the shipment out of our warehouse because of a rush of Christmas orders.

The negative statement can be rephrased to

> Your order for ten microscopes should reach you by December 10. Your bill for the microscopes will reflect a 10% discount to say thank you for accepting a delayed shipment caused by a backlog of Christmas orders.

255

Chapter 7
Memorandums
and Letters:
Sending
Messages

The negative statement rewritten to emphasize "you" becomes a positive statement. The letter writer should follow the admonition of a once-popular song: "Accentuate the positive; eliminate the negative."

Natural Wording

The wording of a letter should be as natural and as direct as possible. Jot down thoughts or organize them in your mind so that they can be presented clearly and naturally. Consider the following wording.

> Pursuant to your request of November 10 that I be in charge of the program at the December DECA meeting, I regret to inform you that my impending departure for a holiday tour of Europe will prevent a positive response.

This sentence probably resulted from a writer trying to sound impressive. The result is stilted, awkward wording. The same information might be clearly and naturally worded.

> I am sorry I cannot be in charge of the DECA program for December. I will be leaving on December 5 for a holiday tour of Europe.

As you write a business letter, keep your reader in mind. How will the letter sound to the reader? Is the information presented in a natural, direct way so that the content is clear?

Form in Letters

Learning about the typical form of a letter includes becoming familiar with standard formats and with the parts of a letter and their placement on a page.

Format

In letter writing, use standard practices concerning paper, keyboarding, appearance, and page layout. Failure to follow these standards shows poor taste, reflects the writer's lack of knowledgeable choices, and invites an unfavorable response. In business correspondence, there is no place for unusual or "cute" stylistic practices. (A specialized type of business letter, the sales letter, sometimes uses attention-getting gimmicks. Follow-up sales letters, however, tend to conform to standard practices.)

Paper
Use good-quality bond (at least 20 pound), standard-size paper. Avoid erasable bond paper because it smudges easily.

Keyboarding
Letters should be keyboarded on a typewriter, a computer, or a word processor to produce letters that look neat and that are easily readable.

Appearance
The general appearance of a letter is very important. A letter that is neat and pleasing to the eye invites reading and consideration much more readily than one that is unbalanced or has noticeable erasures and corrections. The letter should be like a

picture, framed on the page with margins in proportion to the length of the letter. Allow at least a 1 1/2-inch margin at the top and bottom, and at least a 1-inch margin on the sides. Short letters should have wide margins and be appropriately centered on the page. As a general rule, single space within the parts of the letter; double space between the parts of the letter and between paragraphs. It is always a wise investment to spend the extra time to retype or reprint a letter if there is even the slightest doubt about its making a favorable impression on the reader. (See the letter on page 259 for proper spacing.)

Layout Forms

Although there are several standardized page layout forms for letters, the form preferred by most firms is the block (see pages 259, 264, 273, 274, 280, 282, 295). The modified block is an older form (see pages 266, 269, 270, 271, 281). Another layout form gradually gaining favor is the simplified block form.

Block Form. In the block form, *all* parts of the letter are even with the left margin. Open punctuation is sometimes used with the full block form; that is, no punctuation follows the salutation and the complimentary close.

Modified Block Form. In the modified block form, the inside address, salutation, and paragraphs are flush, or even, with the left margin. The heading, complimentary close, and signature are to the right half of the page. Paragraphs may or may not be indented. Open punctuation is sometimes used; that is, no punctuation follows the salutation and the complimentary close.

Simplified Block Form. The parts of the letter in the simplified block form, or the AMS (American Management Society) form, are the heading, the inside address, the subject line, the body, and the signature. The salutation and the complimentary close are omitted. A subject line is always used. Like the block, all parts begin at the left margin.

Parts of a Business Letter

The parts of a business letter follow a standard sequence and arrangement. The six regular parts in the letter include: heading, inside address, salutation, body, complimentary close, and signature (these are illustrated in the letter on page 259). In addition, there may be several special parts in the business letter.

Regular Parts

1. *Heading.* Located at the top of the page, the heading includes the writer's complete mailing address and the date, in that order, as shown below. As elsewhere in standard writing, abbreviations are generally avoided. (Note that the heading does *not* include the writer's name.)

Route 12, Box 758   704 South Pecan Circle
Elmhurst, IL 60126   Hanover, PA 17331
July 25, 1996   9 April 1996

In writing the state name, use the two-letter abbreviation recommended by the postal service.

257

Chapter 7
Memorandums
and Letters:
Sending
Messages

TWO-LETTER ABBREVIATIONS FOR STATES

Alabama	AL	Montana	MT
Alaska	AK	Nebraska	NE
Arizona	AZ	Nevada	NV
Arkansas	AR	New Hampshire	NH
California	CA	New Jersey	NJ
Colorado	CO	New Mexico	NM
Connecticut	CT	New York	NY
Delaware	DE	North Carolina	NC
District of Columbia	DC	North Dakota	ND
Florida	FL	Ohio	OH
Georgia	GA	Oklahoma	OK
Hawaii	HI	Oregon	OR
Idaho	ID	Pennsylvania	PA
Illinois	IL	Rhode Island	RI
Indiana	IN	South Carolina	SC
Iowa	IA	South Dakota	SD
Kansas	KS	Tennessee	TN
Kentucky	KY	Texas	TX
Louisiana	LA	Utah	UT
Maine	ME	Vermont	VT
Maryland	MD	Virginia	VA
Massachusetts	MA	Washington	WA
Michigan	MI	West Virginia	WV
Minnesota	MN	Wisconsin	WI
Mississippi	MS	Wyoming	WY
Missouri	MO		

Note that each abbreviation is written in uppercase (capital) letters and that no periods are used.

Letterhead Stationery. Firms use stationery that has been especially printed for them with their name and address at the top of the page. Some firms have other information added to this letterhead, such as the name of officers, a telephone number, a fax number, or a logo.

On letterhead stationery, the address is already printed on the paper; you add only the date to the heading. Letterhead paper is used for the first page only. (See the letters on pages 264, 266, 270, 271, 273, 274.)

2. *Inside Address.* The inside address is placed flush (even) with the left margin and is usually three spaces below the heading. It contains the full name of the person or firm being written to and the complete mailing address, as in the following illustrations:

Ronald M. Benrey
Electronics Editor, *Popular Science*
355 Lexington Avenue
New York, NY 10017–0127

Kipling Corporation
Department 40A
P.O. Box 127
Beverly Hills, CA 90210

Preface a person's name with a title of respect if you prefer; and when addressing an official of a firm, follow the name with a title or position. Write a firm's name in exactly the same form that the firm itself uses. Although finding out the name of the person to whom a letter should be addressed may be difficult or take some time, it is always better to address a letter to a specific person rather than to a title, office, or firm. In giving the street address, be sure to include the word *Street, Avenue, Circle,* and so on. Remember: Careful writers generally avoid abbreviations.

3. *Salutation.* The salutation, or greeting, is two spaces below the inside address and is flush with the left margin. The salutation typically includes the word *Dear* followed by a title of respect plus the person's last name or by the person's full name: "Dear Ms. Badya:" or "Dear Maron Badya." In addressing a company, acceptable forms include "Dear Davidson, Inc.:" or simply "Davidson, Inc." Some writers use "Dear Sir or Madam."

    Usually the salutation is followed by a colon. Other practices include using a comma if the letter is a combination business-social letter, and, in the modified and full block forms, the option of omitting the mark of punctuation after both the salutation and the complimentary close.

4. *Body.* The body, or the message, of the letter begins two spaces below the salutation. Like any other composition, the body is structured in paragraphs. Generally it is single spaced within paragraphs and double spaced between paragraphs.

    Second Page. For letters longer than one page, observe the same margins as used for the body of the first page. The second-page top margin should be the same as that of the sides of the second page. Be sure to carry over a substantial amount of the body of the letter (at least two lines) to the second page.

    Althought there is no one conventional form for the second-page heading, it should contain (*a*) the name of the addressee (the person to whom the letter is written), (*b*) the page number, and (*c*) the date. The following illustrate two widely used forms:

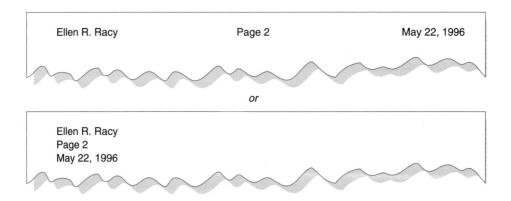

5. *Complimentary Close.* The complimentary close, or closing, is two spaces below the body. It is a conventional expression, indicating the formal close of the letter. "Sincerely" is the commonly used closing. "Cordially" may be used when the writer knows the addressee well. "Respectfully" or "Respectfully yours" indicates that the writer views the addressee as an honored individual or that the addressee is of high rank.

At least 1 1/2-inch
margin at top

80293 Hwy 32 South   HEADING
Elmhurst, IL 60126
25 July 1995   Two or more
                                    keyboard returns

Ronald M. Benrey   INSIDE ADDRESS
Electronics Editor, Popular Science   Content same as
355 Lexington Avenue   address on envelope
New York, NY 10017-0127

Dear Mr. Benrey: Double space
                                    Followed by colon   SALUTATION
                                    Double space

In an electronics laboratory I am taking as a
part of my second-year training at Midwestern   At least 1-inch
Technical College, Elmhurst, Illinois, I have   margin on sides
developed a six-sided stereo speaker system. The   BODY OF LETTER
speaker system is inexpensive (the materials
cost less than $50), lightweight, and quite
simple to construct. The sound reproduction is
excellent.
                        Double space
My electronics instructor believes that other
stereo enthusiasts may be interested in building
such a speaker system. If you think the readers
of Popular Science would like to look at the
plans, I will be happy to send them to you for
reprint in your magazine.
                                    Double space
Sincerely yours,   Capitalize first word,   COMPLIMENTARY
                                    comma after close   CLOSE

Thomas G. Stein
                                    Four keyboard
                                    returns
Thomas G. Stein   SIGNATURE AND
                                                                     TYPED NAME

At least 11/2-inch
margin at bottom

Capitalize only the first word, and follow the complimentary close with a comma. (In using open punctuation, the comma is omitted after both the salutation and the complimentary close.)

6. *Signature.* Every letter should have a legible, handwritten signature in ink. Below this is the keyboarded signature. If the letter is handwritten, print the signature below the handwritten signature.

If a person's given name (such as Dale, Carol, Jerry) does not indicate whether the person is male or female, the person may want to include a title of respect (Ms., Miss, Mrs., Mr.) in the parentheses to the left of the keyboarded signature. In a business letter, a married woman uses her own first name, not her husband's first name. Thus, the wife of Jacob C. Andrews signs her name as Thelma S. Andrews. In addition, she may type her married name in parentheses (Mrs. Jacob C. Andrews) below her own name (see page 266).

The name of a firm as well as the name of the individual writing the letter may appear as the signature (see the letters on pages 266, 271). In this case, responsibility for the letter rests with the name that appears first.

Following the keyboarded signature there may be an identifying title indicating the position of the person signing the letter; for example: Estimator; Buyer, Ladies Apparel; Assistant to the Manager, Food Catering Division. (See pages 264, 270, 271, 273, 274.)

Special Parts

In addition to the six regular parts of the business letter, sometimes special, or optional, parts are necessary. The main ones, in the order in which they would appear in the letter, are the following:

1. *Attention Line.* When a letter is addressed to a company or organization rather than to an individual, an attention line may be given to help in mail delivery. An attention line is never used when the inside address contains a person's name. Typical are attention lines directed to Sales Division, Personnel Manager, Billing Department, Circulation Manager; the attention line may also be an individual's name. The attention line contains the word *Attention* (capitalized and sometimes abbreviated) followed by a colon and name of the office, department, or individual.

   Attention: Personnel Manager *or* Attn: Ms. Robbin Carmichael

   The attention line appears both on the envelope and in the letter, directly above the first line of the address. Since a letter addressed to an individual is usually more effective than one addressed simply to a company, use the attention line sparingly.

2. *Subject, or Reference, Line.* The subject, or reference, line saves time and space. It consists of the word *Subject* or *Re* (a Latin word meaning "concerning") followed by a colon and a word or phrase of specific information, such as a policy number, account number, or model number.

   Subject: Policy No. 10473A *or* Re: Latham VCR Model 926

   The position of the subject line is not standardized. It may appear to the right of the inside address or salutation; it may be centered on the page several spaces below the inside address; it may be flush with the left margin and

261

Chapter 7
Memorandums
and Letters:
Sending
Messages

several spaces below the inside address; it may even be several spaces below the salutation. (See pages 264, 271, 281.)

3. *Identification Line.* When the person whose signature appears on the letter is not the person who keyboarded the letter, there is an identification line. Current practice is to include in the line only the initials (in lowercase) of the keyboard operator. The identification line is two spaces below the signature and is flush with the left margin of the letter. (See pages 264, 266, 271.)

4. *Enclosure.* When you enclose an item (pamphlet, report, check) with the letter, place an enclosure line usually two spaces below the identification line and flush with the left margin. If there is no identification line, the enclosure line is two spaces below the signature and flush with the left margin. The enclosure line is written in various ways and gives varying amounts of information, as illustrated in these four examples.

Enclosure
Enclosures: Inventory of supplies, furniture, and equipment
            Monthly report of absenteeism, sick leave, and vacation leave
Encl: Application of employment form
Encl. (2)

(See pages 266, 269, 273, 280, 281, 282.)

5. *Copy.* When you send a copy of a letter to another person, place the letter *c* (usually lowercase) or the word *copy* followed by a colon and the name of the person or persons to whom you are sending the copy one space below the identification line and flush with the left margin of the letter. (See page 271.) If there is no identification line, the copy notation is two spaces below the signature and even with the left margin.

    cc: Mr. Jay Longman
    copy: Joy Minor

6. *Personal Line.* The word *Personal* or *Confidential* (capitalized and usually underlined) indicates that only the addressee is to read the letter; obviously, this line should appear on the envelope. It usually appears to the left of the last line of the outside address. The personal line may appear in the letter itself, two spaces above the inside address and flush with the left margin.

7. *Mailing Line.* If you send the letter by means other than first class mail, you may want to note this on both the letter and the envelope. The mailing line in the letter is flush with the left margin and appears after all other notations. (See page 269.)

Delivered by Messenger
Certified Mail

The Envelope

The U.S. Postal Service has guidelines for envelope sizes and addresses. These guidelines allow an Optical Character Reader (OCR), an automated device that reads the address and sorts the mail by ZIP code, to operate efficiently.

Regular business envelopes are a minimum of $3^{1}/_{2}$ inches high and 5 inches long and a maximum of $6^{1}/_{8} \times 11^{1}/_{2}$ inches. Mailable thickness is 0.007 to 0.25 inch.

Envelope sizes regularly used by businesses are classified as All-Purpose ($3\frac{5}{8} \times 6\frac{1}{4}$ inches), Executive ($3\frac{7}{8} \times 7\frac{1}{2}$ inches), and Standard ($4\frac{1}{8} \times 9\frac{1}{2}$ inches). Larger envelopes are available in three standard sizes: $6\frac{1}{2} \times 9\frac{1}{2}$ inches, 9×12 inches, and $11\frac{1}{2} \times 14\frac{5}{8}$ inches.

Regular Parts

The two regular parts on the envelope, the outside address and the return address, are illustrated below.

1. *Outside Address.* The content of the outside address on the envelope is identical with the content of the inside address. The postal service prefers single spacing, all uppercase (capital) letters, and no punctuation marks for ease in sorting mail by an OCR. For obvious reasons, the address should be accurate and complete. The postal service encourages using the nine-digit ZIP code to facilitate mail delivery.
2. *Return Address.* Located in the upper left-hand corner of the envelope, not on the back flap, the return address includes the writer's name (without "Ms.," etc.) plus the address as it appears in the heading. The ZIP code should be included.

Regular Parts of the Envelope and Their Spacing

RETURN ADDRESS	THOMAS G STEIN 80293 HWY 32 SOUTH ELMHURST IL 60126	Except for name and date, same as heading in letter
		Begin outside address slightly left of center
OUTSIDE ADDRESS	Content same as in inside address in letter	RONALD M BENREY ELECTRONICS EDITOR POPULAR SCIENCE 355 LEXINGTON AVENUE NEW YORK NY 10017–0127

Special Parts

In addition to the two regular parts on the envelope, sometimes a special part is needed. The main ones are the attention line, the personal line, and mailing directions.

1. *Attention Line.* An attention line may be used when a letter is addressed to a company rather than to an individual. The wording of the attention line on the envelope is the same as that of the attention line in the letter. On the envelope, the attention line is written directly above the first line of the address.
2. *Personal Line.* The word *Personal* or *Confidential* (capitalized and usually underlined) indicates that only the addressee is to read the letter. The personal line, aligned with the left margin of the return address, appears three spaces below the return address.

3. *Special Mailing Directions.* Mailing directions such as SPECIAL DELIVERY, REGISTERED MAIL, CERTIFIED MAIL, or PRIORITY MAIL are typed in all capital letters below the stamp.

263

Chapter 7
Memorandums
and Letters:
Sending
Messages

Types of Letters

There are many, many types of letters. There are numerous books devoted wholly to discussions of letters. This chapter discusses some of the common types of letters written by businesses and individuals: inquiry, order, complaint and adjustment, collection, job application, and application follow-up. For further discussion of letters, see also Chapter 8 Reports, page 309; for the letter of transmittal, see pages 309–310, 327, and 410.

Letter of Inquiry

A letter to the college registrar asking for entrance information, a letter to a firm asking for a copy of its catalog, a letter to a manufacturing plant requesting information on a particular product—each is a letter of inquiry, or request. To write a letter of inquiry, follow these guidelines.

1. State what you want clearly and specifically. If asking for more than two items or bits of information, use an itemized list.
2. Give the reason for the inquiry, if practical. Remember: If you can show clearly a direct benefit to the company or person addressed, you increase your chances of a reply.
3. Include an expression of appreciation for the addressee's consideration of the inquiry. Usually a simple "Thank you" is adequate.
4. Include a self-addressed, stamped envelope with inquiries sent to individuals who would have to pay the postage themselves to send a reply.

On page 264 is a letter of inquiry written from a company handling student loans to a parent regarding the status of her son's application materials.

SOUTHERN EDUCATORS
LIFE INSURANCE COMPANY

4672 Knightsbridge NW   Richmond, Georgia 30093-3239
Telephone: (404) 449-5267   Toll Free: (800) 221-1012

July 26, 1996

Mrs. Patricia A. Woodson
350 Byrd Road
Florence, NC 28560

RE: 78689, Bruce Eugene Woodson, Jr.

Dears Mrs. Woodson:

Recently we sent you forms and instructions for your son's convenience in applying for a Guaranteed Student Loan (Stafford Loan).

You were to fill out certain sections and then send the forms to the college or university that your son will attend. The college or university was to complete the forms and then send them to us for processing.

We have not received your son's completed forms.

If you have filled in the forms and sent them to the college or university, urge the school to send the forms to us immediately.

If we can assist you, please contact us. Our office hours are 8:30 A.M. to 4:45 P.M., Monday through Friday. Our toll free number is (800) 221-1012.

Sincerely,

Lakisa Khanta

Lakisa Khanta
Student Loan Department

GSL/S38

Order Letter

265

Chapter 7
Memorandums
and Letters:
Sending
Messages

The order letter, as the term implies, is a written communication to a seller from a buyer who wishes to make a purchase. For the transaction to be satisfactory, the terms of the sale must be absolutely clear to both. Since in the order letter you the writer (purchaser) request certain merchandise, it is your responsibility to state clearly, completely, and accurately exactly what you want, how you will pay for it, and how you want it delivered.

In writing an order letter, observe the following guidelines:

1. State clearly, completely, and accurately what you want. If ordering two or more different items, use an itemized list format. Include the exact name of the item, the quantity wanted, and other identifying information such as model number, catalog number, size, color, weight, and finish.
2. Give the unit price and total amounts of the merchandise and the method of payment: check, money order, credit card (give name, number, and expiration date), COD, charge to account if credit is established—give account name and number or corporate PO (purchase order) number.
3. Include shipping instructions with the order. Mention desired method for shipment of the merchandise: parcel post, truck freight, railway express, air express, or the like. If the date of shipment is important, say so.

An example of an order letter is shown on page 266.

The Andrews Company

432 Fielding Avenue • Youngstown, Oregon 97386

May 1, 1995

Woodfield Wholesale Company
300 N. State Street
Braxton, CA 90318

Woodfield Wholesale:

Please send by railway express the following items
advertised in your spring sale catalog:

Quantity	Catalog No	Item	Unit Price	Amount
5	77X1628	Chain Saw, 5 H.P. engine	$195.00	$ 975.00
10	32W5602	Bow and Arrow Set	22.00	220.00
100	11R7587	Golf Club Pac	36.00	3,600.00
			Total	$4,795.00

I am enclosing a company check for the merchandise.
Shipping charges and any applicable tax will be paid
within three days after the merchandise arrives.

Sincerely,

THE ANDREWS COMPANY

Thelma S. Andrews

Thelma S. Andrews
(Mrs. Jacob C. Andrews)

msw

Enclosure: Check for $4,795.00

Printed Order Forms

Many companies provide a printed order form for ordering merchandise. These
forms indicate the specific information needed and provide spaces to fill it in, and
they usually give directions for filling in the form. An example of a filled-in order
form is shown on page 267.

267

Chapter 7
Memorandums
and Letters:
Sending
Messagess

MacWAREHOUSE CAtalog 41.0, reg.

MacWAREHOUSE ORDER FORM

3 Easy Ways to Order!

1. PHONE:
1-800-255-6227
24 hours a day, 7 days a week!
INTERNATIONAL ORDERS: 1-908-370-4779

2. FAX:
1-908-905-9279

3. MAIL:
Use this order form and the postage-paid envelope provided.

UPGRADE ORDERS MAY BE FAXED DIRECTLY TO
1-908-363-4834
Remember to include proof of ownership with your upgrade order.

Order by midnight [E] 1-800-255-6227 for Overnight Delivery Only $3.00!

1   ORDERED BY

CODE M54J30
ACCT # 2351177

JERRY CARR
HINDS COMMUNITY COLLEGE
CAINS HALL
RAYMOND MS 39154

2   PLEASE provide daytime phone number.
601 857 3123
(ORDERS WITHOUT PHONE NUMBERS MAY NOT BE PROCESSED)

3   BILLING ADDRESS
(IF DIFFERENT FROM ABOVE)

NAME _____ COMPANY _____

ADDRESS _____

CITY _____ STATE _____ ZIP _____

4   SHIPPING ADDRESS
(IF DIFFERENT FROM ABOVE)

NAME _____ COMPANY _____

ADDRESS _____

CITY _____ STATE _____ ZIP _____

5   METHOD OF PAYMENT
☒ CHECK ☐ MONEY ORDER ☐ C.O.D.
☐ VISA ☐ MASTERCARD ☐ DISCOVER ☐ AMEX
☐ CORPORATE P.O. #
PLEASE INCLUDE TRADE AND BANK REFERENCES IF YOU DO NOT HAVE AN ACCOUNT WITH US.

Month ____ Year ____
EXPIRATION DATE

SIGNATURE _____

6   YOUR SYSTEM [QUANTITY] LC family ____ Mac II family ____
Centris family ____ Quadra family ____ Performa family ____
Power Macintosh family ____ Powerbook family ____

7   ITEMS ORDERED

PRODUCT NUMBER	PAGE #	PRODUCT NAME	QTY.	COST EA.	TOTAL
ACCO756	147	Carrying case	1	$199	$199
ACCO496	135	Command Center Plus	1	79	79
INPO165	125	A3 Mouse	1	74	74

☐ CHECK TO RECEIVE YOUR POWER USERS™ TOOLKIT (POSTAGE & HANDLING ONLY $2.00)   1 $2.00

8. TOTAL MERCHANDISE		352	352
9. SALES TAX (CT, NJ, and OH add applicable sales tax)		
10. SHIPPING		$3.00	3
11. C.O.D. ORDERS (includes shipping)		$6.00	
12. GRAND TOTAL			$355

13. DO YOU CURRENTLY OWN?

PLAN TO PURCHASE? (WITHIN 1 YEAR)

	CD-ROM	FAX MODEM	SCANNER
DO YOU CURRENTLY OWN?	☐	☐	☐
PLAN TO PURCHASE?	☐	☐	☐

Complaint and Adjustment Letters

Complaint and adjustment letters are in some ways the most difficult letters of all to write. The complaint letter describes what is wrong with a product or service; the adjustment letter is the response.

Frequently the writer of a complaint letter is angry or annoyed or extremely dissatisfied, and the first impulse is to express those feelings in a harsh, angry, sarcastic letter. But the purpose of the letter is to bring about positive action that satisfies the complaint. A rude letter that antagonizes the reader is not likely to result in such positive action. Thus, above all in writing a complaint letter, be calm, courteous, and businesslike. Assume that the reader is fair and reasonable and will consider all information presented. Include only factual information, not opinions; and keep the focus on the real issue, not on personalities.

In writing a complaint letter

1. Identify the transaction (what, when, where, etc.). Include copies (not the originals) of substantiating documents—sales receipts, canceled checks, invoices, and the like.
2. Explain specifically what the problem is. Regardless of how angry or inconvenienced you may feel, don't use obscenities, threats, or libelous statements.
3. State the adjustment or action that you want taken, such as repair, replacement, exchange, or refund.
4. Keep a record of all actions. Keep a copy of every letter you send. For persons you talk with over the telephone, record the person's name along with the date, time, and outcome of the conversation.
5. Send the letter by certified mail. Certified mail costs a few dollars extra, but you have proof that the letter was delivered.
6. Remember that reputable companies want to have satisfied customers and that most respond favorably to justifiable complaints.

In writing an adjustment letter

1. Respond to the complaint letter promptly and courteously.
2. Refer to the complaint letter, identifying the transaction.
3. State clearly what action will be taken. If the action differs from that requested in the complaint letter, explain why.
4. Remember to be fair, friendly, and firm.

About one out of four purchases by individuals of services or goods results in a problem, according to the U.S. Office of Consumer Affairs. Therefore, it is wise to be knowledgeable about writing a consumer complaint letter. The sample on page 269 provides a model for such a letter.

Following on page 270 is a complaint letter written in a business setting. The letter illustrates a common complaint: The purchaser believes there has been an oversight or error in the shipment of goods she ordered for the company. An adjustment letter, replying to the preceding complaint letter, follows on page 271.

Model Consumer Complaint Letter in Modified Block Form   **269**

Chapter 7
Memorandums
and Letters:
Sending
Messages

[Your address
  Your city, state, ZIP code
  Date]

Name of contact person
Title
Name of company
Street address or P.O. Box
City, State, ZIP code

Dear [contact person]:

On [date] I bought [product, brand name, model number,
serial number, or service performed]. I bought [product] at
[name of store, location, city, state] and paid by [method
of payment].

Unfortunately, the [product or service] has not been
satisfactory because [explain the problem].

The problem can be resolved if your company will [state the
action you want taken]. Enclosed are copies of [canceled
checks, receipts, warranties, serial numbers, other
documents to verify the terms of purchase].

I look forward to hearing from you and having this problem
resolved within [set a time limit]. Please contact me at
the above address or telephone me at [give area code and
home and office numbers].

                              Sincerely,

                              [Your signature]

                              [Your name]

Enclosures

Letter sent by certified mail

APPLETON'S, INC.

February 20, 1996

Middleton Manufacturing Company
Alton
Illinois 60139

Middleton Manufacturing Company:

Thank you for your prompt delivery of the ten
refrigerators, invoice #479320, that were ordered on 10
February.

Your catalog indicated that each refrigerator contains five
ice trays. None of these was included in the order we
received. Since your invoice does not show a back order or
a separate delivery of the trays, we believe there may have
been an error in filling the order.

Since we are having to delay our sales promotion until we
receive the trays, please ship them by air express.

Sincerely,

Joan A. Manuel, Manager
Appliance Department

4636 MOCKINGBIRD EXPRESSWAY
BATON ROUGE, LOUISIANA 70806

Adjustment Letter in Modified Block Form on Letterhead Stationery   **271**

Chapter 7
Memorandums
and Letters:
Sending
Messages

MIDDLETON MANUFACTURING COMPANY ALTON, ILLINOIS 60139

24 February 1996

Joan A. Manual, Manager
Appliance Department
Appleton's, Inc.
4636 Mockingbird Expressway
Baton Rouge, LA 70806

     Subject: Refrigerators
         Serial Nos. WS-802945 through WS-802954

Dear Joan Manuel:

Your letter of February 20, concerning ice trays, has been
directed to my attention. The trays are being sent to you
by air express today.

We regret the omission of the ice trays when the appliances
were shipped. This was due to a slight difference in the
assembly schedule caused by the new formula coating on the
ice trays, your refrigerators being the first to have these
improved trays.

We apologize for the inconvenience this caused you and we
appreciate your understanding.

                Sincerely,
                MIDDLETON MANUFACTURING COMPANY

                Patrick Beasley

                Patrick Beasley
                Supervisor, Shipping Division

jm
C: J. T. Reeves, Sales Division, Middleton Manufacturing
Company

Collection Letters

Collection letters are used to collect overdue accounts. They must be firm yet friendly to cause the customer to pay up yet keep the customer's goodwill.

Often collection letters are form letters and often they are a series of letters, starting with a reminder letter and moving through various appeal letters to an ultimatum letter, depending on the response of the customer.

A reminder letter, written early in a collection series, might mention that the unpaid balance is overdue and appeal to the customer's pride. It might include the sentence: "If you have already mailed your payment, please disregard this reminder."

An appeal letter, written later in a collection series, might

1. Ask for payment.
2. Ask about possible problems. Suggest the customer call (include a toll-free number) to discuss any such problems.
3. Remind the customer of the value of good credit.
4. Enclose a return envelope to make payment easier.

A sample letter on page 273 illustrates this type of appeal letter.

The ultimatum letter is the last resort. It may

1. State that payment must be received, often by stating a deadline.
2. Mention previous notices sent.
3. State what action will be taken if payment is not received by the deadline, such as turning the account over to a collection agency or a lawyer.

The sample letter on page 274 is an example of the ultimatum letter.

Collection Letter in Block Form on Letterhead Stationery

273

Chapter 7
Memorandums
and Letters:
Sending
Messages

M&P
APPLIANCES

HIGHWAY 15S
ALBANY, GA 31707
912.601.0043
FAX: 912.601.0050

February 8, 1996

Mr. Walter O. Casey
5701 Honda Lane
Forsyth, GA 39129

Dear Mr. Casey:

We are concerned about your overdue account of $161. During
the past 90 days we have sent three statements, reminding
you of this problem. Since you have always paid bills
promptly, we believe some special circumstances may have
caused this delay.

If you need to make special arrangements for paying your
account balance, we will be happy to work out a
satisfactory payment plan. Or, by sending us a check today
for the $161, you can preserve your excellent credit
rating. A good credit rating is of inestimable value.

Please use the enclosed postpaid envelope to mail us your
check or call us at 800-601-0043 to discuss your account.

Sincerely,

Janet W. Malley

Janet W. Malley
Credit Manager

Enc: postpaid envelope

M&P

APPLIANCES

HIGHWAY 15S
ALBANY, GA 31707
912.601.0043
FAX: 912.601.0050

March 11, 1996

Mr. Walter O. Casey
5701 Honda Lane
Forsyth, GA 39129

Dear Mr. Casey:

You have been given every opportunity to pay your overdue
balance of $161. We do not feel that we can permit any
further delay in payment.

If payment is not received immediately, we will be forced
to turn your account over to a collection agency.

Why not preserve your heretofore excellent credit rating
and avoid an unpleasant situation by mailing your check for
$161 today.

Sincerely,

Janet W. Malley

Janet W. Malley
Credit Manager

Letter of Application

275

Chapter 7
Memorandums
and Letters:
Sending
Messages

The letter of application may well be the most important letter you ever write. Perhaps you seek a grant to complete an important job. Or you seek a scholarship to pay for study or research. Or you want to pursue travel/study opportunities. Or, as a college student, you apply for a part-time job or a summer job. As a family breadwinner, you seek employment. Any one of these goals may require a letter of application and a resumé.

The letter of application typically has three sections:

1. Purpose of the letter
2. Background information
3. Request for an interview

The content in each of these three sections is determined by the audience and the purpose of the letter. Consider the guidelines below.

Purpose of the Letter

In the first paragraph, tell why you are writing the letter. Be specific. Identify the grant, the scholarship, travel/study opportunity, or the job you are applying for, and tell how you found out about it. Be specific. If you read information in a newspaper, periodical, or brochure, give the name and date; if you learned about the opportunity from an individual, give the person's name. Explain the reason for wanting the opportunity. If you have special qualifications, state those qualifications.

Background Information

From the resumé, select those facts that qualify you for what you seek. State that a resumé with more complete information is included and make the proper enclosure notation at the end of the letter. Include information to convince the reader of the resumé that you are the person to hire, that your skills and knowledge will be an asset to a company that hires you or will reflect positively on a school or foundation that awards you a grant, scholarship, or travel/study opportunity.

Request for an Interview

In many instances the purpose of the letter of application and the resumé is to get an interview. Any firm interested in employing an applicant will want a personal interview. In the closing paragraph of the letter of application, therefore, request an interview at the prospective employer's convenience. Should there be restrictions regarding time, however, such as classes or work, say so. If distance makes an interview impractical (living in Virginia and applying for a summer job in California), suggest an alternative, such as an interview with a local representative or sending a tape (audio or video) or a disk.

Be sure to include in the closing paragraph how and when you may be reached. For suggestions on the job interview—preparing for an interview, holding an interview, and following up an interview—see Chapter 10 Oral Communication.

Resumé

Resumés are used for a variety of purposes. You may compose a resumé to apply for employment, grants, scholarships, or travel/study opportunities.

Specific background information is given in a resumé. (See pages 000, 000–000, and 000–000 for examples of resumés.) The resumé is a *full* listing of information concerning the applicant. It may also include illustrative work of the applicant. For example, if applying for a job with an advertising agency, the applicant submits a portfolio of art work, perhaps on interactive disk with multimedia components. An architect includes sample designs. If applying for a creative writing workshop, writers include samples of short stories and poems.

Increasingly job applicants are making their resumés interactive and adding multimedia components. Some resumés blend sound, graphics, animation, and text.

In addition to mailing paper resumés to prospective employers—the typical way to submit resumés—persons looking for jobs today are putting resumés on on-line services such as America Online, CompuServe, and the Internet. To use these services, a person must be a subscriber. Job hunters can also submit multimedia resumés by electronic mail to the Internet's Online Career Center, an employment database.

A word of advice. The resumé with multimedia components is not for everyone. The individual needs a good reason to use the multimedia components. And the potential employer or person who receives the resumé must have the equipment and the technical knowledge to review the resumé. The medium for submitting the message is not the most important aspect; the content must be impressive.

The paper resumé, still the most widely used, is an organized listing of qualifications. Included with the resumé is a letter of application, or cover letter. Keep the resumé concise, since busy people are interested only in necessary information. Some employers want only a one-page resumé. With capability of computers to alter point size, you can get many details on one page.

Remember the purpose of the resumé: to show you are the best possible candidate for the job, the grant, the scholarship, or the travel/study. Plan your resumé, the medium and the content, accordingly.

Information in a Resumé

A resumé contains selected details that show you can perform productively in a specified area of employment. The resumé motivates a potential employer (grant or scholarship committee) to interview you. A resumé includes details about such qualifications as skills, personal qualities, education and training, work history, experience, and accomplishments. It typically shows these details in an organized list with guidewords, or headings, to identify the content of each block of information. As you prepare a resumé, select the details that clearly describe you as the person for the job.

Below are headings and descriptions of details that you might include in a resumé. Remember: Select details that show you are the best applicant.

- Resumé Heading. Every resumé includes heading information. The heading information identifies the preparer of the resumé and gives data for contacting the preparer. Heading information always includes the preparer's name, address, and telephone number; other data may include a fax number, a pager number, and an e-mail address. Sample headings appear below.

277

Chapter 7
Memorandums
and Letters:
Sending
Messages

MARLENE KNOWLES

Present address: Permanent address:

312B Mary Nelson Hall   12 Ashwood Road
Mississippi College   Florence, SC 29501
Clinton, MS 39056   (803) 661-6789
(601) 924-1234

TRAN NGYN

124 Pike Street   (303) 576-6789 - pager
Colorado Springs, CO 80906   (303) 576-0001 - work

MARABETH CONNOLLY
MC@cis.com

21 Bridge Avenue ≈≈ Brooklyn, NY 11217 ≈≈ (718) 636-5432 ≈≈ FAX (718) 636-6789

- Objective or Target. State your immediate objective or target. Examples: management position in ready-to-wear apparel, nuclear engineer—research and development, hospital administrator, executive secretary, radiologist, continued study in physical therapy, research in international banking's history and future, social worker.
- Education. Include such facts as the school name, dates of attendance or year of completion, major and minor fields of study, and highest grade completed or degree awarded. If you include several levels of education, begin with the most recent education and list in reverse chronology.
- Work History or Experience. If new to the job market, include voluntary work as well as paid work. If you have held a number of jobs, divide the information into sections such as experience related to the objective stated in the resumé and other jobs. You may decide to include only the jobs that relate directly to the objective.

Details about work history may include only the name of the employer and the dates of employment, or may include the position and a statement or outline of responsibilities.

For work history, begin with current or most recent employment and list in reverse chronology.

- Skills, Capabilities, Accomplishments. In one or more sections, list any special skills. You can use various headings such as Skills, Capabilities, and Achievements to identify the section. Or you can use specific headings such as Management, Communications, Electricity, Graphic Design, Research, Writing, Consulting, Sales, Editing, and so on.
- Special Training, Memberships, Awards, or Other Specialities. If appropriate for the objective of the resumé, include blocks of information listing special training, memberships, awards, or other specialities. For example, list foreign language skills or experience.
- References. Potential employers and selection committees want references. Sometimes they will ask that you include references. Typically references are not included unless requested.

If you are asked to supply references, select three or four individuals who can speak knowledgeably about your qualifications and character. If you have little or no job experience, you might select former teachers or community leaders. Give the name, occupation, business address, and telephone number of each individual. For resumés included with research grants and advanced study applications, select names of professors and others with whom you have studied or conducted research.

Types of Resumés

A resumé follows no one standard form. Once you have jotted down facts under the topics listed above, you can then decide on the headings you will use and the arrangement of the information. You can arrange the sections of a resumé in several different ways with different headings, depending on the purpose of the resumé.

One type of resumé focuses on a clearly stated objective or target. Following the heading information, you state an objective or a target. Then you select headings and information about your capabilities, accomplishments, education, and work experience that specifically relate to the stated objective or target. This type of resumé is a target resumé. An example appears on page 283.

Another arrangement of sections of the resumé emphasizes functions you can perform or skills you have acquired. For example, use headings such as "Communication," "Supervision," "Budget," "Machines," "Personnel," "Research," and the like. Following the headings, give a summary of activities that helped you become competent at performing a function or helped you develop a skill. Or give specific activities related to each heading. If applicable, give work experience that demonstrates your proficiency in these skills. See an example on page 284.

One approach is to select headings such as "Education," "Work Experience," and "Special Qualifications," as illustrated in the sample resumé on page 285. Arrange the headings so that the most impressive details appear near the top of the page. This type of resumé is a traditional or chronological resumé.

An innovative resumé is the multimedia resumé. The multimedia resumé is designed to showcase samples of the sender's work. It may include such media as paper, disks, or audio and video tapes. The multimedia resumés are useful in applying for jobs in computer-related occupations and in occupations where an individual creates a product. A person applying for a job as a newspaper photographer submits a portfolio of journalistic photography. A person applying for a job as a rock band guitarist includes sample guitar work. A person seeking a scholarship to study classical piano sends sample piano playing.

A good rule of thumb: Select the type of resumé and the content and arrange the information on the page to show your qualifications in the best possible way to the intended audience and for the intended purpose.

Examples of Letters of Application

On the following pages are three letters of application. The resumé that accompanies the first letter is on page 283.

279

Chapter 7
Memorandums
and Letters:
Sending
Messages

3621 Bailey Drive
Big Rapids, MI 49307
1 May 1996

Dale Garrett, Office Manager
Carter Insurance Agency
1531 Jayne Avenue
Big Rapids, MI 49307

Dale Garrett:

Through my computer programming instructor, Mr. Tom Lewis,
I have learned that you have an opening in your department
for a computer services department manager.

On 31 May I will complete my degree in Computer
Programming. While a student I have worked for Mr. Lewis,
Director of Data Processing at John Williams Community
College, and as a night supervisor in Midwestern Bell
Telephone's data processing department. Now I am eager to
begin a full-time job in my field.

The enclosed resumé gives a brief outline of my work and
education. In all the computer-related classes I have
taken, I have been ranked as one of the top three students.
As a first-year computer programming major, I received the
award for the outstanding first-year student in
programming. My natural inclination to enjoy work involving
organization and order and my study at John Williams
Community College provide me with a good foundation as a
programmer.

I will be happy to supply any additional information about
myself. Available afternoons and Saturdays for an
interview, I can be reached at (579) 456-2156. I look
forward to hearing from you.

Sincerely,

Alice M. Rydel

Alice M. Rydel

Enclosure: Resumé

Letter of Application in Modified Block Form **281**

Chapter 7
Memorandums
and Letters:
Sending
Messages

500 Northside Drive G-1
Clinton, KS 66813
1 October 1995

Box 8249-P
c/o <u>The Denver Post</u>
1560 Broadway
Denver, CO 80202

Dear Prospective Employer:

SUBJECT: SECRETARY-ADMINISTRATIVE ASSISTANT POSITION

Your advertisement in the 25 September issue of the <u>Denver</u>
<u>Post</u> has sparked my interest in becoming a member of your
team.

My recent college courses in business plus my years of
experience as a secretary qualify me for the position of
secretary-administrative assistant. My business
communication skills, my tact in working with others, and
my computer keyboarding skills enable me to meet any
situation with confidence and success. Enclosed is my
resumé that gives details about my educational
qualifications and work experience.

Could we discuss the possibility of my joining your
progressive and growing company? I look forward to hearing
from you to schedule an appointment.

Sincerely,

Sheri Rayburn

Sheri Rayburn

Enclosure: Resumé

1515 Maria Drive
Jackson, MS 39204
24 September 1995

Ms. Sara Watkins
Director of Pharmacy
Mississippi Baptist Medical Center
1500 North State Street
Jackson, MS 39206

Dear Ms. Watkins:

Through the 22 September 1995 issue of <u>The Clarion-Ledger</u>,
I learned of your opening for a part-time pharmaceutical
technician. Having previously worked in this capacity, I am
very familiar with the responsibilities and duties
required.

Enclosed is a resumé for your consideration. I believe that
my education and five years of experience in a pharmacy
qualify me to be an effective member of the staff at
Mississippi Baptist Medical Center. In the classroom and on
the job, I have constantly proven my abilities and
strengths. As you will note, as a high school honors
graduate, I worked a part-time job and participated in
many extracurricular activities. I believe this proves my
organizational and time management skills and my desire
to achieve.

I would appreciate the opportunity to speak with you in
person to further discuss my qualifications, your business
objectives, and the talents I can bring to your
organization.

Thank you for your time and consideration. I am available
any afternoon for an interview. I can be reached by phone
at (601) 372-5973.

Sincerely,

Kristi L. Baker

Kristi L. Baker

Enclosure: Resumé

Examples of Resumés

Following are three examples of resumés: a target resumé, a skills resumé, and a
traditional resumé. The first resumé accompanies the job application letter on
page 280.

An Example of a Target Resumé　　　　　　　　**283**

Chapter 7
Memorandums
and Letters:
Sending
Messages

ALICE M. RYDEL
3621 Bailey Drive
Big Rapids, MI 49307
(601) 456-2156

JOB TARGET　　　COMPUTER SERVICES DEPARTMENT MANAGER

CAPABILITIES
- Use MacIntosh and IBM computers with ease and efficiency
- Analyze large amounts of data into organized financial statistics
- Use Lotus and other spreadsheet programs and train others
- Use automated accounting system to produce monthly statements
- Manage workers efficiently and effectively
- Keep accurate records of large numbers of accounts

ACCOMPLISHMENTS
- Supervised daily data input in a 12,000-customer billing department
- Set up a simple but efficient file system for record keeping
- Managed a computer lab available to 200 students
- Devised a plan to schedule students for maximum lab use
- Handled inventory of computer lab and submitted requests for materials, equipment, and maintenance

EDUCATION

1994–1996　　　JOHN WILLIAMS COMMUNITY COLLEGE
　　　　　　　　AAS Degree—Computer Programming

1990–1994　　　MURRAH HIGH SCHOOL
　　　　　　　　Diploma—College preparatory and basic business courses

WORK EXPERIENCE

1994–1996　　　John Williams Community College
Part-time　　　Big Rapids, MI

　　　　　　　　Computer Lab Assistant

　　　　　　　　Midwestern Bell Telephone Company
　　　　　　　　Big Rapids, MI　Billing Department

　　　　　　　　Night Supervisor

AWARDS　　　Data Processing Department Award
　　　　　　　　Outstanding First-Year Student in
　　　　　　　　　Programming
　　　　　　　　Citation for excellence in keyboarding
　　　　　　　　　skills

THOMAS D. DAVIS
tddavis@AOL.COM

1045 Drake Place   Ellisville, MA 01047   (521) 363-2371

SUPERVISION: Directed a crew of 20 machinists. Determined work assignments based on priorities. Found solutions to shop production problems.

COMMUNICATION: Orally passed on orders to machinists. Prepared monthly written reports, such as department reports to an immediate supervisor and reports on budget variances to budget control. Prepared daily written reports, such as reports on discrepancies in product conformity.

PERSONNEL: Interviewed and made recommendations for hiring new personnel. Conducted performance evaluations and made recommendations for raises and promotions.

BUDGET: Prepared and monitored the spending of a half-million-dollar department budget.

MACHINE SKILLS: Can operate all common machine shop tools, such as lathes, milling machines, grinding machines. Can use related measuring tools and gauges.

EMPLOYMENT:

1978—Present   Barron Enterprises, Engineering Division, Nye, MA 01047
Fabrication Superintendent
Processing Supervisor

1973—1978   Always Fabrication, Inc., Patterson, IN 47312
Quality Control Checker
Parts Inspector
Layout and Design Assistant

1971—1973   Bickman Manufacturing Company,Cain, IN 47315
Assembly line worker

1969—1971   U.S. Navy
Machinist

EDUCATION: Cain Community College—A.A., Mechanical Technology

SPECIAL TRAINING:
Indiana Technical Institute
- Quality Control with Computers (45 clock hours)
- New Materials in Industry (45 clock hours)
- Production Planning and Problems (45 clock hours)

An Example of a Traditional Resumé   **285**

Chapter 7
Memorandums
and Letters:
Sending
Messages

JAMES E. BROWN, JR.
206 Davis Drive
Monroe, LA 71201

Home telephone: (318) 948-7660
JB276@MONROEVM.BITNET

EMPLOYMENT OBJECTIVE

Supervising technicians and engineers in an electronics
industry or business with possibility for full management
responsibilities

WORK EXPERIENCE

June 1987–   United States Air Force
Sept. 1991   Electronics/Communications
      • Shift Chief, Long-Haul Transmitter Site.
        Supervised three technicians in operating 52
        transmitters and two microwave systems
      • Team Chief, Group Electronics Engineering
        Installation Agency.
        Supervised three technicians in installing
        weather and comunications equipment
      • Technical Writer. Wrote detailed maintenance
        procedures for electronic equipment manuals
      • Instructor in Electronics Fundamentals.
        Continuous three-month classes of 10 persons

Summers and   Manager, Campus Apartments
part-time   Disc jockey, radio stations KXOA and KXRQ
1991-1995   Laboratory Assistant, Broadcast Department
            Technician for theater productions

EDUCATION

B.S. degree in Industrial Management, Louisiana State
University, May 1995

SPECIAL QUALIFICATIONS

- NARTE—Certified Electronics Technician
- Top Secret Clearance for work in U.S. Department of Defense

COLLEGE ACTIVITIES AND HONORS

Member, Society for the Advancement of Management Professionals
Vice-President, Student Audubon Society
Dean's List
Beta Gamma Sigma National Honor Society in Business
   Administration
Associated Students Service and Leadership Award
Listed in <u>Who's Who in American Colleges and Universities</u>

Job Application Form

Many companies provide printed forms, such as the one on pages 287–292, for job applicants. The form is actually a very detailed data sheet. In completing the form, follow directions, be neat, and answer every question. Some companies suggest that applicants put a dash (—), a zero (0), or NA (not applicable) after a question that does not pertain to them. Doing so shows that the applicant has read the question and not overlooked it.

The completed forms of many applicants look very similar. Therefore, it is the accompanying letter that gives you an opportunity to make the application stand out. For such a letter, follow the suggestions for a letter of application given earlier in this chapter.

Following is a typical form used as an application for employment.

Application
For Employment

287

Chapter 7
Memorandums
and Letters:
Sending
Messages

> We consider applicants for all positions without regard to race, color, religion, creed, gender, national origin, age, disability, marital or veteran status, sexual orientation, or any other legally protected status.

(PLEASE PRINT)

Position(s) Applied For	Date of Application

How Did You Learn About Us?

☐ Advertisement   ☐ Friend   ☐ Walk-In
☐ Employment Agency   ☐ Relative   ☐ Other _____

Last Name	First Name	Middle Name

Address	Number	Street	City	State	Zip Code

Telephone Number(s)	Social Security Number

If you are under 18 years of age, can you provide required proof of your eligibility to work? ☐ Yes ☐ No

Have you ever filed an application with us before? ☐ Yes ☐ No

If Yes, give date _____

Have you ever been employed with us before? ☐ Yes ☐ No

If Yes, give date _____

Are you currently employed? ☐ Yes ☐ No

May we contact your present employer? ☐ Yes ☐ No

Are you prevented from lawfully becoming employed in this country because of Visa or Immigration Status?
Proof of citizenship or immigration status will be required upon employment. ☐ Yes ☐ No

On what date would you be available for work? _____

Are you available for work: ☐ Full Time ☐ Part Time ☐ Shift Work ☐ Temporary

Are you currently on "lay-off" status and subject to recall? ☐ Yes ☐ No

Can you travel if a job requires it? ☐ Yes ☐ No

Have you been convicted of a felony within the last 7 years? ☐ Yes ☐ No
Conviction will not necessarily disqualify an applicant from employment.

If Yes, please explain _____

WE ARE AN EQUAL OPPORTUNITY EMPLOYER

Education

	Name and Address of School	Course of Study	Years Completed	Diploma Degree
Elementary School				
High School				
Undergraduate College				
Graduate Professional				
Other (specify)				

Indicate any foreign languages you can speak, read and/or write			
	FLUENT	GOOD	FAIR
SPEAK			
READ			
WRITE			

Describe any specialized training, apprenticeship, skills and extra-curricular activities.

_____

_____

_____

Describe any job-related training received in the United States military.

_____

_____

_____

Employment Experience

289

Chapter 7
Memorandums
and Letters:
Sending
Messages

Start with your present or last job. Include any job-related military service assignments and volunteer activities. You may exclude organizations which indicate race, color, religion, gender, national origin, disabilities or other protected status.

1.

Employer	Dates Employed		Work Performed
	From	To	
Address			
Telephone Number(s)	Hourly Rate/Salary		
	Starting	Final	
Job Title	Supervisor		
Reason for Leaving			

2.

Employer	Dates Employed		Work Performed
	From	To	
Address			
Telephone Number(s)	Hourly Rate/Salary		
	Starting	Final	
Job Title	Supervisor		
Reason for Leaving			

3.

Employer	Dates Employed		Work Performed
	From	To	
Address			
Telephone Number(s)	Hourly Rate/Salary		
	Starting	Final	
Job Title	Supervisor		
Reason for Leaving			

4.

Employer	Dates Employed		Work Performed
	From	To	
Address			
Telephone Number(s)	Hourly Rate/Salary		
	Starting	Final	
Job Title	Supervisor		
Reason for Leaving			

If you need additional space, please continue on a separate sheet of paper.

List professional, trade, business or civic activities and offices held.

You may exclude membership which would reveal gender, race, religion, national origin, age, ancestry, disability or other protected status.

_____

_____

_____

Additional Information

Other Qualifications

Summarize special job-related skills and qualifications acquired from employment or other experience.

_____

_____

_____

Specialized Skills   **Check Skills/Equipment Operated**

		Production/Mobile Machinery (list):	Other (list):
___ CRT	___ Fax		
___ PC	___ Lotus 1-2-3	_____	_____
___ Calculator	___ PBX System	_____	_____
___ Typewriter	___ WordPerfect	_____	_____
		_____	_____

State any additional information you feel may be helpful to us in considering your application.

Note to Applicants: DO NOT ANSWER THIS QUESTION UNLESS YOU HAVE BEEN INFORMED ABOUT THE REQUIREMENTS OF THE JOB FOR WHICH YOU ARE APPLYING.

Are you capable of performing in a reasonable manner the activities involved in the job or occupation for which you have applied?   ___ YES   ___ NO
A description of the activities involved in such a job or occupation is attached.

References

1. _____   (   ) _____
   (Name)   Phone #

   _____
   (Address)

2. _____   (   ) _____
   (Name)   Phone #

   _____
   (Address)

3. _____   (   ) _____
   (Name)   Phone #

   _____
   (Address)

291

Chapter 7
Memorandums
and Letters:
Sending
Messages

NAME:

POSITION:

DATE: ___ / ___ / ___

FOR PERSONNEL DEPARTMENT USE ONLY

Position(s) Applied For Is Open: ☐ Yes   ☐ No

Position(s) Considered For: _____

_____

Date _____

NOTES:

Applicant's Statement

I certify that answers given herein are true and complete to the best of my knowledge.

I authorize investigation of all statements contained in this application for employment as may be necessary in arriving at an employment decision.

This application for employment shall be considered active for a period of time not to exceed 45 days. Any applicant wishing to be considered for employment beyond this time period should inquire as to whether or not applications are being accepted at that time.

I hereby understand and acknowledge that, unless otherwise defined by applicable law, any employment relationship with this organization is of an *"at will"* nature, which means that the Employee may resign at any time and the Employer may discharge Employee at any time with or without cause. It is further understood that this *"at will"* employment relationship may not be changed by any written document or by conduct unless such change is specifically acknowledged in writing by an authorized executive of this organization.

In the event of employment, I understand that false or misleading information given in my application or interview(s) may result in discharge. I understand, also, that I am required to abide by all rules and regulations of the employer.

_____   _____

Signature of Applicant   Date

FOR PERSONNEL DEPARTMENT USE ONLY

Arrange Interview   ☐ Yes   ☐ No

Remarks _____

_____   _____

INTERVIEWER   DATE

Employed   ☐ Yes   ☐ No   Date of Employment _____

Hourly Rate/

Job Title _____ Salary _____ Department _____

By _____   _____

NAME AND TITLE   DATE

NOTES _____

_____

_____

_____

Application Follow-Up Letters

293

Chapter 7
Memorandums
and Letters:
Sending
Messages

Frequently, after writing and sending a letter of application (pages 275, 280, 281, 282), filling out and submitting a job application form (pages 287–292), or completing an interview (pages 275, 448, 450–455), you may write a follow-up letter. The content of the follow-up letter is determined by events occurring after the application or interview.

The follow-up letter may be a request for a response from the prospective employer. This letter may be written if you were promised a response by a certain date but have not received a response by that date. This letter may also be written if you have sent an application for a job you hope will be available. A letter requesting a reponse may resemble the sample letter on page 295.

The letter may simply thank the interviewer, remind the person of your qualifications, and indicate a desire for a positive response. The body of such a letter appears below.

```
    The interview with you on Wednesday, June 10, was
indeed a pleasant, informative experience. Thank you for
helping me feel at ease.
    After hearing the details about the World Bank's
training program and opportunities available to persons who
complete the program, I am eager to be an employee of World
Bank. I believe my experience working with People's Bank
and my associate degree in Banking and Finance Technology
provide a sound background for me as a trainee.
    I look forward to hearing from you that I have been
accepted as a trainee in the World Bank's training program.
```

The follow-up letter may reaffirm the acceptance of a job and your appreciation for the job. The letter may also ask questions that you have thought of since getting the job, mention the date you will begin work, and make a statement about looking forward to working with the company. The body of such a letter may include the following.

```
    Thank you for your offer to hire me as consulting
engineer for Wanner, Clare, and Layshock, Inc. I eagerly
accept the offer. The confidence you expressed in my
abilities to help the firm improve workers' safety
certainly motivates me to be as effective as possible.
    I look forward to beginning work on Tuesday, July 10,
ready to demonstrate that I deserve the confidence
expressed in me.
```

A follow-up letter may be needed if you decide to refuse a job offer. The body of such a letter may be written as follows.

Thank you for offering me a place in the World Bank's training program. The opportunities available upon completion are enticing.

Since my interview with you, the president of People's Bank, where I have worked part-time while completing my associate degree in banking and finance, has offered me permanent employment. The bank will pay for my continued schooling, allow me to work during the summers, and guarantee me full-time employment upon completing an advanced degree.

By accepting this offer from People's Bank, I can live near my aging parents and help care for them.

I appreciate your interest in me and wish continued success for World Bank.

The follow-up application letter is simply a courteous reponse to events following the application or interview. On page 295 is a follow-up letter requesting a response from a prospective employer.

Application Follow-Up Letter in Block Form   **295**

Chapter 7
Memorandums
and Letters:
Sending
Messages

2121 Oak Street
Fort Collins, CO 80521
July 19, 1996

Mr. William Hatton, Personnel Director
World Bank
20 East 53rd Street
New York, NY 10022

Dear Mr. Hatton:

On June 10 I was interviewed by Mr. John Salman, your
representative, for a place in the World Bank's training
program. Mr. Salman told me that I would be notified about
my application by July 1. Although it is the middle of
July, I have not received any response.

Since I must make certain decisions by August 1, could I
please hear from you about my employment possibilities with
World Bank.

I am, of course, eager to become an employee of World Bank
and hope that I will receive a positive response.

Sincerely,

Jayne T. Mannos

Jayne T. Mannos

General Principles in Writing Memorandums and Letters

1. *Memorandums are typically used for written communication between persons in the same company.* Memorandums are used to convey or confirm information.
2. *A memorandum has only two regular parts: the heading and the body.* The heading typically includes the guide words TO, FROM, SUBJECT, and DATE plus the information. The body of the memorandum is the message. The writer usually initials the memo immediately following the typed "From" name; some writers prefer to give their signature at the end of the memo.
3. *An effective business communication is written on good-quality stationery, is neat and pleasing to the eye, and follows a standardized layout form.*
4. *The six regular parts of a business letter are the heading, inside address, salutation, body, complimentary close, and signature.* The parts follow a standard sequence and arrangement.
5. *A letter may require a special part.* Among these are an attention line, a subject line, an identification line, an enclosure line, a copy line, a personal line, and a mailing line.
6. *The envelope has two regular parts: the outside address and the return address.* A special part, such as an attention line or a personal line, may also be needed.
7. *The content of an effective letter is well organized, has the "you" emphasis, stresses a positive approach, uses natural wording, and is concise.*
8. *Among the most common types of letters written by businesses and individuals are letters of inquiry, order, complaint and adjustment, collection, job application, and job application follow-up.* Each type of letter requires special attention as to purpose, inclusion of pertinent information, and consideration of who will be reading the letter and why.
9. *Job application materials present the qualifications of the applicant, including education, work experience, and references.* This information may be presented as a resumé or on a printed job application form.

Applications in Writing Memorandums

Individual and Collaborative Activities

7.1 Make a collection of at least five memorandums and letters of various types and bring them to class. For each communication, answer these questions. In small groups, analyze each memorandum and letter. In each group, select the most effective letter. Share these with the class, explaining why each example is effective.

  a. What layout form is used?
  b. Is a letterhead used? What information is given in the letterhead?
  c. Is the communication neat and pleasing in appearance? Explain your answer.

297

Chapter 7
Memorandums
and Letters:
Sending
Messages

d. What special parts of a memorandum or letter are used?
e. What is the purpose of the communication?
f. Does the communication have the "you" emphasis? Explain your answer.

7.2   Select one of the communications you collected for 7.1 above. Evaluate the item according to the General Principles in Writing Memorandums and Letters, and any special instructions for this type of communication. Hand in both the communication and the evaluation.

Writing Memorandums

In completing Applications 7.3 and 7.4,

a. Write a preliminary draft of the memorandum.
b. Go over the draft in a peer editing group.
c. Revise. See Part III Revising and Editing and the checklists for revising, front and back endpapers.
d. Write the memorandum.

7.3   Assume that you are the president of an organization. The date and the location of the next regular meeting have been changed. You need to send this information in a memo to the members of the organization.

7.4   Assume that you are an employee in a company. You have an idea for improving efficiency that should lead to a larger margin of profit for the company. Present this idea in a memo to your immediate supervisor. State the idea clearly and precisely, and give substantiating data.

Applications in Writing Letters

7.5   Examine the content and form of the following body of a letter addressed to *Popular Science Digest*. Point out items that keep the paragraph from being clear.

In regard to your article a while back on how to make a home fire alarm system in *Popular Science Digest,* which was very interesting. I would like to obtain more information. Would also like to know the names of people who have had good results with same. Give me where they live, too.

7.6   Rewrite the body of the letter shown in 7.5 above.

In completing Applications 7.7–7.15, 7.17, 7.19–7.20,

a. Write a preliminary draft of the letter.
b. Go over the draft in a peer editing group.
c. Revise. See Part III Revising and Editing and the checklists for revising, front and back endpapers.
d. Write the letter.

Inquiry Letters

7.7   Prepare a letter to a person such as a former employer or a former teacher asking permission to use the person's name as a reference in a job application.

7.8   Write a letter to the appropriate official in your college requesting permission to take your final examinations a week earlier than scheduled. Be sure to state your reason or reasons clearly and effectively.

Order Letters

7.9   Prepare an order letter for an item you saw advertised in a newspaper or magazine.

7.10   Assume that you are the instructor in one of your lab courses and you have been given the responsibility of ordering several different pieces of equipment. Write the order letter.

7.11   Find a printed order form and make out an order on it. It is usually wise to write out the information on a sheet of paper and then transfer it to the order form.

Complaint and Adjustment Letters

7.12   Assume that you ordered an item, such as a pair of shoes, a set of wheel covers, a ring, or a bowling ball, and the wrong size was sent to you. Write a claim letter requesting proper adjustment.

7.13   Write a claim letter requesting an adjustment on a piece of equipment, a tool, an appliance, or a similar item that is not giving you satisfactory service. You have owned the item two months and it has a one-year warranty.

7.14   Write an adjustment letter in response to 7.12 or 7.13 above.

Collection Letters

299

Chapter 7
Memorandums
and Letters:
Sending
Messages

7.15 As credit manager of the local college bookstore, write a series of collection letters to be mailed to students with accounts delinquent for varying periods of time.

Resumés

7.16 Prepare a traditional resumé or a skills resumé or both, as directed.

a. Write a preliminary draft of the resumé.
b. Go over the draft in a peer editing group.
c. Revise. See Part III Revising and Editing and the checklists for revising, front and back endpapers.
d. Write the final copy of the resumé.

Job Application Letters

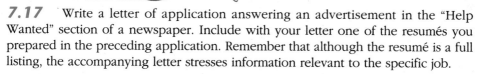

7.17 Write a letter of application answering an advertisement in the "Help Wanted" section of a newspaper. Include with your letter one of the resumés you prepared in the preceding application. Remember that although the resumé is a full listing, the accompanying letter stresses information relevant to the specific job.

7.18 Secure a job application form. (The form used by the U.S. government is especially thorough.) Fill in the form neatly and completely. It is usually wise to write out the information on a separate sheet of paper and then transfer it to the application form.

Job Follow-Up Letters

7.19 Write an application follow-up letter in which you accept the job applied for in Application 7.17.

7.20 Write an application follow-up letter in which you refuse the job offered as a result of the letter in Application 7.17.

Using Part II Selected Readings

7.21 Read "Clear Writing Means Clear Thinking Means . . . ," by Marvin H. Swift, pages 542–547. Under "Suggestions for Response and Reaction," page 547:

a. Answer Number 2.
b. Write Number 3.

7.22 Read "Battle for the Soul of the Internet," by Philip Elmer-Dewitt, pages 500–507. Under "Suggestions for Response and Reaction," page 508, write Number 4.

7.23   Read "Blue Heron," by Chick Wallace, page 550. Under "Suggestions for Response and Reaction," page 551, write Number 7.

7.24   Read "The Effects of Perceived Justice on Complainants' Negative Word-of-Mouth Behavior and Repatronage Intentions," by Jeffrey G. Blodgett, Donald H. Granbois, and Rockney G. Walters, pages 552–562. Under "Suggestions for Response and Reaction," page 563, complete Number 2.

7.25   Read "The Active Document: Making Pages Smarter," by David D. Weinberger, pages 579–584. Under "Suggestions for Response and Reaction," page 584, complete Number 5.

Chapter 8

Reports: Conveying
Needed Information

Outcomes

303

Chapter 8
Reports:
Conveying
Needed
Information

Upon completing this chapter, you should be able to

- Write a definition of reports
- Explain the difference between school reports and professional reports
- Select and use appropriate formats for presenting reports
- Use headings in reports
- Use effective layout and design in reports
- List and identify common types of reports
- Give a periodic report
- Give an observation report
- Give a progress report
- Give a feasibility report
- Give a proposal

Introduction

The word *report* covers numerous communications that fulfill many purposes. You may be concerned about your monthly financial report (a bank statement) or the grade report that you will receive at the end of the term. Or you may be busy polling students about dormitory hours to support your committee's recommendations to the housing council. You may face a due date for a supplementary reading report, or a laboratory report, or a project report.

In the business, industrial, governmental, and corporate worlds, reports are a vital part of communication. A memorandum to the billing department, a weekly production report to the supervisor, a requisition for supplies, a letter to the home office describing the status of bids for a construction project, a performance report on the new computer, a sales report from the housewares department, a report on the availability of land for a housing project, a report to a customer on an estimate for automobile repairs—these are but a few examples of the kinds of reports and the functions they serve.

Scope of Report Writing

Although report writing is a firmly established part of business, industry, government and the corporate world, deciding on the specifics to be taught about report writing is difficult. For one thing, there is no uniformity in report classifications. Depending upon the business, the industry, the particular branch of government, or the textbook, reports may be classified on one or more bases, such as subject matter, purpose, function, length, frequency of compilation, type of format, degree of formality, or method by which the information is gathered. Similarly, there is a lack of uniformity in terminology.

These difficulties underscore the aliveness, the contemporary pace, the elasticity of report writing. Only when report writing has become a relic—serving no practical usefulness—will actual reports fall into well-defined categories.

Below is a report from an architect's firm to a client, Troy Henderson, and a contracting firm, Phoenix Construction Company. The report illustrates the difficulty of classifying reports in neat categories. The sample could be classified as a periodic, observation (field), or progress report.

The report is a periodic report because a similar report is prepared every day to document work done. The report might also be described as an observation (field) report because the writer of the report observed on-site the activities and concerns identified in the report. The report is a progress report because it shows work being done to complete a project—a clubhouse.

Report from an Architect's Firm

Dean/Dale and Dean
architects   a professional association   field report

P.O. Box 4685
Jackson, MS
39216
1301 Mirror Lake Plaza
2829 Lakeland Drive
39208

(601) 939-7717

Project/Project No.   CLUBHOUSE-HINDS COMMUNITY COLLEGE
RAYMOND, MISSISSIPPI
PROJECT NUMBER: 90054

Contact/Supt.:   PHOENIX CONSTRUCTION COMPANY

Date 5 MARCH 1996   Time 7:45 AM   Weather CLOUDY-COOL   Approx. Temp. 60 F

Present at Site   LARRY HOLDEN - BILL DICKEN

Remarks

WORK IN PROGRESS: Concrete

1. Contractor was in the process of pouring final exterior porch slab at north end of building when I arrived.

2. Testing lab personnel had been notified but had not shown up. Finally, got to site approximately 8:30 a.m. Took cylinders from second truck – all appeared per specs.

3. I indicated to Contractor that I had a concern over water soaked spiral columns at bottom. Contractor indicated that he would add additional support to bottom of columns to strengthen this area.

END OF REPORT.

William A. Dicken, Jr.

WILLIAM A, DICKEN, JR.

WADjr : tms

cc: Troy Henderson
Claude Brown

Thus, it is unrealistic to draw sharp boundary lines between types of reports or to try to cover all the situations and problems involved in report writing. However, it is quite realistic and practical for you to:

- Become acquainted with the general nature of report writing
- Develop self-confidence by learning basic principles of report writing. Thus, when you are responsible for writing a report, you can analyze the audience and its need and then present a report to accomplish a stated purpose.
- Study, write, and give orally several common types of reports that you are most likely to encounter as a student and as a future employee.

305

Chapter 8
Reports:
Conveying
Needed
Information

Definition of "Report"

A very basic definition of a report might be this: A report is technical data, collected and analyzed, presented in an organized form. Another definition: A report is an objective, organized presentation of factual information that answers a request or supplies needed data. The report usually serves an immediate, *practical* purpose; it is the basis on which decisions are made. Generally, the report is requested or authorized by one person (such as a teacher or employer) and is prepared for a particular, limited audience.

Reports may be simple or complex; they may be long or short; they may be formal or informal; they may be oral or written. Characteristics such as these are determined mainly by the purpose of the report and the intended audience.

Qualities of Report Content

Reports convey exact, useful information. That information, or content, should be presented with accuracy, clarity, conciseness, and objectivity.

Accuracy
A report must be accurate. If the information presented is factual, it should be verified by tests, research, documentation, authority, or other valid sources. Information that is opinion or probability should be distinguished as such and accompanied by supporting evidence. Dishonesty and carelessness are inexcusable.

Clarity
A report must be clear. If a report is to serve its purpose, the information must be clear and understandable to the reader. The reader should not have to ask: *What does this mean?* or *What is the writer trying to say?* The writer helps to ensure clarity by using exact, specific words in easily readable sentence patterns, by following conventional usage in such mechanical matters as punctuation and grammar, and by organizing the material logically. (See pages 11–18; and Part III Revising and Editing, particularly the sections on Grammatical Usage and Mechanics.)

Conciseness
A report must be concise. Conciseness is "saying much in a few words." Unnecessary wordiness is eliminated, and yet complete information is transmitted. Busy executives appreciate concise, timesaving reports that do not compel them to wade through bogs of words to get to the essence of the matter. Note the example that follows.

WORDY: After all is said and done, it is my honest opinion that the company and all its employees will be better satisfied if the new plan for sick leave is adopted and put into practice.

CONCISE: The company should adopt the new sick leave plan.

Or consider the following report in memorandum format.

Memorandum Lacking Conciseness

```
To:         J. Carraway
From:       T. Jayroe
Subject:    Ideas for revising "Ten Keys to Business Suc-
            cess" 3d ed.
Date:       5 February 1996

It has come to my attention that it is time for us to
consider revising the brochure "Ten Keys to Business Success"
for a 4th edition. Would you be kind enough to take the time
to answer some questions to give me some ideas about the new
edition. Since you have been involved with the brochure since
its first edition, I feel that your ideas are valuable.

First and foremost, do you feel that the 3rd edition has
accomplished its purpose and should we therefore use the
same basic content in the 4th edition? Second, what in your
opinion are the changes we need to make to improve the
brochure? Third and finally, what do you think we should do
about distributing the brochures? Should we continue to
distribute the 3rd edition, or should we wait for a new
edition before distributing any additional brochures?

I need answers to these questions no later than 10 March.
```

Obviously the memo above is wordy. Also the layout of the content could be improved to make reading and comprehension easier. Consider the revised memo below.

Revised Memorandum

```
To:         J. Carraway
From:       T. Jayroe
Subject:    Ideas for revising "Ten Keys to Business
            Success" 3d ed.
Date:       5 February 1996

Please answer the following questions by 10 March.

  1. Should the 4th edition brochure include the same basic
     content?
  2. What changes can we make to improve the brochure?
  3. Should we continue distributing the 3rd edition
     brochures?

Thank you.
```

Note that the revised memo, more concise and more pleasantly designed, conveys the same information as the original memo, but the information can be read more quickly and understood more easily. Revise a report until it contains no more words than those needed for accuracy, clarity, and correctness of expression. For further discussion of conciseness, see pages 14–16.

307

Chapter 8
Reports:
Conveying
Needed
Information

Objectivity

A report must be objective, that is, a report should present data fairly and without bias. Objectivity demands that logic rather than emotion determine both the content of the report and its presentation. The content should be generally impersonal, with no indication of the personal biases and sentiments of the writer. For instance, a report on the comparison of new car warranties for six makes of automobiles ordinarily should not be slanted toward the writer's preferences.

Essential to objectivity is the use of the denotative meaning of words—the meaning that is the same, insofar as possible, to everyone. Denotative meanings of words are found in a dictionary; they are exact and impersonal. Such meanings contrast with the connotative meanings, which permit associated, emotive, or figurative overtones. The distinction between the single denotative meaning and the multiple connotative meanings of a word can be illustrated by examples:

WORD:	war
DENOTATIVE MEANING:	open, armed conflict
CONNOTATIVE MEANINGS:	death, injustice, Vietnam, freedom, cruelty, necessary evil, God vs. Satan, draft, soldiers, high taxes, destruction, bombing, orphans
WORD:	work
DENOTATIVE MEANINGS:	employment, job
CONNOTATIVE MEANINGS:	paying bills, happiness, curse of Adam, accomplishment, 9 to 5, satisfaction, adulthood, alarm clock, fighting the traffic, income, new car, sweat, sitting at a desk, nursing

For further discussion of denotation and connotation, see pages 11–12.

School Reports and Professional Reports

Reports can be classified according to the function they serve: school reports, which are a learning device for students; and professional reports, which are used in business, industry, government, and corporate world.

Both school reports and professional reports are important—but in different ways and for different reasons. School reports, of course, are of more immediate concern to a student because making reports is a widely used learning technique. In the school report it is the student who is important; emphasis is on the student as a learner and as a developer of potential skills. The report shows completion of a unit of study; it reflects the understanding, thoroughness, and intelligence with which he/she has carried out an assignment. The school report also gives students practice in presenting the kinds of reports that their future jobs may require. Such practice can be invaluable in preparing to meet on-the-job demands.

Professional reports—the reports used in business, industry, government, and corporate world—serve a different function from that of school reports. In professional reports, emphasis is on the information that the report contains and on serving the needs of the recipient. The crucial question is this: How well does the report satisfy the needs of the person to whom the report is sent? On the other hand, the instructor who assigns the school report ordinarily neither needs nor will use the information; the instructor is interested in the report only insofar as it reflects the educational progress of the writer.

In both school and professional reports, desirable qualities are accuracy, clarity, conciseness, and objectivity. But you should understand that some of the information given in a school report is not needed in a professional report. For example, in a student investigative report, your instructor may require information on conventional theory and procedure (to be sure that you understand and can explain them). In business, industry, government, and corporate world such information ordinarily would be unnecessary because it is taken for granted that you are knowledgeable about recognized theories and accepted procedures.

Typical Formats for Reports

Most reports (even those presented orally) are put in writing to record the information for future reference and to ensure an accurate, efficient means of transmitting the report when it is to go to several people in different locations. A report may be given in various formats: on a printed form, as a memorandum, as a letter, as nonformal and formal reports, or as an oral report. The format may be prescribed by the person or agency requesting the report; it may be suggested by the nature of the report; or it may be left entirely to the discretion of the writer, who analyzes audience and purpose and then selects an appropriate format.

Printed Form

Printed forms are used for many routine reports, such as sales, purchase requests, production counts, general physical examinations, census information, delivery reports. Printed forms call for information to be reported in a prescribed, uniform manner for the headings remain the same, and the responses—usually numbers or words and phrases—are expected to be of a certain length.

Printed forms are especially timesaving for both the writer and the reader. The writer need not be concerned with structure and organization; the reader knows where specific information is given and need not worry about omission of essential items. However, printed forms lack flexibility: they can deal with only a limited number of situations. Further, they lack a personal touch that provides an opportunity for the writer to express his or her individuality.

In making a report that uses a printed form, the primary considerations are accuracy, legibility, and conciseness. (See the reports on pages 313–314.)

Memorandum

Memorandums and report letters are used in similar circumstances, that is, if the report is short and contains no visual materials. Unlike a report letter, however, the memorandum is used primarily within a firm. (For a detailed discussion of memorandums, see pages 251–253.)

Reports in memorandum formats are illustrated on pages 306, 348–349, 352, and 353–354.

309

Chapter 8
Reports:
Conveying
Needed
Information

Letter

As with the memorandum, the letter is often used for a short report (not more than several pages) that does not include visual materials. The letter is almost always directed to someone *outside* the firm.

The report letter should be as carefully planned and organized as any other piece of writing and should observe basic principles of letter writing. (See Chapter 7 Memorandums and Letters.) The report letter follows conventional letter writing practices concerning heading, inside address, body, salutation, and signature; however, the conventional complimentary close of a report letter is "Respectfully submitted." A subject line is usually included. The report letter often is longer than other business letters and may have internal headings if needed for easier readability. The degree of formality varies, depending on the intended reader and purpose.

Although report letters are widely used, some firms discourage their use to avoid the possible difficulty occasioned if they are filed with ordinary correspondence.

Nonformal and Formal Reports

Conventional report formats include both nonformal and formal reports. Nonformal and formal, vague though often-used terms, refer in actual practice to report length and to the degree to which the report is "dressed up."

Nonformal Reports

The nonformal, or informal, report, usually one page to a few pages in length, is designed for circulation within an organization or for a named reader, and includes only the essential sections of the report proper: an introduction or purpose, the information required for the report, and perhaps conclusions and recommendations.

The nonformal report is far more common than the formal report. Examples of nonformal reports appear on pages 306, 313–314, 324, 336–337, 340–343, 348–350, 353, and 357–360.

Formal Reports

The formal report has a stylized format evolving from the nature of the report and the needs of the reader. Often the formal report is long (from eight to ten up to hundreds of pages), is designed generally for circulation outside an organization, will be read by multiple readers, but will not be read in its entirety by each person who examines it.

Parts. Following is a list of common parts of the formal report and a brief explanation of each part. A writer can combine, omit, or vary these parts to accommodate the intended audience and the purpose of a report.

1. Preliminary Matter
    - Transmittal memorandum or letter—transmits the report from the writer to the person who requested the report or who will act on the report. This

part may include such information as an identification of the report, the reason for the report, how and when the report was requested, problems associated with the report, or reasons for emphasizing certain items. Or this part may communicate a simple message, such as: "Enclosed is the report on customer parking facilities, which you asked me on July 5 to investigate." (For further discussion of letters, see Chapter 7 Memorandums and Letters; for examples of letters of transmittal, see pages 327 and 410.)

- Title page—lists an exact and complete title; usually includes the name of the person or organization for whom the report was prepared, the name of the person making the report, and the date. Arrangement of these items varies. See pages 315, 328, and 411 for sample title pages.
- Table of contents—shows the reader the scope of the report, the specific topics (headings) that the report covers, the organization of the report, and the page numbers. Wording of headings in the table of contents should be exactly as it is in the report. See pages 315 and 413 for sample table of contents.
- List of tables and figures—lists all tables and figures included in the report. Captions for tables and figures in this list are same as those used within the report.
- Summary or abstract—gives the content of the report in a highly condensed form; a brief, objective description of the essential, or central, points of the report. This part may include an explanation of the nature of the problem under investigation; the procedure used in studying the problem; and the results, conclusions, and/or recommendations. The summary is usually no more than 10 percent of the report, but it *represents* the entire report. (For further discussion and examples of abstracts, see pages 223–224, 329, and 414.)

2. Report Proper
   - Introduction—gives an overview of the subject; indicates the general plan and organization of the report, provides background information, and explains the reason for the report. More formal introductions give the name of the person or group authorizing the report, the function the report will serve, the purpose of the investigation, the nature of the problem, the significance of the problem, the scope of the report, the historical background, the plan or organization of the report, a definition or classification of terms, and the methods and materials used in the investigation.
   - Body—presents the information. Includes an explanation of the theory on which the investigation is based, a step-by-step account of the procedure, a description of materials and equipment, the results of the investigation, and an analysis of the results. The body may also include visuals or illustrations.
   - Conclusions and recommendations—deductions or convictions resulting from investigation and suggested future activity. The basis on which conclusions were reached should be fully explained, and conclusions should clearly derive from evidence given in the report. The conclusions are stated positively and specifically and are usually listed numerically in order of importance; the recommendations parallel the conclusions. Conclusions and recommendations may be together in one section, or they may be listed separately under individual headings.

3. End Matter
- Appendix—includes supporting data or technical materials (tables, charts, graphs, questionnaires), information that supplements the text but if given in the text would interrupt continuity of thought.
- Bibliography—lists the references used in writing the report, both published and unpublished material. (For a discussion of bibliography forms, see pages 408–409.)

311

Chapter 8
Reports:
Conveying
Needed
Information

Organization. The various parts of the formal report may be combined or rearranged, depending on the needs of the writer and reader and on the nature of the report. The report is divided into sections, each with a heading. The headings are listed together to form the table of contents.

Examples of formal reports appear on pages 315–319, 327–333, and 410–423.

Oral Report

A student may be assigned a report, such as a reading report, to give as an oral presentation only. However, because of the nature and purpose of reports, most reports will first exist in written form, which allows them to be filed for future reference.

Often reports may be presented in both written and oral forms, the oral presentation emphasizing the major aspects of the report. For example, the treasurer of a small organization might give the latest treasury report orally; the report might include the amount of money in the treasury at the beginning of the previous accounting period, additional income, amount of expenditures, and the total remaining for the next period. The written report, prepared as a record, would probably include these same figures, as well as an itemized listing of sources and amounts of money received and creditors and amounts of money paid out. This report is typically keyboarded. Large corporations send out printed reports, often lengthy and sometimes elaborate, showing profits and losses, dividends paid to stockholders, and decisions regarding future payments of dividends. At the stockholders' meeting, significant aspects of the report may be presented orally by one or more members of the corporation.

Suggestions for giving an oral presentation appear in Chapter 10 Oral Communication.

Visuals

Frequently visuals make information clearer, more easily understood, and more interesting. If a report, therefore, can be made more meaningful for the audience through visual materials, decide what kinds can be used most effectively, and use them.

For a detailed discussion of visuals, see Chapter 11 Visuals.

Headings

- Headings are important in a report. They are an integral part of the page layout and document design and contribute to readability and comprehension of the content. More specifically:

- Headings give the reader a visual impression of major and minor topics and their relation to one another.
- Headings reflect the organization of the report.
- Headings remind the reader of movement from one point to another.
- Headings help the reader retrieve specific data or sections of the report easily.
- Headings make the page more inviting to read by dividing what otherwise would be a solid page of unbroken print.

To be effective, headings must be visually obvious; they must clearly reflect the distinction between major sections and minor supporting sections.

For suggested systems of headings for reports, see pages 9–11. Study the use of headings in the reports on pages 315–319, 324–326, 327–333, 340–343, 348–349, and 353–354. For a discussion of headings in an outline, see pages 357–360.

Layout and Design

Critical to the effectiveness of a report are its page layout and document design. Layout pertains to how the material is laid out, or placed, on the page. There must be ample white space so that the text will be visually appealing. Just as important, material to be emphasized should be indicated through such techniques as uppercase (capital) letters, underlining, boldface, color, boxes, sidebars, and the like. (For detailed discussion, see pages 6–9 and 34–36.)

Design has to do with the principles of visual composition. Careful attention should be given to such matters as report format, choice of visuals, spacing, and kinds and sizes of typefaces. (For further discussion, see pages 9 and 34–36.) The important question is: What layout and design techniques can I use to present my report most effectively for my readers and my purpose?

Common Types of Reports

Among the common types of reports that you may encounter are the periodic report, the observation report, the progress report, the feasibility report, the proposal, and the research report. The following discussion examines each of these types of reports (except the research report, which is discussed separately in the next chapter) as to purpose, uses, main parts, and organizations. Both student and professional reports illustrating the different types are given in several formats (on a printed form, as a memorandum, and in the various conventional report formats).

Periodic Report

Purpose
A periodic report gives information at stated intervals, such as daily, weekly, monthly, or annually, or upon completion of a recurring action. The information may simply be filed as a matter of record or it may be used as a basis for making a decision.

Uses
A daily report of absentees, weekly payroll reports, income tax returns, inventories, a sales report on coverage of a particular geographical area, budget requests, se-

mester grade reports, monthly bank statements, nurses' notes on patients, a company-run report from a fire department unit, an annual inspection of a dam, a bridge, or a building—these are but a few examples of the specific uses that periodic reports serve in practically all phases of business activity.

Most individuals go through the mental process required in periodic reporting in their daily lives. For instance, a parent reviews available clothing for a young child each season. Based upon the review, the parent decides what clothing to buy. Families who are careful with their finances balance their checkbooks and reconcile their bank statements each month to determine if they can make certain purchases.

313

Chapter 8
Reports:
Conveying
Needed
Information

Main Parts

In its simpler, more common forms, the periodic report primarily gives specific measurable quantities, brief responses, and explanations. Data usually are given on a printed form although the report form may vary from a memorandum or a letter to a full-scale, formal report, such as a large corporation's annual report to its stockholders or the United States Corps of Engineers' review after a dam inspection. The periodic report, whatever its form, generally specifies the period of time covered, the subject dealt with, and the pertinent data concerning the subject.

Organization

The organization of a periodic report usually presents little difficulty because of the nature of the material. Typically, the information is presented by categories or chronology (the time sequence in which events occurred). In a firm, the special-purpose periodic report tends to settle into a uniform pattern since it covers the same or similar items each time (thus it permits the use of timesaving printed forms).

Sample Periodic Reports

A periodic report may be as simple and informal as the card below on which medical personnel note periodic changes in a patient's diet.

Periodic Report Form

DIET ORDER AND DIET CHANGES

Patient's Name: _____ Room # _____

Diet Order _____

Change Diet To _____

Serve Late (Meal) _____

Patient Will Be Out: _____

Breakfast ☐ _____ Lunch ☐ _____ Supper ☐ _____

Patient on Pass: From _____ To _____

Move to Room # _____

_____
NURSE'S NAME

_____
DATE

Or it could be a page in a trucker's daily log, a report required of all truckers by the U.S. Department of Transportation. See the following example.

Periodic Report Form

FORM-MCS 59 - Prescribed by the U.S. DEPARTMENT OF TRANS-
PORTATION FEDERAL HIGHWAY ADMINISTRATION Rev. - 67

DRIVER'S DAILY LOG
(ONE CALENDAR DAY –24 HOURS)

Form Approved, Budget Bureau No. 04-R2399
ORIGINAL - File each day at home terminal
DUPLICATE - Driver retains in his possession for one month

RECAP

Day No.

Driving Hrs. Today
Total Line 3

(MONTH) (DAY) (YEAR)   (TOTAL MILEAGE TODAY)

I certify these entries are true and correct: _____
(VEHICLE NUMBERS-SHOW EACH UNIT)

Driving Violation Today

(TOTAL MILES DRIVING TODAY)   (DRIVER'S SIGNATURE IN FULL)

On Duty Hrs. Today
Total Line 3 - 4

(NAME of CARRIER or CARRIERS)   (NAME of CO-DRIVER)

70 HR / 8 DAY DRIVERS

(MAIN OFFICE ADDRESS)   (HOME TERMINAL ADDRESS)

A.

MID-NIGHT 1 2 3 4 5 6 7 8 9 10 11 NOON 1 2 3 4 5 6 7 8 9 10 11   Total Hours

Total Hrs. On Duty Last 7 Days, Incl. Today

1: OFF DUTY

2: SLEEPER BERTH

B.

3: DRIVING

Hrs. Available Tomorrow: 70 Hrs. Minus A

4: ON DUTY (NOT DRIVING)

C.

MID-NIGHT 1 2 3 4 5 6 7 8 9 10 11 NOON 1 2 3 4 5 6 7 8 9 10 11

Total Hrs. On Duty - Last 8 Days, Incl. Today

REMARKS

60 HR / 7 DAY DRIVERS

A.

Total Hrs. On Duty Last 6 Days, Incl. Today

B.

Total Hrs. Available Tomorrow 60 Hrs. Minus A

C.

Shipping document, manifest number, or name of a shipper and commodity. Information required by Section 395.8(o).
Check the time and enter name of place you reported and where released from work and when and where each change of duty occurred.
Explain excess hours - Section 3895.8(o).

Total Hrs. On Duty - Last 7 Days, Incl. Today

FROM: **TO:**

(Starting point or place)   (Destination or turn around point or place)

USE TIME STANDARD AT HOME TERMINAL

© Copyright 1978 & Published by: J.J. Keller & Associates, Inc. – Neenah, Wisc. 54956

The periodic report may also be longer, more detailed, and more formal. Following are selected pages* from a periodic inspection report on the John H. Overton Lock and Dam, located on the Red River and affecting an area from the Mississippi River to Shreveport, Louisiana. The report was prepared by the U.S. Army Corps of Engineers (Vicksburg District). Not included are the appendixes—forty-four pages of supporting reports, nine pages of photographs, and sixty-seven illustrations (plates) detailing every aspect of the dam. Notice in the list of appendixes (see the table of contents page) that an observation report and a trip report are included in the support materials.

*Note: Ellipses (three periods) indicate where original material has been omitted.

U.S. ARMY ENGINEER DISTRICT, VICKSBURG
LOWER MISSISSIPPI VALLEY DIVISION

JOHN H. OVERTON LOCK AND DAM
RED RIVER, LOUISIANA
RED RIVER WATERWAY
LOUISIANA, TEXAS, ARKANSAS, AND OKLAHOMA
MISSISSIPPI RIVER TO SHREVEPORT, LOUISIANA

REPORT OF
PERIODIC INSPECTION NO. 1
MARCH 1989

U.S. ARMY ENGINEER DISTRICT, VICKSBURG
LOWER MISSISSIPPI VALLEY DIVISION
JOHN H. OVERTON LOCK AND DAM
RED RIVER, LOUISIANA
RED RIVER WATERWAY
LOUISIANA, TEXAS, ARKANSAS, AND OKLAHOMA
MISSISSIPPI RIVER TO SHREVEPORT, LOUISIANA
PERIODIC INSPECTION NO. 1
MARCH 1989

Table of Contents

i

JOHN H. OVERTON LOCK AND DAM
RED RIVER, LOUISIANA
PERIODIC INSPECTION NO. 1
16 MARCH 1989

SECTION I – INTRODUCTION

1–01. Authority. Authority for this inspection is contained in ER 1110–2100, "Periodic Inspection and Continuing Evaluation of Completed Civil Works Structures."

1–02. Purpose and Scope. This report presents the results of the first periodic inspection made to ensure the structural integrity and operational adequacy of John H. Overton Lock and Dam. The inspection was held on 16 and 17 March 1989. No features were unwatered for this inspection.

1–03. References. References used in the preparation of this report are listed in Appendix C.

. . .

Periodic Report: Excerpts, cont.

317

Chapter 8
Reports:
Conveying
Needed
Information

SECTION II – PROJECT DESCRIPTION

2–01. <u>General.</u> The project provides for a 9–foot by 200–foot navigation channel extending about 236 miles from the Mississippi River through Old River and Red River to the vicinity of Shreveport, Louisiana. Five locks with dimensions of 84 feet by 685 feet by 14 feet and adjacent dams will provide a lift of about 141 feet. The project also provides for realigning the banks of the Red River from the Mississippi River to Shreveport, Louisiana, by means of dredging, cutoffs, and training works and for stabilizing its banks by means of revetments, dikes, and other methods. Facilities to provide recreation and fish and wildlife development are also an integral part of the project.

2–02. <u>Features Covered by Inspection.</u> The features of the John H. Overton Lock and Dam covered by this inspection were the lock chamber including gate bays and guidewalls, the gated dam, overflow wall, T–wall, and an earthen closure dam.

2–03. <u>Structure Description.</u>

a. <u>General.</u> John H. Overton Lock and Dam was the second one of the locks and dams to be constructed on the Red River. It is located between miles 86.5 and 89.0 (1967 mileage) on the Red River in Rapides Parish, Louisiana, approximately 17 miles downstream from the City of Alexandria, Louisiana. Pertinent data are tabulated in Appendix A.

b. <u>Lock.</u> The lock configuration is a reinforced concrete U–frame supported by a steel H–pile foundation, and consists of an 84–foot by 685–foot (nominal) lock chamber, gate bays with miter gates, culvert type filling and emptying system complete with tainter valves, guardwalls, upper and lower guidewalls, and a control station. The lock guardwall monoliths and the guidewall monoliths are also supported by steel H–piles.

c. <u>Dam.</u> The dam is a reinforced concrete structure consisting of a gated spillway, and adjacent ungated overflow section, and a non–overflow section adjacent to the lock on the landside, all supported by steel H–piles except for the non-overflow section which is supported by prestressed concrete piles. The gated spillway consists of five 60–foot wide by 38–foot high non–submersible tainter gates, individual hoisting equipment for each gate, concrete piers supporting the gates, and a concrete stilling basin with baffle blocks . . . .

d. <u>Closure Dam.</u> An earthen closure dam closes off the former river channel and consists of a semi–compacted clay embankment placed on a hydraulic sand fill base which was placed between two rock dikes.

. . .

<u>SECTION III – DESCRIPTION OF INSPECTION</u>

3–01. <u>Attendance and Agenda.</u>

a. This first periodic inspection of Red River Waterway John H. Overton Lock and Dam was held on 16 March 1989. Members of the inspection team were: [names omitted].

b. Descriptive brochures were prepared and furnished to team members in advance of the inspection.

c. The inspection consisted of briefings covering structure descriptions, problems, and evaluations of observations of engineering measuring devices; a field inspection; and a post–inspection conference discussion. The briefings, field inspection, and conference discussion were all held on 16 March 1989.

d. Findings, evaluations, and proposed actions are presented in the following portions of this report.

e. The upper pool was at elevation 64.0 and the lower pool was at elevation 54.8 on 16 March 1989. The tainter gates were all positioned at an approximate 16–foot gate opening. The structures were not unwatered for this inspection.

3–02. <u>Condition of Structures.</u>

a. Lock. The lock and operating equipment were properly maintained and generally in very good condition (Photos 1–5). Specific items of concern were as follows:

(1) The upstream guidewall has cracks across several shear keys at monolith joints.

(2) There is extensive surface cracking in the cross–over gallery in the upper gate bay monolith and in the top of the riverside lock wall in monolith L–4.

(3) The landside lowering carriage controller is located under the emergency stoplog crane and therefore blocks the view of stoplog movements during emergency operations.

. . .

SECTION IV – CONCLUSIONS AND RECOMMENDATIONS

   4–01. <u>Conclusions.</u> It is concluded from visual inspections, evaluations of engineering measuring devices, and review of design criteria that the Red River Waterway John H. Overton Lock and Dam structures are stable and in safe operating condition subject to the recommendations of paragraph 4–02.

   4–02. <u>Recommendations.</u> The following recommendations were made:

      a. <u>Lock.</u>

         (1) The cracks across the shear keys at monolith joints in the upstream guidewall should be sealed.

         <u>ACTION</u>: Material to seal these cracks has been ordered by project personnel and the sealing of the cracks will be accomplished when the material is received.

         (2) Relocate the landside lowering carriage controller closer to the lock wall.

         <u>ACTION</u>: This has not been done. A work request was initiated for this relocation but was turned down.

         (3) New corrosion protection anodes should be installed on the upstream and downstream miter gates.

         <u>ACTION</u>: Project personnel have new anodes at the jobsite and will install when conditions permit.

         (4) Periodic surveys should be made of the stone protecting the area of the upstream ported guidewall.

         <u>ACTION</u>: Fathometer soundings of this area are made when flow conditions permit and no material has been lost as of yet.

      b. <u>Dam.</u>

         (1) The lightning protection system ground cables on the dam need protection or relocation.

         <u>ACTION</u>: These will be attached by "cadweld" to the bulkhead guides when conditions permit the safe operation of a boat in this area.

         (2) Complete the installation of the automatic tainter gate operation system.

         <u>ACTION</u>: This has been completed.

. . .

4–1

Observation Report

Purpose

The observation report records observable details. It may describe a particular location or site (sometimes called a field report). It may be a collection of information about an existing condition. The report may present results of experimentation, research, or testing. In addition to these activities, major sources of information for the observation report may be personal observation, experience, or knowledgeable people. (See Chapter 10 Oral Communication, pages 455–456, for how to conduct an informational interview.)

Uses

An observation report may be required in any number of areas: business, chemistry, physics, computer science, nursing, electronics, real estate, fire and police science, marketing, agriculture, medicine, construction, environment, forestry and wildlife, and so on. The observation report can be used in many ways. For example, such a report may be important in estimating the value of real estate or the cost of repairing a house; establishing insurance claims for damage from a tornado or a blizzard; improving production methods in a department or firm; choosing a desirable site for a building, highway, lake, or computer lab; or providing an educational experience for a prospective employee or an interested client. The observation report may include the results of tests, such as a blood type test. Some reports may become quite involved, not only describing the experiment but also giving test results and applying the results to specific problems or situations. The observation report may or may not include recommendations.

Main Parts

Since the observation report has a variety of uses and includes various kinds of information, it has no established divisions or format. The report may include such parts as a review of background information, an account of the investigation or description of observed activity or surroundings, an analysis and commentary, and conclusions and recommendations. For a student (or any other interested person), an observation report on a visit to a company might include a description and explanation of physical layout, personnel, materials, and equipment; the individual activities that carry out the major function of the company; and evaluative comments.

An observation report that focuses on experimentation might include such sections (and headings) as "Object" or "Purpose," "Theory" or "Hypothesis," "Method" or "Procedure," "Results," "Discussion of Results" or "Comments," and "Conclusions/Recommendations." A longer report may include "Appendixes" and "Original Data."

Organization

The organization of the report may vary depending on the subject of the report, the purpose, and the audience. The beginning of the report may state the purpose of the report; the specific site, facility, or division observed; and the aspects of the subject to be presented. Next may follow results of the investigation, conclusions, and recommendations. If the report focuses on experimentation, headings such as those listed above in the second paragraph under Main Parts may be combined or rearranged to suit the needs of the writer and the reader.

Sample Observation Reports

321

Chapter 8
Reports:
Conveying
Needed
Information

Two examples of observation reports follow. The first (preceded by the filled-in Plan Sheet) is a report on a visit to an architect's firm. The second report—a more formal report that includes such preliminary matter as a transmittal letter, title page, and summary—concerns the field inspection of a pond dam.

PLAN SHEET
FOR GIVING AN OBSERVATION REPORT

Analysis of Situation Requiring a Report

What is the subject of my report?
observations of offices of Cooke-Douglass-Farr-Lemons, Ltd., Architects-Engineers-Planners

Who is my intended audience?
persons interested in working as architects

How will the intended audience use the report? For what purpose?
to learn about the physical layout, personnel, materials/equipment, and activities in an architectural firm

What format will I use to present the report?
random listing of topics

When is the report due?
15 September 1996

Object or Purpose of the Report

to become familiar with the layout, activities, and equipment and materials in a typical architectural firm

Theory, Hypothesis, Assumption

omit

Method or Procedure

on-site, personal observation

Results; Observations; Comments

Physical layout: attractive reception area, modern telecommunications equipment; glass-walled conference room with modern sofas and coffee table to provide a personal atmosphere for meeting with clients (room also includes boardroom-style meeting table to accommodate up to 18); accounting office for financial operations; specification-writing department where building specifications and construction documents are written and prepared, housed in open rectangular space subdivided into modular partitions of ten-by-twelve-foot offices for designers and drafters, one wall entirely of glass; larger offices for partners and senior architects

Personnel

approximately 30 employees from administrators to senior staff members. Mostly designers and drafters. Also engineers, architects, and support personnel

Materials and equipment

Paper in all sizes, colors, and types for sketching, typing, CAD; special paper for diazo, blue-line, black-line, sepia, and film reproductions; model construction materials such as foamcore board, various weights of cardboard, plywood, styrofoam; assorted glues and adhesives; paint of all colors, types, and textures; all types of writing and drawing instruments plus standard office supplies (staplers, paper clips, rubber bands, etc.); computer-aided drafting equipment, photoimaging equipment, VCR's, computers, word processors, bookbinding machinery, file cabinets, ammonia-aided blue-line drawing machines

Activities

assembly line activity to produce a product, beginning with a senior architect's feasibility report and a schematic design to fill client's need; then a team is assembled including junior architects, civil engineers, electrical engineers, mechanical engineers; architectural designer, drafters, and interior designer; construction inspectors, model builders, and specification writers; and administrative office personnel

Conclusions; Recommendations

omit

Types and Subject Matter of Visuals I Will Include

omit

The Title I May Use

An Architect's Workplace

Sources of Information

visit to the offices of Cooke-Douglass-Farr-Lemons, Ltd.

An Architect's Workplace

James R. Bailey

On 8 September 1996, I visited the offices of Cooke-Douglass-Farr-Lemons, Ltd., Architects-Engineers-Planners in Jacksonville, Florida. As an architecture major, I had arranged with the office manager to tour a place of employment similar to one where I would like to work when I receive my degree in several years. I observed details about the physical layout of the offices, the personnel, the materials and equipment, and the activities.

Physical Layout

The attractive reception area invites clients into the offices of Cooke-Douglass-Farr-Lemons, Ltd. This area has a variety of modern telecommunications equipment. Down the hallway are two typical office spaces. Next is a glass—walled conference room. In this room is a grouping of two modern sofas and a sculpted wood and glass coffee table. In the conference room, informal meetings with clients take place in a more personal atmosphere. The conference room contains a boardroom-style table where up to eighteen people may conduct business and teleconference calls. The next office is the accounting office where daily financial operations are conducted.

Adjacent to the accounting office is the specifications-writing department. In this department all building specifications and construction documents are written and prepared. The work takes place in an open rectangular room that is subdivided with modular partitions into ten-by-twelve-foot work spaces for the designers and drafters. Lining the exterior walls are larger offices for the partners and senior architects. The north exterior wall is glass from ceiling to floor; this gives a feeling of airiness and spaciousness and provides the office with natural light.

Observation Report, Cont.

325

Chapter 8
Reports:
Conveying
Needed
Information

As I was shown the offices, I made a sketch of the physical layout:

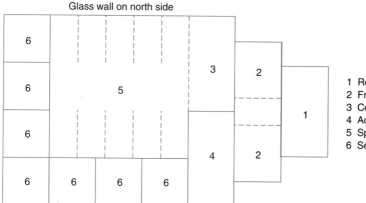

Glass wall on north side

1 Reception area
2 Front offices
3 Conference room
4 Accounting
5 Specifications writing
6 Senior offices

Personnel

Some thirty persons are employed in various capacities. Positions range from office staff to senior partners. The majority of the persons are designers and drafters. The remaining are divided into three groups: architects, engineers, and support personnel.

Materials and Equipment

Within an architect's office is an assortment of materials used on a day-to-day basis. They include paper in all sizes, colors, and types used for sketching, keyboarding, and CAD. Special types of paper are used for various styles of reproduced drawings—i.e., diazo, blue-line, black-line, sepia, and film. Materials used in the construction of architectural models are also available: foamcore board, assorted weights of cardboard, plywood, styrofoam; assorted glues and adhesives; paint of all colors, types, and textures; and an array of other materials to provide realistic touches to the models.

Any type of writing or drawing instrument is available
including computer-aided drafting equipment, photoimaging
equipment, VCRs, computers, word processors, and ammonia-
aided blue-line machines that develop copies of drawings
(blueprints). Other equipment includes VCRs, bookbinding
machines, and large file cabinets. As in any other office,
there are the standard office supplies such as staplers,
paper clips, rubber bands, and the like.

Team Activities

An architect's office is extremely active, with many
individuals working on different sections of a project to
produce a composite result.

Jeffrey Palmer, one of the architects, explained a typical
approach to a project. Just as a car comes off an assembly
line in Detroit, a plan for a multi-storied office building
comes off the "assembly line" in an architect's office. At
the start of the project, the senior architect develops a
feasibility report along with a schematic design to fulfill
a client's needs. Next, the senior architect assembles a
team of key players, including junior architects, civil
engineers, electrical engineers, mechanical engineers, an
architectural designer, several drafters, and an interior
designer. Others involved include various designers,
construction inspectors, model builders, specification
writers, and administrative office personnel.

No project can be completed without the cooperation of all
departments; architecture requires teamwork.

Observation Report, Letter of Transmittal

327

Chapter 8
Reports:
Conveying
Needed
Information

1135 Combs Street
Jackson, MS 39204
6 April 1995

Mr. Harry F. Downing
4261 Marshall Road
Jackson, MS 39212

Dear Mr. Downing:

Attached is the field inspection report of the pond dam on
your property. The dam is considered to be in stable
physical condition although some minor seepage and erosion
were discovered. Recommendations for correcting these are
included in the report.

It has been a pleasure to work with you on this project.

Sincerely yours,

Paul Kennedy

FIELD INSPECTION REPORT
OF
HARRY F. DOWNING DAM

Paul Kennedy
6 April 1995

SUMMARY

On 2 April 1995 an unofficial inspection of the dam of
the Harry F. Downing Pond was conducted by Paul Kennedy.
This dam is considered to be in stable physical condition
although some minor seepage and erosion were discovered.
This conclusion was based on visual observations made on
the date of the inspection.

FIELD INSPECTION REPORT

HARRY F. DOWNING POND
HINDS COUNTY, MISSISSIPPI
PEARL RIVER BASIN
CANY CREEK TRIBUTARY
6 APRIL 1995

PURPOSE

The purpose of this inspection was to evaluate the structural integrity of the dam of the Harry F. Downing Pond, which is identified as MS 1769 by the National Dam Inventory of 1973.

DESCRIPTION OF PROJECT

Location. Downing Pond is located two miles SE of Forest Hill School, Jackson, Mississippi, in Section 23, Township 6, Range 1 East (see Figure 1).

Hazard Classification. The National Dam Inventory lists the location of Downing Pond as a Category 3 (low-risk) classification. Personal observation of areas downstream confirm this classification since only a few acres of farmland would be inundated in the case of a sudden total failure of the dam.

Description of Dam and Appurtenances. The dam is an earth fill embankment approximately 200 feet in length with a crown width of 6 to 10 feet. The height of the dam is estimated to be 16 feet with the crest at Elevation 320.0 M.S.L. (elevations taken from quadrangle maps). Maximum capacity is 47 acre feet. The only discharge outlet for the pond is an uncontrolled overflow spillway ditch in the right (east) abutment. The spillway has an entrance crest elevation of 317.0 M.S.L. and extends approximately 150 feet downstream before reaching Cany Creek. The total intake drainage area for the pond is 30 acres of gently rolling hills.

Observation Report, Cont.

331

Chapter 8
Reports:
Conveying
Needed
Information

Design and Construction History. No design information has been located. City records indicate that the dam was constructed in 1940 to make a pond for recreational purposes.

FINDINGS OF VISUAL INSPECTION

Dam. Apparently the dam was constructed with a 1V or 2H slope. This steep downstream slope is covered with dense vegetation, which includes weeds, brush, and several large trees. These trees range from 15 to 20 feet in height (see Photos 1 and 2). These trees have not likely affected the dam at the present time, but decaying root systems may eventually provide seepage paths.

A normal amount of underseepage was observed about halfway along the toe of the dam (see Photo 3). This seepage was not flowing at the time of the inspection but should be watched closely during high-water periods. The upstream face of the dam has several spots of erosion near the water's edge due to the lack of sod growth. Apparently topsoil was not placed after construction.

Overflow Spillway. The uncontrolled spillway shows no signs of erosion and is adequately covered with sod growth (see Photo 4).

RECOMMENDATIONS

It is recommended that the owner:

1. Periodically inspect the dam (at least once a year).
2. Prevent the growth of future trees on the downstream slope.
3. Install a gage and observe the flow of underseepage as compared to pool levels.
4. Fill areas of erosion and place topsoil and sod to prevent future erosion.

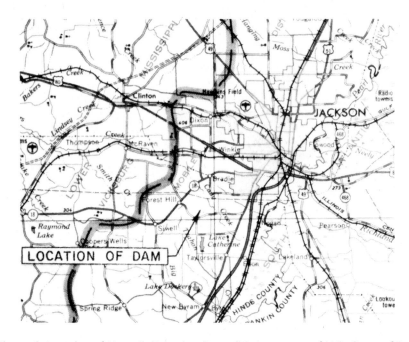

Figure 1. Location of Harry F. Downing Dam. (Map courtesy of U.S. Corps of Engineers Waterways Experiment Station at Vicksburg)

Observation Report, Cont.   **333**

Chapter 8
Reports:
Conveying
Needed
Information

Photo 1. View looking east from left (west) abutment. Note large trees on dam at left.

Photo 2. View looking north at dam from inlet area. Note large trees on dam in background.

Photo 3. View looking west along toe of dam. Note seepage at left.

Photo 4. View looking upstream of un-controlled overflow spillway ditch. Note adequate sod growth.

Progress Report

Purpose

The progress report gives information concerning the status of a project currently under way. In effect, progress reports enable large businesses or organizations to keep up with what is happening within the business or organization. Your instructor may ask you to prepare a progress report as you work on a project.

Uses

Students and employees use a progress report to describe investigations to date, either at the completion of each stage or as requested by a supervisor. For the student, the progress report can signal the teacher that assistance or direction is needed. In industry and business, the progress report keeps supervisory personnel informed so that timely decisions can be made accordingly.

Progress reports answer a variety of questions. How much work toward completing the project has been accomplished? How much money has been spent? How much money is still available? Is it enough to complete the project? Is the project on schedule? Are changes needed? In plan? Performance? Method? Expenditures? Personnel? What unforeseen problems have arisen? In effect, the progress report should answer any questions about the progress (How far have we come?), the status (Where are we now?), and the completion of the project (What remains to be done?).

Main Parts

The progress report introduces the subject, describes work already completed on the project, discusses in detail the specific aspects that are currently being dealt with, and often states plans for the future. Unexpected developments or problems encountered in the investigation may be collected in one section under a heading such as "Problem" or "Unexpected Development" or at the points in other parts of the report where they logically arise. A progress report might include recommendations for change in the plan or procedure. If such recommendations are made, they must be supported by reasons and an explanation of how the changes will affect the project. The recommendations would likely appear under the heading "Recommendations." In a very long report, the recommendations would likely be presented at or near the beginning of the report, since they may be the most important information in the report.

Organization

Obviously a progress report could have a variety of audiences. Supervisors might need to know if more workers will be needed to complete the project. Engineers might need to know if a design should be altered. Students might need to know if they need to allot more time to a major project. A doctor might need to change the medicine or the therapy prescribed for a patient. The sales force for a company might need to change their sales strategy.

The three parts of the report (previous work, current work, future plans) form a natural, sequential order for presenting the information. Indeed, chronology or sequence is commonly used in organizing progress reports (see the report on pages 340–343). Another possible order for presenting progress reports is activity (see the report on pages 336–337).

When progress reports are prepared for use within a business or organization, the content is often presented in memorandum form or on a preprinted form. Other commonly used formats are the business letter and the short report.

Sample Progress Reports

Two examples of progress reports follow. The first describes for subcontractors the status of key aspects of a construction project. Note that this report is titled a Conference Report because the progress on the project was discussed in a meeting or a conference. The second report (preceded by the filled-in Plan Sheet) was written by a student for her major professor, who is recommending students for employment. Notice that both reports include specific data, not generalities.

335

Chapter 8
Reports:
Conveying
Needed
Information

Conference Report

Meeting [X]
Telecom [ ]

Project: GOLF CLUBHOUSE, EAGLE RIDGE GOLF COURSE, HCC-90054

Date of Conference: 17 AUGUST 95 Location: OFFICES OF TROY
HENDERSON,
HINDS COMMUNITY
COLLEGE,
RAYMOND

Purpose: REVIEW CURRENT STATUS PROJECT
DRAWINGS

Participants: TROY HENDERSON
THOMAS WASSON
ANTHONY PRICE
RON HARTLEY
KENNY OUBRE

The following items were discussed on the date above. In
the interest of the project's schedule, if we do not hear
from you within 72 hours, we will proceed on the basis that
this report meets your approval. Ron and Kenny presented
current status plans including: revised schematic site
plans, floor plans, elevations, various building details
and sections, and mechanical and electrical plans. The
following items or actions were noted as a result of review
or discussion:

1 Two (2) handicap parking spaces are adequate.
2 Existing septic tank to be removed.
3 Exterior golf bag racks to be built by Hinds Community
College.
4 No exterior drinking fountain.
5 Construction will be scheduled so that the base for
roads and parking is installed during fall of 1995 and
not in spring of 1996.

Dean/Dale and Dean Architects

Progress Report, Cont.

337

Chapter 8
Reports:
Conveying
Needed
Information

NO.	BY	ITEM	ACTION	BY	DATE
6	TH	Landscape Plan—being constructed by "in-house" forces—not part of bid package.	Contact Thad Owns to determine fees.	RH	
7	TH	Concrete curb & gutter—concern with cost.	Consider asphalt.	DDD	
8	TH	Sewage treatment—location relative to clubhouse and outdoor activities and driving range tee.	Make sure location is sufficient distance from activity areas, but serviceable.	DDD	
9	TH	Eagle weather vane.	Consider adding to roof.	KO	
10	TH	Burglar Alarm—may be required for insurance on retail goods.	Evaluate cost of minimal system of motion detectors.	KO	
11	TH	Bid Alternates—prefer that roof not be an alternate.	Provide alternates in bid form to help assure awardability—provide list to HCC.	DDD	
12	TH	Cost Estimate.	Provide itemized cost estimate as soon as possible.	DDD	
13	TW	Location of clubhouse.	Provide drawing for layout and final approval of building location.	DDD	8/20
			Stake out and review—notify Ron of decision.	TW	8/20
14	TH	Review of HVAC.	Review design for HVAC & mechanical staff	TH	
15		Changes to floor plan or details.	None noted at meeting—present ASAP.	HCC	

END OF REPORT.

RONALD L. HARTLEY, ASLA
RLH: wkb

cc: Participants
        Richard Dean
        Chris Hoffman

                        Dean/Dale and Dean Architects

PLAN SHEET FOR GIVING A PROGRESS REPORT

Analysis of Situation Requiring a Report

What is the subject of my report?
work toward a degree in aircraft mechanics

Who is my intended audience?
instructor of aircraft mechanics

How will the intended audience use the report? For what purpose?
to evaluate my suitability for employment

What format will I use to present the report?
chronological listing of classes, grades, and quality points

When is the report due?
2 October 1995

Purpose of the Report

to show courses, grades, and quality points toward an Associate in Applied Science degree (aircraft mechanics) from Hinds Community College

Work Completed; Comments

1992 Spring Semester

Courses	Hours	Grades	Quality Points
ENG 1113 English Comp I—Technical	*3*	*A*	*12*
PSY 1513 General Psychology I	*3*	*A*	*12*
TBT 1113 Elementary Typing I	*3*	*B*	*9*
TBT 1213 Elementary Shorthand I	*3*	*A*	*12*
TBT 1313 Records Management I	*3*	*A*	*12*
TBT 1413 Business Mathematics	*3*	*A*	*12*
	18		*69*

3.833 GPA
Did not find courses challenging but found myself encouraged by grades; changed major to a two-year aircraft mechanics certificate program.

1993–1994 Semester I

Courses	Hours	Grades	Quality Points
VAP 1138 Powerplants I	*8*	*A*	*32*
VAP 1148 Powerplants II	*8*	*A*	*32*
	16		*64*

Semester II

Courses	Hours	Grades	Quality Points
VAP 2118 Powerplants III	*8*	*A*	*32*
VAP 2128 Airframe I	*8*	*A*	*32*
	16		*64*

Work Completed; Comments, cont.

4.0 GPA. Classes were challenging. Enjoyed Airframe I, working with wood, cloth, glue, and paint.

1994–1995 Semester I

Courses	Hours	Grades	Quality Points
VAP 2138 Airframe II	8	A	32
VAP 2148 Airframe III	8	A	32
	16		64

Semester II

Courses	Hours	Grades	Quality Points
VAP 1118 General Theory and Maintenance I	8	A	32
VAP 1128 General Theory and Maintenance II	8	A	32
	16		64

4.0 GPA. Completed certificate program. Greatest accomplishment.

Work in Progress; Comments

1995–1996 Semester I
Currently enrolled in courses to earn an associate degree.

Courses	Hours
ENG 1123 English Comp II—Technical	3
PHY 2253 Physical Science II	3
Physical Science Lab	
SOC 1113 Introduction to Sociology	3
	9

Work to Be Completed; Comments

Semester II

Courses	Hours
PHI 1113 Old Testament Survey	3
MAT 1213 College Mathematics	3
	6

At semester's end will have the 79 hours and more than the required 158 quality points for the associate degree.

Types and Subject Matter of Visuals I Will Include

table showing work completed for Spring 1992, 1993–1994 Semester I and II, 1994–1995 Semester I and II
table showing work in progress for 1995–1996 Semester I
table showing work to be completed for 1995–1996 Semester II

The Title I May Use

My Work as a Student at Hinds Community College: Purpose, Progress, and Projections

Sources of Information

college catalog; semester grade reports

My Work as a Student at Hinds Community College: Purpose,
Progress, and Projections
Laura McCain
2 October 1995

In the Spring of 1992 I enrolled at Hinds Community
College. Unsure of what I wanted to do, I took general
business courses. In August of 1993 I enrolled in the
aircraft mechanics program to obtain the hours required to
take the FAA mechanics examination and receive a vocational
certificate of completion. The program requires a minimum
of 64 semester hours (128 quality points) or 1,920 clock
hours. After completion of the program, I enrolled in
classes leading to an Associate in Applied Science degree,
specializing in aircraft mechanics. The degree program
requires 15 hours (30 quality points minimum) of general
education in addition to a certificate of completion in a
vocational program. I am in my sixth semester, working on
the last 15 hours required to obtain the associate degree.

1992 Spring Semester

During the spring semester of the 1992 academic year I
completed the following courses with the stated hours,
grades, and quality points.

COURSES	HOURS	GRADES	QUALITY POINTS
ENG 1113 English Composition I— Technical	3	A	12
PSY 1513 General Psychology I	3	A	12
TBT 1113 Elementary Typing I	3	B	9
TBT 1213 Elementary Shorthand I	3	A	12
TBT 1313 Records Management I	3	A	12
TBT 1413 Business Mathematics	3	A	12
	18		69

These courses were the suggested courses for a general
business major. My grade point average (GPA) for the
semester was 3.833 on a 4.0 scale. The success of my first
semester encouraged me to continue my education, but I was
not challenged by the general business major. I then
enrolled in the aircraft mechanics program.

Progress Report, Cont.

341

Chapter 8
Reports:
Conveying
Needed
Information

1993–1994

I completed these courses with the following hours, grades, and quality points.

Semester I

COURSES	HOURS	GRADES	QUALITY POINTS
VAP 1138 Powerplant I	8	A	32
VAP 1148 Powerplant II	8	A	32
	16		64

Semester II

COURSES	HOURS	GRADES	QUALITY POINTS
VAP 2118 Powerplant III	8	A	32
VAP 2128 Airframe I	8	A	32
	16		64

These were the required courses for the first year of the two-year certificate program in aircraft mechanics. My quality point average for the year was 4.0. These classes were very challenging but my mechanical aptitude helped me succeed. I enjoyed the Airframe I course more than the Powerplants III course because it required working with wood, cloth, glue, and paint. I had not had much experience with these materials before the course.

1994–1995

During the first semester of the 1994–1995 academic year I completed the required courses with the following hours, grades, and quality points.

Semester I

COURSES	HOURS	GRADES	QUALITY POINTS
VAP 2138 Airframe II	8	A	32
VAP 2148 Airframe III	8	A	32
	16		64

Semester II

COURSES	HOURS	GRADES	QUALITY POINTS
VAP 1118 General Theory and Maintenance I	8	A	32
VAP 1128 General Theory and Maintenance II	_8_	A	_32_
	16		64

At the end of this school year I had earned a certificate in the aircraft mechanics program with an overall 4.0 quality point average. The completion of this certificate program is one of my greatest accomplishments.

1995–1996

During the 1995–1996 academic year I am currently enrolled in courses to earn an Associate in Applied Science degree, specializing in aircraft mechanics. I am taking the following courses.

Semester I

COURSES	HOURS
ENG 1123 English Composition II—Technical	3
PHY 2253 Physical Science II and Physical Science Lab	3
SOC 1113 Introduction to Sociology	_3_
	9

During the second semester of the 1995–1996 academic year I plan to take the additional courses required for an associate degree.

Semester II

COURSES	HOURS
PHI 1113 Old Testament Survey	3
MAT 1213 College Mathematics	_3_
	6

Progress Report, Cont.

343

Chapter 8
Reports:
Conveying
Needed
Information

Upon completion of this semester, I plan to have more than the required 79 hours and the 158 quality points to obtain an associate degree.

My Future

I intend to graduate from Hinds Community College as an aircraft mechanics specialist with an Associate in Applied Science degree. When I complete my degree, I will be qualified to take the FAA mechanics exam and to enter the aviation maintenance field in one of several positions.

Feasibility Report

Purpose

The feasibility report is a systematic analysis of what is possible and practical, of what can be accomplished or brought about. The feasibility report offers answers to such questions as these: Should we do this? Which of these choices should I select? Analysis for a feasibility report is very similar to analysis in an effect-cause or comparison-contrast situation. (See Chapter 5 Analysis Through Effect-Cause and Comparison-Contrast. See especially Problem Solving, pages 191–192.)

Uses

Feasibility reports are frequently used in business, industry, government, and corporate world. Their data serve as the basis for significant decisions. Should our company move its central headquarters from this geographical area to another part of the country? Should we rent, lease, or buy our equipment? What is the best location in this city for my fast-food franchise? Which investment company can meet the needs of my client? Should we spend the estimated $20,000 to repair our company cars, or should we buy a new fleet? Which marketing strategy should I adopt? Whatever the situation, the feasibility report gives a complete, accurate analysis of the possibilities and presents recommendations.

Main Parts

Typically the main parts of the feasibility report are these:

> Summary
>
> Conclusions
>
> Recommendations
>
> Introduction: background, purpose, definition, scope (criteria)
>
> Method for gathering information
>
> Discussion

The main parts, however, and thus the headings, may vary, depending on the subject, audience, and particular considerations.

Organization

The summary, conclusions, and recommendations may be combined. They usually come first in a feasbility study because these items—particularly the recommendations—are the focus of the report. Then typically follows a section that gives background information, explains the purpose of the study, defines and describes the subject, and explains the scope of the study. The scope of the study lists the criteria—that is, the standards by which an item is judged or a choice is made.

The lengthiest section of the report is the discussion; here full data are given with analysis and commentary on each possible solution or option, and substantiation of all recommendations. If you are asked to recommend a suitable site for dumping hazardous waste, for example, your recommendation will have far-reaching implications—financial, ecological, environmental, and human. Your analysis, therefore, of the feasibility of locating the dump site in a particular area must be thorough. A casual look at the areas is not a sufficient basis for a recommendation.

345

Chapter 8
Reports:
Conveying
Needed
Information

You must collect as much relevant data as possible. Are there other similar sites used for dumping hazardous waste? How have persons living in the area been affected? How long have the sites been used for dumping? What are the similarities/differences between these sites and the proposed site? Is the proposed site accessible for transporting the waste at a minimum level of danger? At what cost? Is the land available? At what cost? You would have to answer these and many other questions before you could justifiably recommend for or against the proposed dumping site.

The discussion section of the feasibility report may be organized by one of the three main comparison-contrast patterns (point by point, subject by subject, similarities/differences). For discussion of these organizational patterns, see pages 200–203. The most usual pattern is point by point, that is, criterion by criterion (a criterion is a standard by which an item is judged). The criteria, for instance, for selecting a new car might be cost, gasoline mileage, standard equipment, and warranty coverage. These criteria might then become headings in the report, each followed by a discussion of data relating to each criterion.

Visuals can enhance the meaning and clarity of the feasibility report, and various kinds of visuals can be used effectively. If the report recommends purchasing new office furniture to complement the existing carpet and walls, color photographs would reinforce the recommendation. More often, perhaps, tables and graphs will be effective to compare facts and features. A table could be used to show clearly the criteria—cost, gasoline mileage, standard equipment, and warranty coverage—as each applies to individual cars considered for purchase. See Chapter 11 Visuals.

Sample Feasibility Report

An example of a feasibility report follows. The report is preceded by the filled-in Plan Sheet. In the study of available VCRs, note how the data are concisely presented in a table.

PLAN SHEET
FOR GIVING A FEASIBILITY REPORT

Analysis of Situation Requiring a Report

What is the subject of my report?
data on 4-head remote VHS VCRs for possible purchase

Who is my intended audience?
director of housing

How will the intended audience use the report? For what purpose?
to accept or reject recommendation to purchase a VCR

What format will I use to present the report?
memorandum

When is the report due?
9 December 1995

Background, Purpose, Definition, Scope of the Report

Hardy-Puryear Residence Hall Council wants to add a VCR to the lounge. Criteria established:
- *4 heads for special effects*
- *remote on-screen programming*
- *at least 155-channel cable capability*
- *8-event/1-year timer*
- *automatic programming*
- *index search*
- *cost within $350*

Method or Procedure

visited Wal-Mart, K-Mart, Hooper Sound, and Electric City; looked for ads in <u>The Clarion-Ledger</u> during November; looked at a catalog from Service Merchandise

Results; Observations; Comments

Found several suitable VCRs, different makes and models:
Goldstar Model GHV-9000M
GE Model VG-4216
Magnavox VR 9035 AT
Teac Model MC-560
Emerson Model VCR 964

Each VCR had various desirable features.

Goldstar — *4 heads for special effects, remote on-screen programming, 181-channel cable capability, 8-event/1-year timer, automatic programming, $299.02*

GE — *4 heads, remote on-screen programming, 181-channel cable capability, 8-event/1-year timer, automatic programming, index search, $349.94*

Magnavox — *4 heads, remote on-screen programming, 155-channel cable capability, 4-event/1-year timer, $269.92*

Teac — *4 heads, remote on-screen programming, 155-channel cable capability, 8-event/1-month timer, $379.86*

Emerson — *4 heads, remote on-screen programming, 110-channel cable capability, 4-event/28-day timer, index search, $216.76*

Conclusions; Recommendations

Purchase of the GE Model VG-4216 at a cost of $349.94

Types and Subject Matter of Visuals I Will Include

table showing each model and characteristics based upon established criteria

The Title I May Use

memo subject line replaces the title

Sources of Information

Wal-Mart, K-Mart, Hooper Sound, Electric City, <u>The Clarion Ledger</u>, <u>Service Merchandise</u> catalog

To: Terri May, Director of Housing
From: Lucy Majers, Hardy-Puryear Residents Council
                Representative
Date: 9 December 1995
Subject: FEASIBILITY REPORT OF AVAILABLE 4-HEAD REMOTE VHS
            VCR'S

<u>Recommendation</u>: While any of the VCRs looked at would be acceptable, I recommend the GE Model VG-4216. It meets all of the criteria and even exceeds the criteria in the channel cable capability with a 181-channel capability. The price of $349.94 is also within the cost limit.

<u>Purpose</u>: The purpose of this feasibility study is to locate and purchase a 4-head remote VHS VCR for viewing movies and recording programs. The VCR will be used in the lounge of Hardy-Puryear Hall. The criteria established by the residence hall council for the VCR are the following:

- 4 heads for special effects
- remote on-screen programming
- at least 155-channel cable capability
- 8-event/1-year timer
- automatic programming
- index search
- cost within $350

<u>Method</u>: I visited four local stores: Wal-Mart, K-Mart, Hooper Sound, and Electric City, and I looked at ads in the daily paper, <u>The Clarion-Ledger</u>, for the month of November. I also looked at a catalog from Service Merchandise. I found many possible VCRs for purchase, but I limited my selection to five for serious consideration.

Feasibility Report, Cont.

349

Chapter 8
Reports:
Conveying
Needed
Information

Data: The five VCRs considered include a Goldstar 4-Head
Remote Hi-Fi Stereo Model GHV-900M, a GE 4-Head Remote Hi-
Fi Stereo Model VG-4216, a Magnavox 4-Head Remote Model VR
9035 AT, a Teac 4-Head Remote Hi-Fi Stereo Model MC-560,
and an Emerson 4-Head Remote Model VCR 964.

Data for the five VCR's are shown in the accompanying
table.

Each of the VCRs has four heads for special effects and
remote on-screen programming. The Goldstar and the GE
models have 181-channel cable capability, the Magnavox and
the Teac have 155, and the Emerson only 110. The Goldstar
and the GE models have an 8-event/1-year timer, the Teac
has an 8-event/1-year timer, the Emerson has only a 4-
event/28-day timer, and the Magnavox has a 4-event/1-year
timer. The Goldstar and GE models have automatic
programming; the other models do not. The GE and Emerson
models have an index search system; the other models do
not. Prices range from a high of $379.86 for the Teac VCR
to a low of $216.76 for the Emerson.

VCRs for Possible Purchase

Brand Name and Model	4 Heads for Special Effects	Remote On-Screen Pro-gramming	Channel Cable Capability	Events and Time Span Timer	Automatic Pro-gramming	Index Search	Cost
Goldstar Model GHV–9000M	Yes	Yes	181	8-event 1-year	Yes	No	$299.92
GE Model VG–4216	Yes	Yes	181	8-event 1-year	Yes	Yes	$349.94
Magnavox VR 9035 AT	Yes	Yes	155	4-event 1-year	No	No	$269.92
Teac Model MC–560	Yes	Yes	155	8-event 1-month	No	No	$379.86
Emerson Model VCR 964	Yes	Yes	110	4-event 28-day	No	Yes	$216.76

Proposal

Purpose

A proposal is a well-thought-out plan. The plan may be to start something new, to change the way things are now, or to sell a product or service. The idea suggested in the proposal, a plan for action, might come from an employee who perceives a need and independently thinks up a solution; or the idea may come from an employer or manager who requests that a proposal be submitted. In either case, this type of proposal may be referred to as an *internal* proposal—that is, a proposal that is initiated within and applies to a problem within a company or organization.

Foundations, corporations, businesses, or agencies may request the submission of a proposal from another organization or an individual. Such a proposal may be referred to as an *external* proposal—that is, it is initiated outside the organization. Proposals may also be referred to as *solicited*—that is, a proposal written in response to a Request for Proposal (RFP); or *unsolicited*—that is, a proposal written because of a perceived problem, the need for a change or a service, or a study of existing conditions. Requests for proposals usually include specific guidelines for submitting the proposal; the writer must follow these guidelines exactly.

Below is an example of a Request for Proposal from the *Commerce Business Daily,* a daily listing of U.S. government procurement invitations, contract awards, subcontracting leads, sales of surplus property, and foreign business opportunities.

NASA-Ames Research Center, Mail Stop 241–1, Moffett Field, CA 94035–1000

15—FABRICATION OF F–18 RADOMES SOL RFP2–34946(JDJ) POC Contact Point, Jim Johnson, 415/604–3010, Contracting Officer, Susan Martin, 415/604–5787. This is a RFP synopsis for the construction of three (3) full scale F–18 radomes. NASA Ames Research Center is conducting a full scale F–18 aircraft wind tunnel test in their 80 × 120 foot wind tunnel as part of the NASA High Angle of Attack Research Program. One of the test objectives will be to study the effects of forebody vortex control devices. Two methods of pneumatic forebody vortex control will be tested: (1) slot blowing; (2) discrete jets. These forebody vortex control devices will be located on the radome of the F–18. The contractor will fabricate the F–18 radome models for wind tunnel testing using a mold from an F–18 production radome. The task includes fabricating the radome models with the forebody vortex control device built in, installing forebody vortex control hardware, installing pressure orifices and tubing, and painting the fabricated radome models. This procurement will be conducted as a Small Business Set Aside. All firms which meet the small business size standard for SIC Code 3728 are invited to submit a proposal. It is anticipated that this will be a firm fixed price contract with a delivery date of March 31, 1992. The anticipated release date of the RFP is September 27, 1991. No telephone requests for the RFP will be accepted. All responsible sources may submit a proposal which shall be considered by the Agency. (0241)

Readers of such RFPs in the *Commerce Business Daily* are representatives of firms and individuals interested in submitting proposals for the particular jobs or equipment.

Uses

Proposals may have such divergent uses as a letter from a student government association to the college president setting forth proposed changes in the class attendance policy; a memo from a sales associate to the department manager detailing the need for additional salespersons; a multivolume document from an arms manu-facturer bidding on a contract for the Department of Defense; a proposal from a professor to a foundation for a grant to research visual problems in preschool children.

351

Chapter 8
Reports:
Conveying
Needed
Information

Main Parts

Since the proposal has a variety of uses, since it may be solicited or unsolicited, since it may be formal or nonformal, and since it may be written by one person or a group, it has no set format or divisions. Depending on the purpose of the pro-posal and the audience, however, you might include these parts:

> Purpose
> Problem
> Proposed solution or plan (Who? What? How much? When?)
> Methods or procedure
> Qualifications of the writer (if needed)
> Recommendations or permission to implement the recommendations
> Conclusion

Longer, more formal proposals may include such sections (and headings) as "Letter of Transmittal," "Abstract," "Summary," "Title Page," "Table of Contents," and "List of Illus-trations," as well as "Budget," "Personnel," "Schedule," "Evaluation," and "Appendixes."

Organization

At the onset, consider the audience, who will probably be more than one reader. Who will read the proposal? Do the readers recognize that a problem exists, or must the proposal convince them? How much do the readers know about the back-ground of the problem or circumstance? Who implements the recommendations?

After gathering information, identifying readers, and analyzing readers' needs, select an appropriate organizational format. For a short, internal proposal, you might use a memorandum. For a short, external proposal, you might use a business letter. In some instances the form may be dictated by your instructor, your em-ployer, your company, or by RFP (Request for Proposal) specifications.

As previously noted, in most proposals you will organize the information in blocks or sections with appropriate headings to make reading and understanding easier. Use visuals or graphics wherever they would enhance meaning.

Give an overview of the situation with a statement of the problem; describe, define, explain the problem in detail. Include specific facts. Clearly identify the cause of the problem or situation. Show how and why the current situation or cir-cumstance is unacceptable. Convince the readers that you have thoroughly investi-gated the situation. Depending on the situation, you may need to show that you have looked at significant records, observed activities, talked to appropriate per-sons, researched any needed background information, and so on.

Finally, recommend that the solution be implemented. If appropriate, ask for permission to implement the proposed solution.

Sample Proposals

Following are examples of proposals. First is a memo to an architect proposing work to be done to correct problems in a building project. Second is a memo from a student analyzing a proposed research report. Last is a student-written proposal (preceded by the filled-in Plan Sheet) for remodeling a family restaurant.

roof tops

A SUBSIDIARY OF INDEPENDENT ROOFERS, INC.

STANDING SEAM ROOF

934 DOUG STREET   JACKSON, MS 39202
601/555-8336   FAX: 601/555-3864

To: W. A. (Bill) Dickens, Jr.
From: T. Steppe
Reference: Hinds Community College Club House
Date: September 11, 1995

After considering our discussion this morning, I would like
to propose the following:

First Contract: (Separate from all others.)
Seam the remaining portion of the building as needed to
complete. $4,500.00 payment on completion. This price does
not include any work on front portion of building, or ridge
rows, or caps.

Contract 2-A: Attempt to repair to a reasonable completion,
the front portion of the Club House. I accept no
responsibility if this cannot be done satisfactorily. If I
can repair the roof under this section I will expect
payment of $6000.00 plus whatever materials have to be
purchased to do so. I anticipate these not to exceed
$2000.00. This would also be payment on completion.

Part 2B: If I cannot complete Section A of this agreement I
feel that removal and replacement will be absolutely
necessary. I would hope that this will not be necessary.
Price would be $11,600.00 to remove, replace, and complete.

Because of other commitments in the near future and the
fact that this project has to have my personal hands-on
attention, I would appreciate a quick reply.

If at any time in the future I can help you in any way in
preplanning a project, please call.

Proposal from a Student (Memorandum Format)   **353**

Chapter 8
Reports:
Conveying
Needed
Information

TO: Dr. Nell Ann Pickett
FROM: Janet Thomas, ENG 1123-Technical
DATE: April 20, 1995
SUBJECT: Topic for library research report

Statement of Problem

For my report I would like to explore Alzheimer's disease
from the standpoint of the person who provides care to a
family patient at home. The caretaker is usually a spouse,
a son or a daughter, or some other close relative.

While there seems to be ample material about what is
known and not known about Alzheimer's disease itself, I
have found few articles dealing with how family members
cope with caring for a patient at home.

Reason for Choice of Topic

As a nursing major who plans to specialize in geriatrics,
I have a keen interest in dysfunctions and diseases of the
elderly. Although Alzheimer's disease is not restricted to
the elderly, it is certainly a debilitating illness for
many older people.

In addition to my professional interest, I have a
personal interest in the disease and especially in the role
of the caretaker of the diseased person. My grandfather has
Alzheimer's disease; for three years I have observed my
grandmother, who, with the help of my father and two aunts,
has taken care of my grandfather. I would like to research
this topic so that I can better understand the emotional
and physical stress and frustration that my grandmother and
other family members have experienced as they have tried to
care for my grandfather.

Procedure

I will research this topic in the college library and in
the library at the Nursing/Allied Health Center. Also I
will structure a list of questions and then interview these
people (and perhaps others): a home health nurse who weekly
visits several homes with Alzheimer's disease patients,
several nonrelative caretakers of Alzheimer's disease
patients, my grandfather's doctor, the director of the
county mental health association, an appropriate person at
the state health department, and a social worker at the
regional hospital.

Possible Problems

There should be no problem in finding numerous articles
and other information on Alzheimer's disease itself. But
sorting out the information about the caretaker and the
changes in her/his life-style and attitude toward life will
probably take a lot of time.

At this time, I foresee no problems in meeting the
deadline of May 11 for the report or in meeting the interim
deadlines for working outline, working bibliography notes,
and preliminary draft.

Need for Consultation with Instructor

I would like for you to look over my interview questions
when I have them made out. I'm thinking about taking a tape
recorder to record each interview, but I'm not sure how
that would work out (especially with elderly caretakers).

PLAN SHEET
FOR GIVING A PROPOSAL

Analysis of Situation Requiring a Proposal

What is the subject of my report?
remodeling Chamoun's Restaurant, Clarksdale, MO

Who is my intended audience?
The Chamoun family

How will the intended audience use the report? For what purpose?
to decide whether or not to remodel the existing restaurant

What format will I use to present the report?
conventional report format—informal

When is the report due?
4 Nov. 1995

Purpose of the Report
to present remodeling and expansion ideas for the restaurant to convince owners that such a move is desirable

Problem or Situation

Small grocery and deli established in 1967. Popularity of food (Lebanese, American, and Italian specialties) has led to requests for evening dining. To expand hours (currently the store and small deli are open from 8:00 AM to 7:00 PM Monday through Saturday) would require expanding space for diners. Currently only seven small tables are available, and a carryout service.

Proposed Solution or Plan, Including Method or Procedure

Set hours for restaurant operation from 11:00 AM to 10:00 PM Monday through Saturday and 11:00 AM to 2:00 PM on Sundays.
Add a 20´ × 20´ dining room area for 65 to 85 customers.
Enclose the partially exposed grill-kitchen area and add industrial fittings and appliances.
Remove gasoline pumps; add a waiting area by enclosing porch area with glass.

Proposed Solution, cont.:

Add a parking lot for approximately 40–55 cars.
Remodel existing dining area to accommodate a larger lunch crowd.
Add Middle Eastern furnishings to accentuate the restaurant's name.
Install intercom stereo system for customer listening.
Brick the outside walls and landscape the grounds.
Cost: $50,000–$70,000.

Conclusion

Well-established business and top-quality food well known in the Clarksdale area.
Remodeling should prove profitable.

Other Information

none

Types and Subject Matter of Visuals I Will Include

diagram showing the present layout of Chamoun's; diagram showing the proposed layout

The Title I May Use

Proposal for Remodeling Chamoun's Restaurant, Clarksdale, Missouri

Sources of Information

family members, local building contractor

Proposal

357

Chapter 8
Reports:
Conveying
Needed
Information

Proposal for Remodeling Chamoun's Restaurant
Clarksdale, Missouri
Mary Louise Chamoun
November 4, 1995

Purpose

The purpose of this proposal is to present remodeling and expansion ideas for Chamoun's Restaurant, in Clarksdale, Missouri.

Present Conditions

Chamoun's is a family owned and operated grocery and delicatessen. The business was established in 1967 by Chabik and Louise Chamoun as a grocery and has since become a deli serving a variety of Lebanese, American, and Italian specialties. With the many requests for evening dining, the question of remodeling and expanding has become dominant in family conversation.

Chamoun's is presently open from 8:00 A.M. to 7:00 P.M. Monday through Saturday. The store operates a small deli containing seven small tables and a carryout service with the small grocery business. Figure 1 shows the present floor plan.

Remodeling Recommendations

An informal estimated cost of remodeling and expanding the restaurant is $50,000 to $70,000. The restaurant hours would be 11:00 A.M. to 10:00 P.M. Monday through Saturday and 11:00 A.M. to 2:00 P.M. on Sundays.

My recommendations for expansion and remodeling (see Figure 2) are as follows:

1. Add a 20′ × 20′ dining room area to accommodate 65–85 customers.
2. Enclose and furnish with industrial fittings and appliances the partially exposed grill/kitchen area.
3. Remove the gasoline pumps and replace the porch area with an enclosed glass waiting area.

4. Add a parking lot to accommodate 40-55 cars.
5. Remodel the present dining area for the large lunch crowd.
6. Brick the outside of the restaurant, and landscape the grounds.
7. Decorate the interior with Middle Eastern furnishings, to accentuate the name of the restaurant.
8. Install an intercom stereo system for customer dining pleasure.

The business is well established and the food is top quality and well known in Clarksdale and the surrounding area. With the demand in Clarksdale for good restaurants, it would be quite profitable to remodel and to offer evening dining.

Proposal, cont.

359

Chapter 8
Reports:
Conveying
Needed
Information

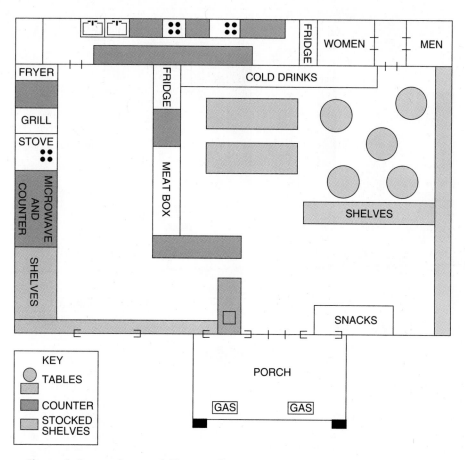

Figure 1. Present layout of Chamoun's.

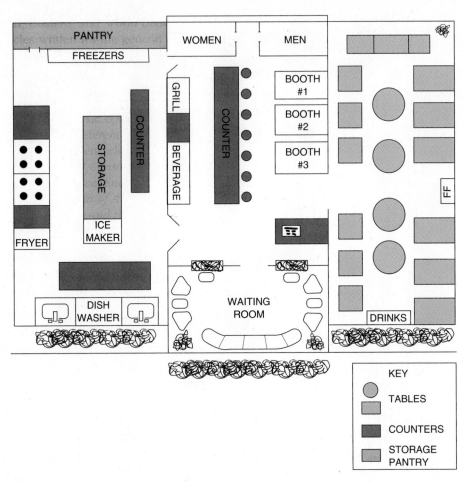

Figure 2. Future layout of Chamoun's.

RESEARCH REPORT

361

Chapter 8
Reports:
Conveying
Needed
Information

The research report, prepared after a careful investigation of all available resources, is similar to what some students refer to as the research paper or term paper. The research report requires note-taking, documentation, and the like. For a full discussion, see Chapter 9 The Research Report.

General Principles in Giving a Report

1. *A report must be accurate, clear, concise, and objective.*
2. *Reports may be classified according to the function they serve.* They may be school reports, which are a testing device for students. They may be professional reports, which are used in business, industry, government, and corporate world.
3. *Reports may be given in various formats.* These formats include printed forms, memorandums, letters, conventional report formats (nonformal or formal), and oral presentations.
4. *Reports frequently include visuals.* Visuals may make information clearer, more easily understood, and more interesting.
5. *Reports usually include headings.* Headings reflect the major points of the report and their supporting data.
6. *Page layout and document design are integral aspects of effective reports.*
7. *Common types of reports include the periodic report, the observation report, the progress report, the feasibility report, the proposal, and the research report.*

Procedure for Giving a Report

1. Determine the audience and the purpose of the report. It is essential to know who will be reading or listening to the report and why. Plan the report accordingly.
2. Review the common types of reports. Think about whether the report you are preparing is a periodic report, an observation report, a progress report, a feasibility report, a proposal, a research report, or a combination of two or more of these types.
3. Investigate and research the subject, being careful to take accurate notes and to list sources. Make use of all available sources—print and nonprint materials, knowledgeable people, experiments, observation.
4. Select an appropriate format for the report. Decide whether the report will be oral or written or both. Decide whether to use a printed form, a memorandum, a letter, or a conventional report format.
5. Organize the report, using headings. Arrange the material in a logical order, as determined by audience, purpose, and subject matter.
6. Plan and prepare visuals. Decide which information should be given in visuals and what type of visuals to use. Then carefully prepare the visuals and place them in the report where they are most effective.
7. Plan the page layout and document design. Think through how the text and the visuals can be most effectively placed on the page, and think through

how you want each page and the report as a whole to appeal visually and intellectually to the reader.

8. Remember that a report should be accurate, clear, concise, and objective.

Applications in Giving Reports

Individual and Collaborative Activities

8.1   In a standard desk dictionary, look up the word *report;* list all the meanings that are given. Indicate the meanings that you think are related to report writing.

8.2   Interview at least two people who are employed in business, industry, government, or corporate world. Find out the kinds of writing they must do. Determine which of these kinds of writing could be classified as report writing. Present your findings in a one-page report.

8.3   Working in teams of three or four, make a collection of at least three to five reports. Using the criteria of audience, purpose, accuracy, clarity, conciseness, and objectivity, analyze the reports. Present your analysis in a one-page report. Choose one member of the team to show the sample reports to the other class members and discuss your team's analysis of the reports. Or have individual team members discuss one report each.

Writing Reports

In completing applications 8.4–8.13,

a. Fill in the appropriate Plan Sheet on pages 365–374, or one like it.
b. Write a preliminary draft.
c. Go over the draft in a peer editing group.
d. Revise. See Part III Revising and Editing and the checklists for revising, front and back endpapers.
e. Write the final report.

Periodic Reports

8.4   Write a daily periodic report on your activities in your office, field, shop, or laboratory. Then expand this report to a weekly periodic report on your activities.

8.5   Assume that you have an annual scholarship of $2,500 from a company that regularly hires employees from your major field of study. Every semester (or quarter) you must report how you have spent any or all of the money and how you intend to spend any remaining portion. Write the report in the form of a letter.

363

Chapter 8
Reports:
Conveying
Needed
Information

Observation Reports

8.6   Write an observation report on an experiment or test.

8.7   Assume that you have been experimenting with three brands of the same product or piece of equipment in an effort to decide which brand name you should select for your home, shop, or lab. Write an investigative report of your observations.

8.8   Visit a business or industry related to your major field of study, a governmental office or department, or an office or division of your college, to survey its general operation. Write an observation report on your visit.

8.9   Write an observation report including conclusions and recommendations on one of the topics below or on a similar topic of your own choosing.

a. Parking problems on campus
b. On-the-job training for laboratory technicians or persons in other health-related occupations
c. Condition of a structure (building, dam, fire tower) or piece of equipment
d. Efficiency of the registration process on campus
e. Conditions at a local jail, prison, hospital, rehabilitation facility, mental institution, or other public institution
f. The present work situation of last year's graduates in your major field of study
g. Employment opportunities in your major field of study
h. Services by an organization such as the Better Business Bureau or the Chamber of Commerce

Progress Reports

8.10   Write a report showing your progress toward completing a degree or receiving a certificate in your major field.

8.11   Write a weekly progress report on your work in a specific course or project (math, science, research, specific technical subject). Using the same course or project, write a monthly progress report on your work.

Feasibility Reports

8.12   Write a feasibility report including conclusions and recommendations on one of the topics below or on a similar topic of your own choosing.

a. Purchase or renting of an item, such as a car, computer, or apartment
b. Selection of a site for a business or an industry

Proposals

8.13   Write a proposal on one of the topics below or on a similar topic of your choosing.

a. Need for purchasing new equipment for a laboratory or shop
b. Rearrangement of equipment in a laboratory or shop
c. A color scheme and furnishings for lounges in a college union building
d. A landscaping plan, a type of air conditioning system, an interior decorating plan, or some other topic related to your major field of study, for a housing project
e. A system of emergency exit patterns and directions for several campus buildings
f. A situation that needs change (unsafe buildings, too large classes, lack of facilities for the handicapped, and so on)

Oral Reports

8.14   As directed by your instructor, adapt for oral presentations the reports you prepared in the applications above. Ask your classmates to evaluate your speech by filling in the Evaluation of Oral Presentations on page 459, or one like it.

Using Part II Selected Readings

8.15   Read "Battle for the Soul of the Internet," by Philip Elmer-Dewitt, pages 500–507. Under "Suggestions for Response and Reaction," page 508, answer Number 5.

8.16   Read "Clear Writing Means Clear Thinking Means … ," by Marvin H. Swift, pages 542–547. What points in this article are important in writing reports?

8.17   Read the poem "Blue Heron," by Chick Wallace, page 550. Under "Suggestions for Response and Reaction," page 551, answer Number 8.

8.18   Read the report "The Effects of Perceived Justice on Complainants' Negative Word-of-Mouth Behavior and Repatronage Intentions," by Jeffrey G. Blodgett, Donald H. Granbois, and Rockney G. Walters, pages 552–562. Under "Suggestions for Response and Reaction," page 563, answer Number 5.

8.19   Read "Journeying Through the Cosmos," by Carl Sagan, pages 570–575. The last seven paragraphs describe the observations of Eratosthenes that led him to believe that the Earth is round. Recast this material in the form of an observation report by Eratosthenes.

PLAN SHEET FOR GIVING A PERIODIC REPORT

Analysis of Situation Requiring a Report

What is the subject of my report?

Who is my intended audience?

How will the intended audience use the report? For what purpose?

What format will I use to present the report?

When is the report due?

Type of Periodic Report (daily, weekly, monthly, etc.)

Dates covered:

Information to Be Reported

Work Date/Time: Work Completed:

Types and Subject Matter of Visuals I Will Include

The Title I May Use

Sources of Information

PLAN SHEET
FOR GIVING AN OBSERVATION REPORT

Analysis of Situation Requiring a Report

What is the subject of my report?

Who is my intended audience?

How will the intended audience use the report? For what purpose?

What format will I use to present the report?

When is the report due?

Object or Purpose of the Report

Theory, Hypothesis, Assumption

Method or Procedure

Results; Observations; Comments

Personnel

Materials and equipment

Activities

Conclusions; Recommendations

Types and Subject Matter of Visuals I Will Include

The Title I May Use

Sources of Information

PLAN SHEET FOR GIVING A PROGRESS REPORT

Analysis of Situation Requiring a Report

What is the subject of my report?

Who is my intended audience?

How will the intended audience use the report? For what purpose?

What format will I use to present the report?

When is the report due?

Purpose of the Report

Work Completed; Comments

Work Completed; Comments, cont.

Work in Progress; Comments

Work to Be Completed; Comments

Types and Subject Matter of Visuals I Will Include

The Title I May Use

Sources of Information

PLAN SHEET
FOR GIVING A FEASIBILITY REPORT

Analysis of Situation Requiring a Report

What is the subject of my report?

Who is my intended audience?

How will the intended audience use the report? For what purpose?

What format will I use to present the report?

When is the report due?

Background, Purpose, Definition, Scope of the Report

Method or Procedure

Results; Observations; Comments

Conclusions; Recommendations

Types and Subject Matter of Visuals I Will Include

The Title I May Use

Sources of Information

PLAN SHEET
FOR GIVING A PROPOSAL

Analysis of Situation Requiring a Proposal

What is the subject of my report?

Who is my intended audience?

How will the intended audience use the report? For what purpose?

What format will I use to present the report?

When is the report due?

Purpose of the Report

Problem or Situation

Proposed Solution or Plan, Including Method or Procedure

Proposed Solution, cont.:

Conclusion

Other Information

Types and Subject Matter of Visuals I Will Include

The Title I May Use

Sources of Information

Chapter 9

The Research Report: Becoming Acquainted with Resource Materials

Outcomes

Upon completing this chapter, you should be able to

- Select an appropriate subject for a research report
- State the general problem investigated in a report
- Divide the general problem into specific problems or questions (preliminary outline)
- Make a working bibliography
- Use the public access catalog
- Use periodical indexes
- Locate source materials for a research report
- Write a direct quotation note card
- Write a paraphrase note card
- Write a summary note card
- Formulate the central idea of a research report
- Make a table of contents for a research report
- Document a research report
- Write a complete research report

377

Chapter 9
The Research
Report:
Becoming
Acquainted with
Resource
Materials

Introduction

Some of the writing and speaking you will do as a student—term papers, book reviews, reports on various topics, speeches—will require research. Also, assignments you may be given as an employee in business or industry—feasibility reports, proposals, process explanations, or even a speech to a local civic club—may require research. The information gathered through research can be organized into a final presentation to accomplish a stated purpose for a specific audience.

To research a topic thoroughly, you need to know how to find the possible sources of information and how to use these sources effectively.

This chapter provides a general guide to library research, one type of research that can be used to prepare and write a research report. It does not attempt, however, to deal with the more intricate points in library research or with the methods used in pure research or in scientific investigation, two other types of research. It does not attempt to present every acceptable method and form for writing a research report. This chapter does present in a logical, step-by-step sequence a procedure for researching and writing a successful report.

The first part of the chapter explains basic library research, and the second part outlines and explains the procedure for writing a research report.

Locating Materials

Materials about a subject are usually located by a systematic, organized search of available sources. Therefore, to begin research, you must first have some idea of possible sources of information and know how to make a systematic search of the sources.

Sources of Information

Possible sources of information can be classified as:

- Personal observation or experience
- Personal interviews
- Free or inexpensive materials
- Library materials

An investigation of these general sources will help you to compile a list of specific sources—books, periodicals, people, agencies, companies—that can supply needed information about a subject or topic.

Personal Observation or Experience

Personal observation or experience can make writing more realistic and vivid, and sometimes this observation or experience is essential. If writing, for instance, about the need for more up-to-date laboratory equipment in local technical education centers or in a medical research facility, personal inspections would be quite helpful.

Do not be misled, however, by surface appearances. Remember that the same conditions and facts may be interpreted in an entirely different way by another person. For this reason, do not depend completely on personal observations or personal experiences for information. Every statement given as fact in a research presentation must unquestionably be based on validated information.

Personal Interviews

Interviewing persons who are knowledgeable about a topic lends human interest to the research project. Talking with such people may also prevent chasing up blind alleys and may supply information unobtainable elsewhere.

Since methods of getting facts through interviews require almost as much thought as the methods involved in library research, interviewing should be carried out systematically. The first thing to remember is courtesy. In seeking an appointment for an interview, request cooperation tactfully. Second, give thoughtful preparation to the interview. Be businesslike in manner and prepare your questions in advance. Finally, keep careful notes of the interview. In quoting the person interviewed, be sure to use his or her exact words and intended meaning. And remember, although the personal interview is a valuable source of information, it offers only one person's opinions. (See Chapter 10 Oral Communication, pages 455–456, for a discussion of how to conduct an informational interview. The material on job interviews, pages 448, 450–455, also may be helpful.)

Free or Inexpensive Materials

Many agencies distribute free or inexpensive pamphlets, documents, and reports that contain much valuable information. The U.S. Government Printing Office, the various departments of the United States government, state and local agencies, industries, insurance companies, labor unions, and professional organizations are just a few of the sources that may supply excellent material on a subject.

379

Chapter 9
The Research
Report:
Becoming
Acquainted with
Resource
Materials

Not at all unusual is the case of a student who, writing about Teflon, requested information from the DuPont Company. In her letter she stated that as a student in mechanical technology she was writing a library research report about industrial uses of Teflon. Within a few days the student received a packet of extremely helpful materials. Some of the material was so new that it was months before it began to appear in periodicals. Some of the material consisted of reprints of magazine articles that the student did not have access to. And some of the individual pieces of material she received would never appear in magazine or book form.

In seeking material from various agencies, make the request specific. For instance, a person researching aluminum as a structural building material might be disappointed in the response to a request to Alcoa for information about aluminum. Such a request is difficult if not impossible for the company to fill because the particular need is not specified. A request for information about aluminum as a structural building material, however, specifies the subject and thus encourages a more satisfactory response.

In requesting materials, consider the time element. Do not assume that all materials will arrive immediately or that all or even most of the information for a research report will come from these requested materials. Depending too much on material to arrive before the due date of the report could be disastrous.

Library Materials

Much research is carried out in a library, whether a school library, a public library, or a specialized library such as a medical library, a business library, or a law library. The library contains a wealth of information in printed materials—books, pamphlets, newspapers, and other periodicals—and in nonprint materials—tapes, films, microforms, computer disks, compact disks. But this information is useless to the researcher until it is found and used to accomplish a purpose.

Public Access Catalog

The single most useful item in the library is the public access catalog, formerly called the card catalog. The public access catalog (PAC), once available only in the form of 3-by–5-inch cards, is increasingly being made available in such electronic forms as a computer-output-microform (COM) catalog and/or on-line computer system, known as an on-line public access catalog (OPAC).

Microfiche, usually referred to as fiche, is a sheet of microfilm typically 4 by 6 inches and capable of accommodating and preserving a considerable number of pages in reduced form. One 4-by–6-inch fiche can, on the average, accommodate the information from 1,600 3-by–5-inch cards in the card catalog. The COM catalog is produced in this way: The information contained in the public access catalog is input and stored in a computer to which new acquisitions are routinely added. Periodically, a computer tape of the entire catalog is retrieved and reproduced on either microfiche or microfilm.

The public access catalog entries, called bib (bibliographic) records, are arranged alphabetically according to the first important word in the heading (first line) on the entry. Most items listed in the public access catalog have a minimum of three entries: author entry, title entry, and one or more subject entries. The heading of each bib record is determined by the type of entry: author, title, or subject. Only the heading for each entry differs; the basic bibliographic information for each entry is the same.

A typical entry in the public access catalog contains the following information (see the corresponding colored numbers on the following sample entries):

1. Heading—subject, author, or title
2. Author (or editor) listed first on the book
3. Complete title of the book
4. Listing of all authors (or editors)
5. City of publication (the state is also given for less well-known cities)
6. Publishing company
7. Year of publication
8. Number of introductory pages
9. Number of pages
10. Illustrations
11. Height in centimeters (a centimeter is 0.4 inch)
12. International Standard Book Number (ISBN)
13. Headings by which the book is cataloged
14. Call number (the designation for classifying and shelving the book in the library.)

Subject Entry from Public Access Catalog
(Numbers in color added to correspond to preceding list)

1   NURSING RECORDS
   2 Eggland, Ellen Thomas
      3 Nursing documentation : charting, recording, and reporting / 4 Ellen Thomas Eggland,
      Denise Skelly Heinemann. 5 Philadelphia : 6 J.B. Lippincott, 7 c1994.
      8 xi, 260 p. : ill. ; 28 cm.   9   10   11
   12   0397550103 (pbk.)
      13 1. Nursing records. 2. Communication in nursing.
         I. Heinemann, Denise Skelly. II. Title.
      14   610.73 Eg3n

As you begin looking in the public access catalog for books and other materials containing information on your subject, you may not know any authors or titles to consult. Therefore, look for subject entries. Suppose you are researching for a report entitled "Medical Applications of the Laser." If you do not know any title or author concerning the subject, first look in the "L" section for "Laser" and "Lasers." However, do not stop after looking under these two subject headings; look also under other related subject headings. Some of your most important information might be catalogued under such subjects as "Medicine" or "Surgery" or "Physics."

Entry from On-Line Public Access Catalog

381

Chapter 9
The Research
Report:
Becoming
Acquainted with
Resource
Materials

```
  To pick a new button, press TAB or button's first letter.
  To view the next item(s) you found, press SEND now.
  HELP        GOBACK        STARTOVER    PRINT      NEWCOMMAND
  LIKE        OPTIONS
  FORWARD     BACKWARD      MARK
  - - - - - - - - - - - - - - - - - - - - - - - - - - - - - -
        THIS IS RECORD NUMBER 226 OF THE 395 YOU FOUND IN THE CATALOG
  364.973 C86M 1994–95

              3 Title:  Criminal justice 94/95 / John J. Sullivan, Joseph L.
                        Victor, editors. 4
             Edition:  18th ed.
    Publication info:  Guilford, Conn. : Dushkin Pub. Group, c1994. 5  6  7
      Physical desc:   240 [3] p. : ill. ; 28 cm. 9  10  11
             Series:   Annual editions
       General note:   1561342688 12
          | Subject:   Criminal justice, Administration of United States.

                        (Displaying 1 of 1 volumes)

     RAYMOND CALL NUMBER  14             COPY MATERIAL   LOCATION
       1) 364.973 C86M 1994–95            1 NONFICTION OPENSTACKS
     PAGE 1                 VERIFY      RECEIPT       (c) Sirsi Corporation 62n
```

An on-line computer system (or on-line public access catalog) makes available quickly and accurately millions of items of information stored in a central computer. A terminal links the user to the computer. The on-line public access catalog (OPAC) provides highly efficient access as you search by the traditional types of entry (subject, author, title) as well as by the following:

- Publication date
- Publisher
- International Standard Book Number (ISBN)
- Language (English, Spanish, French, and so on)
- Format (films, filmstrips, slides, compact disks, and so on)

On some systems you may search by key word, which means that you can search the information in each bibliographic record (entry) in the computer for a meaningful word, term, or concept. Searching by key word is particularly valuable if you want to know everything available by or about a particular person, such as Albert Einstein, or if you are not sure of the correct subject heading to use.

Many on-line public access catalogs include several modules or units that are integrated or work together. These modules may include circulation, academic reserves, requests for materials, acquisitions, serials, and bibliographic interfaces. When the modules are integrated in a system, the information on each item includes where the item is located, if it is checked out, if it has been placed on reserve, or if it is on order.

Some advantages of the OPACs are their capability for multiple users to access the system simultaneously; for remote access from home, dorm, or office via dial-up access or off-site terminals; for use as gateways to search other information retrieval systems, such as some of the periodical indexes that will be discussed later; and for printers to be connected.

Further explanation of on-line systems can be found on page 388.

Library of Congress Subject Headings

Particularly useful in looking up subject entries is the *Library of Congress Subject Headings* (and its supplements). This book gives subject descriptors used by the Library of Congress, and subsequently by most other libraries. Following is an excerpt indicating some of the subject descriptors for "Laser" and "Lasers."

EXCERPT FROM *LIBRARY OF CONGRESS SUBJECT HEADINGS*

Laser anemometer
  USE   Laser Doppler velocimeter
Laser angioplasty   *(May Subd Geog)*
    BT   Lasers in surgery
       Thermal angioplasty
Laser automobile
    ₍TL215.L₎
    BT   Plymouth automobile
Laser beam cutting
    BT   Machining
       Metal-cutting
Laser-beam recording
  USE   Laser recording
Laser beams
    UF   Beams, Laser
       Laser radiation
    NT   Laser-plasma interactions
       Laser plasmas
       Laser pulses, Ultrashort
  — **Atmospheric effects**
      ₍QC976.L36₎
    UF   Atmospheric effects on laser beams
    BT   Meteorological optics
  — **Diffraction**
    BT   Diffraction
  — **Scattering**
    NT   Laser speckle
Laser cell counting
  USE   Flow cytometry
Laser cell separation
  USE   Flow cytometry
Laser cell sorting
  USE   Flow cytometry
Laser chemistry
  USE   Laser photochemistry
Laser coagulation
    UF   Laser photocoagulation
    BT   Light coagulation
    RT   Lasers in surgery

Laser communication systems
    UF   Coherent light communication systems
       Communication systems, Optical
         (Laser-based)
       Light communication systems, Laser-
         based
       Optical communication systems, Laser-
         based
    BT   Integrated optics
       Optical communications
       Telecommunication systems
    NT   Astronautics—Optical communication
         systems
       Optical radar
Laser communication systems industry
    (May Subd Geog)
      ₍HD9696.L35-HD9696.L354₎
    BT   Telecommunication equipment
         industry
Laser-controlled land grading
    (May Subd Geog)
    UF   Laser grading
    BT   Earthwork
Laser cooling   *(May Subd Geog)*
    BT   Cooling
Laser dentistry
  USE   Lasers in dentistry
       . . .

┌─────────────────────────────┐

SYMBOLS

UF   Used For
BT   Broader Topic
RT   Related Topic
SA   See Also
NT   Narrower Topic

└─────────────────────────────┘

As you begin looking in the public access catalog for sources of information, check to see if the alphabetizing is letter by letter or word by word. Most dictionaries alphabetize letter by letter, but encyclopedias and indexes may use either alphabetical order or some other order such as chronological, tabular, regional. Knowing this information is essential; otherwise you might incorrectly assume the library has no materials on a subject. The following example illustrates two methods of alphabetizing:

383

Chapter 9
The Research
Report:
Becoming
Acquainted with
Resource
Materials

Letter by Letter	*Word by Word*
art ballad	art ballad
art epic	art epic
article	art lyric
art lyric	art theatre
art theatre	Arthurian legend
Arthurian legend	article
artificial comedy	artificial comedy
artificiality	artificiality

Indexes

Indexes are to periodicals (magazines, newspapers) what the public access catalog is to books. By consulting indexes to periodicals, specific sources of information on a subject can be found without looking through hundreds of thousands of magazines.

Much of any needed information on a technical or business subject is likely to come from magazines. For one thing, a great deal of information in magazines is never published in book form. Furthermore, because the writing and publishing of a book usually takes at least a year, the information in magazines is likely to be much more current.

At the beginning of any index are directions for use, a key to abbreviations, and a list of periodicals indexed.

Computerized Periodical-Index Systems

Many libraries provide electronic systems for finding citations to periodical materials on specific subjects. The user sits at a workstation typically composed of a microcomputer with keyboard, a monitor, a disk drive, and a printer plus the software, usually a tape or a disk, on which the information is stored. The user enters the subject words, makes selections from the options that appear on the screen, and chooses which citations to be printed for the user to look up.

Some systems have abstracts of the indexed articles. The abstracts may be printed and used as sources if the original article is unavailable. See page 385 for more information on indexes containing abstracts. Other systems, as explained on page 388, have access to a database of the complete text of the cited material. The complete text, which may be available on-line, on compact disk (CD), or in a microform, may be retrieved, read on the screen or on a reader/printer, and printed if needed.

Often the electronic systems contain rolling indexes. As current citations and abstracts are added, the space on the system is filled and the older entries are removed. These older entries are often retained in a backfile database available on-line, on a compact disk, or in microform. Backfiles can be excellent sources when historical information is needed. To provide additional access to electronic periodical indexes, some libraries are networking computers so that several terminals may access the same electronic index simultaneously. Often, several indexes are available on a LAN (local area network), so research in more than one periodical index may be accomplished at one computer terminal. Some indexes are available on tapes that may be loaded with the on-line public access catalog (OPAC) and accessed through the same computer terminals as the library's OPAC.

General Periodicals Index

Among electronic periodical indexes particularly helpful to students is the *General Periodicals Index.* It comes in four editions: Select, Library, Research I, and Research II. Research II edition indexes approximately 1,600 general interest and scholarly publications in such subject areas as social sciences, general sciences, humanities, business, management, economics, and current affairs. Updated monthly, it indexes data for the current four years and the current six months of two newspapers, *The New York Times* and *The Wall Street Journal.* It also contains full text coverage of over three hundred of the periodicals indexed.

The following excerpt is from another electronic periodical-index, *Magazine Index Plus.**

> InfoTrac $ Magazine Index Plus 1991 – Sep 1994
>
> Heading: LASER BEAMS
>   -Usage
> 5. Catch a ride on a laser beam. (pulsed-detonated
>   engines) by Tim Stevens il v242 Industry Week May 17
>   '93 p45(1)
>   69C0446
>   ABSTRACT / HEADINGS

The entry (one of nine entries under the heading LASER BEAMS—Usage) gives this information.

> "Catch a Ride on a Laser Beam": title of the article
> (Pulsed-detonated engines): additional information about the subject
> Tim Stevens: author
> il: illustrations included in the article
> 242: volume number
> *Industry Week:* magazine title
> May 17, 1993: date
> p45(1): page number and length of the article

The location code 69C0446 near the end of the entry indicates that the article is available in the Magazine Collection, which contains the full text of selected magazines on microfilm cartridges. The article is on frame 446 of cartridge 69C. ABSTRACT/HEADINGS indicates that a brief summary of the article is available on the index and all the headings under which the article is listed are given. The headings often suggest other terms to use for locating additional citations on a topic.

385

Chapter 9
The Research
Report:
Becoming
Acquainted with
Resource
Materials

Readers' Guide to Periodical Literature.　The *Readers' Guide* is a general index of almost two hundred fifty leading popular magazines published from 1900 to the present. It is published monthly with cumulative issues every three months and a bound cumulation each year. This cumulation saves the researcher from having to look in so many different issues and keeps the index up to date.

The main body of the *Readers' Guide* is a listing, by subject and author, of periodical articles. Each entry gives the title of the article, the author (if known), the name of the magazine, the volume, the page number, and additional notations for such items as bibliography, illustration, or portrait. Following the main body of the index is an author listing of citations to book reviews.

The excerpt below from the August 1994 issue of the *Readers' Guide to Periodical Literature* is from the listing for the subject "Lasers."*

Medical use
Curing snoring [use of lasers to trim tissue obstructing air flow] R. Robinson. il *American Health* v12 p13 D '93
The perforated heart [transmyocardial revascularization; work of Mahmood Mirhoseini] S. Frandzel. il *American Health* v12 p11 D '93
Three who can see [radial keratotomy and excimer laser surgery] L. E. Ma. il *Health* (San Francisco, Calif.: 1992) v7 p76 N/D '93
Military use
　See also
　High-Energy Laser Systems Test Facility (U.S.)
Boeing, Rockwell confront airborne laser challenges] S. Evers.] *Aviation Week & Space Technolgy* v140 p75 My 30 '94
Surgical use
　See Lasers—Medical use

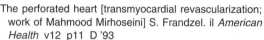

The first entry under "Medical Use" gives this information.

"Curing Snoring": title of article
[use of lasers to trim tissue obtructing air flow]: additional information about the subject
R. Robinson: author
il: illustrations
American Health: magazine title
12: volume number
13: page number
December 1993: date

Readers' Guide Abstracts.　The *Readers' Guide Abstracts* each year adds about 60,000 abstracts of the articles indexed in *Readers' Guide* and of articles in *The New York Times*. Each abstract is a concise summary that presents the major points, facts, and opinions of the original article. *Abstracts* is updated monthly in the CD-ROM format and twice a week on-line. Indexing coverage is January 1983 to the present, and abstracting coverage is September 1984 to the present.

Similar in format to *Readers' Guide Abstracts* are the Wilson Abstracts (Business, Applied Science & Technology, General Science, Art, Humanities, and Social Sciences), NewsBank's *Index to Periodicals,* and the ProQuest Abstracts (Newspaper Abstracts, Periodical Abstracts, ABI/INFORM), and Health Index.

Applied Science & Technology Index.   The *Applied Science & Technology Index* indexes nearly 400 English language periodicals in the fields of aeronautics and space science, atmospheric sciences, chemistry, computer technology and applications, construction industry, energy resources and research engineering, engineering, fire and fire prevention, food and food industry, geology, machinery, mathematics, metallurgy, mineralogy, oceanography, petroleum and gas, physics, plastics, textile industry and fabrics, transportation, and other industrial and mechanical arts. The main body of the *Index* lists subject entries to periodical articles. Additionally, there is an author listing of citations to book reviews and a product review section. The *Index* is issued monthly except July and has an annual cumulation.

The arrangement of entries is similar to that in the *Readers' Guide to Periodical Literature.*

Business Periodicals Index.   This is an index to over 350 English language periodicals in the fields of accounting, advertising and marketing, agriculture, banking, building, chemical industry, communications, computer technology and applications, drug and cosmetic industries, economics, electronics, finance and investments, industrial relations, insurance, international business, management, personnel administration, occupational health and safety, paper and pulp industries, petroleum and gas industries, printing and publishing, public relations, public utilities, real estate, regulation of industry, retailing, taxation, transportation, and other specific businesses, industries, and trades.

The main body of the *Index* lists subject entries to business periodical articles. Additionally, there is an author listing of citations to book reviews.

The *Business Periodicals Index* is published monthly except August and has quarterly and annual cumulations. It is very similar in format and arrangement to the *Readers' Guide* and the *Applied Science & Technology Index.*

Other Indexes to Professional Journals

In addition to *General Periodicals Index, Readers' Guide to Periodical Literature, Applied Science & Technology Index,* and *Business Periodicals Index,* the following indexes to periodicals may be useful:

Biological and Agricultural Index (since 1916; formerly *Agricultural Index*)

Business Index (since 1979)

Cumulative Index to Nursing and Allied Health Literature (since 1956; formerly *Cumulative Index to Nursing Literature*)

Engineering Index (since 1884)

General Science Index (since 1980)

Hospital Literature Index (1945)

PAIS International In Print (since 1915; formerly *Public Affairs Information Service Bulletin*)

Social Sciences Index (since 1974; formerly part of *Social Sciences and Humanities Index,* 1965–1974, and of *International Index,* 1907–1965)

Indexes to Newspapers

Newspaper indexes include:

387

Chapter 9
The Research
Report:
Becoming
Acquainted with
Resource
Materials

- *Bell and Howell's Index to the Christian Science Monitor* (since 1960, monthly with annual cumulations; formerly *Index to the Christian Science Monitor*)
- *The London Times Index* (since 1906, monthly with annual cumulation)
- *National News Index* covering *The New York Times, The Christian Science Monitor, The Wall Street Journal,* the *Los Angeles Times,* and *The Washington Post* (since 1979, monthly, "rolling" cumulation index). Covers the most current four years of data
- *The New York Times Index* (since 1851, semimonthly, annual cumulation)
- *Newspaper Abstracts* has two editions, Complete and National. The Complete Edition covers *The New York Times, The Wall Street Journal, The Christian Science Monitor, USA Today,* and five regional newspapers (since 1989 monthly cumulative index). Includes backfiles (1985–1988) for some papers.
- *The Wall Street Journal Index* (since 1950, monthly, annual volume)

The New York Times Index is especially useful. The *Times* covers all major news events, both national and international. Because of its wide scope and relative completeness, the *Times Index* provides a wealth of information. It is frequently used even without viewing the individual paper of the date cited. A brief abstract of the news story is included with each entry. Thus, someone seeking a single fact, such as the date of an event or the name of a person, may often find all that is needed in the *Index.* In addition, since all material is dated, the *Times Index* serves as an entry into other, unindexed newspapers and magazines. *The New York Times Index* is arranged in dictionary form with cross-references to names and related topics.

Indexes to periodicals are available in a variety of formats, which include print (on paper), microfilm, CD-ROM, and on-line. Some indexes are available in several of these formats.

Indexes to Government Publications

The U.S. government prints all kinds of books, reports, pamphlets, and periodicals for audiences from the least to the most knowledgeable. The best-known index to government documents is the *Monthly Catalog of United States Government Publications,* published by the U.S. Superintendent of Documents. Each monthly issue lists the documents published that month.

U.S. Government Books is a selective list of government publications of interest to the general public.

Also available from the U.S. government are some 175 subject bibliographies of government publications. These are listed in the *Subject Bibliographies Index;* subjects of bibliographies range from business to education, health, science, and history.

A guide to reports from research is *Government Reports Announcements & Index,* published twice a month. It has an annual, cumulative index titled *Government Reports Annual Index.*

Government Reference Books is an annotated guide that provides access to important reference works issued by agencies of the United States government. The guide is published every two years; the entries are arranged by subject.

Essay Index

It is often difficult to locate essays and miscellaneous articles in books. The *Essay and General Literature Index* is an index by author, subject, and some titles of essays and articles published in books since 1900. The index is kept up to date by supplements. Subjects covered include social and political sciences, economics, law, education, science, history, the various arts, and literature.

Other "Helps" in Library Research

In addition to the public access catalog and the indexes, other tools or sources of information may be valuable in research.

On-Line Computer Search

In an on-line computer search, various databases produced by the government, by nonprofit organizations, or by private companies are accessed through a computer terminal.

Using the computer terminal to access the database is a two- or three-step procedure. First, contact is established with the computer containing the database via telephone lines. Some computers contain only one database; some contain many. The database needed must be accessed. Then, using carefully formulated search logic, the computer searches the database for the desired materials.

Search strategies should be formed, using one or more key words before going "on-line" (on-line is the time actually spent interacting with the computer). A thesaurus or a subject heading list, if available for the index, should be used. Each has "see" and "see also" terms to help in selecting the proper subject headings to use.

A computer search is fast (particularly for searches covering several years), current (some information is available on-line weeks or months before the printed version), accurate, thorough (more in-depth coverage since there is more access to the information), and convenient.

Computer searches require the assistance of a librarian and during busy times may need to be scheduled in advance. In addition, computer searches may be expensive, although the cost of using databases varies greatly. The user is charged for on-line time (cost per minute of time the user is "connected" to the database) and for each citation and each page of text printed.

Traditionally, database searches have been used for locating citations for journal and newspaper articles on a subject. Now with entire works—periodical articles, books, and even encyclopedias on-line—databases may be used for locating detailed financial data and directory listings on companies, statistics, biographical information, news-wire stories, information on colleges and universities, and quotations. The information itself is becoming on-line, not just the citations to it. See pages 383–384 for further explanation of on-line systems.

The Internet

The Internet is a global computer network of interconnected educational, scientific, business, and governmental networks. Through a computer, modem, and telephone line, anyone on the Internet can communicate with any other person or group of people who are also connected to it.

A growing variety of resources are becoming available on the Internet. These include on-line databases such as full-text periodicals, books, and government

389

Chapter 9
The Research
Report:
Becoming
Acquainted with
Resource
Materials

documents; statistics; business and travel information; and indexes. Also accessible are the on-line catalogs of many academic and research libraries. Software, pictures, and even sounds can be located and downloaded from the Internet.

Many libraries provide the Internet for their faculty and students to use. While there are connect fees and phone line charges, most of the resources on the Internet do not have user fees for searching or downloading information; however, some resources are available through businesses and involve fees and pass words or account numbers to access data.

NewsBank

NewsBank Reference Library is a microfiche newspaper reference service covering social, health, legal, political, international, arts, consumer, education, economic, and scientific fields. Articles are collected from more than 450 leading newspapers all over the United States and indexed by subject. The printed indexes are published monthly and are cumulated quarterly in March, June, and September, with an annual cumulation in December. The articles, reproduced on microfiche, are provided monthly with the index. NewsBank is particularly useful because of the on-the-spot availability of all indexed articles.

Given below is an excerpt from the April-June 1994 *Index* with an explanation of how to use the *Index*.*

COMPUTER CRIMES
   hackers
      computer security–1994 BUS 21:B9

COMPUTER INDUSTRY
    See also **Computer Crimes; Electronic Equipment and
    Supplies; Semiconductor Industry**
   competition
     data storage
       charts–1994 BUS 36:C12–14
     impact–1994 BUS 21:B12–13
     studies and reports–1994 BUS 21:B14–C1
   employee benefits
     health benefit plans, retirees
       Unisys–1994 EMP 23:C8
     life insurance
       Digital Equipment–1994 EMP 29:C11
   employment market
     layoffs
       Digital Equipment–1994 EMP 23:C9–10, 23:C11
       Massachusetts–1994 EMP 23:C12–14
   exports
     supercomputers–1994 BUS 21:C2
   industry conditions
     flat-panel displays–1994 BUS 28:D1–2
   manufacturers
     consumer services–1994 BUS 36:D1–2
   new products
     computer animation–1994 BUS 36:D3–4

*_NewsBank Index._ Copyright 1994 by NewsBank, Inc. Material reproduced by permission of the publisher.

studies and reports
  keyboards–1994 BUS 21:C3
trade shows
  Comdex
    Georgia: Atlanta–1994 BUS 36:D5

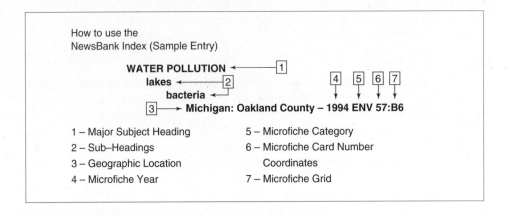

The *Newsbank Reference Service PLUS,* an electronic retrieval system, uses CD-ROM (compact disk-read only memory) and the personal computer. This system allows the user to quickly and easily access newspaper articles in *NewsBank* (1981 to the present), *Names in the News* (1981 to the present), and *Review of the Arts* (1980 to the present). Since 1990, the system has included abstracts for the articles in *News-Bank.* Also used with *NewsBank Reference Service PLUS* is the NewsBank *Index to Periodicals,* which indexes articles in 100 general periodicals (1988 to the present).

Also published by NewsBank are *CD NewsBank* and *STAT FACT. CD NewsBank* contains full text articles on issues of national or international importance from over 100 local newspapers in the United States and articles from national and international wire services. Over 40,000 articles are added annually. Updated monthly, it uses the same searching techniques as *NewsBank Reference Service PLUS. STAT FACT* contains state and national statistical data from over 250 resources. Searchable by subject, each statistic includes the source.

Business NewsBank, another NewsBank product, is similar to NewsBank. It provides a printed index to business articles from over 600 newspapers and regional business journals in the United States. The articles pertain to companies, industries, products, and personnel. *Business NewsBank PLUS,* on CD, includes an electronic index to *Business NewsBank* (1985 to the present) and since 1994 has included the articles in full text.

Social Issues Resources Series

A valuable help in library research is the Social Issues Resources Series (SIRS). SIRS is a selection of articles reprinted from newspapers, magazines, journals, and government publications. The articles are organized into volumes with each volume, in a loose-leaf notebook format, a different social issue. Among the titles for the volumes are Aging, Alcohol, Communication, Consumerism, Corrections, Defense, Energy, Family, Food, Health, Human Rights, Mental Health, Pollution, Privacy, Sports, Technology, and Women. Each volume contains a minimum of 20 articles and is updated with an annual supplement of another 20 articles. (Each volume eventually contains 100 articles.) The articles are selected to represent various reading levels, differing points of view, and many aspects of the issue.

The SIRS articles, tables of contents, and indexes are available in these formats: print (paper), microfiche, and CD-ROM.

SIRS Science is a series of five volumes covering Earth Science, Physical Science, Life Science, Medical Science, and Applied Science. Similar to the regular SIRS volumes, these also contain articles selected from various sources held in loose-leaf notebooks. Each of the volumes contains 70 indexed articles in chronological order. The volumes are annual.

SIRS Critical Issues contains volumes on current, significant topics. Similar to the regular SIRS volumes, the Critical Issues began in 1987 with *The AIDS Crisis;* and in 1989 *The Atmosphere Crisis* was added.

In 1991, SIRS added the series SIRS Global Perspectives, which contains four volumes: *History, Government, Economics,* and *World Affairs.* Similar to the SIRS Science series, each annual volume contains 70 articles.

Two full text databases on CD-ROM available from SIRS are *SIRS Researcher* and *SIRS Government Reporter. SIRS Researcher* contains selected full-text articles from over 800 newspapers, magazines, journals, and government publications from 1988 to the present. *SIRS Government Reporter* contains selected full-text documents from federal departments and agencies including tables and charts of census data. Summaries of the documents are included. Both databases are searchable by subject and key word.

Nonprint Materials

Most colleges have a nonprint department, a media or learning resources center, or a library department that supervises audiovisual materials. Check with the librarian or person in charge concerning the cataloging of such materials as audiotape, videotapes, computer disks, films, filmstrips, and slides. These items may be alphabetized in the public access catalog or in a separate catalog. They may contain valuable information on a research topic.

Periodicals Holdings List

The periodicals holdings list is a catalog of all the periodicals in a library. The alphabetical list gives the name of each magazine and newspaper and the dates of the available issues. The list may also indicate whether the issues are bound or unbound, the format, and where in the library the periodicals are located.

The periodicals holdings list may be in various forms, such as a drawer of 3-by-5-inch cards, a visible index file, a typed sheet, a microform, or a computer printout, or it may be available on-line. Some libraries combine the periodicals holdings list with the public access catalog.

It is essential to know where the periodicals holdings list in a library is located. When in the various periodical indexes you find titles of articles that seem usable, you need to know if the library has the specified magazines. The periodicals holdings list will give you this information; you can save time by checking the list yourself.

Reference Works

Libraries are filled with all kinds of reference books. Common among these are encyclopedias, dictionaries, books of statistics, almanacs, bibliographies, handbooks, and yearbooks.

391

Chapter 9
The Research
Report:
Becoming
Acquainted with
Resource
Materials

Encyclopedias. Encyclopedias may be general encyclopedias or specialized encyclopedias. General encyclopedias may be a good starting point for locating information about a subject and for gaining an overall view of it. *Americana, Britannica, Collier's,* and *World Book* are well-known general encyclopedias that provide articles written for the general reader.

A valuable specialized encyclopedia is the *McGraw-Hill Encyclopedia of Science and Technology,* an international reference work in twenty volumes including an index. The articles arranged in alphabetical order (word by word), cover pertinent information for every area of modern science and technology. Each article includes the basic concepts of a subject, a definition, background material, and multiple cross-references.

Specialized encyclopedias are numerous. Included are the *Encyclopedia of American Forest and Conservation History, Encyclopedia of Psychology, Encyclopedia of Crime and Justice, Encyclopedia of Textiles, Encyclopedia of Banking and Finance, The Encyclopedia of Alcoholism, Encyclopedia of Physics, Encyclopedia of American Business History and Biography,* and *Macmillan Encyclopedia of Architects.*

Dictionaries. General-use dictionaries include *The American Heritage College Dictionary, Merriam Webster's Collegiate Dictionary, The Random House Dictionary of the English Language,* and *Webster's New World Dictionary of American English.* Specialized dictionaries include those of computer languages, architecture, welding, decorative arts, technical terms, astronomy, U.S. military terms, Christian ethics, and economics.

Books of Statistics. Statistics are often an important part of a report because they support conclusions, show trends, and lend validity to statements. Since statistics change frequently, currency is usually essential.

Up-to-date statistics on a subject may be located in such sources as almanacs, yearbooks, and encyclopedias as well as recent newspaper and magazine articles on the topic. Statistical information on the past, such as wages in the 1930s, may be located in *Historical Statistics of the United States, Colonial Times to 1970.*

Statistics for the United States and other countries in such areas as industry, business, society, and education are found in *Statistics Sources.*

For annual updates of statistical data, see the *Statistical Abstract of the United States* (since 1878) and the *Statistical Yearbook* (since 1949), which covers yearly events for about 150 countries.

Two monthly indexes that provide access to a large area of statistical information are *American Statistics Index* (ASI) and *Statistical Reference Index* (SRI). ASI (since 1973) indexes most statistical sources published by the federal government and SRI (since 1980) indexes statistical information in U.S. sources other than those published by the federal government. Both of these complementing indexes have annual cumulations.

For statistical works held by the library, look in the public access catalog under the subject headings "Statistics" and "United States—Statistics." For statistical works just on a topic, also look under the topic for the subheading "Statistics," for example "Agriculture—Statistics."

Almanacs. Perhaps the best-known general almanac is *The World Almanac and Book of Facts.* It includes such diverse facts as winners of Academy Awards, accidents and deaths on railroads, civilian consumption of major food commodities per person, National Football League champions, notable tall buildings in North American cities, and the U.S. military pay scale.

Other almanacs include *Information Please Almanac, The Universal Almanac,* and *Whitaker's Almanac.*

Bibliographies. Bibliographies are guides to sources of information on specific subjects. They do not generally provide answers but rather list sources, such as periodicals, books, pamphlets, and nonprint materials, which contain the needed information.

Most bibliographies contain specific citations to information on a subject; however, others, such as the *Encyclopedia of Business Info Sources,* lists by subject such types of sources as encyclopedias and dictionaries, handbooks and manuals, periodicals, statistical sources, and almanacs and yearbooks. This type of work is particularly useful when researching a topic about which little is known or few sources seem to be available.

For bibliographies available in the library on a specific subject, consult the public access catalog. Look under the subject for the subheading "Bibliography," for example, "Business—Bibliography," "Marketing—Bibliography," or "Nursing—Bibliography."

Handbooks. Handbooks contain specialized information for specific fields. The variety of handbooks is illustrated by the following examples. There are handbooks for word processor users, secretaries, electronics engineers, photographers, and construction superintendents, as well as handbooks for the study of suicide, transistors, air conditioning systems design, food additives, simplified television service, and tropical fish, to list a few.

Persons interested in information about job opportunities will find the *Occupational Outlook Handbook* quite useful. Published biennially by the U.S. Bureau of Labor Statistics, it describes about 250 occupations and for each gives the outlook, the number of persons currently employed, the possible salary, educational requirements, possibility for advancement, and so on.

Yearbooks. Yearbooks, as the name suggests, cover events in a given year. An encyclopedia yearbook updates material covered in a basic set—this is an example of a general yearbook.

Specialized yearbooks include such various titles as the *Yearbook of Agriculture, Yearbook of American and Canadian Churches, Yearbook of Drug Therapy, Yearbook of Astronomy, Yearbook of Sports Medicine,* and *Yearbook of Labor Statistics.*

Note that the above examples of encyclopedias, dictionaries, books of statistics, almanacs, bibliographies, handbooks, and yearbooks illustrate the number and variety of these reference works. A quick glance at *Books in Print,* under these categories, further emphasizes the number and variety.

Guides to Reference Works

For a quick review of what reference materials exist, the following guides are useful. If you find in these guides materials not available in your library, ask the librarian about an interlibrary loan. But remember that interlibrary loans take time.

Periodicals. *Ulrich's International Periodicals Directory* includes an alphabetical listing (both subject and title) of in-print periodicals, both American and foreign, and lists some of the works that index the periodicals.

393

Chapter 9
The Research
Report:
Becoming
Acquainted with
Resource
Materials

Books.   Sheehy's *Guide to Reference Books* lists approximately 10,000 reference titles of both general reference works and those in the humanities, social sciences, history, and the pure and applied sciences. This guide and its supplements also include valuable information on how to use reference works.

The information in Sheehy's *Guide* may be updated by using *American Reference Books Annual* (ARBA), which lists by subject the reference books published each year in the United States.

The *Cumulative Book Index* lists books printed in the English language each year. *Books in Print* is a listing of the majority of the books in print in America.

CD-ROMs

CD-ROMs In Print provides information on CD-ROM products and producers worldwide. The title directory lists titles alphabetically giving title, description, subject areas, provider, and hardware/software requirements. It also includes a subject index.

Vertical File

Much printed information exists in other than book and magazine forms. Pamphlets, booklets, bulletins, clippings, and other miscellaneous unbound materials are usually in a collection called the vertical file. These materials are filed or cataloged by subject. The vertical file and the *Vertical File Index,* which lists current pamphlets, may be valuable sources of information.

Planning and Writing a Research Report

Now that you are familiar with the four general sources of information—personal observation or experience, personal interviews, free or inexpensive materials from various agencies, and library materials and their use—you are ready to begin planning and writing a research report.

A research report is the written result of an organized investigation of a specific topic. It requires systematic searching out and bringing together information from various sources. It requires that the gathered information be presented in a conventional, easy-to-follow form.

Writing a research report can be profitable and rewarding. Knowing how and where to search for information and then how to present that information in a practical, understandable, and logical manner are real accomplishments.

The procedure for writing a research report centers around four major steps:

1. Selecting a subject and defining the problem
2. Finding the facts
3. Recording and organizing the facts
4. Reporting the facts

Selecting a Subject and Defining the Problem

Selecting a Subject

The success of a research project depends on a wise choice of subject. If a specific subject is assigned, of course the selection problem is solved. However, if you choose the subject, consider the following guidelines.

1. *Choose a subject that fulfills a need.* Doing research should not be a chore or just another assignment—it should be an adventure that fulfills a need to know. Investigating some aspect of a future vocation, following up a question or a statement in a class, finding out more about an invention or a discovery, wanting to know the functions and uses of a particular mechanism—whatever the subject, let it be something that appeals to you because you want and need to know the information.

2. *Choose a subject that can be treated satisfactorily in the allotted time.* No one expects a college student to make an earthshaking contribution to human knowledge by presenting the result of months and years of research and study. Library research gives the student experience in finding, organizing, and reporting information on a specific subject. Therefore, the topic chosen should be sufficiently limited for adequate treatment in the few weeks allotted for the paper. The subject "Dogs," for instance, is much too broad to try to cover in a few weeks. Even "Hunting Dogs" would require far more time than is available. But "The Care and Training of Bird Dogs" is a specific topic that could reasonably be investigated in the usual time designated for a research report.

3. *Choose a subject on which there is sufficient material available.* Before deciding definitely on a subject, check with instructors and the library to be sure that the topic has sufficient accessible material. Generally, it is wise to choose another topic if the principal library you use has little or no information on the proposed topic. If the topic is highly specialized or has information in only very select journals, locate several references before deciding definitely to use the topic. Sometimes a news broadcast or a current event may suggest a subject. Again, be sure that enough material is available for an effective report.

395

Chapter 9
The Research
Report:
Becoming
Acquainted with
Resource
Materials

Defining the General Problem

After selecting the subject for a research report, state specifically the central problem or idea being investigated. The formulation of this central problem, or basic question, is important because it determines the kind of information to look for. In writing a report, the idea is not simply to gather information that happens to fall under a general heading; the idea is to gather information that relates directly to the subject and that can give it form and meaning. For the subject "The Effects of Alcohol on People," for example, the central problem might be defined as the following: "How does alcohol affect people: How does it affect their bodies, behavior, relationships with other people?" This question gives direction to the researcher's reading and to the investigation as a whole.

Defining the Specific Problems

As a guide in searching for information, divide the central problem into specific problems or questions. These questions serve as a preliminary outline for more selective reading. Again, take the subject "The Effects of Alcohol on People" with the central problem, "How does alcohol affect people: How does it affect their bodies, behavior, relationships with other people?" Divide this general question into smaller ones, such as these:

1. Does alcohol kill brain cells?
2. Why do people drink?

3. Why do some people become silly and giggle a lot and others become morose and withdrawn when they drink?
4. What organs of the body does alcohol affect?
5. How does excessive drinking affect family relationships?
6. How much alcohol does it take to have an effect on the body?
7. Why can some people take or leave alcohol while others become addicted to it?

Other questions, of course, might be added to these.

As the investigation proceeds and the body of collected data increases, change or drop or add to the questions in the preliminary outline.

Finding the Facts

Finding the facts for a research report requires a search of all available resource materials on the subject of the report. As discussed earlier in this chapter, the possible sources of information can be classified as: (1) personal observation, (2) personal interviews, (3) free or inexpensive materials from various agencies, and (4) library materials.

Make a thorough, systematic search of all available materials. A good place to start is a general encyclopedia. From there move to the public access catalog, remembering to check every possible topic you can think of that is related to the subject of the report. Then check indexes to periodicals; if the subject requires current information, check only the issues of current date. If, however, the date of publication is not a factor, check in all available issues. Remember to check in more than one index. You might start with the *General Periodicals Index* or the *Readers' Guide* and then the *Applied Science & Technology Index* or the *Biological and Agricultural Index* or one of the other specialized indexes.

A check of these basic resources is a good start in finding the facts. Then you can go to any other available resources.

Recording and Organizing the Facts

Recording and organizing information is a key step in writing a successful research report. This step involves evaluating resource material, compiling a working bibliography, taking notes, stating the central idea of a report and contructing an outline or table of contents, and arranging the note cards to fit the outline or table of contents.

Evaluating Resource Material

You must *evaluate* resource material for all sources are not equally useful or reliable. To evaluate material carefully, you need to know the difference between primary and secondary sources, to separate fact from opinion, and to use the non-textual qualities of a source in evaluating it.

Primary and Secondary Sources. Primary sources are those giving firsthand accounts of an event or condition. Material written by Dr. Jonas Salk on the development of the Salk polio vaccine would be a primary source. Secondary sources are works written about firsthand accounts. A magazine article about Dr. Salk would be a secondary source. Although secondary sources are generally more readily available, use primary sources whenever possible.

Fact and Opinion. Separating fact from opinion is essential in evaluating a source. A fact is an item of information that can be proved, such as "The book has 875 pages" or "Hereford is a breed of cattle." An opinion is an interpretation of a fact, or an idea about a fact, such as "This book has 875 *exciting* pages" or "Hereford is the *most profitable* breed of cattle." The researcher must be able to distinguish facts from the writer's opinions.

397

Chapter 9
The Research
Report:
Becoming
Acquainted with
Resource
Materials

Nontextual Qualities. The nontextual qualities of a source may be helpful in evaluation. The reputation of the author and of the publishing company, date of publication, table of contents, introductory material, bibliography, index—each gives an indication of the usefulness and reliability of a source. An article about the surface of the moon written in 1968 would not be as reliable as an article written by Armstrong and Aldrin after their moon landing in 1969 or one written by astronauts who visited the moon later.

Compiling a Working Bibliography

Once you have selected a subject and defined the problem to be researched, the next step is to begin research. As you locate possible specific sources, list these sources (books, periodicals, and so on) to compile a working bibliography or tentative bibliography.

You will probably compile your working bibliography through a combination of computer printouts of citations, print copies of citations from a microform (such as microfiche and microfilm), photocopied citations from print indexes, and citations or sources that you record in your handwriting.

If you handwrite citations, a workable way to record them for efficient use later is to record the information in the same order and form to be used for the final bibliography or list of works cited. You will notice that the form is different from the form used in a public access catalog, in a computer printout, or in an index.

A good plan to follow is to record complete bibliographical information as you go. Record bibliographical information on 3-by-5- or 4-by-6-inch cards, one card for each source. The cards are easier to handle than pieces of paper and simplify alphabetizing the final bibliography. Later as you examine the book, article, or other material, you will check to be sure that the bibliographical information you have recorded is accurate and complete. Desperate last-minute dashes to the library can take hours of time.

The following samples illustrate a standard form for recording on cards the bibliographical information for a book, a journal article, and a popular magazine article. Forms for other citations are given on pages 405–409.

Bibliography Card for a Book. A bibliography card for a book includes the following items:

1. Author's or editor's name (last name first for easy alphabetizing)
2. Title of the book (underlined)
3. Edition (if other than the first)
4. Volume number (if the work is in several volumes)
5. City of publication (followed by a colon)
6. Publishing company
7. Year of publication
8. Call number

In addition, it may be helpful to write the following information on the bibliography card.

9. Notation as to the contents of the book, such as textbook
10. Library where the book is located, if several libraries are used. Place the name of the library below the call number.

Sample Card for a Book

> *Lewis, Thomas E. <u>BASIC Programming</u>.*
> *Dubuque: Brown, 1994.*
>
> *Illustrations*
>
> *Textbook*
>
> *0.01.6*
> *L53q*
> *college library*

Bibliography Card for an Article in a Journal. On each bibliography card for an article from a journal, record:

1. Author's name (last name first)
2. Title of the article (in quotation marks)
3. Title of the journal (underlined)
4. Volume number
5. Date of publication (in parentheses)
6. Page numbers

Sample Card for an Article in a Journal

> *Heywood, John S. "Unionization and the*
> *Pattern of Nonunion Wage Supplements."*
> *<u>Southern</u> <u>Economic</u> <u>Journal</u> 60 (April 1994):*
> *1020–1029.*

Bibliography Card for an Article in a Popular Magazine. On each bibliography card for an article from a popular magazine, record

1. Author's name (last name first), if given
2. Title of the article (in quotation marks)
3. Title of the magazine (underlined)
4. Date of publication
5. Page numbers

399

Chapter 9
The Research
Report:
Becoming
Acquainted with
Resource
Materials

Sample Card for an Article in a Popular Magazine

Levinson, Marc. "Running Low on Gas."
<u>*Newsweek*</u> *29 August 1994: 38–39.*

Taking Notes

While investigating each source in the working bibliography, save time by first consulting the table of contents and the index for the exact pages that may contain helpful information. After locating the information, scan it to get the general idea. Then, if the information is of value, read it carefully. If certain material may be specifically used in the paper, make notes. Sometimes students photocopy sources and use a highlighter to mark pertinent passages. Be wary of trying to use such pages as notes because you will tend to give reports on individual sources rather than integrate material from all sources throughout the paper.

General Directions for Taking Notes. Regardless of the shortcuts you may think you have devised for taking notes, following these directions will be the shortest shortcut.

1. Take notes in ink on either 3-by-5- or 4-by-6-inch cards. Ink is more legible than pencil, and cards are easier to handle than pieces of paper.
2. If it is difficult to decide when to take notes and when not to, stop *trying* to take notes. Read a half dozen or so of the sources, taking no notes. Study the preliminary outline or table of contents. (Do not hesitate to revise the outline— it is purely for your use.) Then return to the sources with a clearer idea of what facts are needed.
3. Write one item of information on a note card. Write on only one side of the card. Cards can then be rearranged easily.
4. Always write notes from different sources on different cards.
5. Take brief notes—passages that may possibly be quoted in the paper, statistics and other such specific data, proper names, dates, and only enough other information to jog the memory. Writing naturally or well cannot be done if the

paper is based solely on notes rather than on an overall knowledge and understanding. One caution: Write enough so that you can use the information after it is "cold."

6. Write a key word or phrase indicating the content of each note in the upper left corner for quick reference. This key word or phrase will be especially helpful in arranging cards according to topic; often these key words or phrases become headings in the final outline.

7. In the upper right corner, identify the source (author's name and, if necessary, title of the work) and give the specific page number from which the material comes. This information in the right corner keys the note card to the bibliography card, needed for writing notes and bibliography.

8. Remember that you must give proper credit in the paper for all borrowed material—quotations, exact figures, or information and ideas that are not widely known among educated people. Remember, too, that an idea stated in one's own words is just as much borrowed as an idea quoted in the author's exact words. Therefore, take notes carefully so that you know which notes are direct quotes and which are paraphrases or summaries.

Types of Notes.   Notes may be direct quotations, paraphrases, or summaries.

- *Direct Quotation.* Some notes will be direct quotations, that is, an author's or speaker's exact words.

    There may be passages in which an author has used particularly concise and skillful wording, and you may want to quote them; in this case, the author's exact words are written on the note card. Generally, however, take down few direct quotations—they create a tendency to rely on someone else's phrasing and organization for the report. If it is desirable to leave out part of a quotation, use ellipsis points, that is, three spaced dots (. . .) or if the omission includes an end period, the end period plus three spaced dots (. . . .). Be sure to put quotation marks around all quoted material.

Sample Note Card: Direct Quotation

Noise as health hazard—statistic   *"Now Hear This" p. 50*

"Increasingly, the racket that surrounds us is being recognized not only as an environmental nuisance but also as a severe health hazard. About 28 million Americans, 11%, suffer serious hearing loss, and more than a third of the cases result from too much exposure to loud noise."

- *Paraphrase and Summary.* Most of the notes for a library report will be paraphrases and summaries rather than direct quotations. A paraphrase is a restatement in different words of someone else's statement. A summary is the gist—it includes the main points—of a passage or article. In reading material, listening to a speaker, or viewing visual material, get main ideas and record them in your own words. Jot down key phrases and sentences that summarize ideas clearly. Do not bother taking down isolated details and useless illustrative material. It is not necessary to write notes in complete sentences, but each note should be sensible, factual, and legible—and meaningful after two weeks.

Following are examples of paraphrase and summary note cards.

401

Chapter 9
The Research
Report:
Becoming
Acquainted with
Resource
Materials

Sample Note Card: Paraphrase

Noise as health hazard—statistic   *"Now Hear This" p. 50*

Loud noise—serious health hazard to millions of Americans. Accounts for the hearing impairment of more than a third of some 28 million hard-of-hearing people in our country.

Sample Note Card: Summary

Noise as health hazard   *"Now Hear This" pp. 50–51*

Americans do not seem to be concerned about hearing loss caused by loud noise at work or in their leisure activities. But loud noise is causing severe hearing loss for millions.

Constructing a Formal Outline or Table of Contents

At the very beginning of the research project you formulated a basic question or problem; then you divided this general question into specific questions, or aspects, as a preliminary outline. As investigation has progressed, you probably have added to, changed, and omitted some of the original questions. Thus, the present list of questions looks quite different from the original list. (You will need to revise and bring up to date the outline or table of contents after you have taken notes from source materials.)

By this stage in the research you should have clearly in mind the central idea or central question of the report. Write this central idea or question at the top of what will be the working outline page or table of contents page.

From the central idea, the list of questions or divisions in the revised preliminary outline or table of contents, and key words in the upper left-hand corner of the note cards, structure a formal outline or table of contents to serve as your working plan for the first draft of the report. Each major division in the outline or table of contents should reflect a major aspect of the central idea. Taken as a whole, all of the major divisions should cover the central idea and adequately explain it.

Most formal reports in business and industry have a table of contents. The table of contents is a listing, with page numbers, of all the headings in the report. (On pages 25, 315, and 413 and on the first page of each chapter in this book—pages 1, 26, 74, 105, and so on, are examples of tables of contents.) The table of contents and the outline serve similar purposes: They both show at a glance the content of the report, and they use indenting to distinguish major headings, divisions, and subdivisions.

The formal outline may be either a sentence outline or a topic outline. Remember: In a sentence outline all the headings are complete statements (no heading should be a question), and in a topic outline all the headings are phrases or single words. The sentence outline is more complete and helps to clarify the writer's thinking on each point. The topic outline is briefer and shows at a glance the divisions and subdivisions of the subject. Whichever type of outline you choose, be consistent; do not mix topics and sentences. (For further treatment of outlining, see pages 139–140 and 404.)

The table of contents or the working outline will serve as a work plan in writing the first draft of the report; it will let you know where you are and will provide a systematic organization of the information.

Arranging the Note Cards to Fit the Outline

When you have completed the table of contents or the working outline, mark each note card with a roman numeral, letter of the alphabet, or arabic numeral to show to what section and subsection of the outline the note card corresponds or write on the card the table of contents heading. Then rearrange the note cards accordingly. If the note cards are insufficient, look up additional material; if there are irrelevant note cards, lay them aside. When you have enough material to cover adequately each point in the table of contents or in the outline, you are ready to begin writing the report. Each main heading in the table of contents or in the outline should correspond to a major section of the report and each subheading to a supporting section.

Reporting the Facts

Writing the First Draft

With the table of contents or the working outline for a guide and the note cards for content, write the first draft of the report rapidly and freely. Concentrate on getting thoughts down on paper in logical order; take care of grammar, punctuation, and mechanical points later. In composing the first draft, you will have to add short transitional passages to connect the material from the note cards. Also, you will likely rephrase notes (except for direct quotations) to suit the exact thought and style of the report as it evolves. Add headings for each section, using key words and phrases from each section of the table of contents or the outline. (See pages 9–11 and 311 for a discussion of headings.) Think through the page layout and design of the report (see pages 6–9, 34–36).

Introduction and Conclusion

In the first paragraph or in a section headed "Introduction," state the central idea of the report. Let the reader know what to expect, to what extent the subject will be explored, as indicated in the major headings of the outline. The introduction, designed to attract the reader's attention, serves as a contract between writer and reader. It may be a single paragraph, several paragraphs, or even several pages. In it, the writer commits the report to a subject, a purpose, and an organization of material. (See also Chapter 8 Reports, page 310, for further discussion of the introduction.)

In the introduction, make clear the subject to be covered and the extent of the coverage. Stating the central idea usually is sufficient to identify the subject; a sentence summarizing the major headings of the outline will show the extent of coverage.

Supply any information needed to help the reader understand the material to follow; this information might include a definition of the subject itself or of terms related to the subject. Explain the significance of your investigation and reporting.

After completing the body of the paper, close with a short section that is a summary, a climax, conclusions or recommendations, or suggestions for further study drawn from the material that has been presented.

403

Chapter 9
The Research
Report:
Becoming
Acquainted with
Resource
Materials

Quotations and Paraphrases

If the exact words of another writer are used, indicate that you are quoting. Enclose short quotations in quotation marks. If the quotation is four lines or more in length, set it off; instead of using quotation marks, indent each line ten spaces from the left margin.

Incorporate paraphrased matter and indirect quotations into the body of the report, and do not use quotation marks. However, document *all* borrowed information, whether quoted directly or paraphrased. See Types of Notes earlier in this chapter, pages 400–401. (Documentation is discussed in detail later in this chapter.)

A Note About Plagiarism.   Plagiarism is implying that another person's ideas or words are your own ideas or words. Sometimes plagiarism is unintentional, sometimes intentional. To avoid the problem of plagiarism,

1. Place quotation marks around any material that you yourself did not write—direct quotations—and credit the source. Indicate material copied on notes by using quotation marks. In the report you may wish to introduce the material by using the name of the person who wrote it or the title of the work in which it appeared.
2. Write paraphrased material in your own style and language, and credit the source. Simply rearranging the words in a sentence or changing a few words does not paraphrase.

Plagiarism is a serious offense, punishable by law. When you use another person's ideas or wording—and do not give credit—you are plagiarizing. There is no excuse for plagiarizing. A careful handling of information will help you avoid this problem.

Documenting Borrowed Information

Documenting information means supplying references to support assertions and to acknowledge ideas and material that you borrow. Documentation is usually done by inserting in the text a shortened parenthetical identification or a number that corresponds to a work identified fully in a footnote at the bottom of the page, in a note at the end of the report, or, most often, in a list of "Works Cited" at the end of the report. Sometimes included is a bibliography listing all the sources used in writing the report, whether specifically referred to or not. (Documentation practices are discussed in detail below and continuing through page 409.)

Revising the Outline and First Draft

After writing the first draft of the report, study it carefully for accuracy of content, mechanical correctness, logical organization, clarity, proportion, and general effectiveness. Check to see if you have kept the intended reader in mind. The writer may understand perfectly well what is written, but the real test of writing skill is whether or not the reader will understand the report.

Make needed revisions in the outline and the first draft. Make sure that the headings in the report correspond exactly to the headings in the outline. (For a further discussion of headings, see pages 9–11 and 311–312.) Plan and prepare visuals. Think through the kinds of visuals that will be the most effective and decide where to place the visuals. (For a detailed discussion, see Chapter 11 Visuals.)

The revising process may involve making several drafts of the table of contents or outline and the report. But after the hours, days, and weeks already invested in this project, a few more hours spent polishing the report are worth the effort. Careful revision may mean the difference between an excellent and a mediocre report. (See Part III Revising and Editing and the checklists for revising, front and back endpapers.)

Preparing the Final Copy

When you have completed the table of contents or the outline, the text of the report, and the documentation, prepare a letter of transmittal, a title page, and an abstract (see pages 309–310). For the entire report, follow the general format directions for writing, on the inside front cover of this book. Double-space the keyboarded report.

Proofread the report thoroughly. Clip the pages together in this order: letter of transmittal, title page, outline or table of contents, abstract, body of the report, and works cited page. The research report is finished at last and ready to be handed in! (For a detailed discussion of reports, see Chapter 8 Reports.)

Documentation

Documentation provides a systematic method for identifying sources of references used in a research report.

Documentation styles vary from discipline to discipline and even within disciplines. Societies or organizations of persons in a specific discipline may provide style manuals of their own. Examples include the American Mathematical Society's *Manual for Authors of Mathematical Papers,* the Council of Biology Editors' *CBE Style Manual,* the American Chemical Society's *The ACS Style Guide: A Manual for Authors and Editors,* the American Institute of Physics' *AIP Style Manual,* the International Committee of Medical Journal Editors' "Uniform Requirements for Manuscripts Submitted to Biomedical Journals," the *American Medical Association Manual of Style,* and *A Uniform System of Citation* distributed by the Harvard Law Review Association.

Then there are basic manuals used by various disciplines, such as *The MLA Style Manual* (New York: MLA, 1985); *The Chicago Manual of Style* (14th ed.; Chicago: U of Chicago P, 1993); *Publication Manual of the American Psychological Association* (4th ed.; Washington: American Psychological Assn., 1994); Turabian's *Manual for Writers of Term Papers, Theses, and Dissertations* (5th ed.; Chicago: U of Chicago, 1987); Campbell, Ballou, and Slade's *Form and Style: Research Papers, Reports, Theses* (9th ed.; Boston: Houghton, 1994); or the *United States Government Printing Office Style Manual* (Washington: GPO, 1984).

405

Chapter 9
The Research
Report:
Becoming
Acquainted with
Resource
Materials

Although documentation practices in literature, history, and the arts are somewhat uniform, documentation in other areas follows various practices. Among the science disciplines, for instance, there is little uniformity of procedure.

While the *reason* for documentation is the same in all areas—citation of sources for statements presented or to acknowledge borrowed matter—the *procedures* for citing sources are somewhat different.

The following discussion is presented with these reservations and with this advice: For documentation practices, consult with the instructor assigning the report, consult with instructors in the subject area of the report, and study the practices in pertinent professional journals. Whatever format you choose, be consistent.

Four Styles of Documentation

In documenting information, the important thing is to cite the source of the information. The source can be given in the text:

> According to Robert W. Van Giezen ("Occupational Wages in the Fast-Food Restaurant Industry," *Monthly Labor Review* 117 [1994]: 24–30), the majority of the employees in fast-food restaurants are 20 years of age or younger.

Such a citation within the text, however, detracts from the emphasis in the sentence; and several such citations within the text are likely to annoy the reader and hinder readability.

To avoid the distracting, full internal inclusion of sources, writers use various styles or systems of documentation. Four styles dominate, referred to here as Style A, Style B, Style C, and Style D.

Style A: MLA

Style A, recommended by the Modern Language Association (MLA), is distinguished by a textual parenthetical listing of the reference (author's last name and the page number) and at the end of the report, a full citation of the reference in the alphabetical list of works cited.

Example:

> The majority of the employees in fast-food restaurants are 20 years of age or young (Van Giezen 24).

The full citation from the "Works Cited" page:

> Van Giezen, Robert W. "Occupation Wages in the Fast-Foot Restaurant Industry." *Monthly Labor Review* 117 (1994): 24–30.

See pages 411–423 for a report using Style A documentation.

Style B: APA and Chicago

Style B is recommended by *The Publication Manual of the American Psychological Association* (APA) and *The Chicago Manual of Style* (Chicago). It is distinguished by a textual parenthetical listing (author's last name and the year of publication) and at the end of the report, a full citation of the reference in the list of works cited.

Example:

> The majority of the employees in fast-food restaurants are 20 years of age or younger (Van Giezen, 1994). [Chicago omits the comma between author and year.]

The full citation in the list of references:

[handwritten: No CAPS only first word]

> APA: Van Giezen, R. W. (1994). Occupational Wages in the Fast-Food Restaurant Industry. *Monthly Labor Review,* 117(8), 24–30.

> Chicago: Van Giezen, Robert W. Occupational Wages in the Fast-Foot Restaurant Industry. *Monthly Labor Review* 117, no. 8 (1994): 24–30.

Another example of Style B:

Text of Paper:

> According to Broomhall and Johnson (1994), the better students move to pursue higher paying jobs resulting in reduced productivity of the local workforce and discouraging new employers from locating in the community.

Entry from "References" at the End of the Paper:

> Broomhall, David E., and Thomas G. Johnson. 1994. "Economic Factors that Influence Educational Performance in Rural Schools." *American Journal of Agricultural Economics* 76:557–567.

Style C: Number System

Style C is a number system. A number is usually placed in parentheses or in brackets (sometimes the number is written as a superscript—slightly above the regular line of type) in the text. This number refers to an entry in the numbered list of references at the end of the paper.

Example:

Text of Paper:

> According to recent studies, some species of whales are increasing in numbers (11, 12).

Entries from "Literature Cited" at the end of the paper:

> 11. Schmidt, K. *Science* 1994, *263*, 25–26.
> 12. Goodyear, J. D. *J Wildl Manage* 1993, *57*, 503–513.

The list of references at the end of the paper may be given (a) in alphabetical order by author's name or (b) in the order in which the sources are mentioned in the text of the paper (as in the above example).

References in science papers may be written in a number of ways; there is no one uniform format. Given below is a sampling of notes using various formats. (Remember, whatever format you choose for your paper, be consistent.)

Examples (each example from a different paper):

(4) Sprent, Jonathan; Tough, David F. *Science,* 265:1395–1400 (1994).

17. Ammons BS, Smalley DL, Madison BM: Evaluation of a New Rapid Latex Test for the Detection of Heterophil Antibody. *Clin Lab Sci* 7:245–249, 1994.

9. Puttré, M., "CAD for Pen-Based Systems," *Mechanical Engineering,* August 1994, pp. 62–63.

6. Love, L. C., Principles of Metallurgy (Reston, VA: Reston Publishing Company, 1991), chap. 2. p. 40.

407

Chapter 9
The Research
Report:
Becoming
Acquainted with
Resource
Materials

Style D: Traditional

Style D is the traditional humanities system of documenting through footnotes or endnotes and with or without a list of works cited or a bibliography. A superscript number in the text is keyed to a full citation at the bottom of the page or, more commonly, at the end of the report.

Example:

Funding received by the Centers for Disease Control for infectious disease research was lower in 1992 (when adjusted for inflation) than funds received 40 years prior.[1]

The full citation as a footnote or endnote:

[1]Philip E. Ross, "A New Black Death?" *Forbes* 12 Sept. 1994: 241.

If there is a list of references, the citation is as follows:

Ross, Philip E. "A New Black Death?" *Forbes* 12 Sept. 1994: 240–243, 246, 250.

In Style D the first reference to a source is a complete citation, as illustrated in the example above. Frequently, reference is made to a source a second, third, or more times. Any source reference after the first usually gives the author's last name and the page number(s). The reference may also include an intelligible short title if necessary to make clear which work is cited.

First reference:	[2]Charles G. Ramsey, *Ramsey/Sleeper Architectural Graphic Standards,* ed. John Ray Hoke, Jr., 9th ed. (New York: J. Wiley, 1994), p. 229.
Second reference:	[3]....
	[4]....
	[5]Ramsey, p. 708.

Notice that footnotes/endnotes are indented as if they were paragraphs and that the items of information are separated by commas, not periods.

Citations List and Bibliography

A citations list or bibliography typically concludes a library research report. A citations list gives author, title, and publishing information about all source materials referred to in the text. The citations list is headed "Works Cited," "Works Consulted," "References," or a similar title.

The citations list is usually in alphabetical order. If Style C documentation (a number system) is used, however, the numbered list of citations may be either alphabetical or in the order in which the sources are mentioned in the text.

A bibliography is an alphabetical listing of all source materials—whether specifically referred to or not—used in preparing and writing a research report.

An annotated bibliography includes descriptive or evaluative comments about each entry. (See Chapter 6 The Summary, especially page 224.)

The citations list and bibliography observe hanging indentation, that is, in every entry each line after the first is indented.

Examples of Citations

Given below are examples of citations to various kinds of references: books, periodicals, and nonprint materials. These examples follow Style A documentation (author-page parenthetical insertion in the text and an end-of-report citations list). The sample research report that concludes this chapter follows Style A (MLA) documentation.

Remember that although documentation styles differ, they differ only in details—details of punctuation and abbreviation and of order in the placement of information. All documentation styles give the essential information for locating a particular source.

Citations for Books

One author	Neel, James V. *Physician to the Gene Pool: Genetic Lessons and Other Stories.* New York: Wiley, 1994.
Two or more authors	Silverman, Jay, Elaine Hughes, and Diana Roberts Wienbroer. *Rules of Thumb: A Guide for Writers.* 3rd ed. New York: McGraw, 1996.
Three or more authors	Bowen, H. Kent, et al. *The Perpetual Enterprise Machine: High Performance Product Development in the 1990's.* New York: Oxford UP, 1994.
Corporate author	National Safety Council Staff. *First Aid Guide.* 2nd ed. Boston: Jones, 1993.
	United States. Department of Agriculture. *Nutrition, Eating for Good Health.* Agriculture Information Bulletin 685. Washington: GPO, 1993.
Editor	Zerwekh, JoAnn, and Jo Carol Claborn, eds. *Nursing Today: Transition and Trends.* Philadelphia: Saunders, 1994.
Essay	Mohrman, Kathryn. "The Public Interest in Liberal Education." In David H. Finifter and Arthur M. Hauptman (Eds.), *New Directions for Higher Education,* #85. San Francisco: Jossey-Bass, 1994. 21–30.
	Fetterley, Judith. "The Sanctioned Rebel." *Studies in the Novel* 3 (Fall 1971): 293–304. Rpt. in *Critical Essays on The Adventures of Tom Sawyer.* Ed. Gary Scharnhorst. New York: Hall, 1993. 119–29.

409

Chapter 9
The Research
Report:
Becoming
Acquainted with
Resource
Materials

Republished book	Finck, Henry. *Romantic Love and Personal Beauty*. 1891. Havertown: West, 1973.
Encyclopedia article	"Field Effect Transistor." *McGraw-Hill Encyclopedia of Science and Technology*. 1992 ed.
	Kornblum, Zvi C. "Photochemistry." *Encyclopedia Americana*. 1994 ed.

Citations for Articles in Periodicals

Article in a journal	Hoge, Charles W., et al. "An Epidemic of Pneumococcal Disease in an Overcrowded, Inadequately Ventilated Jail." *The New England Journal of Medicine* 331 (1994): 643–48.
Article in a popular magazine	Spragins, Ellyn. "Much Ado About . . . Not Much." *Newsweek* 5 Sept. 1994: 48–50.
	Scott, Dan. "Designer Catalysts." *Chem Matters* Apr. 1994: 13–15.
Article reprinted in another source	Crawford, Kimberly A. "Surreptitious Recording of Suspects' Conversations." *FBI Law Enforcement Bulletin*. Dept. 1993: 26–32. Rpt. in Social Issues Resources Series, *Privacy* 5: art. 15.
Item in a newspaper	Trumbull, Mark. "Voice-Processing Services Link Telephones with PCs." *The Christian Science Monitor* 15 Sept. 1994: 9.
	"The Power of One." Editorial. *The Wall Street Journal* 20 Sept. 1994, southwest ed.: A20.

Citations for Nonprint Materials

NewsBank (microfiche)	Rohan, Barry. "Answering the Call." *Detroit Free Press* [MI] 18 July 1994. NewsBank, Health, 1994, fiche 46, grids E5-E6.
Filmstrip (or film or videotape)	*Body Language Videorecording: An Introduction to Non-Verbal Communication*. Videocassette. Learning Seed, 1993. 25 min.
Computer program	*Wordperfect*. Version 3.1. Computer software. System 6.0.7 or later. MacPlus or higher. 2 MB RAM. 11 MB hard disk space. WordPerfect Corp., 1993.
Interview	Tenhet, Toby. Personal interview. 22 Aug. 1995.

A Sample Research Report

Following is a sample research report prepared by a student concerned about human fetal tissue implantation. Included are the letter of transmittal, the title page, the outline (plus an alternate table of contents), an abstract, the report, and the list of works cited. Note the use of headings throughout the report and the use of visuals. The report uses Style A (MLA) documentation.

P.O. Box 8021
Jackson, MS 39284-8021
7 May 1992

Dr. Nell Ann Pickett
Department of English
Hinds Community College
Raymond, MS 39154-9799

Dear Dr. Pickett:

Enclosed is the report entitled "The Ethical Use of Human
Fetal Tissue in Transplantation" as required in English
1123-Technical Approach. It discusses a new treatment for
certain diseases, ethical issues concerning this type of
treatment, and some possible solutions to this dilemma.

All research for this report was done at the McLendon
Library. As a nursing major, this subject is of particular
importance to me. It addresses certain issues that I may
face in the future as an R.N.

Thank you for all your wonderful assistance and
encouragement to me during this research.

Sincerely,

Judy A Smith

Judy A. Smith

Enclosure: Research report

Research Report, Title Page

411

Chapter 9
The Research
Report:
Becoming
Acquainted with
Resource
Materials

<u>**THE ETHICAL USE OF HUMAN FETAL TISSUE**</u>
<u>**IN TRANSPLANTATION**</u>

by
Judy A. Smith

English 1123-BE
Hinds Community College
Raymond, Mississippi
7 May 1992

THE ETHICAL USE OF HUMAN FETAL TISSUE
IN TRANSPLANTATION

The purpose of this paper is to present the political and
ethical issues in the use of human fetal tissue in
transplantation and to suggest possible solutions.

  I. Discovery analysis

     A. Early-century fetal tissue use
     B. Recent fetal tissue use

  II. Benefits of using fetal tissue

     A. Unique properties
     B. Valuable natural resource

  III. Factors in the use of fetal tissue in transplantation

     A. Moratorium of federal funding of research
     B. Lack of federal supervision
     C. Pressure to find cures for diseases
     D. Issues for debate
     E. Immoral act

  IV. Possible solutions

     A. Recombinant DNA
     B. Cell lines

Note: A research report in the workplace typically includes a table of contents rather than an outline. Below is an example of a table of contents that could replace the outline on the preceding page.

Your report should have an outline *or* a table of contents—not both.

Research Report, Table of Contents

413

Chapter 9
The Research
Report:
Becoming
Acquainted with
Resource
Materials

TABLE OF CONTENTS

2

ABSTRACT

Medical science may have discovered a revolutionary way to
treat such diseases as Parkinson's disease, Alzheimer's
disease, diabetes, cancer, AIDS, and certain spinal cord
injuries through the use of human fetal tissue
transplantation. But this discovery has led to a heated
political and ethical debate between research scientists
and conservative groups. Solutions to this dilemma are
possible through further types of research—in recombinant
DNA and cell lines—that could neutralize the controversial
issues.

3

Research Report   **415**

Chapter 9
The Research
Report:
Becoming
Acquainted with
Resource
Materials

<u>**THE ETHICAL USE OF HUMAN FETAL TISSUE
IN TRANSPLANTATION**</u>

In a nation dependent on medical research for
discoveries of treatments and cures for diseases and
ailments, scientists have opened up an ethical Pandora's
box by using human fetal tissue from induced abortions for
the purpose of therapeutic transplantation. Each day in our
nation, over four thousand abortions take place (Orenstein
298). To many people in our society, abortion for any
reason is immoral.

Can it be ethical to use human fetal tissue for
transplantation? That is the question that is the basis of
this report.

<u>**DISCOVERY ANALYSIS**</u>

Discoveries through fetal research have taken place
since the nineteenth century (Maugh grid F12). Human fetal
tissue has proven to be a valuable resource in the world of
research (Greely et al. 1093).

<u>**Early-Century Fetal Tissue Use**</u>

Since the 1930s, research with human fetal tissue has
led to the production of vaccines for polio and smallpox,
certain treatments for diseases of the fetus, and
biochemical treatment of cancer cells.

Recent Fetal Tissue Use

    The most recent use of fetal tissue is to treat and possibly cure diseases such as Parkinson's disease, Huntington's chorea, Alzheimer's disease, diabetes, AIDS, cancer, and several conditions such as spinal cord injuries and quadriplegia. Figure 1 illustrates how fetal tissue can be used to treat a victim of Parkinson's disease (Clark 62).

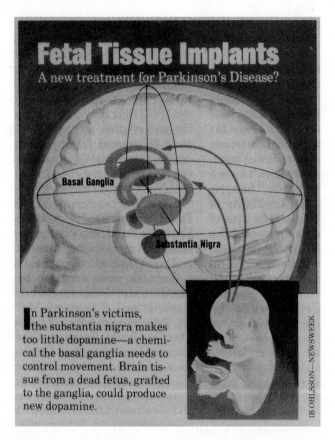

Figure 1. Fetal tissue implantation in a victim of Parkinson's disease.

Research Report, Cont.

417

Chapter 9
The Research
Report:
Becoming
Acquainted with
Resource
Materials

BENEFITS OF USING FETAL TISSUE

The reason fetal tissue is so beneficial for transplantation purposes is that it has exceptional abilities to regenerate expediently. Because of its unique properties and its availability as a valuable natural resource, human fetal tissue is in great demand for research.

Unique Properties

Fetal tissue has several unique properties that make it useful for transplantation. One property is that fetal tissue is highly compatible; it adapts readily and easily to the area of transplantation. This minimizes the problem of typing and cross-matching tissues. A second property is that fetal tissue stimulates growth and can tolerate long-term storage. The third property is that there is little or no rejection by the recipient (Weiss 296).

Valuable Natural Resource

Fetal tissue is viewed among scientists as a valuable natural resource (Orenstein 300). Because of the moral disputes in progress, however, the supply of fetal tissue is in short supply (Maugh grid F13).

FACTORS IN THE USE OF FETAL TISSUE IN TRANSPLANTATION

The use of fetal tissue for transplantation has ignited strong debates between certain groups. The areas in dispute are federal funding, federal supervision, pressure for disease cures, issues for debate, and immoral act.

Federal Funding of Research

Because of the tremendous arguments over the ethical implications of this issue, a moratorium was imposed in May 1988 (Levine 112) and has since been extended (Elmer-Dewitt 52). This moratorium denied "federal funding of research utilizing human fetal tissue, obtained from induced abortions, for therapeutic transplantations'" (Levine 112).

This moratorium does not restrict private funding and private experimenting with fetal tissue for transplantation.

Lack of Federal Supervision

A potential problem resulting from the lack of federal support is that research funded by the private sector and fetal surgery can be performed without federal supervision or possibly ethical scrutiny (Orenstein 308). This creates risks of abuse to women and the possibility of poor quality control. An example of this is transmitting AIDS to a healthy person from an infected tissue sampling.

Despite the controversy over the abortion issue at hand, many people believe some good should come from the end results of an abortion; that is, the remains should be used for therapeutic transplants (Slovut grid B14).

Pressure to Find Cures for Diseases

Many patients and their families suffer from the complications of serious illness and diseases. Thus, society has placed a great demand on finding cures for these ailments.

As Barbara J. Culliton states in her report on fetal research, "A [continued] ban on fetal research would be devastating to scientific experimentation in a number of areas, including Parkinson's disease, Alzheimer's, diabetes, cancer, and AIDS" (1423). Actual experiments, so far, have been tried on patients with Parkinson's disease and on patients who are diabetic.

Research Report, Cont.

419

Chapter 9
The Research
Report:
Becoming
Acquainted with
Resource
Materials

Issues for Debate

The flip side of the debating issue has to do with the arguments that anti-abortion advocates project.

First, they protest the use of fetal tissue from induced abortions for the purpose of transplantation, saying this type of research will encourage more women to have abortions (Foreman grid C7). But some of these advocates seemingly contradict themselves. An example is one woman who wanted to get pregnant, have an abortion, and then donate the remains to her father, who suffered from Alzheimer's, to be transplanted. Another woman wanted to do the same thing. But her tissue would be used to treat her own diabetic condition (Clark 63).

Second, these advocates also suggest that tissue used in transplantation should come from spontaneous abortions. Researchers have concluded, however, that tissue from spontaneous abortions is of little use. The reason is that this tissue (spontaneously aborted) is abnormal or infected (Annas and Elias 1081). Too, if fetal tissue is not retrieved and stored immediately after an abortion, it deteriorates rapidly, depleting usefulness.

Immoral Act

Anti-abortion advocates believe that abortion is wrong; therefore it would be further unscrupulous to use tissues from aborted fetuses for the treatment of otherwise untreatable disorders (Krimsky et al.).

Both anti- and pro-abortion groups do agree, however, that women and clinics should not be paid for tissue samplings (Orenstein 302).

POSSIBLE SOLUTIONS

Like everything else in life, there are people who abuse the system. In order for fetal tissue transplantations to be ethical as well as legal, guidelines should be set—guidelines similar to those for organ transplants. In the end not everyone will be satisfied, but hopefully a compromising line can be drawn between opposing forces.

8

Currently, the Department of Health and Human Services (DHHS) still holds a moratorium on the use of fetal tissue obtained from induced abortions, for the purpose of transplantation. However, one of its sections—the National Institutes of Health (NIH)—has concluded it is acceptable to use human fetal tissue from induced abortion, for therapeutic transplantations. With the moratorium still in effect, this policy remains a problem (Krimsky).

Two possible solutions to this dilemma are recombinant DNA techniques and developing cell lines.

Recombinant DNA

One solution to these concerns is the future use of recombinant DNA techniques. Through these techniques, growing large quantities of various types of fetal cells would be possible (Culliton 1423).

Cell Lines

Another resolution might be the production of various cell lines of human fetal cells, such as W138 (Wehrwein). W138 was a piece of lung tissue retrieved from a fetus in Sweden, around 1958 or 1959. A cell line descended from W138 is still being reproduced. As illustrated in Figure 2, researchers have already been successful in growing fetal-cell cultures in the laboratory (Weiss 297).

Research Report, Cont.

421

Chapter 9
The Research
Report:
Becoming
Acquainted with
Resource
Materials

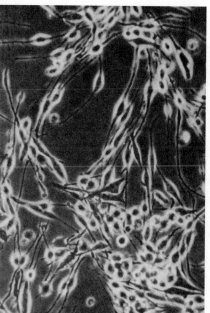

Eventually, nerve cells grown in culture may lessen or eliminate the need for fetal cells as transplants. Scientists have already implanted into animals cultured neuroblastoma cells (left), which are grown from tumors and resemble the precursor cells of neurons. Under proper conditions, laboratory neuroblastoma cells can differentiate into cells (below) that resemble mature neurons, able to make, store, and release neurotransmitters and generate electrical potentials. Their long fibers can make specific connections to muscles or to other nerve cells.

Figure 2. Laboratory production of cell lines.

10

These two solutions would be highly agreeable to opposing sides, thus eliminating the need for actual fetuses. How much time it would take to yield these kinds of results is unknown.

CONCLUSION

The question concerning the ethical use of fetal tissue for transplantation is one that has to be answered in each individual's mind. Although the U.S. Senate and House of Representatives lifted the ban of funding research on fetal tissue transplantation, ethical concerns still abound.

Research Report, Works Cited Page

423

Chapter 9
The Research
Report:
Becoming
Acquainted with
Resource
Materials

WORKS CITED

Annas, George J., and Sherman Elias. "The Politics of Transplantation of Human Fetal Tissue." The New England Journal of Medicine 320 (20 Apr. 1989): 1079–82.

Clark, Matt. "Should Medicine Use the Unborn?" Newsweek 14 Sept. 1987: 62–63.

Culliton, Barbara J. "White House Wants Fetal Research Ban." Science 16 Sept. 1988: 1423.

Elmer-Dewitt, Philip. "The Doctors Take on Bush." Time 5 Aug. 1991: 52–53.

Foreman, Judy. "Ethicists Back Fetal Tissue in Transplants." Boston Globe 20 Apr. 1989. NewsBank, Health, 1989, fiche 50, grid C7.

Greely, Henry T., et al. "The Ethical Use of Human Fetal Tissue in Medicine." The New England Journal of Medicine 320 (20 Apr. 1989): 1093–96.

Krimsky, Sheldon, Ruth Hubbard, and Colin Gracey. "Fetal Research in the United States: A Historical and Ethical Perspective." Genewatch July/Oct. 1988: 1+. Rpt in Social Issues Resources Series, Medical Science, 1989: Art. 13.

Levine, Robert J. "Fetal Research: The Underlying Issue." Scientific American Aug. 1988: 112.

Maugh, Thomas H., II. "Use of Fetal Tissue Stirs Hot Debate." Los Angeles Times 16 Apr. 1988. NewsBank, Health, 1988. fiche 55, grids F12–14.

Orenstein, Peggy. "The Use of Aborted Fetal Tissue in Medical Research Is as Controversial as Abortion Itself." Vogue Oct. 1989: 298–302, 308.

Slovut, Gordon. "Major 'U' Study Analyzes Fetal Tissue Transplants." Minneapolis Star and Tribune 14 Feb. 1990. NewsBank, Science, 1990, fiche 7, grids B14–C1.

Wehrwein, Peter. "Fetal Tissue Research Promising, Controversial." Times (Albany, NY) Union 3 Dec. 1989. NewsBank, Health, 1989, fiche 156, grids A8–9.

Weiss, Rick. "Forbidding Fruits of Fetal-Cell Research." Science News 5 Nov. 1988: 296–98.

Applications in Writing a Research Report

9.1   Answer the following questions concerning your college library.

a. What are the library hours?
b. What services does the library provide for (a) photocopying material, (b) paper printout of microform materials, (c) paper printout of materials in electronic formats?
c. What is the procedure for checking out a book?
d. For how long a period of time may books be checked out?
e. Where are reserve books shelved?
f. What is the procedure for requesting a magazine?
g. Where are current magazines located?
h. Where are nonprint materials located?
i. Which print and nonprint periodical indexes does the library have?

9.2   Draw a floor plan of your college library. Indicate the location of the following: public access catalog; periodical indexes; periodicals holdings list; general encyclopedias; technical encyclopedias, dictionaries, and guides; biographical reference works; and works in your major field.

9.3   In a periodical index (such as *Magazine Index/Plus, Reader's Guide to Periodical Literature,* or *Applied Science & Technology Index*) look up a general subject, such as hydraulics, nutrition, computers, radio receivers, metals, recreation, stock market, seawater, space research, antipollution laws, soils, glass, fibers, eye, electronic circuits, wood, internal-combustion engines, and so on. For the general subject, list at least five subclassifications sufficiently limited for a research report.

9.4   Choose one of the following topics (or pick a topic of your own). Formulate a possible basic question, or basic problem, to be investigated in a research report. Then divide the basic question into a list of specific questions (preliminary outline).

a. Problems in developing an electric automobile
b. Early results of the discovery of penicillin
c. New interest in cottage industries
d. Benjamin Franklin's minor inventions
e. Architectural problems in designing an underground home
f. Ethics and the Internet
g. July 20, 1969
h. Cooking food electronically
i. The next ten years in prosthetic development
j. The business office—100 years ago and today
k. The laser in eye surgery
l. Effects of disappearing farms on the U.S. economy

m. Changes in telemarketing techniques in the last 15 years

n. Videos—yesterday and today

o. Topic of your own choosing

425

Chapter 9
The Research
Report:
Becoming
Acquainted with
Resource
Materials

9.5   Explain each of the items on the following author citation from a public access catalog.

```
657

SI a    Siegel, Joel G
            Accounting handbook [by] Joel G. Siegel
            [and] Jae K. Shim. New York, Barron's [1990]
            836p. 24cm.

            1. Accounting-Handbooks, manuals, etc.
            2. Shim, Jae K., joint author
```

9.6   Turn to page 385. Look at the last three entries from the *Readers' Guide to Periodical Literature*. Explain the items in each of the entires.

9.7   Select a topic that is sufficiently restricted for a research report. Find and list at least two magazine articles, two books, and one encyclopedia article that contain information on the topic. Use conventional bibliographical forms. (See pages 408–409 for examples.)

INDEX CARDS FRIDAY –

9.8   Using the article on pages 585–587, "The Power of Mistakes," complete the following:

a. Make a bibliography card for the article.

b. Make a direct-quotation note card for the first sentence in paragraph 3.

c. Make a paraphrase note card of the information in the second and third sentences of paragraph 3.

d. Make a direct-quotation note card for the second sentence in paragraph 6, omitting the phrase "that can be turned."

e. Make a summary note of paragraph 7.

f. Of the information on the preceding notes, which would require documentation in a report? Why?

g. Show documentation for the summary note written in e above.

h. Write a paragraph about how you have learned from a mistake you made. Refer at least twice to this article, using parenthetical documentation. Write the entry for the Works Cited page.

9.9 Arrange the following information as if it were the final bibliography for a report.

Risk Management for Software Projects, by Alex Down and others, published in 1994 by McGraw-Hill, Inc., of New York City.

Just-in-Time Systems for Computing Environments, by Ralph L. Kliem and Irwin S. Ludin, published in 1994 by Greenwood of Westport, Connecticut.

"The 486: A Final Requiem," by Michael J. Miller, in *PC Magazine,* the 17 May 1994 issue, vol. 1, pp. 277–288.

"The Superhighway Patrol," by Laurie Flynn, in *The New York Times,* National edition, Section 3, page 8, columns 5–6, on 11 September 1994.

Software Development Tools: A Source Book, edited by Stephen J. Andriole, published in 1993 by Petrocelli Books, Princeton, New Jersey.

Legal Aspects of Computer Use, by William J. Luddy and Stuart R. Wolk, published in 1993 by Lansing Books in Boston.

Oral Presentation

9.10 As directed, adapt your research report for oral presentation. Ask your classmates to evaluate your speech by filling in an Evaluation of Oral Presentations on page 459, or one like it.

Using Part II Selected Readings

9.11 Read "The Powershift Era," by Alvin Toffler, pages 509–514. Under "Suggestions for Response and Reaction," page 515, complete Number 3.

9.12 Read "Quality," by John Gallsworthy, pages 521–525. Under "Suggestions for Response and Reaction," page 525, write Number 6.

9.13 Read the excerpt from "Non-Medical Lasers: Everyday Life in a New Light," by Rebecca D. Williams, pages 564–569. Under "Suggestions for Response and Reaction," page 569, write Number 4.

9.14 Read "Journeying Through the Cosmos," by Carl Sagan, pages 570–575. Under "Suggestions for Response and Reaction," page 576, write Number 3.

Schedule

9.15 Write a research report. Use the schedule as a checklist of steps in writing the report. Fill in the appropriate Plan Sheets from pages 428–434 before beginning each major step.

ASSIGNMENT	DATE DUE
Choice of general subject	_____
Choice of specific topic	_____
Statement of basic problem (basic question)	_____
Statement of specific questions (preliminary outline)	_____
Working bibliography	_____
Note cards	_____
Working outline	_____
Purpose statement or central idea	_____
First draft with correct documentation	_____
Final report, including outline or table of contents and documentation	_____
(Your instruction may also require that bibliography cards, note cards, computer printouts, outlines, preliminary drafts, copies of photocopied materials, etc., be turned in with the final report.)	_____

Chapter 9
The Research
Report:
Becoming
Acquainted with
Resource
Materials

PLAN SHEET 1
FOR WRITING A RESEARCH REPORT:
SELECTING A SUBJECT

Technical Field

My technical field is:

Allotted Time

Today's date is:

The completed report is due on:

That means I have _____ weeks to work on this report.

Possible Subjects

Five to ten possible subjects for my research report are the following:

1.   6.

2.   7.

3.   8.

4.   9.

5.   10.

Specific Topics

From the preceding list of possible subjects, the one that interests me the most is:

I can narrow this subject to at least three topics that I could treat satisfactorily in the allotted time:

1.

2.

3.

Chosen Topic

After checking in the library and with my instructors to be sure there is sufficient available material, I have chosen the following as the topic for my research report:

Questions

I need to ask my instructor these questions:

PLAN SHEET 2
FOR WRITING A RESEARCH REPORT:
DEFINING THE PROBLEM

Subject

The topic for my research report is:

General Problem

I can state the general problem or idea that I am investigating as this question:

Specific Problems

I can divide the general problem into these specific problems or questions:

Questions

I need to ask my instructor these questions:

PLAN SHEET 3
FOR WRITING A RESEARCH REPORT:
FINDING THE FACTS

Subject

The subject for my research report is:

Sources to Be Researched

Personal Observation or Experience   In gathering information for writing, I will need to visit the following places to observe conditions:

Personal Interviews   In gathering information for writing, I may want to interview the following people:

Library Materials   The library (or libraries) I will use is (are):

A tentative bibliography of available materials includes:

Free and Inexpensive Materials   In gathering information for writing, I may write or call the following companies or agencies:

Questions

I need to ask my instructor these questions:

PLAN SHEET 4
FOR WRITING A RESEARCH REPORT:
RECORDING AND ORGANIZING THE FACTS

Subject

The subject for my research report is:

Purpose Statement or Central Idea

I can sum up the purpose or the main idea of my report in the following sentence:

Formal Outline

The formal outline is a _____ (sentence or topic) outline. The outline is as follows:

Introduction

The introductory material for the report is as follows:

Closing

The closing material for the report is as follows:

Visuals

Kinds of visuals and what they will show are as follows:

Questions

I need to ask my instructor these questions:

Chapter 10

Oral Communication: Saying It Clearly

Outcomes

437

Chapter 10
Oral
Communication:
Saying It
Clearly

Upon completing this chapter, you should be able to

- List ways in which an oral presentation differs from a written presentation
- List and identify the modes of delivery for a formal speech
- Classify oral presentations according to general purpose
- Prepare and use visuals in oral presentations
- Prepare and deliver to a specific audience an oral presentation that meets a clearly stated purpose
- Evaluate an oral presentation
- Demonstrate knowledge and understanding of how to hold a job interview
- Demonstrate knowledge and understanding of how to hold an informational interview

Introduction

In school or on the job, you probably speak thousands of times as often as you communicate in any other way. Even a moment's consideration of how people make a living will point to the importance of being able to express ideas orally.

Every time you speak, you convey something of who you are and what you think. Your vocabulary, pronunciation, grammatical usage, phrasing, and expressed ideas are aspects of speech that make an impression on your listeners who then form opinions about you. Unfortunately, sometimes the impression you make can be a negative or poor one simply because you cannot orally express what you want to say. Such a situation shows a lack of self-confidence and knowledge in handling oral communication.

This chapter will acquaint you with various group and individual communication situations and will give suggestions to help you develop the self-confidence and knowledge needed for effective oral expression.

Speaking Differs from Writing

Speaking and writing have much in common because they are both forms of communication based on language. Speaking differs, however, in several important ways:

1. *Level of diction.* Speaking, typically, requires a simpler vocabulary and shorter sentences of less involved structure.
2. *Amount of repetition.* More repetition is needed in speaking to emphasize and to summarize important points.
3. *Kind of transitions.* Transitions from one point to another must be more obvious in speaking. Such transitions as *first, second,* and *next* signal movement often conveyed on the printed page through paragraphing and headings.

4. *Kind and size of visuals.* Speaking lends itself to the use of exhibits and projected materials; some kinds of flat materials such as charts, drawings, and maps must be constructed on a large scale.

Study these features in the speech "Self-Motivation: Ten Techniques for Activating or Freeing the Spirit," by Richard L. Weaver II, on pages 589–595.

Classification as Informal and Formal

Oral communication might be broadly classified as informal and formal. The term *informal* describes nonprepared speech and *formal* describes well-planned, rehearsed speech.

Informal Presentations

Most of us spend a large percentage of each day in informal communication: We talk with friends, parents, fellow workers, neighbors, other family members. Often, you may be asked to share views or knowledge about events, people, places, or things with a group, such as a service club, a professional organization, or a class. You may be asked at the meeting to respond impromptu; or you may be asked ahead of time but make little or no preparation. In both cases, you share information through informal oral communication.

Formal Presentations

Formal oral communication involves a great deal of preparation and attention to delivery. Professional and learned people are often asked to share their views and knowledge of their field. The deliberate, planned, carefully organized and rehearsed presentation of ideas and information for a specific audience and purpose constitutes formal communication.

Formal presentations may be categorized according to the speaker's mode of delivery as:

- Extemporaneous
- Memorized
- Read from a manuscript

In the extemporaneous mode—the most often used of the three—the speaker refers to brief notes or an outline, or simply recalls from memory the points to be made. In this way, the speaker is able to interact with the audience and convey sincerity and self-assurance. In the memorized mode, the speaker has written out the speech and committed it to memory, word for word. The memorized speech is typically lacking in spontaneity and an at-ease tone; too, the speaker has the very real possibility of forgetting what comes next. In the third mode of delivery, the speaker reads from a manuscript. While this type of delivery may be needed when exact wording is required in a structured situation, the manuscript speech has serious limitations. It is difficult for the speaker to show enthusiasm and to interact with the audience, delivery is usually stilted, and the audience may soon become inattentive.

The effective speaker considers the occasion, the audience, and the purpose of the speech in determining the mode of delivery.

439

Chapter 10
Oral
Communication:
Saying It
Clearly

Classification According to General Purpose

In addition to informal and formal, another way to classify oral communication is according to its general purpose: to entertain, to persuade, or to inform. Any one of these types of communication could be presented formally or informally.

To Entertain

Oral communication that is meant to entertain is intended to provide enjoyment for those who listen. Probably few individuals would have many occasions to communicate solely for the purpose of entertaining except on an informal basis with friends and relatives.

To Persuade

The goal of communication that is meant to persuade is to affect the listeners' beliefs or actions. On the job you may well find yourself responsible for persuading supervisory personnel or customers or employees to change a method or a procedure, to hire additional personnel, to buy a certain piece of machinery or equipment, and so forth. Whether presenting an idea, promoting a plan, or selling a product, the same basic principles of persuasion are invovled.

The art of persuasion can be summed up in two sentences: Present a need, want, or desire of the customer (buyer). Show the customer how your idea, service, or product can satisfy that need, want, or desire.

Basic Considerations in Persuasion

If you are to prepare and present a persuasive speech effectively, you must be aware of several factors: needs, wants, and desires of people; and kinds of appeals.

Needs, Wants, and Desires of People. The actual needs of people in order to exist are few: food, clothing, and shelter. In addition to these physical needs, people have numerous wants and desires. Among these are the following:

- *Economic security.* This includes a means of livelihood and ownership of property and material things.
- *Recognition.* People want social and professional approval. They want to be successful.
- *Protection of self and loved ones.* Safety and physical well-being of self and of family and friends are important.
- *Aesthetic satisfaction.* Pleasant surroundings and pleasing the senses can be very satisfying.

Consideration of people's needs, wants, and desires is essential in order to present effectively an idea, plan, or product. For example, if you were to try to persuade your employer to purchase new computers, you probably would appeal to the employer's economic and aesthetic desires; that is, you probably would emphasize the time that could be saved and the improved appearance of material. Or if you were a supervisor impressing upon a worker the importance of following dress regulations, you would stress protection (safety) of self and the possible loss of economic security for the worker and his or her family should injury or accident occur. Or if you were to persuade a co-worker to take college courses in the evening, you would point out the recognition and economic security aspects— gaining recognition and approval for furthering education and skills and the possible subsequent financial rewards.

Appeals—Emotional and Rational, Direct and Indirect. Persuasion is the process of using combined emotional and rational appeals and principles. Emotional appeals are directed toward feelings, inclinations, and senses; rational appeals are directed toward reasoning, logic, or intellect. Undoubtedly, many times emotional appeals carry more weight than rational appeals.

The most satisfying persuasion occurs when people make up their own minds or direct their own feelings toward a positive reception of the idea, plan, or product—but without being told to do so. Thus, indirect appeals, suggestions, and questions are usually much more effective than a direct statement followed by proof.

The Persuasive Presentation

The persuasive presentation involves four steps or stages: opening, need and desire intensified, supporting proof, and close. For timing in moving from one step to another, you must use your judgment by constantly analyzing conditions and audience mood.

In approaching a single listener or a small group of listeners, be sincere and cordial. A firm handshake should set a tone of friendliness. It is natural also to exchange a few brief pleasantries (How are you?—How is business?—Beautiful weather we are having) before getting down to business. With a large group of listeners, such a personal approach is impossible. You can, however, be sincere and cordial.

Opening. In the opening, the listener's attention is aroused. Thus the opening should immediately strike the listener's interest and should present the best selling points. This may be done directly or indirectly.

DIRECT: Our new Top Quality razor blades give a closer, cleaner shave than any others on the market.

INDIRECT: Do your customers ask for razor blades that give a closer, cleaner shave?

Need and Desire Intensified. When the audience's attention is aroused, each main selling point is then developed with explanatory details.

Both emotional and logical appeals show how the proposal will help satisfy one or more of these basic desires: economic security, recognition, protection of self and loved ones, and aesthetic satisfaction.

At this stage and throughout the proposal presentation, the listener may raise objections. The best way to handle these is to be a step ahead of the listener; that is, be aware of all possible objections, prepare effective responses, and incorporate them into your presentation.

Supporting Proof. Description and explanation must be supported by evidence or proof. Visual exhibits, demonstrations, testimony of users, examples of

441

Chapter 10
Oral
Communication:
Saying It
Clearly

experience with and uses of the product, and statistics showing specifications and increased productivity are methods to prove the worth of your proposal.

      Close. In closing a presentation, you usually are wise to assume the positive attitude that the audience will accept the idea or plan or will buy the product. The following suggestions reflect such an attitude: *When may we begin using this procedure? Which model do you prefer?* Reaffirmation of how the proposal will enhance the listener's business often helps to conclude the deal and to reinforce his or her satisfaction. If you detect a negative attitude, avoid a definite "no" by suggesting further consideration of the proposal or a trial use of the product and another meeting at a later date.

To Inform (Emphasis of Remainder of Chapter)

Of the three general purposes of oral communication—to entertain, to persuade, and to inform—the informative purpose is most frequently employed on the job. In communicating instructions and processes, descriptions of mechanisms, definitions, analyses through classification and partition, analyses through effect and cause and through comparison and contrast, summaries, and reports (Chapters 1–6 and 8 in Part I), the speaker has as a major goal *informing* the listeners.

      Giving oral informative presentations is a very significant aspect of an employee's communication responsibilities. The next three sections of this chapter—Preparation of an Oral Presentation, Delivery, and Visual Materials—are geared to the informative speech. The basic principles discussed in these sections are, of course, applicable to any speech situation. For instance, the steps in preparing a speech are essentially the same, whether the purpose of the speech is to entertain, to persuade, or to inform.

      As you study this chapter, review the oral presentations "How Alexander Fleming Discovered Penicillin," pages 54–55; "Description: The Winsome Professional Quiet 1800 Blow Dryer," pages 89–90; and the speech "Self-Motivation: Ten Techniques for Activating or Freeing the Spirit," by Richard L. Weaver II, pages 589–595.

Preparation of an Oral Presentation

Preparing an oral presentation includes these steps:

- Determine the specific purpose
- Analyze the type of audience
- Gather the material
- Organize the material
- Determine the mode of delivery
- Outline the speech, writing it out if necessary
- Prepare visual materials
- Rehearse

Determine the Specific Purpose

The general purpose of an employee's speech typically is to inform; sometimes the general purpose is to persuade, or, occasionally, to entertain. The *specific* purpose, however, must be determined if the speech is to be effective. The reason for the speech and those who will use the information must be established. Data that completely, accurately, and clearly present the subject must be given, analyzed, and interpreted thoroughly and honestly. Recommendations can then be made accordingly.

Analyze the Type of Audience

Your speech, if it is to be effective, must be designed especially for the knowledge and interest level of the intended audience or listeners. Adapt the vocabulary and style to the particular audience. For instance, if you were to report on recent applications of the laser, your report to a group of nurses, to a group of engineers, to a college freshman class of physics students, or to a junior high science club would differ considerably. Each group represents a different level of knowledge and a different partisan interest.

Gather the Material

You gather material primarily from three sources: interviews and reading, field investigation, and laboratory research.

The extent to which you use one or more of these sources depends on the nature of the speech. A report in history, for instance, may simply call for the reading of certain material in a book. An investigation of parking facilities in a particular location may call for personal interviews plus on-site visits. Or an analysis of the hardiness of certain shrubs when exposed to sudden temperature changes may involve both field investigation and experimental observation.

Organize the Material

To organize the material, select the main ideas; do not exceed three or four. (Remember that your audience is listening, not reading.) Arrange supporting data under each main idea. Use only the supporting data necessary to develop each main idea clearly and completely.

After the main body of material is organized, plan the introduction. Let the audience know the reason for the speech, the purpose, the sources of data, and the method or procedure for gathering the data. Then state the main ideas to be presented. The function of the introduction is to set an objective framework in which the audience will accept the information as accurate and as significant.

Plan the conclusion. It should contain a summary of the data, a summary of the significance or of the interpretation of the data, and conclusions and recommendations for action or further study.

A suggested outline for the introduction, body, and conclusion is given in the section Outline the Speech, Writing It Out if Necessary.

(The organization of a persuasive speech is treated earlier in the chapter.)

Determine the Mode of Delivery

Once you have analyzed the speaking situation and gathered and organized the material, you can determine the appropriate mode of delivery. Is it more appropriate to speak extemporaneously, to recite a memorized speech, or to read from a manuscript? (Of course, you may have been told which mode to use; thus the decision has already been made for you.) The memorized speech is most appropriate in such situations as competing in an oratorical contest or welcoming an important visting dignitary. Reading from a script is most appropriate if presenting a highly technical scientific report, giving a policy speech, or the like. For most other situations, extemporaneous speaking is the most appropriate.

443

Chapter 10
Oral
Communication:
Saying It
Clearly

Outline the Speech, Writing It Out If Necessary

Outline your speech. A suggested outline form is as follows:

Introduction
  I. Reason for the speech
     A. Who asked for it?
     B. Why?
 II. Purpose of the speech
III. Sources of data
IV. Method or procedure for gathering the data
 V. Statement of main ideas to be presented

Body
  I. First main idea
     A. Sub-idea
        1., 2., etc. Data
     B. etc. Sub-ideas
        1., 2., etc. Data
II., III., etc. Second, third, etc., main ideas
     A., B., etc. Sub-ideas

Conclusion
  I. Summary of the data
 II. Summary of the significance or of the interpretation of the data
III. Conclusions and recommendations for action or further study

If you plan to present a memorized speech or read from a script, write out the speech. Give special care to manuscript form and to the construction of visuals if you are to distribute copies of the speech (copies should be distributed *after* the oral presentation, not before or during it).

Prepare Visual Materials

Carefully select and prepare visuals to help clarify information and to crystallize ideas. See the section that follows entitled Visual Materials, pages 445–448.

Rehearse

For an extemporaneous speech: From your outline, make a note card (3-by-5-inch, narrow sides up and down) of the main points that you want to make. Indicate on the card where you plan to use visuals. Rehearse the entire speech several times, using only the note card (not the full outline). Get fixed in your mind the ideas and supporting data and the order in which you want to present them.

For a memorized speech: Commit to memory the exact wording of the script. As you practice the speech, put some feeling into the words; avoid a canned, artificial sound.

For a speech read from a manuscript: Just because you are to read a speech doesn't mean you shouldn't practice it. Go over the speech until you know it so thoroughly that you can look at your audience almost as much as you look at the script. Number the pages so that they can be kept in order easily. Leave the pages loose (do not clip or staple them together); you can then unobtrusively slide a finished page to the back of the stack.

Some speakers find it helpful to tape-record their speech once or twice while rehearsing; then they play back the recording for an objective analysis of their strengths and weaknesses.

Rehearsing your presentation several times is very important; it gives you self-confidence and it prepares you to stay within the time allotted for the speech.

Delivery

A major factor in oral communication is effective delivery, or *how* you say what you say. When giving a speech, observe the following suggestions.

• Walk to the Podium with Poise and Self-Confidence

From the moment the audience first sees you, give a positive impression. Even if you are nervous, the appearance of self-confidence impresses the audience and helps you to relax.

• Capture the Audience's Attention and Interest

Begin your speech forcefully. Opening techniques include asking a question, stating a little-known fact, and making a startling assertion (all, of course, should pertain directly to the subject at hand).

• Look at the Audience

Interact with the audience through eye-to-eye contact, but without special attention to particular individuals. You should not overdo looking at your notes, the floor, the ceiling, over the heads of your audience, or out the window.

• Stick to an Appropriate Mode of Delivery

If, for instance, your speech should be extemporaneous, don't read a script to the audience.

• Put Some Zest in Your Expression

Relax; be alive; show enthusiasm for your subject. Avoid a monotonous or "memorized" tone and robot image. Have a pleasant look on your face.

• Get Your Words Out Clearly and Distinctly

445

Chapter 10
Oral
Communication:
Saying It
Clearly

Make sure that each person in the audience can hear you. Follow the natural pitches and stresses of the spoken language. Speak firmly, dynamically, and sincerely. Enunciate distinctly, pronounce words correctly, use acceptable grammar, and speak on a language level appropriate for the audience and the subject matter.

• Adjust the Volume and Pitch of Your Voice

This adjusting may be necessary for emphasis of main points and because of distance between speaker and audience, size of audience, size of room, and outside noises. Be certain everyone can hear you.

• Vary Your Rate of Speaking to Enhance Meaning

Don't be afraid to pause; pauses may allow time for an idea to become clear to the audience or may give emphasis to an important point.

• Stand Naturally

Stand in an easy, natural position, with your weight distributed evenly on both feet. Bodily movements and gestures should be natural; well-timed, they contribute immeasurably to a successful presentation.

• Avoid Mannerisms

Mannerisms detract. Avoid such mannerisms as toying with a necklace or pin, jangling change, or repeatedly using an expression such as "You know" or "Uh."

• Show Visuals with Natural Ease

For specific suggestions, see Showing Visuals on pages 447–448.

• Close—Don't Just Stop Speaking

Your speech should be a rounded whole, and the close may be indicated through voice modulation and a simple "Thank you" or "Are there any questions?"

Visual Materials

Visual materials can significantly enhance your oral presentation. Impressions are likely to be more vivid when you use visuals. In general, they are more accurate than the spoken word. Showing rather than telling an audience something is often clearer and more efficient. And showing *and* telling may be more successful than either method by itself. For instance, a graph, a diagram, or a demonstration may present ideas and information more quickly and simpler than can words alone.

In brief, visual materials are helpful in several ways. They can convey information, supplement verbal information, minimize verbal explanation, and add interest.

See Chapter 11 Visuals for a discussion of all types of visual materials for both oral and written communication.

Types of Visuals in Oral Presentations

Visuals for use with oral presentations can be grouped into three types: flat materials, exhibits, and projected materials. A brief survey of these can help you determine which visuals are most appropriate for your needs.

Flat Materials

Included in flat materials are two-dimensional materials such as the chalkboard, bulletin board, flannel board, magnetic boards, handout sheets, posters (page 89), cartoons (page 187), charts, maps (page 332), and scale drawings.

Although these are usually prepared in advance and revealed at the appropriate time (as in the poster used in the oral presentation of the description of the Winsome blow dryer, pages 89–90), sometimes they are created spontaneously during the presentation (as in outlining steps on a chalkboard or dry erase board). A chalkboard or easel and paper (a pad of newsprint is excellent) serve beautifully. Ideally, the visuals should be created in advance and reproduced from memory or notes during the presentation.

In using printed handout material, give careful attention to its time and manner of distribution. The main thing that you should guard against is competing with your own handout material—the audience reading when it should be listening.

An easel is almost essential in displaying pictures, posters, cartoons, charts, maps, scale drawings, and other flat materials. Various lettering sets, tracing and template outfits, and graphic supplies purchased in hobby or art supply stores—as well as computers—can facilitate a neat visual.

For a more detailed discussion of flat materials, see pages 468–482 in Chapter 11 Visuals.

Exhibits

Visual materials such as demonstrations, displays, dramatizations, models, mock-ups, dioramas, laboratory equipment, and real objects comprise exhibits. These are usually shown on a table or stand.

Undoubtedly the demonstration is one of the best aids in an oral presentation. In fact, at times the entire presentation can be in the form of a demonstration. When performing a demonstration, be sure that all equipment is flawlessly operable and that everyone in the audience can see; if practical, allow the audience to participate actively. (The oral presentations on penicillin, pages 54–55, and the Winsome dryer, pages 89–90, use demonstration, real objects, and flat materials.)

For a more detailed discussing of exhibits, see pages 482–484 in Chapter 11 Visuals.

Projected Materials

Projected materials are those shown on a screen by use of a projector: pictures, slides, films, filmstrips, and transparencies. (A transparency is used in the oral description of the Winsome dryer, pages 89–90.) When using projected materials, a long pointer is essential, and an assistant often is needed to operate the machine.

For a more detailed discussion of projected materials, see pages 484–485 in Chapter 11 Visuals.

Preparing and Showing Visuals

447

Chapter 10
Oral
Communication:
Saying It
Clearly

The most effective use of visual materials occurs when you select the most appropriate kinds of visuals and when you prepare and show visuals well.

Preparing Visuals

Once you have chosen specific kinds of visuals from the general types of flat materials, exhibits, and projected materials, give careful attention to their presentation. The following should assist you.

1. Determine the purpose of the visual. Select visuals that will help the audience understand the subject. Adapt them to your overall objective and to your audience.
2. Organize the visual. Gear the information and its arrangement to quick visual comprehension.
3. Consider the visibility of the aid: size, colors, and typography. The size of the visual aid is determined largely by the size of the presentation room and the size of the audience. Visuals should be large enough to be seen by the entire audience.
4. Keep the visual simple. Do not include too much information.
5. In general, portray only one concept or idea in each visual.
6. Make the visual neat and pleasing to the eye. Clean, bold lines and an uncrowded appearance contribute to the visual's attractiveness.
7. Select and test needed equipment. If you need equipment to show your visuals—overhead projector, liquid crystal display pad, slide projector, VCR—select the equipment and test it to be sure it is operable. Check the room for locations and types of electrical outlets; these may affect the placement of the visual equipment. You may need a long extension cord. Determining needs and setting up equipment ahead of time allow you to make your presentation in a calm, controlled manner.

Showing Visuals

Show visual materials with natural ease, avoiding awkwardness. This is the basic principle in showing visuals in any kind of oral communication. The following suggestions are aspects of that basic principle.

1. Place the visual so that everyone in the audience can see it.
2. Present the visual at precisely the correct time. If you need an assistant, rehearse with the assistant. Showing a visual near the beginning of a presentation often helps you to relax and to establish contact with the audience.
3. Face the audience, not the visual, when talking. In using a chalkboard or dry erase board, for instance, be sure to talk to the audience, not the board.
4. Keep the visual covered or out of sight until needed. After use, cover or remove the visual, if possible. Exposed drawings, charts, and the like are distracting to the audience.
5. Correlate the visual with the verbal explanation. Make the relationship of visual and spoken words explicit.
6. When pointing, use the arm and hand next to the visual, rather than reaching across the body. Point with the index finger, with the other fingers loosely curled under the thumb; keep the palm of the hand toward the audience.
7. Use a pointer as needed, but don't make it a plaything.

Visuals should not be a substitute for the speaker, or a prop, or a camouflage for the speaker's inadequacies. Further, the use of visuals should not constitute a show, obviating the talk.

Appropriately used, visuals can decidedly enhance an oral presentation.

Evaluating an Oral Presentation

On the following page is a class evaluation form for oral presentations. The vertical spaces across the top are for student names. The evaluative criteria are listed under two headings, "Delivery" and "Content & Organization"; to these is added "Overall Effectiveness." The members of the class are to evaluate one another on each criterion, using this scale:

4 = Outstanding

3 = Good

2 = Fair

1 = Needs improvement

Then the total number of points for each speaker is tabulated. Total scores can range from a high of 64 to a low of 16.

The evaluation procedure can be simplified, if desired, by using only the Overall Effectiveness criterion. The highest number of points for a speaker would then be 4 and the lowest, 1.

Job Interview

Whether or not an applicant gets the job is usually a direct result of the interview. Certainly, the information in the application letter and resumé is important, but the *person* behind that information is the real focus (see pages 275–285 in Chapter 7 Memorandums and Letters, for a discussion of the application letter and resumé). The personal circumstance of the job interview allows the applicant to be more than written data. The impression that the applicant makes is often the deciding factor in whether he or she gets the job.

Following the suggestions below will help ensure your making the right impression. The suggestions are grouped according to the three steps in the job interview process: preparing for an interview, holding an interview, and following up an interview.

Preparing for an Interview

Give careful attention to preparing for an interview. This involves acquaintance with the job field, company analysis, job analysis, interviewer analysis, personal analysis, and preparation of a resumé. A procedure for rehearsing an interview is suggested at the end of this section.

449

Chapter 10
Oral
Communication:
Saying It
Clearly

EVALUATION OF ORAL PRESENTATIONS

(See page 448 for directions)

Course and Section _____   Date _____   Evaluated by (Name) _____

Students' Names

DELIVERY

Forceful introduction

Poise

Eye contact

Sticking to mode of delivery

Zest (enthusiasm)

Voice control

Acceptable pronunciation and grammar

Avoidance of mannerisms

Ease in showing visuals

Clear-cut closing

Sticking to specified length

CONTENT & ORGANIZATION

Stating of main points at outset

Development of main points

Needed repetition and transitions

Effective kinds and sizes of visuals

OVERALL EFFECTIVENESS

TOTAL POINTS

4 = Outstanding   2 = Fair
3 = Good   1 = Needs improvement

COMMENTS:

Acquaintance with the Job Field

Part of preparing for a job interview involves learning as much as you can about the field in which you seek a job. Become familiar with the possible career choices, job opportunities, advancement possibilities, salaries, and the like. An excellent way of learning about your field is talking with persons who are currently employed in that field, especially in the kind of job that interests you.

Another excellent way of becoming acquainted with your job field is by reading in occupational guides, such as the *Occupational Outlook Handbook,* or by consulting brochures published by various professional societies. (All these reading materials usually are available in a college or public library or in a counselor's office.)

Company Analysis

Find out as much as you can about the company by which you are to be interviewed. Your investigation may lead you to conclude that you really don't want to work for that particular company. More likely, however, your investigation will provide you with information useful for your letter of application, for your interview, and for later if you get the job.

Your analysis of the company should take into account such items as the following:

Background: How old is the company? Who established it? What are the main factors or steps in its development?

Organization and management. Who are the chief executives? What are the main divisions?

Product (or service). What product does it manufacture or handle? What are its manufacturing processes? What raw materials are used and where do they come from? Who uses the product? Is there keen competition? How does the quality of the product compare with that of other companies?

Personnel. How many persons does the company employ? What is the rate of turnover? What is the range of skills required for the total work force? What are company policies concerning hiring, sick leave, vacation, overtime, retirement? What kinds of in-service training are provided? What is the salary range? Are there opportunities for advancement?

Obtaining answers to these and other pertinent questions may require consulting various sources. Remember that the enterprising applicant who really wants the job will overcome whatever difficulties or expend whatever energy is required to find needed information.

This information can be obtained from the Chamber of Commerce, trade and industrial organizations, local newspapers, interviews with company personnel, inspection tours through the company, company publications, and correspondence with the company.

Job Analysis

Just as you analyze the various aspects of the company with which you will interview, you also should analyze the particular job for which you will interview. That is, you should be as knowledgeable as possible about such job factors as the following:

Educational requirements
Necessary skills
Significance of experience
On-the-job responsibilities
Desirable personal qualities
Promotion possibilities
Salary range
Special requirements

451

Chapter 10
Oral
Communication:
Saying It
Clearly

This analysis can help you to see yourself and your qualifications more objectively and thereby contribute to self-confidence.

Sources of information concerning a particular job are likely to be the same as those for the company analysis.

Interviewer Analysis

Learn something about the person who will interview you; it usually is well worth the effort. Of course, the communication that gained you an interview provided some information about the interviewer; but more information would relieve undue anxiety and help to smooth the way and establish rapport.

Just as the salesperson analyzes a prospective customer—his or her wants, needs, and interests—before making a big sales effort, so you should analyze the person who will interview you.

Personal Analysis

In preparing for the job interview, analyze yourself. Think through your attitudes, your qualifications, and your career goals. Be prepared to answer such questions as the following:

Why did I apply for this job?
Do I really want this job?
Have I applied for jobs with other companies?
What do I consider my primary qualifications for this job?
Why should I be hired over someone else with similar qualifications?
What do I consider my greatest accomplishment?
Can I take criticism?
(If you have held other positions) Why did I leave other jobs?
Do I prefer to work with people or with objects?
How do I spend my leisure time?
What are my ambitions in life?
What are my salary needs?

Preparation of a Resumé

If your letter of application did not include a resumé, or vita, prepare one to take with you to the interview. This orderly listing of information about yourself will help you organize your qualifications. Further, having the resumé with you at the interview will help you to present all significant information and will provide the interviewer, at a glance, with an outline of pertinent information about your abilities.

See pages 276–279, 283–285 in Chapter 7 Memorandums and Letters for specific directions in preparing a resumé.

Rehearsing an Interview

With the help of friends, practice an interview. Set up an office situation with desk and chairs. Ask a friend to assume the role of the interviewer and to ask you questions about yourself (see the questions listed in the section on Personal Analysis) and questions about the information on your resumé. Ask your friends to comment honestly on the rehearsal. Then swap roles; you be the interviewer and a friend, the applicant. If possible, tape the interview and play it back for critical analysis.

Another way to rehearse is by yourself, in front of a full-length mirror. Dress in the attire you will wear for the interview. Study yourself impartially; go over aloud the points you plan to discuss in the interview. Keep your gestures and facial expressions appropriate.

Rehearsals can help you gain self-confidence and organize your thoughts.

Holding an Interview

Ordinarily, you cannot know exactly how an interview will be conducted. Thus it is impossible to be prepared for every situation that can arise. It is possible, however, to become acquainted with the usual procedure so that you can more easily adapt to the particular situation.

The usual procedure for an interview includes establishing the purpose of the interview, observance of business etiquette, questions and answers, and closing the interview.

Establishing the Purpose of the Interview

At the outset of the interview, establish why you are there. If you seek a specific person, say so. If you seek a job within a general area of a company, let that be known. Be flexible, but give the interviewer a clear idea of your job preferences.

Following is a list of job interview tips from the *Occupational Outlook Handbook.**

Job Interview Tips

Preparation:
- Learn about the organization.
- Have a specific job or jobs in mind.
- Review your qualifications for the job.
- Prepare answers to broad questions about yourself.
- Review your resume.
- Practice an interview with a friend or relative.
- Arrive before the scheduled time of your interview.

---

*Reprinted by permission from the *Occupational Outlook Handbook,* 1994–1995 ed. U.S. Department of Labor, Bureau of Labor Statistics.

Personal Appearance:
- Be well groomed.
- Dress appropriately.
- Do not chew gum or smoke.

453

Chapter 10
Oral
Communication:
Saying It
Clearly

The Interview:
- Answer each question concisely.
- Respond promptly.
- Use good manners. Learn the name of your interviewer and shake hands as you meet.
- Use proper English and avoid slang.
- Be cooperative and enthusiastic.
- Ask questions about the position and the organization.
- Thank the interviewer, and follow up with a letter.

Test (if employer gives one):
- Listen closely to instructions.
- Read each question carefully.
- Write legibly and clearly.
- Budget your time wisely and don't dwell on one question.

Information to Bring to an Interview:
- Social Security number.
- Driver's license number.
- Resumé. Although not all employers require applicants to bring a resumé, you should be able to furnish the interviewer with information about your education, training, and previous employment.
- Usually an employer requires three references. Get permission from people before using their names, and make sure they will give you a good reference. Try to avoid using relatives. For each reference, provide the following information: Name, address, telephone number, and job title.

Observance of Business Etiquette

Dressing appropriately is extremely important. Good business manners, if not common sense, demand that you arrive on time, be pleasant and friendly but businesslike, avoid annoying actions (such as chewing gum or tapping your feet), let the interviewer take the lead, and listen attentively.

Questions and Answers

During most of the inteview, the interviewer will ask questions to which you respond. Your responses should be frank, brief, and to the point—yet complete. Following is a list of common job interview questions.*

*Reprinted by permission from "Rehearsing for Success: Mock Interviews Build Confidence," by Debra J. Housel, in the Fall 1991 issue of *The Balance Sheet,* p. 27.

Common Job Interview Questions

1. What are your long-range career objectives? How will you achieve them?
2. What are your other goals?
3. What would you like to be doing in five years? Ten years?
4. What rewards do you expect from your career?
5. What do you think your salary will be in five years?
6. Why did you choose this career?
7. Which would you prefer: excellent pay or job satisfaction?
8. Describe yourself.
9. How would your best friend describe you?
10. What makes you put forth your greatest effort?
11. How has your education prepared you for this career?
12. Why should I hire you?
13. Tell me your specific qualifications that make you the best candidate for this job.
14. Define success.
15. What do you think it takes to be successful in our company?
16. How can you contribute to our organization?
17. Describe the ideal supervisor-subordinate relationship.
18. What is your proudest accomplishment? Why?
19. If you were hiring a person for this position, what qualities would you look for in an applicant?
20. Do your grades reflect your scholastic achievement?
21. What subject was your favorite? Why?
22. What subject did you dislike? Why?
23. Do you have plans for additional education?
24. What have you learned from your extracurricular activities?
25. Do you work well under pressure?
26. Will you accept part-time employment or job sharing? Why/why not?
27. Tell me what you know about this organization.
28. Why do you want to work for us?
29. What criteria are you using to evaluate this company as a potential employer?
30. Would you relocate if necessary? If not, why not?
31. Are you willing to travel?
32. What major problem have you encountered? How did you resolve it?
33. What have you learned from your mistakes?
34. Describe your greatest strength.
35. Explain your greatest weakness.

Be honest in discussing your qualifications, neither exaggerating nor minimizing them. The interviewer's questions and comments can help you determine the type of employee sought; then you can emphasize your suitable qualifications. For instance, if you apply for a business position and the interviewer mentions that the job requires some customer contact, present your qualifications that show you have dealt with many people. That is, emphasize any work, experience, or courses that pertain to direct contact work.

In the course of the interview, you will likely be asked if you have any questions. Don't be afraid to ask what you need to know concerning the company or the job and its duties (such as employee insurance programs, vacation policy, overtime work, travel).

Closing the Interview

Watch the interviewer for clues that it is time to end the interview. Express appreciation for the time and courtesies given you, say good-bye, and leave. Lingering or prolonging the interview usually is an annoyance to the interviewer.

At the close of the interview, you may be told whether or not you have the job; or the interviewer may tell you that a decision will be made within a few days. If the interviewer does not definitely offer you a job or indicate when you will be informed about the job, ask when you may telephone to learn the decision.

Following Up an Interview

Following up an interview reflects good manners and good business. Whether the follow-up is in the form of a telephone call, a letter, or another interview usually depends on whether you got the job, you did not get the job, or no decision has been made.

If you got the job, a telephone call or a letter can serve as confirmation of the job, its responsibilities, and the time for reporting to work. Too, the communicator can express appreciation for the opportunity to become connected with the company.

If you did not get the job, thank the interviewer by telephone or by letter for his or her time and courtesy. This kind of goodwill is essential to business success.

Perhaps a final decision has not yet been reached concerning your employment. A favorable decision may well hinge on a wisely executed follow-up. If the interview ended with "Keep in touch with us" or "Check with us in a few days," you have the go-ahead to request another interview. A telephone call or a letter would be less effective, for neither is as forceful in maintaining the good impression that has been made. If the interview ended with "We'll keep your application on file" or "We may need a person with your qualifications a little later," follow up the lead. After a few days, write a letter. Mention the interview, express appreciation for it, include any additional credentials or emphasize credentials that since the interview seem to be particular significant, and state your continued interest in the position.

Informational Interview

A student writing a report about careers in sociology visits a sociologist, a social worker, and an occupational counselor. An employee who has been promoted to a higher position talks with associates concerning new job responsibilities. A member of a service organization polls the membership for suggested projects to undertake. The prospective buyer of a used car questions the owner about its condition. All

455

Chapter 10
Oral
Communication:
Saying It
Clearly

these situations call for informational interviews—that is, conversing with another person to gain needed information. In addition, the preparer of an observation report (see pages 320–333 in Chapter 8 Reports) and of a research report (see Chapter 9 The Research Report) frequently derives information from knowledgeable people. In seeking such information, especially from those not obliged to give it, you must prepare carefully for the interview.

Preparing for an Interview

Knowledgeable people usually are willing to share their knowledge, provided the time required to do so seems worthwhile to them. Follow these steps to ensure that time is well spent for both interviewee and interviewer:

1. Ask for an interview. A frank, informal request usually is sufficient. Identify yourself, and explain briefly why the interview is important. State the kind of information being sought.
2. Set a convenient time and place for the interview. Accommodation of the person who must grant the interview is especially important.
3. Carefully plan the questions to be asked. Think through the reason for the interview and the kind of information desired.
4. The parties to the interview should be aware of each other's knowledge and resources so that a valuable exchange of information results.

The Interview

After making the arrangements for the interview and planning specifically your contribution, you are prepared for the interview itself.

Follow this guide for a more satisfactory interview:

1. Be persuasive in explaining the significance of the interview and the value of the information sought. Make the other person feel that the time and knowledge shared can be of value to both parties.
2. Explain how you will use the information. Assure the interviewee that the information will be treated honestly and that confidential information will remain confidential.
3. Ask well-planned questions. Refer to your notes if necessary.
4. Take notes on what is said. One technique is to bring to the interview written questions. Space can be left to write in the interviewee's responses. Instead, if the interviewee does not object, you may want to tape the interview.
5. Review the information, clarifying where necessary. Especially if you intend to quote, read the quotation back to the interviewee as a check for accuracy.
6. Close the interview by sincerely thanking the person giving the information.

Courtesy Follow-Up

After the interview, write a brief letter of appreciation to the person who granted the interview. This courteous gesture is both good manners and good business.

General Principles in Oral Communication

457

Chapter 10
Oral
Communication:
Saying It
Clearly

1. *An oral presentation differs from a written presentation.* Differences occur in level of diction, amount of repetition, kind of transitions, and kind and size of visuals.
2. *Modes of delivery include extemporaneous, memorized, and read from a manuscript.* The speaker selects a mode appropriate for a given situation.
3. *The general purpose of a presentation may be to entertain, to persuade, or to inform.*
4. *An effective presentation requires preparation.* The steps in preparing for a presentation include determining the specific purpose; analyzing the type of audience; gathering the material; organizing the material; determining the mode of delivery; outlining the speech, writing it out of necessary; preparing visual materials; and rehearsing.
5. *The delivery of the presentation is very important.* Effective delivery includes walking to the podium with poise and self-confidence, capturing the audience's attention and interest, looking at the audience, sticking to an appropriate mode of delivery, putting some zest in expression, getting words out clearly and distinctly, adjusting the volume and pitch of voice, varying the rate of speaking to enhance meaning, standing naturally, avoiding mannerisms, showing visuals with natural ease, and closing appropriately.
6. *Visuals for oral presentations include flat materials, exhibits, and projected materials.*
7. *Evaluating a presentation requires considering the purpose and audience, the delivery, and the content and organization of the presentation.*

Applications in Oral Communication

Individual and Collaborative Activities

10.1   Select five advertisements and bring them to class. To what needs, wants, and desires do the ads appeal? Decide for each one whether the major appeal is emotional or rational.

10.2   Team up with a classmate to practice job interviews. Alternate the roles of interviewer and interviewee. Be constructively critical of each other. If you have access to a tape recorder or a camcorder, record your practice sessions and play them back for study.

10.3   Assume that your instructor is the personnel manager of a company you would like to work for.

a. Prepare for the interview by making a written job field analysis, company analysis, job analysis, interviewer analysis, personal analysis, and resumé.

b. Write a letter of application, including a resumé, to the personnel manager. See pages 275–285. If the letter is successful, you will be granted an interview.
c. Attend the interview at the scheduled time in the office of the personnel manager.
d. Use an appropriate follow-up.

10.4   Arrange for an interview (conference) with an instructor in whose class you would like to be making better grades. Then write a one-page report on the interview.

Oral Presentations

In completing applications 10.5–10.7,

a. Gather the supporting material.
b. Organize the material.
c. Outline the presentation.
d. Go over the outline in a peer editing group.
e. Prepare appropriate visuals.
f. Rehearse.
g. Give the presentation.
h. Ask your classmates to evaluate your speech by filling in the Evaluation of Oral Presentations on page 459, or one like it.

10.5   Assume that you are in *one* of the situations below. What would you say?

a. You have been employed in a firm for six months. You are asking your employer for a raise.
b. You have a plan for facilitating the flow of traffic for campus events. Present your plan to the local police chief (or other appropriate official).
c. You have an automobile (or some other piece of personal property) for sale. A prospective buyer is coming to talk with you.
d. You are a football coach getting ready to talk to your team at halftime. Your team is losing 17–0.
e. You want your parents to buy an automobile (or some other expensive item) for you.
f. Select any other situation in which you present an idea, plan, or product.

10.6   As directed by your instructor, prepare one of the topics below for an oral presentation.

a. A set of instructions or description of a process (review Chapter 1 in Part I)
b. A description of a mechanism (review Chapter 2 in Part I)
c. An extended definition (review Chapter 3 in Part I)
d. An analysis through classification or partition (review Chapter 4 in Part I)

459

Chapter 10
Oral
Communication:
Saying It
Clearly

NOTE: You may need to photocopy this form or make one like it. Ask your instructor for directions.

EVALUATION OF ORAL PRESENTATIONS
(See page 448 for directions)

Course and Section _____ Date _____ Evaluated by (Name) _____

Students' Names

DELIVERY

Forceful introduction

Poise

Eye contact

Sticking to mode of delivery

Zest (enthusiasm)

Voice control

Acceptable pronunciation and grammar

Avoidance of mannerisms

Ease in showing visuals

Clear-cut closing

Sticking to specified length

CONTENT & ORGANIZATION

Stating of main points at outset

Development of main points

Needed repetition and transitions

Effective kinds and sizes of visuals

OVERALL EFFECTIVENESS

TOTAL POINTS

4 = Outstanding   2 = Fair
3 = Good   1 = Needs improvement

COMMENTS:

e. An analysis through effect and cause or comparison and contrast (review Chapter 5 in Part I)
f. A report (review Chapter 8 in Part I)
g. A request from an employer or supervisor on the job

10.7   Arrange for an interview with your program adviser, a school counselor, an employment counselor, or some other person knowledgeable in your major field of study. Try to determine the availability of jobs, pay scale, job requirements, promotion opportunities, and other such pertinent information for the type of work you are preparing for. Present your findings in an oral report.

Using Part II Selected Readings

10.8   Read the speech "Self-Motivation: Ten Techniques for Activating or Freeing the Spirit," by Richard L. Weaver II, pages 589–595.

a. Under "Suggestions for Response and Reaction," page 596, respond to Number 5.
b. As directed by your instructor, answer the other questions following the selection.

10.9   Read "Are You Alive?," by Stuart Chase, pages 537–540. Under "Suggestions for Response and Reaction," present your response to Number 1c as a speech. Ask your classmates to evaluate your speech by filling in the Evaluation of Oral Presentations on page 459, or one like it.

10.10   Read "The Effects of Perceived Justice on Complainants' Negative Word-of-Mouth Behavior and Repatronage Intentions," by Jeffrey G. Blodgett, Donald H. Granbois, and Rockney G. Walters, pages 552–562. How are informational interviews (pages 455–456) used in the study?

Chapter 11

Visuals: Seeing Is Convincing

Outcomes

Upon completing this chapter, you should be able to

- List ways in which visuals can be helpful
- Explain how to use visuals effectively
- Explain how computers are influencing the use and the production of visuals
- Explain the special characteristics of tables, charts, graphs, photographs, drawings, diagrams, exhibits, and projected materials
- Use visuals in written and in oral presentations

Introduction

Visuals can help immeasurably in your communication needs. Both in written and in oral presentations, visuals can clarify the information and impress it upon the minds of your audience.

The various kinds of visuals have unique qualities that make each type the "right" one to use in certain circumstances. All kinds require careful planning and preparation. Some are usable only in written or only in oral presentations, although most types are adaptable to either; some require a great deal of technical knowledge to prepare—others can be rather easily constructed by an amateur. With computers and available software, a variety of effective graphics can be created with relative speed and ease. Computer-generated graphics range from simple to highly technical.

To select and show the visuals that best serve your purpose, you should consider both the reasons for using visuals and the best way to use them effectively. Further, you should become familiar with the most frequently used types (tables, charts, graphs, photographs, drawings, diagrams, exhibits, and projected materials), their specific contributions, and their special characteristics. (See also pages 445–448 in Chapter 10 Oral Communication.)

Advantages of Visuals

Visuals can be helpful in several ways.

1. Visuals can capitalize on *seeing*. For most people, the sense of sight—more so than hearing, smell, touch, or taste—is the most highly developed of the senses.
2. Visuals can convey some kinds of messages better than words can. Ideas or information difficult or impossible to express in words may be communicated more easily through visuals.
3. Visuals can simplify or considerably reduce textual explanation. Accompanying visuals often clarify words.
4. Visuals can add interest and focus attention.

Of course, visuals can work against you as well as for you. So guard against an overreliance on visuals, against poorly planned visuals, and, most of all, against snafus in timing or presentation.

Remember, however, that appropriately used, visuals can richly enhance your presentation.

Using Visuals Effectively

Visuals can be a simple, effective way of presenting information that will make a lasting, positive impression on your audience. Following are suggestions that will help ensure your using visuals to the best advantage.

1. *Study the use of visuals by others.* Analyze their use in books and periodicals and by speakers and lecturers, especially in your field of study. Note such things as intended audience, the kinds of information presented or supplemented, the kind of visual selected for a particular purpose, the design and layout of the visual, the amount of accompanying textual explanation, and the overall effectiveness of the visual.
2. *Select the kinds of visuals that are most suitable.* Consider the purpose of your presentation, the needs of your audience, and the specific information or idea to be presented.
3. *Prepare the visual carefully.* Organize information logically, accurately, completely, and consistently. Include all needed labels, symbols, titles, and headings.
   a. *Do not include too much in a visual.* Plan one overall focus. Information should be easy to grasp visually and intellectually.
   b. *Make the visual pleasing to the eye.* It should be neat, uncrowded, attractive, and should have sufficient margins on all sides.
   c. *Use lettering to good advantage.* Whether constructing the visual on a computer or by hand, be consistent. Avoid carelessly mixing styles of lettering or typefaces and mixing uppercase (capital) letters with lowercase ones. Space consistently between letters and between words. Suggestions for lettering by hand: Using a pencil, lightly rule guidelines and block out the words. Then do the actual lettering. After the lettering has dried, erase the pencil lines with art gum.
   d. *Give each visual a caption.* State clearly and concisely what the viewer is looking at. For tables, place the caption above the visual; in all other instances, place the caption below the visual.
4. *Decide whether to make the visual "run on" in the text or to separate it from the text.* The run-on visual is a part of the natural sequence of information within a paragraph. It is not set apart with a title, a number, or lines (rules). Usually the run-on visual is short and the information contained is uncomplicated. The separate visual is "dressed up" with a number (unless the communications contains only one visual), a caption, and rule lines. Such a visual usually is more complicated and requires more space than the run-in visual. In addition, the separate visual is movable; that is, it can be located other than at the place where it is mentioned in the text (see number 5 below).

5. *Determine where to place the visual.* Ideally a visual is placed within the text at the point where it is discussed. However, some other placement may be more practical (such as on a following page, on a separate page, or at the end of the communication) if:
   * The space following the textual reference is not sufficient to accommodate the visual,
   * A visual that merely supplements verbal explanation interferes with reader comprehension.
   * A number of visuals are used and seem to break the content flow of a presentation.

   When visuals form a significant part of a report, they are *listed,* together with page numbers, under a heading such as "List of Illustrations." This list appears on a separate page immediately following the table of contents.

6. *Refer to the visual in the textual explanation.* The audience should never be left to wonder: "Why is this visual included?" It is essential that you establish a proper relationship between the visual and the text. The extent of textual explanation is determined largely by the complexity of the subject matter, the purpose of the visual, and the completeness of labels on the visual. In referring to the visual, use such pointers as "See Figure 1," "as illustrated in the following diagram," or "Table 3 indicates the pertinent factors."

7. *Use correct terminology in referring to visuals.* Tables are referred to as tables; all other visuals are usually referred to as figures. Examples: "Study the amounts of salary increases shown in Table 2." "Note the position of the automobile in Figure 6." "As the graph in Figure 4 indicates. . . ."

8. *Give credit for borrowed material.* The credit line, usually in parentheses, typically is placed immediately following the title of the visual or just below the visual. For bibliographical forms other than those illustrated below, see pages 408–409 in Chapter 9 The Research Report.

   a. *If you have borrowed or copied an entire visual, give the source.* Although there is no one standard format for giving the source, the following examples use acceptable formats.

   EXAMPLE 1

   TABLE 4
   CAUSES OF INDUSTRIAL ACCIDENTS
   [The table goes here.]
   Source: John R. Barnes, *Industry on Trial* (New York: HarperCollins, 1993), p. 221.

   EXAMPLE 2

   Figure 6. Automobile troubleshooting using the charging system
   (Source: Ford Division-Ford Motor Co.)

   EXAMPLE 3

   Table 2. Projected College Enrollments for 1995 (*Education Almanac* [New York: Educational Associates, 1992], p. 76.)

   (Note the brackets where parentheses ordinarily are used; another pair of parentheses would be confusing.) For other examples, see pages 33, 134, and 153, and, in this chapter, pages 466, 467, 469, 472, 473, 475, 476, 478, and 479.

b. *If you have devised the visual yourself but gotten the information from another source, give the source.* The following examples use acceptable formats.

EXAMPLE 1

Figure 5. Per Capita Income in Selected States. (Source of Information: U.S. Dept. of Urban Affairs)

EXAMPLE 2

Figure 1. Average size of American families. (Data from U.S. Bureau of the Census)

For another example, see Figure 7 in this chapter (page 472).

9. *If necessary, mount the visual.* Photographs, maps, and other visuals smaller than the regular page should be mounted. Attach the visual with dry-mounting tissue, spray adhesive, or rubber cement (glue tends to wrinkle the paper).

Computer Graphics

The computer is revolutionizing both the role and the production of visuals in communication. This computer revolution can be largely attributed to three conditions: (1) the increasing availability of terminals for mainframe systems, (2) the widespread availability and downward pricing of microcomputers, and (3) the continuous additions to a growing array of graphic software, that is, commercially produced programs for easily and swiftly transforming data into various kinds of graphics. For the novice or intermediate-level computer user, the software aspect is of most interest.

Graphics software provides programs for creating basic kinds of visuals and programs for creating more complex visuals.

Basic Visuals—Charting

Persons in business and industry—regardless of the size of the company—are using computer graphics to enlarge upon the time-proved positive difference that visuals make. The general term for producing basic visuals on the computer is *charting*.

The computer user can create line graphs, pie graphs (pie charts), bar graphs (bar charts), consumer marketing maps, forecasting charts, and a host of other basic visuals. These can be produced in black-and-white or color hard copies on a sheet of paper or a transparency; with a color plotter (a device that acts as a mechanical arm and that "draws" with a pen), the visuals can be produced in multiple colors and on clear polyester film for overhead projection. The computer user can prepare and present slides and other graphics with available presentation software.

What this means in terms of preparing a business report is that the data in the computer can be programmed to produce such visuals as a pie chart of percentages of sales for particular products, a graph displaying a sales matrix, a line graph depicting total sales and fixed expenses, or bar charts reflecting sales, gross profits, and overhead expenses. The visuals are generated as an integral aspect of the report.

Given in Figures 1–4 are examples of computer-generated graphs.

Four Examples of Computer-Generated Graphs

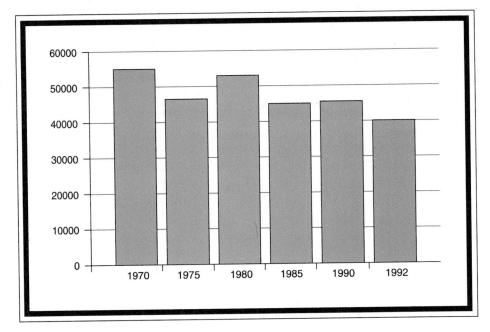

Figure 1. Automobile deaths in thousands.
(Data from the *1994 World Almanac*)

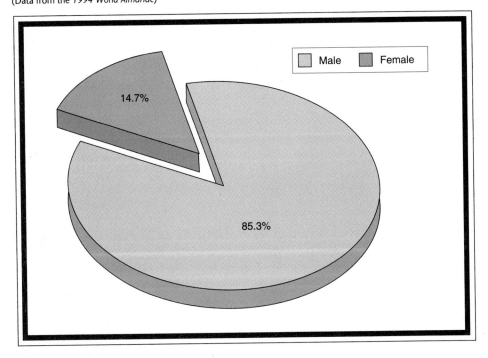

Figure 2. Deaths involving firearms 1990, by gender.
(Source of Information: *1994 World Almanac*)

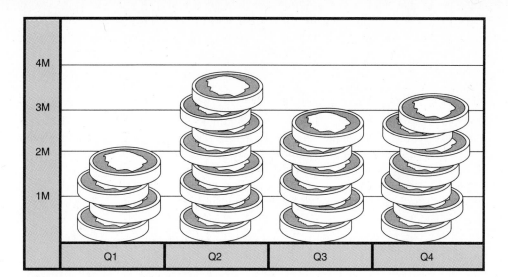

Figure 3. Quarterly profit in millions for XYZ corporation 1993.

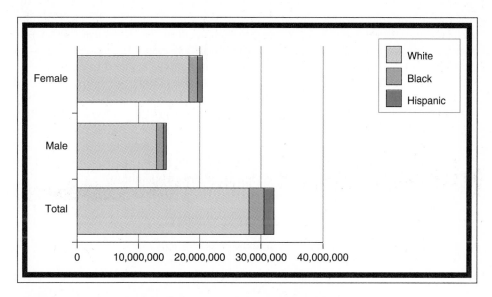

Figure 4. U.S. population age 65 and older, in millions, 1990.
(Source of Information: *1994 World Almanac*)

More Complex Visuals

In addition to charting, other software programs provide more complex, more sophisticated kinds of visuals.

One kind of such presentation is the slide-show program. This program permits
the user to select the computerized charts for demonstration; the user sets up the
charts in sequence, times the needed intervals, and programs the charts for projection on a screen. Some slide-show programs offer animation and graphic editing as
well as such design features as borders, color, and multiple typefaces.

Other presentation software allows the splicing of such media as videotape, film, off-air broadcasts, and interactive computer programs and the producing of three-dimensional computerized graphics.

Effect on Manually Produced Visuals

Computer graphics are rapidly replacing manually produced graphs and charts. Manually produced visuals—whether by hand, with a rapidograph or other such aids, or through a company's graphic arts department—are increasingly being replaced by electronically produced visuals.

The computer-generated visuals can be produced quickly, accurately, and economically. Furthermore, they can be easily updated.

Tables

Tables are an excellent form for presenting large amounts of data concisely. (See the tables on pages 36, 153, 206, 208, 340–342, and 349.) Although tables lack the eye appeal and interest-arousing dramatization of such visuals as charts or graphs, they are unexcelled as a method of organizing and depicting statistical information compiled through research. In fact, information in most other visuals showing numerical amounts and figures derives from data originally calculated in tables.

Study the table on the following page, noting the quantity of information and its arrangement.

Tables may be classified as informal (such as those on pages 206 and 340–342) or as formal (such as the table on page 469). Informal tables are incorporated as an integral part of a paragraph and thus are not given an identifying number or title. Formal tables are set up as a separate entity (with identifying number and title) but are referred to and explained as needed in the pertinent paragraphs.

General Directions for Constructing Tables

As you prepare tables, observe the following basic practices:

1. Number each table (the number may be omitted if only one table is included) and give it a descriptive caption (title); number and caption are omitted in informal tables. Center the number and the caption *above* the table. Often the word table is in all capitals, the number in arabic numerals, and this label centered above the caption in all capitals.

TABLE 2
LEADING CAUSES OF DEATH AMONG AMERICANS

Other acceptable practices include giving the table number in roman numerals or as a decimal sequence.

Table II. Leading Causes of Death among Americans
Table 2.1. Leading causes of death among Americans

No. 714. Money Income of Families—Percent Distribution, by Income Level, Race, and Hispanic Origin, in Constant (1992) Dollars: 1970 to 1992

[Constant dollars based on CPI-U-X1 deflator. Families as of March of following year. Beginning with 1980, based on householder concept and restricted to primary families. Based on Current Population Survey; see text, sections 1 and 14, and Appendix III. For definition of median, see Guide to Tabular Presentation. See also *Historical Statistics, Colonial Times to 1970,* series G 1-8, G 16-23, G 190-192, and G 197-199]

YEAR	Number of families (1,000)	Under $10,000	$10,000-$14,999	$15,000-$24,999	$25,000-$34,999	$35,000-$49,999	$50,000-$74,999	$75,000 and over	Median income (dollars)
ALL FAMILIES[1]									
1970	52,227	8.6	7.5	17.4	20.2	23.1	16.3	6.8	33,519
1975	56,245	8.2	8.3	17.4	17.8	23.1	17.5	7.6	34,249
1980	60,309	8.3	7.4	16.5	16.8	21.8	19.2	10.0	35,839
1985[2]	63,558	9.3	7.2	16.1	15.7	20.1	19.3	12.2	36,164
1989[3]	66,090	8.3	7.1	14.8	14.6	19.7	20.1	15.4	38,710
1990	66,322	8.6	6.7	15.4	15.2	20.0	19.4	14.8	37,950
1991	67,173	9.4	7.0	15.6	15.2	19.6	19.2	14.0	37,021
1992	68,144	9.5	7.3	15.5	15.0	19.2	19.6	13.9	36,812
WHITE									
1970	46,535	7.3	6.9	16.7	20.5	24.1	17.2	7.3	34,773
1975	49,873	6.8	7.6	16.9	18.1	23.9	18.5	8.2	35,619
1980	52,710	6.7	6.6	16.0	17.1	22.7	20.2	10.7	37,341
1985[2]	54,991	7.5	6.5	15.6	16.0	20.8	20.3	13.3	38,011
1989[3]	56,590	6.4	6.3	14.4	14.8	20.5	21.1	16.5	40,704
1990	56,803	6.5	6.2	14.9	15.4	20.7	20.5	15.7	39,626
1991	57,224	7.1	6.4	15.3	15.4	20.4	20.3	15.1	38,920
1992	57,858	7.2	6.6	15.2	15.3	20.0	20.8	14.9	38,909
BLACK									
1970	4,928	20.8	13.5	24.1	17.4	14.0	8.5	1.7	21,330
1975	5,586	20.6	14.7	21.3	16.5	16.0	8.8	1.9	21,916
1980	6,317	21.5	13.7	21.3	14.8	14.8	10.5	3.2	21,606
1985[2]	6,921	23.7	11.9	21.2	13.6	14.8	10.9	3.9	21,887
1989[3]	7,470	22.3	13.0	18.5	13.7	14.9	11.9	5.9	22,866
1990	7,471	23.8	11.2	18.9	13.6	15.3	11.2	6.1	22,997
1991	7,716	25.9	11.0	18.2	14.2	14.9	11.0	5.0	22,197
1992	7,888	26.3	11.8	18.8	13.0	14.0	10.8	5.2	21,161
HISPANIC[4]									
1975	2,499	16.0	13.1	24.2	18.7	17.5	8.1	2.4	23,844
1980	3,235	15.1	12.5	22.7	17.8	16.6	11.6	3.8	25,087
1985[2]	4,206	17.4	13.4	19.9	16.7	15.8	11.9	5.0	24,809
1989[3]	4,840	16.3	11.0	20.6	15.9	16.6	13.4	6.3	26,528
1990	4,981	16.8	12.3	20.7	16.2	17.0	10.9	6.1	25,152
1991	5,177	18.2	11.6	21.2	16.3	15.7	11.1	5.9	24,614
1992	5,318	17.7	12.6	21.7	16.2	15.1	11.3	5.4	23,901

[1] Includes other races not shown separately. [2] Beginning 1983, data based on revised Hispanic population controls and not directly comparable with prior years. [3] Beginning 1987, data based on revised processing procedures and not directly comparable with prior years. See text, section 14, and source. [4] Persons of Hispanic origin may be of any race.

Source: *Statistical Abstract of the United States,* 1994.

2. Label each column accurately and concisely. If a column shows amounts, indicate the unit in which the amounts are expressed; example: Wheat (in metric tons).

3. To save space, use standard symbols and abbreviations. If items need clarification, use footnotes, placed immediately below the table. (Table footnotes are separate from ordinary footnotes placed at the bottom of the page.)

4. Generally use decimals instead of fractions, unless it is customary to use fractions (as in the size of drill bits or hats).

5. Include all factors or information that affect the data. For instance, omission of wheat production in a table "Production of Chief United States Crops" would make the table misleading.

6. Use ample spacing and rule lines (straight lines) to enhance the clarity and readability of the table. Caution: Generally, use as few rule lines as necessary.
7. If a table is divided for continuation, repeat the column headings.
8. If a table is more than a page long and must be continued to another page, use the word *continued* at the bottom of each page to be continued and at the top of each continuation page. If column totals are given at the end of the table, give subtotals at the end of each page; and at the top of each continuation page, repeat the subtotals from the preceding page.

Charts

Although the term *chart* is often used as a synonym for *graph,* a chart is distinguished by the various shapes it can take, by its use of pictures and diagrams, and by its capacity to show nonstatistical as well as statistical relationships (see pages 133, 134). More importantly, a chart can show relationships better than other types of visuals can. Frequently used types of charts are the pie chart, the bar chart, the organization chart, and the flowchart.

General Directions for Constructing Charts

Manually constructing charts, as well as other visuals, requires careful attention to details, a bit of arithmetic, and a few basic materials: ruler, pen or pencil, and paper.
In the construction process:

1. Number each chart as Figure 1 or Fig. 1, Figure 2, and so on or Figure 1.1, Figure 1.2, and so on (this number may be omitted if only one visual is included), and give the chart a descriptive caption (title). Center the number and caption (both on the same line) *below* the chart.
2. Label each segment concisely and clearly.
3. Use lines or arrows if necessary to link labels to segments.
4. Place all labels and other information horizontally for ease in reading.

Pie Chart

The pie chart is a circle representing 100 percent. It is divided into segments, or slices, that represent amounts or proportions. The pie chart is especially popular for showing monetary proportions and is often used to show proportions of expenditures, income, or taxes. Although not the most accurate form for presenting information, it has strong pictorial impact. More than any other kind of visual, the pie chart permits simultaneous comparison of the parts to one another and comparison of one part to the whole.

Constructing a pie chart is relatively simple if you will follow the general directions given above and these additional suggestions.

1. Typically begin the largest segment in the "twelve o'clock" position; then, going clockwise, give the next largest segment, and so on. Note: Some computer programs for constructing pie charts do not allow this positioning of segments.

2. Lump together, if practical, items that individually would occupy very small segments. Label the segment "Other," "Miscellaneous," or a similar title, and place this segment last.
3. Put the label and the percentage or amount on or near each segment.

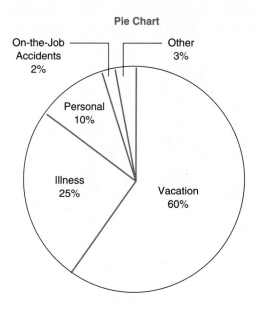

Figure 5. Work loss for all employees.

Bar Chart

The bar chart, also called column chart or bar graph, is one of the simplest and most useful visuals, for it allows the immediate comparison of amounts. (Bar charts are shown on pages 146 and 472–473.)

The bar chart consists of one or more vertical or horizontal bars of equal width, scaled in length to represent amounts. (When the bar is vertical, the visual is called a column chart; when the bar is horizontal, the visual is called a bar chart.) The bars are often separated to improve appearance and readability.

To give multiple data, a bar may be subdivided, or multiple bars may be used, with crosshatching, colors, or shading to indicate different divisions.

Note the difference in the three accompanying examples of bar charts, Figure 6, Figure 7, and Figure 8.

Vertical Bar Chart with Shading and Crosshatching

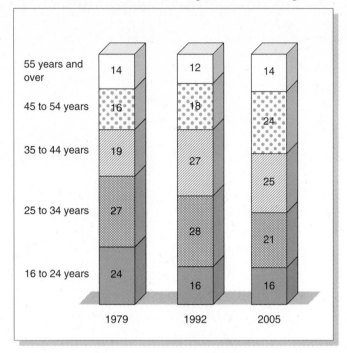

Figure 6. Percentage distribution by age of the civilian labor force.
(*Source:* Bureau of Labor Statistics)

Horizontal Bar Chart with Multiple Bars

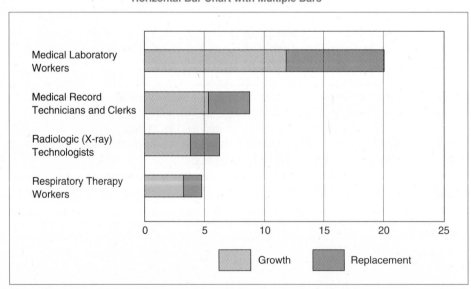

Figure 7. Average annual openings, 1980–1989 (in thousands), selected medical technologist, technician, and assistant occupations.
(*Source:* Bureau of Labor Statistics)

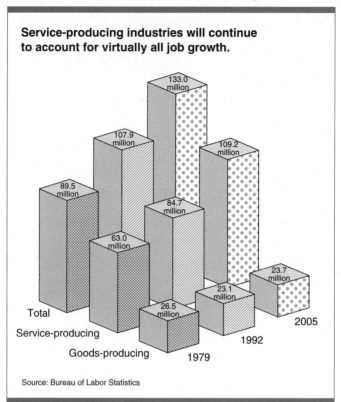

Figure 8. Non-farm wage and salary employment.
(*Source:* Bureau of Labor Statistics)

Organization Chart

The organization chart is effective in showing the structure of such organizations as businesses, institutions, and governmental agencies. (An organization chart appears in Figure 9 on page 474 and in a sample report in Chapter 4, page 166.) Unlike most other charts, the organization chart does not present statistical information. Rather, it reflects lines of authority, levels of responsibility, and kinds of working relationships.

Organization charts depict the interrelationships of (1) staff, that is, the personnel; (2) administrative units, such as offices or departments; or (3) functions, such as sales, production, and purchasing.

A staff organization chart shows the position of each individual in the organization, to whom each is responsible, over whom each has control, and the relationship to others in the same or different divisions of the organization. The administrative unit organization chart shows the various divisions and subdivisions. The administrative units of a large supermarket, for instance, include the produce department, the meat department, the grocery department, and the interrelationships of different activities, operations, and responsibilities. A college organization chart, for instance, might show the structure of the college by functions: teaching, community service, research, and the like.

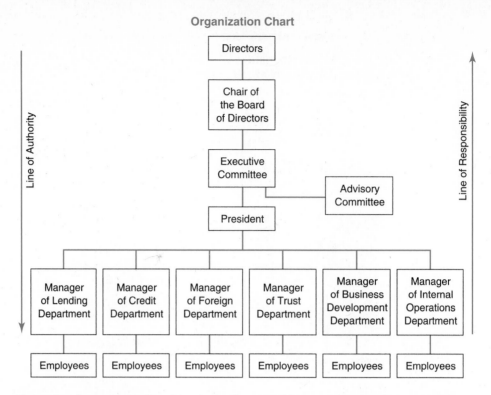

Figure 9. A line organization plan of a medium-sized bank. In this plan, the managers must perform highly specialized functions and at the same time direct or supervise other employees.

An organization chart must be internally consistent; that is, it should not jump randomly from, say, depicting personnel to depicting functions.

Blocks or circles containing labels are connected by lines to indicate the organizational arrangement. Heavier lines are often used to show chain of authority, while broken lines may show coordination, liaison, or consultation. Blocks on the same level generally suggest the same level of authority.

Flowchart

The flowchart, or flow sheet, illustrated in Figure 10 and Figure 11, shows the flow or sequence of related actions. The flowchart pictorially presents events, procedures, activities, or factors and shows how they are related. (Sample presentations, pages 54 and 57, contain flowcharts.)

The flowchart is an effective visual for showing the flow of a product from its beginning as raw material to its complete form, or the movement of persons in a process, or the steps in the execution of a computer operation. Labeled blocks, triangles, circles, and the like (or simply labels) represent the steps, although sometimes simplified drawings that suggest the actual appearance of machines and equipment indicate the various steps. Usually, arrowhead lines show the direction in which the activity or product moves.

The flowchart in Figure 10 below uses labeled blocks and a circle, arrowhead lines, and screened and unscreened lettering to explain the work process in a laundry and dry cleaning plant.

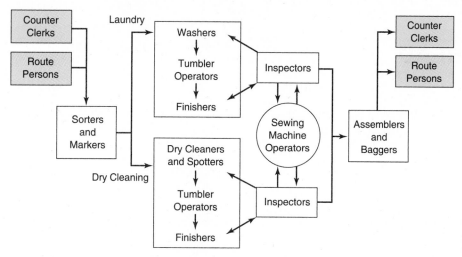

Figure 10. How work flows through a laundry and dry cleaning plant.
(Source: Bureau of Labor Statistics)

The flowchart in Figure 11 depicts a natural process: nerve supply from the brain to the teeth.

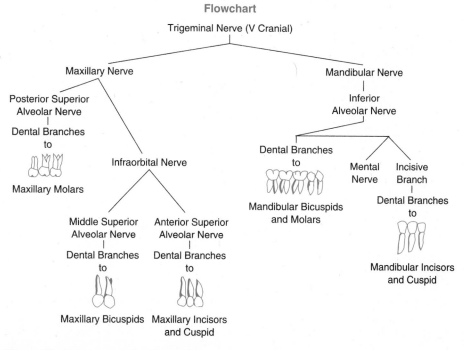

Figure 11. Nerve supply to the teeth.
(Redrawn by permission from Russell C. Wheeler, *Dental Anatomy and Physiology* [Philadelphia: W. B. Saunders, 1985], p. 94)

The following flowchart* guides a reader to make choices about what treatment to pursue for a head injury. Through a series of questions, answers, and arrows, the flowchart directs the reader to choose a treatment based on observed effects and other factors. The accompanying home treatment column tells the reader what to do if home treatment is the suggested treatment.

THINGS TO KNOW

HEAD INJURIES

• **The main concern in a head injury,** where the skull is not clearly damaged, is that there may be bleeding *inside* the skull. The accumulation of blood may eventually put pressure on the brain and cause brain damage.

Head injuries are often not serious, but brain injuries can be.

HOME TREATMENT

• **Apply ice to the bruised area** to minimize the swelling. A bump ("goose egg") often develops, in any case.

The size of the bump does *not* indicate the severity of the injury (i.e., a small bump may be serious, and a large bump may mean only a minor injury.)

• **Observe the victim carefully.** Symptoms of bleeding inside the head usually occur within the first 24 to 72 hours.

• **A typical *minor* head injury** occurs when a victim runs into someone or something and bangs his or her head. The victim is initially stunned, and a bump begins to form.

The person may vomit once or twice in the first few hours. He or she may nap after all the excitement, but is easily aroused.

Neither pupil is enlarged. Within eight hours, the person is back to normal, except for the prominent "goose egg" swelling.

Source: Adapted with permission from Take Care of Yourself *by David M. Vickery, MD and James F. Fries, MD.*

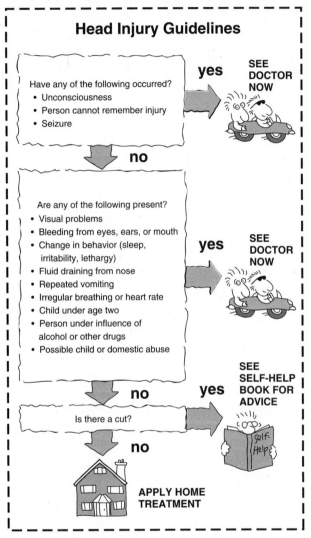

From the *Hope Health Letter,* May 1994, p. 2.

Graphs

Graphs present numerical data in easy-to-read form. They are often essential in communicating statistical information, as a glance through a business periodical, a report, or an industrial publication will attest.

*Reprinted by permission of The Hope Heart Institute, Seattle, WA.

Graphs are especially helpful in identifying trends, movements, relationships, and cycles. Production or sales graphs, temperature and rainfall curves (see Figure 12), and fever charts are common examples. Graphs simplify data and make their interpretation easier. But whatever purpose a specific graph may serve, all graphs emphasize *change* rather than actual amounts.

Consider, for instance, the graph in Figure 13, showing the changes over a ten-year period in the number and size of farms.

At a glance you can see the change over the years. Presented verbally or in another visual form, such as in a table, the information would be less dramatic and would require more time for study and analysis. But presented in a graph, the information is immediately impressed upon the eyes and the mind.

General Directions for Constructing Graphs

Consider the following as you prepare graphs:

1. A graph is labeled as Figure 1 or Fig. 1, Figure 2, or Figure 1.1, Figure 2.1, and so on (this label may be omitted if only one visual is included). A graph is given a descriptive caption, or title. The label and caption (same line) are placed below the graph.
2. A graph has a horizontal and a vertical scale. The vertical scale usually appears on the left side (the same scale may also appear on the right side if the graph is large), and the horizontal scale appears underneath the graph.
3. Generally, the independent variable (time, distance, etc.) is shown on the horizontal scale; the dependent variable (money, temperature, number of people, etc.) is shown on the vertical scale.
4. The horizontal scale increases from left to right. If this scale indicates a value other than time, labeling is necessary. The vertical scale increases from bottom to top; it should always be labeled. Often this scale starts at zero, but it may start at any amount appropriate to the data being presented.
5. The scales on a graph should be planned so that the line or curve creates an accurate impression, an impression justified by facts. To the viewer, sharp rises and falls in the line mean significant changes. Yet the angle at which the line goes up or down is controlled by the scales. If a change is important, the line indicating that change should climb or drop sharply; if a change is unimportant, the line should climb or drop less sharply. See the graph in Figure 13.
6. A graph may have more than one line. The lines should be amply separated to be easily recognized and distinguished, yet close enough for clear comparison. Each line should be clearly identified either above or to the side of the line or in a legend. Often, variations of the solid unbroken line are used, such as a dotted line or a dot/dash line. Note the kinds of lines and their identification in Figure 12, Figure 13, and Figure 14.
7. The line connecting two plotted points may be either straight, as in Figure 13, or smoothed out (faired, curved), as in Figure 12. A straight line is usually desired if the graph shows changes that occur at stated intervals; a faired line, if the graph shows changes that occur continuously.

Graph

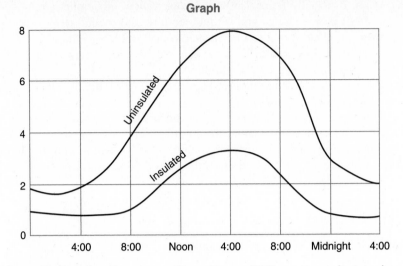

Figure 12. Heat flow into air-conditioned house (BTU per square foot per hour).

Graph

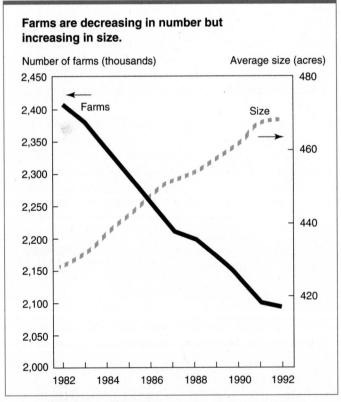

Figure 13. Changes in farm numbers and sizes.
Source U.S. Department of Agriculture

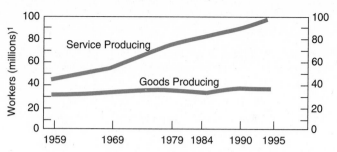

¹Includes wage and salary workers, the self-employed, and unpaid family workers.

Figure 14. Industries providing services will continue to employ many more people than those providing goods.
(Source: Bureau of Labor Statistics)

Photographs

Photographs provide a more exact impression of actual appearance than other visual aids. Photographs on pages 81, 85, 86, 117, and 333 are used to validate reports. They supply far more concreteness and realism than drawings. Though helpful in supplementing verbal description and for giving information, photographs are of the greatest value as evidence in proving or showing what something is.

Photographs have certain limitations, however. Since they present only appearance (except, of course, for such specialities as X-ray photography or holography), internal or below-the-surface exposure is impossible, and drawings or diagrams might be necessary. Further, unless retouched (or cropped) or taken with proper layout for a given purpose, photographs may unavoidably present both significant and insignificant elements of appearance with equal emphasis; or they may even miss or misrepresent important details.

The photograph in Figure 15 was used in filing an insurance claim.

Figure 15. Damage to a car.

Drawings and Diagrams

Drawings

Drawings are especially helpful in all kinds of technical communication. (Among the many drawings in the text are those on pages 29, 30, 33, 40, 41, 50, 59, 84, 92, 95, 115, 116, 121, 143–145, 148, 159, and 210.)

A drawing, though sometimes suggestive or interpretive, ordinarily portrays the actual appearance of an object. Like a photograph, it can picture what something looks like; but unlike a photograph—and herein lies one of its chief values—it can picture the interior as well as the exterior. A drawing makes it possible to place the emphasis where needed and thus omit the insignificant. Furthermore, a drawing can show details and relationships that might be obscured in a photograph. And a drawing can be tailored to fit the need of the user; it can show, for example, a cutaway view (see pages 50, 82, 83, 92, 95), an exploded view (see pages 81, 82, 95, and 159), or an enlarged view of a particular part (see Figure 16 below).

Drawing

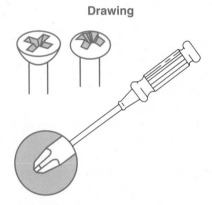

Figure 16. Phillips screwdriver.

Making a simple drawing is relatively uncomplicated and usually is much easier and less expensive than preparing a photograph. If the object being drawn has more than one part, the parts should be proportionate in size (unless enlargement is indicated). The name of each part that is significant in the drawing should be clearly given, either on the part or near it, connected to it by a line or arrow. If the drawing is complex and shows a number of parts, symbols (either letters or numbers) may be used with an accompanying key.

Diagrams

A diagram is a plan, sketch, or outline, consisting primarily of lines and symbols. A diagram is designed to demonstrate or explain a process, object, or area, or to clarify the relationship existing among the parts of a whole. Diagrams are especially valuable for showing the shape and relative location of items (see the floorplans on pages 160, 162, 325, 359–360) and configurations (see pages 159 and 162). Too, diagrams are helpful in explaining a concept (as in defining horsepower, page 121).

Diagrams are indispensable in modern construction, engineering, and manufacturing. A typical example is the design for a fireplace in Figure 17 and the design for an evaporative container, Figure 18.

Diagram

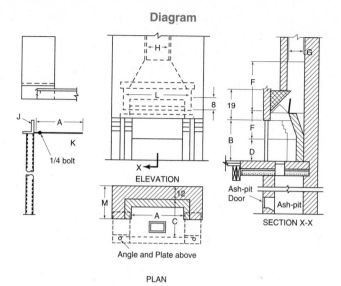

Figure 17. Proven design for a three-way, conventionally built fireplace.
(Courtesy of Donley Brothers Co.)

Diagram

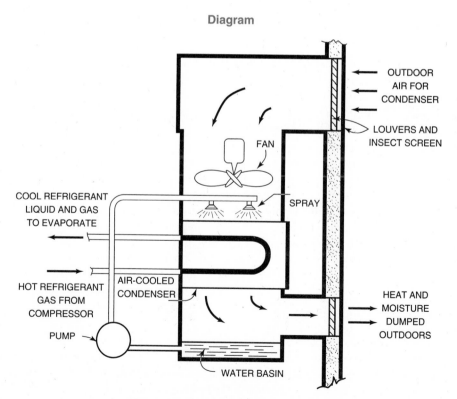

Figure 18. An evaporative condenser.

Schematic Diagram

The schematic diagram, a specialized diagram, is an invaluable aid in various mechanical fields, particularly in electronics. As with all visuals, the schematic diagram, like the one below of an electronic device, has standard symbols, terminology, and procedures that should be followed in its preparation.

Schematic Diagram

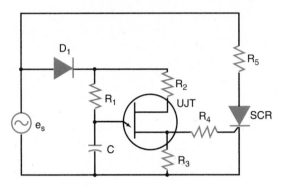

Figure 19. Basic circuit of an SCR controller.

Exhibits

Exhibits, particularly effective in oral presentations, are designed as learning experiences that involve people, objects, or representations presented in orderly sequence. Among the most frequently used kinds of exhibits are demonstrations, displays, real objects, models, dioramas, posters, and the dry erase board or chalkboard. (See Chapter 10 Oral Communication, especially page 446.)

Demonstration

A demonstration describes, explains, or illustrates a procedure or idea. A demonstration provides realism, for it is an enactment of actual steps or aspects using real objects. For instance, an explanation of how to apply a tourniquet, a description of what happens when acid is applied to copper, or a discussion of Newton's law of gravity is more realistic to an audience if it is demonstrated. In the oral explanation of the discovery of penicillin on pages 54–55 and in the oral description of a hair dryer on pages 89–90, the speaker uses demonstration to good advantage.

Display

A display is an arrangement of materials, such as photographs, news clippings, mobiles, or three-dimensional objects, designed to dramatize significant information or ideas. Often displays are presented on a bulletin board, flannel board, table, or similar area.

For an effective showing of displays, observe the following suggestions:

1. Plan the display around a specific theme.
2. Arrange the material so that it tells a story.
3. Provide clearly readable captions and labels, as needed.
4. Show the material in an attractive setting with pleasing backgrounds and accessories.

Real Object

Real objects are especially effective in a presentation. More so than pictures, models, or other representations, real objects give immediacy. Both animate and inanimate, real objects provide an opportunity to show actual size, weight, sound, movement, and texture. (Real objects are used in the oral presentation of the discovery of penicillin, pages 54–55, and the description of a hair dryer, pages 89–90.)

Model

A model is a three-dimensional representation of a real object. A model can be used when the real object is not available, is too large or too small, or is otherwise unsuitable for use with the presentation. Although models are of various types, generally they permit easy handling and convenient observation; they can provide interior views of objects; they can be stripped of some details so that other details can be easily observed; and they can be disassembled and put together to show the interrelationship of parts.

Diorama

A diorama is a three-dimensional scene of proportionately scaled objects and figures in a natural setting. A diorama is framed by a box, pieces of cardboard fastened together at right angles, or other such delineating devices. The diorama gives an in-depth, realistic view.

Dioramas are commonly used in museums, advertising displays, and instructional materials.

Poster

A poster is a very versatile aid, for it can contain a wide range of information and include a number of other visuals such as charts, photographs, pictures, drawings, and diagrams. In addition, anyone can design a good poster. (A poster is used in the oral presentation describing the Winsome Professional Blow Dryer, pages 89–90.)

In preparing a poster, review Using Visuals Effectively, pages 463–465, especially number 3, on preparing a visual. Remember that everything on a poster must be large enough for *all* viewers to see.

Chalkboard or Dry Erase Board

The chalkboard or dry erase board, though not really an exhibit in itself, is one of the most convenient means for visually communicating information. The board can be erased quickly, and new material can be added as the learning sequence progresses. (The chalkboard is used in the outlining of steps in the oral presentation of how Alexander Fleming discovered penicillin, pages 54–55.)

When using a board, be sure to write large enough for all material to be visible from the rear of the room; develop one point at a time; remove or cover distracting material; and stand to one side when pointing out material.

Projected Materials

Projected materials are uniquely effective in oral presentations. (See Chapter 10 Oral Communication, especially page 446.) Among the most common projected materials are films and videos, filmstrips, slides, and transparencies. These are projected onto a screen or some other appropriate surface such as a wall. A slide projector, a liquid crystal display pad, a 16-millimeter projector, a filmstrip projector, an overhead projector, a 1/2-inch or a 3/4-inch tape player or similar equipment is required.

Computers also provide effective ways of presenting projected materials. Using computers, you can project onto a screen materials including pictures, animation, and sound from a laser disk or a compact disk. The content of video tapes, filmstrips, and slides can be stored on laser and compact disks; however, to transfer from one medium to the other requires a special services laboratory. And commercially prepared laser and compact disks can be purchased.

Films and Videos

Films (more accurately motion pictures) and videos are excellent for portraying the action and movement inherent in a subject. They present a sense of continuity and logical progression. And sound can be added easily. However, since the preparation of films and videos requires specialized knowledge and is expensive in terms of time, equipment, and materials, commercially prepared films and videos are often borrowed, rented, or purchased.

Filmstrip

A filmstrip is a series of still pictures photographed in sequence on 35-millimeter film. They may be supplemented by captions on the frames (pictures), recorded narration, or script reading. Filmstrips are compact, easily handled, and, unlike slides, always in proper sequence. Although rather difficult to prepare locally, filmstrips from commercial producers are inexpensive and cover a wide array of topics.

Slide

Slides, usually 2 inches by 2 inches, are taken with a simple 35-millimeter camera; they provide colorful, realistic reproductions of original subjects. Exposed film is sent to a processing laboratory, which returns the slides mounted and ready for projection. Also, sets of slides on particular topics can be purchased; these often are accompanied by taped narrations.

Slides are quite flexible. They are easily rearranged, revised, handled, and stored; and automatic and remotely controlled projectors are available for greater efficiency and effectiveness. If handled individually, however, slides can get out of order, be misplaced, or even be projected incorrectly.

Transparency

Transparencies have become quite popular in conveying information to groups of people. They are easy and inexpensive to prepare and to use, and the projector is simple to operate. (A transparency is used in the oral presentation describing the Winsome Professional Blow Dryer, pages 89–90.)

The overhead projector permits a speaker to stand facing an audience and project transparencies (sheets of acetate usually $8\frac{1}{2}$ inches by 10 inches) onto a screen behind the speaker. The projection may be enlarged to fit the screen, so the speaker may more easily point to features or mark on the projected image. Room light is at a moderate level.

In preparing a transparency, you may write on the film with a grease pencil, India ink, or special acetate ink; or you may use a special copying machine or some types of photocopy machines.

General Principles in Using Visuals

1. *Visuals can be extremely helpful.* They can capitalize on seeing, convey information, reduce textual explanation, and add interest.
2. *Effective use of visuals requires careful planning.* Appropriate selection, placement, and reference to visuals are crucial to their effective use. The visual should not include too much, should be pleasing to the eye, and should use lettering to good advantage.
3. *Credit should be given for borrowed material.* Whether an entire visual or simply the information is borrowed, credit should be given for the source.
4. *Layout of text on the page can visually aid the reader.* Such layout devices as headings, paragraphing, indentation, list formats, double columns, and ample white space can enhance a report.
5. *Computers have both simplified and expanded the production of graphic illustrations.*
6. *Tables present large amounts of data concisely.* Although tables lack eye appeal and interest-arousing dramatization, they are unexcelled as a method of organizing and depicting research data.
7. *Charts show relationships.* Common types of charts are the pie chart, the bar chart, the organization chart, and the flowchart.
8. *Graphs show change.* Graphs portray numerical data helpful in identifying trends, movements, and cycles.
9. *Photographs show actual appearance.* Photographs are of particular value as evidence in proving or showing what something is.
10. *Drawings and diagrams show isolated or interior views, exact details, and relationships.* Drawings and diagrams can be easily tailored to fit the needs of the user.
11. *Exhibits dramatize a concept or objects before an audience.* Frequently used kinds of exhibits are demonstrations, displays, real objects, models, dioramas, posters, and the chalk and dry erase boards.

12. *Projected materials show pictures and other forms of information on a screen before an audience.* Common projected materials include films and videos, filmstrips, slides, and overhead-projected transparencies.

Applications in Using Visuals

11.1 From periodicals, reports, brochures, government publications, and the like, make a collection of ten visuals pertaining to your major field of study. If necessary, photocopy the visual from the source, and write a two- to three-sentence comment concerning each visual.

11.2 Present the following information in an appropriate visual or visuals: Government service, one of the nation's largest fields of employment, provided jobs for 15 million civilian workers in 1992—about one out of six persons employed in the United States. Nearly four-fifths of these workers were employed by state or local governments, and more than one-fifth worked for the federal government.

11.3 Select information from the table on page 469, or from the table below. Present the selected information in a nontable visual.

Improper Driving as Factor in Accidents, 1990

Kind of Improper Driving	Fatal Accidents			Injury Accidents			All Accidents[a]		
	Total	Urban	Rural	Total	Urban	Rural	Total	Urban	Rural
Improper driving	61.2	62.3	60.6	67.8	70.5	63.6	66.6	67.7	64.5
Speed too fast[b]	30.5	31.5	30.2	24.4	20.7	30.2	21.0	17.6	28.1
Right of way	12.1	17.8	9.6	24.3	30.7	14.3	23.7	27.6	15.7
Drove left of center	11.8	6.0	14.3	4.2	2.4	7.1	3.5	2.1	6.3
Improper overtaking	1.5	1.0	1.6	1.4	1.1	1.9	2.1	1.7	2.7
Made improper turn	0.7	0.8	0.6	2.2	2.5	1.8	3.7	4.2	2.7
Followed too closely	0.5	0.7	0.4	6.2	7.8	3.6	6.5	7.6	4.4
Other improper driving	4.1	4.5	3.9	5.1	5.3	4.7	6.1	6.9	4.6
Total	**100.0%**	**100.0%**	**100.0%**	**100.0%**	**100.0%**	**100.0%**	**100.0%**	**100.0%**	**100.0%**

[a]Principally property damage accidents, but also includes fatal and injury accidents.
[b]Includes "speed too fast for conditions."
Source: Urban and rural reports from ten state traffic authorities to National Safety Council.
Note: Figures are latest available.

11.4 Select any fifteen consecutive pages in this book. Analyze the visuals (including page layout and design of the visuals).

a. List the selected page numbers.
b. List the various techniques used in page layout and design of the visuals.
c. List the kinds of visuals included.
d. Write a paragraph evaluating the effectiveness of the page layout and design of the visuals.

11.5   Locate material on your major field of study in the latest edition of *Occupational Outlook Handbook* (published biennially by the U.S. Department of Labor's Bureau of Labor Statistics). From the information given, prepare a report, using visuals where appropriate. Create your own visuals; do not merely copy the ones in the *Handbook*.

Tables

11.6   Make a survey of twenty persons near your age concerning their preferences in automobiles: make, color, accessories. Present your findings in a table.

11.7   From a bank or home finance agency, obtain the following information concerning a $50,000 house loan, a $70,000 house loan, and a $90,000 house loan:

Interest rate
Total monthly payment for a 30-year loan
Total interest for 30 years
Total cost of the house after 30 years
Cost of insurance for 30 years
Estimated taxes for 30 years

Present the information in a table.

Pie Charts

11.8   Present the following information in the form of a pie chart.

From Steer to Steak

Choice steer on hoof	1,000 lbs
Dresses out 61.5%	615 lbs
Less fat, bone, and loss	183 lbs
Salable beef	432 lbs

11.9   Prepare a pie chart depicting your expenses at college for a term. Indicate the total amount in dollars. Show the categories of expenses in percentages.

Bar Charts

11.10   Prepare a bar chart showing the total amount in dollars paid annually in McGowan's retirement plan and the individual amounts paid by McGowan, by Denis County, and by the state.

Roy McGowan is transportation maintenance supervisor for Denis County Public Schools. His salary is $40,000 per year. McGowan participates in a retirement plan into which annually he pays 5 percent of his salary, Denis County pays 4 percent of his salary, and the state pays 2 percent of his salary.

11.11   Make a bar chart showing the sources of income and areas of expenditure for a household, your college or some other institution, or a firm.

Organization Charts

11.12   Make an organization chart of the administrative personnel of your college or some other institution or of the personnel in a firm.

11.13   Make an organization chart of a club or organization with which you are familiar.

Flowcharts

11.14   Make a flowchart of the registration procedures or some other procedure in your college or of a procedure in your place of employment.

11.15   Make a flowchart depicting from beginning to completion the flow of a process or product in your field of study.

Graphs

11.16   Construct a line graph depicting your growth in height *or* weight, or someone else's growth, for a ten-year period. (Make estimates if actual amounts are unknown.)

11.17   Prepare a multiple-line graph showing the enrollment for the past ten years in your college and in another college in your area. If you are unable to obtain the exact enrollment figures, use your own estimates.

Photographs

11.18   Find two photographs (from your own collection; in newspapers, periodicals, etc.) that present evidence as only a photograph can. Write a brief comment about each photograph.

11.19   Visualize a situation pertinent to your major field of study in which a photograph would be essential. Then find or make a suitable photograph. Finally, write a paragraph that describes the situation and include the photograph.

Drawings and Diagrams

11.20   Make a drawing of a piece of equipment used in your major field of study and indicate the major parts. Then write a paragraph identifying the piece of equipment and include the drawing.

11.21 Make a drawing or diagram that illustrates this concept:

The ancient Greek philosopher Aristotle thought that the heavier something is, the faster it falls to the ground. The Renaissance physicist Galileo discovered that this is not true. Rather, everything falls to the ground at the same rate; for example, a baseball and a lead cannon ball dropped from a tower would both hit the ground at the same time.

11.22 Make a floor plan (diagram) of your living quarters, including the location of all pieces of furniture.

Exhibits and Projected Materials

11.23 Prepare an oral presentation in which you use exhibits and/or projected materials.

Using Part II Selected Readings

11.24 Read "Are You Alive?," by Stuart Chase, pages 537–540. Under "Suggestions for Response and Reaction," page 541, prepare a pie chart as directed in Number 2.

11.25 Read "The Effects of Perceived Justice on Complainants' Negative Word-of-Mouth Behavior and Repatronage Intentions," by Jeffrey G. Blodgett, Donald H. Granbois, and Rockney G. Walters, pages 552–562. Under "Suggestions for Response and Reaction," page 563, construct charts as directed in Number 3.

11.26 Read the transcript of the speech "Self-Motivation: Ten Techniques for Activating or Freeing the Spirit," by Richard L. Weaver II, pages 589–595. Under "Suggestions for Response and Reaction," page 596:

a. Answer Number 5.
b. Suggest at least three visuals that the speaker might have used in presenting the ten main points.

Part II
Selected Readings

Part I suggests ways to improve writing and speaking, two language usages. Part II suggests ways to improve reading, a third language usage.

Part II of *Technical English* is a collection of readings included for several reasons. First, the readings are intended to complement the material in Part I by illustrating, in part or as a whole, the kinds of writing discussed in Part I. They are intended to stimulate thinking and writing, with assignments for each reading correlated to specific chapters in Part I.

The collection is also intended to stimulate reading, to interest you as student and worker, and to encourage you to read additional materials.

Without question, reading skill is essential. In a recent yearbook of the Association for Supervision and Curriculum Development, the following comments appeared:

> Reading is . . . the process of interrelating many varied experiences, drawing meanings with symbols that are almost infinitely varied in their combinations and permutations. It is, therefore, not a simple process that is mastered once and for all. As students move into the organized bodies of knowledge with their own technical terminologies and special vocabularies, in short their languages, they must to a degree learn to read again. Each special field has its own language and one who would succeed in the field must learn its language.

In the process of communication in business, industry, and service institutions, everybody has to read and assimilate larger and larger amounts of information. Many corporations and companies with multiple locations require a division to prepare print and nonprint communications for distribution to company personnel throughout the country and around the world. These communications vary; they may be statements of company policy on such subjects as compensation, working hours, holidays, and benefits; reports on research into new methods, techniques, processes, and uses of material; or periodic bulletins about company personnel who, for example, have made significant suggestions for improving or increasing production. Intercompany communications are also numerous and varied, the number and the variety typically increasing proportionally with the size of the company.

Reading skill, whether the material is on a computer screen or a printed page, is essential to keep up with what is happening in this age of technology. Institutions and organizations publish numerous periodicals and professional journals for almost all possible areas of interest. Some of these are available to the general public; others are available only through membership in professional organizations. Reading such periodicals and journals keeps the skilled worker aware of current developments.

You will find that you cannot possibly read all the available materials related to your professional interest; you must be selective. The following study-reading method offers one way to help you decide whether an article is worthy of careful reading and study.

This method will help you read and study assigned material more efficiently and carefully. By following each step as you read the selections in Part II, you will be better prepared to carry out the assignments following each selection.

The procedure includes scanning, reading, reviewing, responding, and reacting—activities to help you become familiar with the content of a selection and thus provide you with a basis for selective or assigned reading.

Scanning

Scanning (prereading, previewing, surveying) a selection should reveal pertinent information, such as the subject or topic, tone or mood, and depth or detail of content. You may discover such information by noting specific parts of a selection, such as its title and subtitle, headings, annotation, visuals, introduction and conclusion, and questions or problems.

Title and Subtitle

Note the title and whether it has a subtitle. It may indicate the subject or the topic. For example, the titles of some selections—"Self-Motivation: Ten Techniques for Activating or Freeing the Spirit," "Excerpt from Non-Medical Lasers: Everyday Life in a New Light," and "The Active Document: Making Pages Smarter" very clearly identify the subjects. Titles such as "Quality," and "Are You Alive?" require the reader to find more information before learning exactly what the selections are about.

The title or subtitle may also indicate the tone or the mood of the selection—humorous, sarcastic, persuasive, and so forth.

Headings

Note internal headings that may be used throughout a selection. These identify the subjects of the sections within a selection, and the headings help you to see the overall outline of a selection more quickly. The following articles use headings: "Battle for the Soul of the Internet," "The Powershift Era," "Hardest Lessons for First-Time Managers," "Clear Writing Means Clear Thinking Means . . . ," "The Effect of Perceived Justice on Complainants' Negative Word-of-Mouth Behavior and Repatronage Intentions," "Non-Medical Lasers: Everyday Life in a New Light," "Journeying Through the Cosmos" from *The Shores of the Cosmic Ocean,* "The Active Document: Making Pages Smarter," and "Self-Motivation: Ten Techniques for Activating or Freeing the Spirit."

Annotation

Sometimes an annotation or a headnote suggests the subject, content, or main point of a selection. Read any such annotation for the information on content it may give. Or it may give information about the author: background, education, experience.

Each of the selections in Part II has an annotation.

Visuals

Look at any visuals included in a selection—pictures, sketches, graphs, diagrams, and the like. (Review Chapter 11 in Part I.) Visuals may give more specific clues about content. Note the use of visuals in the article, "Non-Medical Lasers: Everyday Life in a New Light," "Journeying Through the Cosmos" from *The Shores of the Cosmic Ocean,* and "The Active Document: Making Pages Smarter."

Introduction and Conclusion

Read the first paragraph (or first few paragraphs); it usually previews the content of the selection. The first section may be entitled "Introduction," such as in the formal report, "The Effects of Perceived Justice on Complainants' Negative Word-of-Mouth Behavior and Repatronage Intentions."

Read the concluding paragraphs (or paragraphs); it may contain various kinds of information. The author may restate the central idea (see pages 607–608) or relist the main points used in developing or explaining the central idea. The author may summarize the content or reach a conclusion(s) supported or justified by the content.

Sometimes sections are labeled "Summary" or "Conclusion." For example, the formal report "The Effects of Perceived Justice on Complainants' Negative Word-of-Mouth Behavior and Repatronage Intentions" includes a separate section, an abstract (see page 554), at the beginning of the report. The formal report also includes an executive summary and a closing summary of the report. A reading of these sections identifies details the author considers important; look for these details as you read the entire selection.

Questions or Problems

Sometimes questions or problems will follow a selection. Read these carefully. They emphasize ideas for reflection and response. They direct you to main ideas and concepts.

Reading

On the basis of information gained through scanning a selection, you may decide to read it—if, of course, the decision to read or not to read is yours to make. As a student or a worker you may be assigned a selection; scanning the selection, then, provides a background for your reading.

Initial Reading

Read a selection through quickly for an overall view.

Further Reading

Read a selection a second (and, if necessary, a third and fourth) time, reading carefully and thoroughly for details.

Vocabulary

Sometimes in reading a selection you find a word that is unfamiliar. You may be able to determine the meaning of such a word from the context—other words in the sentence and in surrounding sentences. Or you may have to look up the word in a dictionary. There you will probably find several meanings; of these, you will have to decide which meaning is more logical as the word is used in the selection.

In the following selections, vocabulary study words are printed in **boldface** type.

Reviewing

After reading a selection, think about what you have read. List the main ideas presented and summarize each briefly. You may want to list and summarize these ideas as an oral review of your reading, or you may want to write the review. Possibly you will do both an oral and a written review, especially if a selection contains information of immediate or future importance.

Think about the relationship among the sections of the article. Try to understand the relationship of the sections to the article as a whole (see pages 219, 224–230).

Review words that are new to you. Define the words according to their use in the selection; then try using the words in your own sentences.

Responding and Reacting

Rarely does anyone read anything without responding or reacting in some way. The reaction may be simply a fleeting thought such as "I don't believe this," "I don't like this," or "Pretty good article." Or it may be more involved, depending on the reason you read the material, your interest in it, or your need for the information. (See Evaluative Summary, pages 233–236.)

Procedure for Comprehending

1. Scan a selection. Note specifically:
   a. Title and subtitle(s)
   b. Headings
   c. Annotation
   d. Visuals
   e. Introduction and conclusion
   f. Questions or problems
2. Read a selection at least twice, once quickly for an overall view and again carefully and thoroughly for details. Learn the meanings of unfamiliar words in the selection.
3. Review what you have read. If necessary, write down main ideas and summarize each briefly; then review them orally. Think about the relationship among the sections of the article.
4. Respond and react, either as you choose or as you are directed.

Clear Only If Known

EDGAR DALE

Edgar Dale of Ohio State University answers the question: Why do people give directions poorly and sometimes follow excellent directions inadequately?

For years I have puzzled over the **inept** communication of simple directions, especially those given me when touring. I ask such seemingly easy questions as "Where do I turn off Route 40 for the by-pass around St. Louis? How do I get to the planetarium? Is this the way to the Federal Security Building?" The individual whom I hail for directions either replies, "I'm a stranger here myself" or gives you in kindly fashion the directions you request. He finishes by saying pleasantly, "You can't miss it."

But about half the time you do miss it. You turn at High Street instead of Ohio Street. It was six blocks to the turn, not seven. Many persons who give directions tell you turn to right when they mean left. You carefully count the indicated five stoplights before the turn and discover that your guide meant that blinkers should be counted as stoplights. Some of the directions exactly followed turn out to be inaccurate. Your guide himself didn't know how to get there.

Now education is the problem of getting our bearings, of developing orientation, of discovering in what direction to go and how to get there. An inquiry into the problem of giving and receiving directions may help us discover something important about the educational process itself. Why do people give directions poorly and sometimes follow excellent directions inadequately?

First of all, people who give directions do not always understand the complexity of what they are communicating. They think it a simple matter to get to the Hayden Planetarium because it is simple for them. When someone says, "You can't miss it," he really means, "I can't miss it." He is suffering from what has been called the COIK fallacy—Clear Only If Known. It's easy to get to the place you are inquiring about if you already know how to get there.

We all suffer from the COIK fallacy. For example, during a World Series game a recording was made of a conversation between a rabid Brooklyn baseball fan and an Englishman seeing a baseball game for the first time.

The Englishman asked, "What is a pitcher?"

"He's the man down there pitching the ball to the catcher."

"But," said the Englishman, "all of the players pitch the ball and all of them catch the ball. There aren't just two persons who pitch and catch."

Later the Englishman asked, "How many strikes do you get before you are out?"

The Brooklyn fan said, "Three."

"But," replied the Englishman, "that man struck at the ball five times before he was out."

Reprinted by permission of Edgar Dale.

These directions about baseball, when given to the uninitiated, are clear only if known.

Try the experiment sometime of handing a person a coat and ask him to explain how to put it on. He must assume that you have lived in the tropics, have never seen a coat worn or put on, and that he is to tell you *verbally* how to do it. For example, he may say, "Pick it up by the collar." This you cannot do, since you do not know what a collar is. He may tell you to put your arm in the sleeve or to button up the coat. But you can't follow these directions because you have no previous experience with either a sleeve or a button.

The communication of teachers to pupils suffers from the COIK fallacy. An uninitiated person may think that the decimal system is easy to understand. It is—if you already know it. Some idea of the complexity of the decimal system can be gained by teachers who are asked by an instructor to understand his explanation of the duo-decimal system—a system which some mathematicians will say is even simpler than the decimal system. It is not easy to understand with just one verbal explanation, I assure you.

A teacher of my acquaintance once presented a group of parents of first-grade children with the shorthand equivalents of the first-grade reader and asked them to read this material. It was a frustrating experience. But these parents no longer thought it was such a simple matter to learn how to read in the first grade. Reading, of course, is easy if you already know how to do it.

Sometimes our directions are over-complex and introduce unnecessary elements. They do not follow the law of **parsimony**. Any unnecessary element mentioned when giving directions may prove to be a distraction. Think of the directions given for solving problems in arithmetic or for making a piece of furniture or for operating a camera. Have all unrelated and unnecessary items been eliminated? Every unnecessary step or statement is likely to increase the difficulty of reading and understanding the directions. There is no need to overelaborate or **labor** the obvious.

In giving directions it is also easy to overestimate the experience of our questioner. It is hard indeed for a Philadelphian to understand that anyone doesn't know where the City Hall is. Certainly if you go down Broad Street, you can't miss it. We know where it is: why doesn't our questioner? . . .

Another frequent reason for failure in the communication of directions is that explanations are more technical than necessary. Thus a plumber once wrote to a research bureau pointing out that he had used hydrochloric acid to clean out sewer pipes and inquired, "Was there any possible harm?" The first reply was as follows: "The **efficacy** of hydrochloric acid is indisputable, but the corrosive residue is incompatible with metallic permanence." The plumber then thanked them for the information approving his procedure. The dismayed research bureau tried again, saying, "We cannot assume responsibility for the production of toxic and noxious residue with hydrochloric acid and suggest you use an alternative procedure." Once more the plumber thanked them for their approval. Finally, the bureau, worried about the New York sewers, called in a third scientist who wrote: "Don't use hydrochloric acid. It eats hell out of the pipes."

Some words are not understood and others are misunderstood. For example, a woman confided to a friend that the doctor told her she had "very close veins." A patient was puzzled as to how she could take two pills three times a day. A little girl told her mother that the superintendent of the Sunday school said he would drop them into the furnace if they missed three Sundays in succession. He had said that he would drop them from the register.

We know the vast difference between knowing how to do something and being able to communicate that knowledge to others, or being able to verbalize it. We know how to tie a bow knot but have trouble telling others how to do it.

Another difficulty in communicating directions lies in the unwillingness of a person to say that he doesn't know. Someone drives up and asks you where Oxford Road is. You realize that Oxford Road is somewhere in the vicinity and feel a sense of guilt about not even knowing the streets in your own town. So you tend to give poor directions instead of admitting that you don't know.

Sometimes we use the wrong medium for communicating our directions. We make them entirely verbal, and the person is thus required to keep them in mind until he has followed out each of the parts of the directions. Think, for example, how hard it is to remember Hanford 6–7249 merely long enough to dial it after looking it up.

A crudely drawn map, of course, would serve the purpose. Some indication of distance would also help, although many people seem unable to give adequate estimates of distances in terms of miles. A chart or a graph can often give us an idea in a glance that is communicated verbally only with great difficulty.

But we must not put too much of the blame for inadequate directions on those who give them. Sometimes the persons who ask for help are also at fault. Communication, we must remember, is a two-way process.

Sometimes an individual doesn't understand directions but thinks he does. Only when he has lost his way does he realize that he wasn't careful enough to make sure that he really did understand. How often we let a speaker or instructor get by with such mouth-filling expressions as "emotional security," "audiovisual materials," "self-realization," without asking the questions which might clear them up for us. Even apparently simple terms like "needs" or "interests" have hidden many confusions. Our desire not to appear dumb, to be presumed "in the know," prevents us from really understanding what has been said.

We are often in too much of a hurry when we ask for directions. Like many tourists we want to get to our destination quickly so that we can hurry back home. We don't bother to savor the trip or the scenery. So we impatiently rush off before our informant has really had time to catch his breath and make sure that we understand.

Similarly, we hurry through school subjects, getting a bird's-eye view of everything and a closeup of nothing. We aim to cover the ground when we should be uncovering it, looking for what is underneath the surface.

It is not easy to give directions for finding one's way around in a world whose values and directions are changing. Ancient landmarks have disappeared. What appears to be a lighthouse on the horizon turns out to be a mirage. But those who do have genuine expertness, those who possess tested, authoritative data, have an obligation to be clear in their explanations. Whether the issue is that of atomic energy, UNESCO, the UN, or conservation of human and natural resources, clarity in the presentation of ideas is a necessity.

We must neither overestimate nor underestimate the knowledge of the inquiring traveler. We must avoid the COIK fallacy, realize that many of our communications are clear only if already known.

Questions on Content

1. What is the COIK fallacy?
2. List five of the author's reasons why people give directions poorly.
3. The author puts some of the blame for lack of communication involving directions on the person who requests help. What two reasons does he describe?

Vocabulary

Define the following words as they are used in the article (the words are printed in **boldface** type):

1. inept
2. parsimony
3. labor
4. efficacy

Suggestions for Response and Reaction

1. Relate an experience when you failed to give or receive instructions or directions clearly.
2. Write a descriptive (see pages 220–224) and an informative (see pages 224–233) summary of this article.
3. Write instructions (see pages 27–44) for putting on a coat, tying shoelaces, or some other "simple" procedure.

Battle for the Soul of the Internet

PHILIP ELMER-DEWITT

The world's largest computer network, once the playground of scientists, hackers, and gearheads, is being overrun by lawyers, merchants, and millions of new users. Is there room for everyone?

There was nothing very special about the message that made Laurence Canter and Martha Siegel the most hated couple in cyberspace. It was a relatively straight-forward advertisement offering the services of their husband-and-wife law firm to aliens interested in getting a green card—proof of permanent-resident status in the U.S. The computer that sent the message was a perfectly ordinary one as well: an IBM-type PC parked in the spare bedroom of their ranch-style house in Scottsdale, Arizona. But on the Internet, even a single computer can wield enormous power, and last April this one, with only a tap on the enter key, stirred up an international controversy that continues to this day.

The Internet, for those who are still a little fuzzy about these things, is the world's largest computer network and the nearest thing to a working prototype of the information superhighway. It's actually a global network of networks that links together the large commercial computer-communications services (like CompuServe, Prodigy and America Online) as well as tens of thousands of smaller university, government and corporate networks. . . . According to the Reston, Virginia-based Internet Society, a private group that tracks the growth of the Net, it reaches nearly 25 million computers users . . . and is doubling every year.

Now, just when it seems almost ready for prime time, the Net is being buffeted by forces that threaten to destroy the very qualities that fueled its growth. It's being pulled from all sides: by commercial interests eager to make money on it, by veteran users who want to protect it, by governments that want to control it, by pornographers who want to exploit its freedoms, by parents and teachers who want to make it a safe and useful place for kids. The Canter-and-Siegel affair, say Net observers, was just the opening **skirmish** in the larger battle for the soul of the Internet.

What the Arizona lawyers did that fateful April day was to "Spam" the Net, a colorful bit of Internet jargon meant to evoke the effect of dropping a can of Spam into a fan and filling the surrounding space with meat. They wrote a program called Masspost that put the little ad into almost every active bulletin board on the Net—some 5,500 in all—thus ensuring that it would be seen by millions of Internet users, not just once but over and over again. Hoverer Rheingold, author of *The Virtual Community,* compares the experience with opening the mailbox and finding "a letter, two bills and 60,000 pieces of junk mail."

In the eyes of many Internet regulars, it was a **provocation** so bald-faced and deliberate that it could not be ignored. And all over the world, Internet users

Reprinted with permission from *Time,* 25 June 1994, pages 50–56.

501

Philip
Elmer-Dewitt
Battle of the Soul
of the Internet

responded spontaneously by answering the Spammers with angry electronic-mail messages called "flames." Within minutes, the flames—filled with unprintable epithets—began pouring into Canter and Siegel's Internet mailbox, first by the dozen, then by the hundreds, then by the thousands. A user in Australia sent in 1,000 phony requests for information every day. A 16-year-old threatened to visit the couple's "crappy law firm" and "burn it to the ground." The volume of traffic grew so heavy that the computer delivering the E-mail crashed repeatedly under the load. After three days, Internet Direct of Phoenix, the company that provided the lawyers with access to the Net, pulled the plug on their account.

Even at that point, all might have been forgiven. For this kind of thing, believe it or not, happens all the time on the Internet—although not usually on this scale. People make mistakes. Their errors are pointed out. The underlying issues are thrashed out. And either a consensus is reached or the combatants exhaust themselves and retire from the field.

But Canter and Siegel refused to give ground. They declared the experiment "a tremendous success," claiming to have generated $100,000 in new business. They threatened to sue Internet Direct for cutting them off from even more business (although the suit never materialized). And they gave an unrepentant interview to the *New York Times*. "We will definitely advertise on the Internet again," they promised.

It was like a declaration of war, and as if on cue, the **harassment** surged anew. The lawyers' fax machine spewing out page after page of blank paper. Hundreds of bogus magazine subscriptions began showing up on their doorstep. And technicians began devising tools that would prevent Canter and Siegel from making good their threat. The most ingenious: a piece of software written by a Norwegian programmer that came to be known as the "cancelbot"—a sort of information-seeking robot that roams the Internet looking for Canter and Siegel mass mailings and deletes them before they spread.

The Green Card Incident, as the Canter-and-Siegel affair came to be known, brought to the surface issues that had been lurking largely unexamined beneath the Net's explosive growth. It was not designed for doing commerce, and it does not gracefully accommodate new arrivals—especially those who don't bother to learn its strange language or customs or, worse still, openly defy them.

The Internet evolved from a computer system built 25 years ago by the Defense Department to enable academic and military researchers to continue to do government work even if part of the network were taken out in a nuclear attack. It eventually linked universities, government facilities and corporations around the world, and they all shared the costs and technical work of running the system.

The scientists who were given free Internet access quickly discovered that the network was good for more than official business. They used it to send each other private messages (E-mail) and to post news and information on public electronic bulletin boards (known as Usenet newsgroups). Over the years the Internet became a favorite haunt of graduate students and computer hackers, who loved nothing better than to stay up all night exploring its weblike connections and devising new and interesting things for people to do. They constructed elaborate fantasy worlds with Dungeons & Dragons themes. They built tools for navigating the Net—like the University of Minnesota's Gopher, which makes it easy for Internet explorers to tunnel from one place on the Internet to another. Or like the programs whimsically named Archie, Jughead and Veronica, which allow users to locate a particular word or program from vast libraries of data available to net users. More and more news-groups were added, until the bulletin-board system had grown into a dense tangle of discussion topics with bizarre computer-coded titles like alt.tasteless.jokes, rec.arts.erotica and alt.barney.dinosaur.die.die.die.

Until quite recently it was painfully difficult for ordinary computer users to reach the Internet. Not only did they need a PC, a **modem** to connect it to the phone line and a passing familiarity with something called Unix, but they could get on only with the cooperation of a university or government research lab.

In the past year, most of those **impediments** have disappeared. There are now dozens of small businesses that will sell access to the Net starting at $10 to $30 a month. And in the past few months, mainstream computer services like America Online have started to make it possible for their subscribers to reach parts of the Internet through standard, easy-to-use menus.

But with floods of new arrivals have come new issues and conflicts. Part of the problem is technical. To withstand a nuclear blast and keep on ticking, the Net was built without a central command authority. That means that nobody owns it, no-body runs it, nobody has the power to kick anybody off for good. There isn't even a master switch that can shut it down in case of emergency. "It's the closest thing to true anarchy that ever existed," says Clifford Stoll, a Berkeley astronomer famous on the Internet for having trapped a German spy who was trying to use it to break into U.S. military computers.

But a large part of the problem is cultural. The rules that govern behavior on the Net were set by computer hackers who largely eschew formal rules. Instead, most computer wizards subscribe to a sort of anarchistic ethic, stated most suc-cinctly in Steven Levy's *Hackers*. Among its tenets:

FAQs (Frequently Asked Questions)

What is the Internet?

The Internet is a vast international network of networks that enables computers of all kinds to share services and communicate directly, as if they were part of one gi-ant, seamless, global computing machine.

How do I get connected?

That depends on how connected you want to be.

- If you have an account at CompuServe or Prodigy, you can already send and re-ceive E-mail through the Internet.
- If you have an America Online account, you can also use other Internet services, like the electronic bulletin boards (called newsgroups).
- If you have an account at Delphi or any one of dozens of smaller commercial op-erations, you can get access to even more of the Internet—but still indirectly, through a dial-up modem.
- For you to be directly plugged into the Internet and use all its services, your com-puter must have what is inelegantly called a TCP/IP (for Transmission Control Protocol/Internet Protocol) connection. To set that up, you would probably need the help of a professional—or better still, a teenager with a high-speed modem.

- Access to computers should be unlimited and total.
- All information should be free.
- Mistrust authority and promote **decentralization.**

503

Philip
Elmer-Dewitt
Battle for the
Soul of the
Internet

The Internet was built up by people who lived and breathed the hacker ethic—students at Berkeley and M.I.T., researchers at AT&T Bell Laboratories, computer designers at companies such as Apple and Sun Microsystems. "If there is a soul of the Internet, it is in that community," says Mark Stahlman, president of New Media Associates, a research firm in New York City.

As long as the community was relatively small, it could be self-policing. Anybody who got out of line was shouted down or shunned. But now that the population of the Net is larger than that of most European countries, those informal rules of behavior are starting to break down. The Internet is becoming Balkanized, and where the mainstream culture and hacker culture clash, open battles are breaking out. Canter and Siegel may head the most-hated list, but they are hardly alone.

Here Come the Newbies!

Tensions between old-timers and new arrivals—or "newbies"—flare up every September as a new crop of college freshmen (armed with their first Internet accounts) are loosed upon the network. But the annual hazing given clueless freshmen pales besides the welcome America Online users received last March, when the Vienna, Virginia-based company opened the doors of the Internet to nearly 1 million customers. It was bad enough that America Online users, clearly identifiable by the @aol.com attached to their user IDs, were making all the usual mistakes—asking dumb questions, posting messages in the wrong place and generally behaving like

---

What are those Internet services?

- E-mail, which is like the post office (only faster)
- Talk, which is like the telephone (except that you have to type out everything you want to say)
- Internet Relay Chat (IRC) which is like CB radio—noisy and confusing
- File Transfer Protocol (FTP), for fetching programs and big documents from remote computers
- Tienet, to operate those remote computers from your own desktop.
- Archie, Veronica, Jughead and WAIS (Wide Area Information Servers), tools for searching the huge libraries of information stored on the Net
- Gopher, for tunneling quickly from one place on the Net to another
- The World Wide Web, a more advanced navigation system that organizes its contents by subject matter
- Mosaic, a kind of onscreen control panel that enables you to drive through the Web by pointing and clicking your electronic mouse
- Internet Talk Radio, which broadcasts sound recordings (like the popular interview show *Geek of the Week*)
- CUSeeMe, an Internet video conferencing system that enables up to eight users to see and hear each other on their computer screens

boorish tourists. But because of a temporary bug in AOL software, every message they wrote was duplicated eight times—magnifying their errors and making the AOL folks sitting targets for locals already disposed to resent their presence on the Net.

The result was a verbal **conflagration** that dominated the newsgroups for weeks and is still smoldering four months later. . . . Things deteriorated when the AOL crowd began to give as good as they got, hinting that the old-timers ought to make way for people who actually paid for their Internet services. Feelings are still raw on both sides and are not likely to be salved until the next wave of newbies arrives—probably from CompuServe, as early as August. If history is any guide, the loudest complaints about the new immigrants will come from those who immediately preceded them—the next-to-new-comers from America Online.

Sex and the Net

For those interested in pornography, there's plenty of it on the Internet. It comes in all forms: hot chat, erotic stories, explicit pictures, even XXX-rated film clips. Every night brings a fresh crop, and the newsgroups that carry it (alt.sex, alt.binaries.pictures.erotica, etc.) are among the top four or five most popular. . . .

For purely technical reasons, it is impossible to censor the Internet at present. "It's designed to work around censorship and blockage," explains Stoll. "If you try to cut something, it self-repairs." But some antipornography activists have found a clever way to cope with that. From time to time, they will appear in newsgroups devoted to X-rated picture files and start posting messages with titles like "YOU WILL ALL BURN IN HELL!" These typically provoke flurries of angry responses—until it dawns on the pornography lovers that by filling the message board with their rejoinders, they are pushing out the sexy items they came to enjoy.

Keeping Secrets

No battle on the Internet has been as public as the one waged over the Clipper Chip—the U.S. Government-designed **encryption** system for encoding and decod-

---

What is Usenet?

Usenet is a collection of electronic bulletin boards (called newsgroups) set up by subject matter and covering just about every conceivable topic, from molecular biology to nude sunbathing. The newsgroups are organized into hierarchies, such as science (sci), recreation (rec), society (soc) and the miscellaneous category called alternate (alt). A sampling:

- sci.astro.hubble—astronomical data from the Hubble Space Telescope
- rec.arts.books—where bookworms gather to discuss their favorite authors
- comp.risks—a digest of brief reports about computers run amuck
- soc.culture.bosna.herzgvna—where the war is fought with words, not mortars
- alt.best.of.internet—a place where people re-post choice tidbits found on the Net
- alt.fan.lemurs—celebrating the legend, lore and humor of Madagascar's most famous animals

ing phone calls and E-mail so that they are protected from snooping by everyone but the government itself. The information-should-be-free types on the Internet were strongly opposed to Clipper from the start, not because they were against encryption ironically, but because they wanted a stronger form of encryption—encryption for which the government doesn't have a back-door key, as it intends to have with the Clipper system.

In the ensuing debate—much of which took place over the Net—government officials maintained that they needed Clipper to be able to interpret and **decipher** messages from mobsters, drug dealers and terrorists. Not so, claim critics. "Clipper is not about child molesters or the Mafia but about the Internal Revenue Service," argues Bruce Fancher, proprietor of a New York City Internet service provider called Mindvox. "Clipper just doesn't make sense any other way." As more and more commerce takes place on the Internet, contends Fancher, the IRS is going to need a surefire way to track the flow of cyberbucks—and to collect its share.

Who Needs the Press?

If it is true, as A.J. Liebling once wrote, that "freedom of the press is guaranteed only to those who own one," then the Internet may represent journalism's ultimate liberation. On the Net, anyone with a computer and a modem can be his own reporter, editor and publisher—spreading news and views to millions of readers around the world. . . .

But publishing on the Internet has its risks, as Brock Meeks learned. Meeks, a reporter by day for *Communications Daily* in Washington, by night publishes an electronic broadsheet called *CyberWire Dispatch,* in which he tells readers what he thinks is *really* going on. Last April he investigated an Internet advertisement offering $500 or more just for receiving junk E-mail and uncovered what he called a bait-and-switch scheme operated by "a slick direct-mail baron" in Ohio. He wrote a

---

How do I find the good stuff?

That depends on what you mean by good. If you are a professional, ask a colleague for the name of the newsgroups or mailing lists devoted to your specialty. Otherwise your best bet is to buy one of the dozens of Internet guidebooks published in the past year and start exploring. In no particular order:

- *The Internet: Complete Reference;* Harley Hahn and Rick Stout; Osborne McGraw-Hill; $29.95
- *The Internet Navigator;* Paul Gilster; John Wiley; $24.95
- *The Whole Internet User's Guide and Catalog;* Ed Krol; O'Reilly; $24.95
- *Netguide;* Peter Rutten, Albert Bayers and Kelly Maloni; Random House; $19
- *Internet Starter Kit;* Adam Engst; Hayden; $29.95.
- *Navigating the Internet;* Mark Gibbs and Richard Smith; Sams Publishing; $24.95
- *Cruising Online;* Lawrence Magid; Random House; $25
- *Internet Guide for New Users;* Daniel Dern; McGraw-Hill; $27.95
- *How the Internet Works;* Joshua Eddings; Ziff-Davis; $24.95
- *On Internet 94;* Meckler-media; $45

505

Philip
Elmer-Dewitt
Battle for the
Soul of the
Internet

story headlined JACKING IN FROM THE P.T. BARNUM PORT and dispatched it to the Net. He was promptly sued for libel. Whatever the truth of the story—or the merit of the suit—Meeks now faces a $25,000 legal bill that, because he was working on his behalf, not his employer's, he must pay out of his own pocket. It was a pointed reminder to reporters—and would-be reporters on the Internet—that the laws of libel don't stop at the borders of cyberspace. . . .

Traditional journalism flows from the top down: the editor decides what to cover, the reporters gather the facts, and the news is packed into a story and distributed to the masses. News on the Net, by contrast, is bottom up: it bubbles from newsgroups whenever anyone has anything to report. Much of it may be bogus, error-ridden or just plain wrong. But when writers report on their area of expertise— as they often do—it carries information that is frequently closer to the source than what is found in newspapers.

In this **paradigm** shift lie the seeds of revolutionary change. The Internet is a two-way medium. Although it is delivered on a glowing screen, it isn't at all like television. It's not one-to-many, like traditional media, but many-to-many. It doesn't work in couch-potato mode. And as Canter and Siegel discovered, it doesn't take kindly to in-your-face advertising.

But it does represent a new and fast-growing market. For better or worse, the Internet is filled with bright, well-educated, upwardly mobile people—a demographic that makes it particularly attractive to those with things to sell. And while the green-card lawyers were creating a diversion, hundreds of businesses were quietly staking out the territory. Silicon Graphics, a computer manufacturer, uses the Internet to distribute software and answer customer questions. Joe Boxer, a San Francisco design firm that makes colorful and off-beat men's briefs, invites customers to submit "underwear stories" to its Internet address (joeboxer@boxer.com).

Is there such a thing as proper "etiquette"?

Yes! (And thanks for asking.) A few tips for making friends and avoiding unnecessary flames:

- When you arrive at a new newsgroup, spend a couple of weeks lurking (reading messages without posting your own) to get a feel for the place before adding your two cents.
- Keep your posts brief and to the point.
- Stick to the subject of that particular newsgroup.
- If you're responding to a message, quote the relevant passages or summarize it for those who may have missed it.
- Don't start a "flame war" unless you're willing to take the heat.
- Never publish private E-mail without permission.
- Don't post test messages or clutter newsgroups with "I agree" and "Me too!" messages.
- Don't type in all caps. (IT'S LIKE SHOUTING!)
- Don't E-mail unsolicited advertisements.
- Don't flame people for bad grammar or spelling errors.
- Read your FAQS and don't ask stupid questions.

507

Philip
Elmer-Dewitt
Battle for the
Soul of the
Internet

"I think the market is huge," says Martin Nisenholt, an advertising executive at Ogilvy & Mather who has drawn up a set of guidelines for marketing to the Net. (Rule No. 1: Intrusive E-mail is unwelcome.) He insists there's a place for advertising on the network. It's O.K. to post an ad for a used computer, for example, in a newsgroup called comp.system.mac.wanted, or to sell flowers in a corner of the Net marked florist.com. Global Network Navigator, one of the first Internet publishers to include advertising in its offerings, now has 45 online clients, including Lonely Planet Publications, an international publisher of travel guides. . . .

While the Net is not entirely ready for business, the pieces are falling into place. A system that will enable merchants to take credit-card numbers over the Internet and verify their customers' signatures, for instance, is expected to be up and running before the end of the year. Right now the hot product is a program called Mosaic, which gives the Internet what the Macintosh gave the personal computer: a navigation system that can be understood at a glance by anybody who can point and click a mouse. Hundreds of companies are using Mosaic to establish an easy-to-find presence on the Net. Last year there were a handful of these Mosaic "sites"; today there are more than 10,000, including such blatantly commercial ventures as the California Yellow Pages and the Internet Shopping Network.

And what about the folks who settled the Internet when it was still a frontier town? Some have left, preferring to spend time with their family and friends. Most are bracing for the next wave of homesteaders. Dave Farber, a University of Pennsylvania computer-science professor, has developed what he calls "New York City filters"—techniques for surviving in a densely populated network and for sorting E-mail that arrives at the rate of 400 pieces a day. Others use "bozo filters" and "kill files"—lists of individuals whose past behavior has convinced Internet users that their lives will be richer and much saner if they never read another word those bozos write.

The Internet has grown too large to think of it as a single place, says Esther Dyson, a board member of the Electronic Frontier Foundation, an Internet watchdog group. "It needs to be subdivided into smaller neighborhoods. There should be high-class neighborhoods. There should be places that parents feel are safe for their kids."

San Francisco's Whole Earth 'Lectronic Link (WELL) is perhaps the most famous of these new **virtual communities.** It is connected to the Net but protected by a "gate" that won't open without a password or a credit card. Stacy Horn, a former WELL user, built a similar system on the West Coast with this twist: she offered free accounts to women, hoping they would provide a "civilizing force" to counterbalance the Internet's testosterone-heavy demographics. It turned out to be a successful formula, and Horn has plans to build similar services in six U.S. cities, including Boston, Minneapoolis and Los Angeles.

The danger, if this trend continues, is that people will withdraw within their walled communities and never venture again into the Internet's public spaces. It's a process similar to the one that created the suburbs and replaced the great cities with shopping malls and urban sprawl. The magic of the Net is that it thrusts people together in a strange new world, one in which they get to rub virtual shoulders with characters they might otherwise never meet. The challenge for the citizens of cyberspace—as the battles to control the Internet are joined and waged—will be to carve out safe, pleasant places to work, play and raise their kids without losing touch with the freewheeling, untamable souls that attracted them to the Net in the first place.

Questions on Content

1. a. Who are Laurence Canter and Margaret Siegel?
   b. How did they anger other users of the Internet?
2. How does the article classify those who threaten to destroy the positive qualities that generated the Internet's growth?
3. Who built the Internet?
4. What has caused the flood of new issues and conflicts surrounding the Internet?

Vocabulary

Define the following words as they are used in the article (the words are printed in
boldface type):

1. skirmish   6. decentralization   11. virtual communities
2. provocation   7. conflagration
3. harassment   8. encryption
4. modem   9. decipher
5. impediments   10. paradigm

Suggestions for Response and Reaction

1. What use does the article make of definition (see pages 109–118) and comparison (see pages 200–212)? Cite specific examples.
2. In the article, find examples of jargon (see page 13).
3. Using information from the article, write an extended definition (see pages
   112–118) of the Internet.
4. Write a letter (see pages 254–262 and page 309) to your congressional representative supporting one of the following statements:
   a. The Internet use should be censored.
   b. The Internet use should not be censored.
5. Write a research report (see pages 394–409) on the advantages of the Internet.
   Use information in the article, your own knowledge if you are an Internet user,
   and information gained through interviewing an Internet user and reading other
   articles.

The Powershift Era

ALVIN TOFFLER

Out of the massive restructuring of power systems at every level throughout the world emerges a rare event in human history—a revelation of the very nature of power.

This is a book about power at the edge of the 21st century. It deals with violence, wealth, and knowledge and the roles they play in our lives. It is about the new paths to power opened by a world in upheaval.

Despite the bad odor that clings to the very notion of power because of the misuses to which it has been put, power in itself is neither good nor bad. It is an inescapable aspect of every human relationship, and it influences everything from our sexual relations to the jobs we hold, the cars we drive, the television we watch, the hopes we pursue. To a greater degree than most imagine, we are the products of power.

Yet of all the aspects of our lives, power remains one of the least understood and most important—especially for our generation.

For this is the dawn of the Powershift Era. We live at a moment when the entire structure of power that held the world together is now disintegrating. A radically different structure of power is taking form. And this is happening at every level of human society.

In the office, in the supermarket, at the bank, in the executive suite, in our churches, hospitals, schools, and homes, old patterns of power are fracturing along strange new lines. Campuses are stirring from Berkeley to Rome and Tapiei, preparing to explode. Ethnic and racial clashes are multiplying.

In the business world we see giant corporations taken apart and put back together, their CEOs often dumped, along with thousands of their employees. A "golden parachute" or goodbye package of money and benefits may soften the shock of landing for a top manager but gone are the **appurtenances** of power: the corporate jet, the limousine, the conferences at glamorous golf resorts, and above all, the secret thrill that many feel in the sheer exercise of power.

Power isn't just shifting at the pinnacle of corporate life. The office manager and the supervisor on the plant floor are both discovering that workers no longer take orders blindly, as many once did. They ask questions and demand answers. Military officers are learning the same thing about their troops. Police chiefs about their cops. Teachers, increasingly, about their students.

This crackup of old-style authority and power in business and daily life is accelerating at the very moment when global power structures are disintegrating as well.

Ever since the end of World War II, two superpowers have straddled the earth like colossi. Each had its allies, satellites, and cheering section. Each balanced the other, missile for missile, tank for tank, spy for spy. Today, the course, the balancing act is over.

As a result, "black holes" are already opening up in the world system: great sucking power vacuums, in Eastern Europe for example, that could sweep nations and peoples into strange new—or, for that matter, ancient—alliances and collisions. Power is shifting at so astonishing a rate that world leaders are being swept along by events, rather than imposing order on them.

There is strong reason to believe that the forces now shaking power at every level of the human system will become more intense and pervasive in the years immediately ahead.

Out of this massive restructuring of power relationships, like the shifting and grinding of **tectonic plates** in advance of an earthquake, will come one of the rarest events in human history: a revolution in the very nature of power.

A "powershift" does not merely transfer power. It transforms it.

The End of Empire

The entire world watched awestruck as a half-century-old empire based on Soviet power in Eastern Europe suddenly came unglued in 1989. Desperate for the Western technology needed to energize its **rust-belt** economy, the Soviet Union itself plunged into a period of near-chaotic change.

Slower and less dramatically, the world's other superpower also went into relative decline. So much has been written about America's loss of global power that it bears no repetition here. Even more striking, however, have been the many shifts of power away from its once-dominant domestic institutions.

Twenty years ago General Motors was regarded as the world's premier manufacturing company, a gleaming model for managers in countries around the world and a political powerhouse in Washington. Today, says a high GM official, "We are running for our lives." We may well see, in the years ahead, the actual breakup of GM.

Twenty years ago IBM had only the feeblest competition and the United States probably had more computers than the rest of the world combined. Today computer power has spread rapidly around the world, the U.S. share has sagged, and IBM faces stiff competition from companies like NEC, Hitachi, and Fujitsu in Japan; Groupe Bull in France; ICL in Britain, and many others. Industry analysts speculate about the post-IBM era.

Nor is all this a result of foreign competition. Twenty years ago three television networks, ABC, CBS, and NBC, dominated the American airwaves. They faced no foreign competition at all. Yet today they are shrinking so fast, their very survival is in doubt.

Twenty years ago, to choose a different kind of example, medical doctors in the United States were white-coated gods. Patients typically accepted their word as law. Physicians virtually controlled the entire American health system. Their political clout was enormous.

Today, by contrast, American doctors are under siege. Patients talk back. They sue for malpractice. Nurses demand responsibility and respect. Pharmaceutical companies are less deferential. And it is insurance companies, "managed care groups," and government, not doctors, who now control the American health system.

Across the board, then, some of the most powerful institutions and professions inside the most powerful of nations saw their dominance decline in the same twenty-year period that saw America's external power, relative to other nations, sink.

Lest these immense shake-ups in the distribution of power seem a disease of the aging superpowers, a look elsewhere proves otherwise.

While U.S. economic power faded, Japan's skyrocketed. But success, too, can trigger significant power shifts. Just as in the United States, Japan's most powerful Second Wave or rust-belt industries declined in importance as new Third Wave industries rose. Even as Japan's economic heft increased, however, the three institutions perhaps most responsible for its growth saw their own power plummet. The first was the governing Liberal-Democratic Party (LDP). The second was the Ministry of International Trade and Industry (MITI), arguably the brain behind the Japanese economic miracle. The third was Keidanren, Japan's most politically potent business federation.

Today the LDP is in retreat, its elderly male leaders embarrassed by financial and sexual scandals. It is faced, for the first time, by outraged and increasingly active women voters, by consumers, taxpayers, and farmers who formerly supported it. To retain the power it has held since 1955, it will be compelled to shift its base from rural to urban voters, and deal with a far more **heterogeneous** population than ever before. For Japan, like all the high-tech nations, is becoming a de-massified society, with many more actors arriving on the political scene. Whether the LDP can make this long-term switch is at issue. What is not at issue is that significant power has shifted away from the LDP.

As for MITI, even now many American academics and politicians urge the United States to adopt MITI-style planning as a model. Yet today, MITI itself is in trouble. Japan's biggest corporations once danced attendance on its bureaucrats and, willing or not, usually followed its "guidelines." Today MITI is a fast-fading power as the corporations themselves have grown strong enough to thumb their noses at it. Japan remains economically powerful in the outside world but politically weak at home. Immense economic weight pivots around a shaky political base.

Even more pronounced has been the decline in the strength of Keidaren, still dominated by the hierarchies of the fast-fading smokestack industries.

Even those dreadnoughts of Japanese fiscal power, the Bank of Japan and the Ministry of Finance, whose controls guided Japan through the high-growth period, the oil shock, the stock market crash, and the yen rise, now find themselves impotent against the turbulent market forces destabilizing the economy.

Still more striking shifts of power are changing the face of Western Europe. Thus power has shifted away from London, Paris, and Rome as the German economy has outstripped all the rest. Today, as East and West Germany progressively fuse their economies, all Europe once more fears German domination of the continent.

To protect themselves, France and other West European nations, with the exception of Britain, are hastily trying to integrate the European community politically as well as economically. But the more successful they become, the more of their national power is transfused into the veins of the Brussels-based European Community, which has progressively stripped away bigger and bigger chunks of their sovereignty.

The nations of Western Europe thus are caught between Bonn and Berlin on the one side and Brussels on the other. Here, too, power is shifting rapidly away from its established centers.

The lists of such global and domestic power shifts could be extended indefinitely. They represent a remarkable series of changes for so brief a peacetime period. Of course, some power shifting is normal at any time.

Yet only rarely does an entire globe-girdling *system* of power fly apart in this fashion. It is an even rarer moment in history when all the rules of the power game change at once, and the very nature of power is revolutionized.

Yet that is exactly what is happening today. Power, which to a large extent defines us as individuals and as nations, is itself being redefined.

God-in-a-White-Coat

A clue to this redefinition emerges when we look more closely at the above list of apparently unrelated changes. For we discover that they are not as random as they seem. Whether it is Japan's meteoric rise, GM's embarrassing decline, or the American doctor's fall from grace, a single common thread unites them.

Take the punctured power of the god-in-a-white-coat.

Throughout the heyday of doctor-dominance in America, physicians kept a tight choke-hold on medical knowledge. Prescriptions were written in Latin, providing the profession with a semi-secret code, as it were, which kept most patients in ignorance. Medical journals and texts were restricted to professional readers. Medical conferences were closed to the laity. Doctors controlled medical-school curricula and enrollments.

Contrast this with the situation today, when patients have astonishing access to medical knowledge. With a personal computer and a modem, anyone from home can access data bases like Index Medicus, and obtain scientific papers on everything from Addison's disease to zygomycosis, and, in fact, collect more information about a specific ailment or treatment than the ordinary doctor has time to read.

Copies of the 2,354-page book known as the PDR or *Physicians' Desk Reference* are also readily available to anyone. Once a week on the Lifetime cable network, any televiewer can watch twelve uninterrupted hours of highly technical television programming designed specifically to educate doctors. Many of these programs carry a disclaimer to the effect that "some of this material may not be suited to a general audience." But that is for the viewer to decide.

The rest of the week, hardly a single newscast is aired in America without a medical story or segment. A video version of material from the *Journal of the American Medical Association* is now broadcast by three hundred stations on Thursday nights. The press reports on medical malpractice cases. Inexpensive paperbacks tell ordinary readers what drug side effects to watch for, what drugs not to mix, how to raise or lower cholesterol levels through diet. In addition, major medical breakthroughs, even if first published in medical journals, are reported on the evening television news almost before the M.D. has even taken his subscription copy of the journal out of the in-box.

In short, the knowledge monopoly of the medical profession has been thoroughly smashed. And the doctor is no longer a god.

This case of the dethroned doctor is, however, only one small example of a more general process changing the entire relationship of knowledge to power in the high-tech nations.

In many other fields, too, closely held specialists' knowledge is slipping out of control and reaching ordinary citizens. Similarly, inside major corporations, employees are winning access to knowledge once monopolized by management. And as knowledge is redistributed, so, too, is the power based on it.

Bombarded by the Future

There is, however, a much larger sense in which changes in knowledge are causing or contributing to enormous power shifts. The most important economic development of our lifetime has been the rise of a new system for creating wealth, based no longer on muscle but on mind. Labor in the advanced economy no longer consists of working on "things," writes historian Mark Poster of the University of California (Irvine), but of "men and women acting on other men and women, or . . . people acting on information and information acting on people."

The substitution of information or knowledge for brute labor, in fact, lies behind the troubles of General Motors and the rise of Japan as well. For while GM still thought the earth was flat, Japan was exploring its edges and discovering otherwise.

As early as 1970, when American business leaders still thought their smokestack world secure, Japan's business leaders, and even the general public, were being bombarded by books, newspaper articles, and television programs heralding the arrival of the "information age" and focusing on the 21st century. While the end-of-industrialism concept was dismissed with a shrug in the United States, it was welcomed and embraced by Japanese decision-makers in business, politics, and the media. Knowledge, they concluded, was the key to economic growth in the 21st century.

It was hardly surprising, therefore, that even though the United States started computerizing earlier, Japan moved more quickly to substitute the knowledge-based technologies of the Third Wave for the brute muscle technologies of the Second Wave past.

Robots proliferated. Sophisticated manufacturing methods, heavily dependent on computers and information, began turning out products whose quality could not be easily matched in world markets. Moreover, recognizing that its old smokestack technologies were ultimately doomed, Japan took steps to facilitate the transition to the new and to buffer itself against the dislocations entailed in such a strategy. The contrast with General Motors—and American policy in general—could not have been sharper.

If we also look closely at many of the other power shifts cited above, it will become apparent that in these cases, too, the changed role of knowledge—the rise of the new wealth-creation system—either caused or contributed to major shifts of power.

The spread of this new knowledge economy is, in fact, the explosive new force that has hurled the advanced economies into bitter global competition, confronted the socialist nations with their hopeless **obsolescence,** forced many "developing nations" to scrap their traditional economic strategies, and is now profoundly dislocating power relationships in both personal and public spheres.

In a **prescient** remark, Winston Churchill once said that "empires of the future are empires of the mind." Today that observation has come true. What has not yet been appreciated is the degree to which raw, elemental power—at the level of private life as well as at the level of empire—will be transformed in the decades ahead as a result of the new role of "mind."

The Making of a Shabby Gentility

A revolutionary new system for creating wealth cannot spread without triggering personal, political, and international conflict. Change the way wealth is made and you immediately collide with all the entrenched interests whose power arose from the prior wealth-system. Bitter conflicts erupt as each side fights for control of the future.

It is this conflict, spreading around the world today, that helps explain the present power shake-up. To anticipate what might lie ahead for us, therefore, it is helpful to glance briefly backward at the last such global conflict.

Three hundred years ago the industrial revolution also brought a new system of wealth creation into being. Smokestacks speared the skies where fields once were cultivated. Factories proliferated. These "dark Satanic mills" brought with them a totally new way of life—and a new system of power.

Peasants freed from near-servitude on the land turned into urban workers subordinated to private or public employers. With this change came changes in power relations in the home as well. Agrarian families, several generations under a single roof, all ruled by a bearded patriarch, gave way to stripped-down nuclear families from which the elderly were soon extruded or reduced in prestige and influence. The family itself, as an institution, lost much of its social power as many of its functions were transferred to other institutions—education to the school, for example.

Sooner or later, too, wherever steam engines and smokestacks multiplied, vast political changes followed. Monarchies collapsed or shriveled into tourist attractions. New political forms were introduced.

If they were clever and farsighted enough, rural landowners, once dominant in their regions, moved into the cities to ride the wave of industrial expansion, their sons becoming stockbrokers or captains of industry. Most of the landed gentry who clung to their rural way of life wound up as shabby gentility, their mansions eventually turned into museums or into money-raising lion parks.

Against their fading power, however, new elites arose: corporate chieftains, bureaucrats, media **moguls.** Mass production, mass distribution, mass education, and mass communication were accompanied by mass democracy, or dictatorships claiming to be democratic.

These internal changes were matched by gigantic shifts in global power, too, as the industrialized nations colonized, conquered, or dominated much of the rest of the world, creating a hierarchy of world power that still exists in some regions.

In short, the appearance of a new system for creating wealth undermined every pillar of the old power system, ultimately transforming family life, business, politics, the nation-state, and the structure of the global power itself.

Those who fought for control of the future made use of violence, wealth, and knowledge. Today a similar, though far more accelerated, upheaval has started. The changes we have recently seen in business, the economy, politics, and at the global level are only the first skirmishes of far bigger power struggles to come. For we stand at the edge of the deepest powershift in human history.

Questions on Content

1. List at least five areas where there has been a striking change in power in the last twenty years.
2. What changes have taken place in Americans' views of medical doctors?
3. What does Toffler mean by this statement: "The most important economic development of our lifetime has been the rise of a new system for creating wealth, based no longer on muscle but on mind" (second sentence under "Bombarded by the Future").

Vocabulary

Define the following words as they are used in the piece (the words are printed in **boldface** type):

1. appurtenances
2. tectonic plates
3. rust-belt
4. heterogeneous
5. obsolescence
6. prescient
7. moguls

Suggestions for Response and Reaction

1. How do you define power? To you, who are powerful people?
2. Interview your grandparents or persons of a similar age concerning who made decisions "then" and who makes decisions "now" in the family, in the workplace, in the voting booth, and in other significant aspects of life. Write a paper comparing and contrasting (see pages 200–212) who made decisions then and who makes decisions now. Arrive at conclusions that reach beyond your family, your associates, your locale.
3. Select a sentence from "The Powershift Era" that invites research (see pages 394–409). Amend the sentence to make it your own assertion that requires much further reading and supporting documentation. Write the research report, as directed by your instructor.

Hardest Lessons for First-Time Managers

LINDA A. HILL

A Harvard Business School professor points out tools for success, whether you supervise a rookie, work for one, or just got promoted yourself.

A new manager's initial concept of what it means to manage typically focuses on the rights, privileges and authority that come with the job: "I thought, 'Finally I'll have the power to do things right around here,'" one first-timer recalled. He found out it wasn't so simple.

In accepting their promotions, new managers are making the commitment to form a new professional identity. They have to manage *people,* not just tasks, and they must master three critical lessons: to switch from relying on formal authority to establishing **credibility,** from striving for subordinate control to building subordinate commitment and from managing individuals to leading the team. The transformation challenges the best of them, both intellectually and emotionally.

Understanding the Job

Rookie managers first have to let go of the deeply held attitudes and habits they developed when they were responsible only for their own performance. Many think managing means doing what they've been doing all along, only with more power and control. One new sales manager described his role as that of the "lead salesperson with the final authority and accountability," and complained that "if people would just leave me alone, I could do my job."

Instead, to use the analogy of an orchestra, a new manager must move from being a violinist, who concentrates on playing one instrument, to being the conductor, who coordinates the efforts of many musicians and must know about every instrument. She must also find new ways of deriving satisfaction from her work and measuring her success, a mental switch that can be traumatic. In the words of one new manager, "I never knew a promotion could be so painful."

Most first-timers soon discover that *task* learning (including building key relationships) is only part of the story. They are surprised by how much *personal* learning they must undergo. One manager, who admitted that she preferred action to analysis, said she had never believed that adults could change themselves very much. But in her first year as a manager, she found herself "growing and discovering new sides" of herself, learning to manage her weaknesses and play to her strengths. "I don't need to be liked," she said. "[The staff knows] I just want to get the job done. I don't have to play any games to protect my ego, and so I give it to them straight." She began finding new satisfaction in directing her efforts to helping subordinates "fulfill their dreams."

Reprinted by permission from *Working Woman,* February 1994, pages 18–21.

A new manager's first challenge is to adjust to the daily reality of managerial life, which is often pressured, hectic and fragmented. She must learn to respond to the unending, confusing and sometimes conflicting demands of subordinates, bosses and peers. One new manager declared, "I'm not the boss; I'm a hostage. And there are loads of terrorists around here." Rookies quickly learn that their new position is one of dependence as much as authority, since their success now depends largely on the performance of subordinates. In particular, those who have been star performers have to face up to the fact that many subordinates fall short of their expectations, yet they must figure out how to get them to deliver.

In the beginning, many first-time managers, eager to implement their ideas, use a hands-on, **autocratic** approach. But they soon find out that to be effective in their new role, they have to develop the capacity to exercise influence without relying simply on the power of their position. After six months or so, just when they think they are getting a handle on managing their team, they begin to recognize a new dependence on those over whom they have no formal authority—their bosses and peers. Rookies tend to learn the hard way that unless they understand the power dynamics of the organization and its impact on their unit, and unless they actively manage these external relationship, their own success is jeopardized. Subordinates of new managers frequently find themselves having to meet unreasonable performance objects with limited resources because their bosses have not devoted enough time and energy to managing up and laterally.

Developing Judgment

In managing individual performance, new managers have to learn to balance three sets of **paradoxical** tensions: treating subordinates fairly, but as individuals; holding them accountable, yet tolerating their mistakes and deficiencies; and maintaining control, while also providing autonomy. They discover that the real trick is not only managing individuals, but also providing leadership to cultivate a high-performance culture.

As they work with people day by day, new managers begin to create personal maps for diagnosing the inevitable problems and conflicts they encounter and develop rules of thumb for resolving them. By experimenting with different ways of handling these situations, they build up a kind of "personal case law" as a set of guiding principles.

First-time struggles with delegation, a fundamental managerial skill, form a good case in point. Managers have to decide upon just the right length of rope to give subordinates—too short and they feel over-controlled; too long and they feel unsupported and neglected. In time, new managers must also come to terms with another basic principle that feels intuitively illogical: To treat people fairly is to treat them differently, since the right amount of delegation for one employee may be grossly inappropriate for another.

In this regard, most new managers are more comfortable working with less experienced subordinates than with more experienced ones. One manager had the insight to realize that he found the less experienced employees easier to work with because they seemed to depend on him as much as he did on them. In contrast, his more experienced people, especially the most successful, were forever complaining that they got too little respect and autonomy and too much attention. In fact, many experienced subordinates refer to "new-manager syndrome," the tendency to over-manage.

Three primary factors often keep first-time managers from becoming effective delegators. First, they have to accept the distinction between managing the people and managing the task. For many, it is a slow and painful process to be "weaned off the task side to get over to the management side," as one manager lamented. The second inhibitor is their personal preference: They themselves tend to be good at technical tasks and like doing them. They fear that their skills will become obsolete as they spend less time on details. The third and perhaps most significant inhibitor is insecurity. New managers do not know whom to trust or how to assess a subordinate's competence, personal integrity and motivation. "This job demands that you rely on somebody else to get it done," observed one new manager. "Before, *you* were the master of your fate. You had no one to blame but yourself, and no one to take credit but yourself."

How do managers eventually learn to delegate? Mostly, circumstances force them to: They come to realize that their jobs are simply too big and complex to handle alone. Eight months into her new job, one manager admitted, "It was a simple cost-benefit analysis. I couldn't do everything I was in charge of. I was working 70-hour weeks. I just couldn't keep up any longer."

Learning How to Learn

To be successful, new managers have to become self-directed learners. They must force themselves to be periodically introspective, collecting feedback, analyzing it and altering their behavior when necessary. They also have to learn how to manage themselves and their emotions. Inevitable feelings of stress and anxiety can profoundly affect daily functioning. Many first-timers will find themselves all too willing to revert to the familiar, comfortable "doer" role when faced with messy managerial responsibilities. Only when managers understand *why* they are regressing can they begin to combat this tendency.

When they are promoted, new managers may find it useful to make a list of whom they are dependent on to get the job done—subordinates, superiors, peers and associates both inside and outside the organization. They should seek out these constituencies and learn about their expectations, priorities, concerns and preferred work styles. By so doing, new managers can begin to size up various tradeoffs to be made as work situations require them to juggle competing needs and expectations.

New managers may also want to keep a record of how they spend their time to make sure that they are allocating it appropriately: how much time they are spending on *doing* as opposed to *managing*. Creating and regularly updating a personal-development report can also be useful. One manager got into the habit of taking 10 minutes every Friday before he left the office to note "big mistakes made and key lessons learned." Another kept a detailed journal of her experiences and impressions the first three months on the job. Such simple tools can help managers confront their misconceptions about managerial work and learn from their experience sooner. In addition, they help identify a set of benchmarks by which to gauge their own progress.

Finally, it is important that new managers recognize that the challenges of becoming a manager cannot be handled alone. Those who have a more extensive and varied network and are willing to ask for help find it easier to cope the first year. Talking with peers, both past and current, can be invaluable, since such discussions tend to be more informal and managers can feel free to explore ideas and disclose problems. Unfortunately, new managers are often reluctant to ask for help; it doesn't fit their conception of the boss as expert. One manager explained that she avoided participating in the company training program offered to new managers because "if they realize how little I know, they will think I am a promotion mistake."

In response to such anxieties, new managers often embark on a futile search for the perfect mentor to help them navigate their new assignment. Instead, the first-time manager's goal should be to cultivate a network of relationships (superior and peer, internal and external) for different types of support, feedback and advice. Such relationships can be fruitful only if the new managers are willing to take some risks and open themselves to constructive criticism.

Managing a New Manager

A new manager, by definition, is a work-in-progress, and bosses have to be realistic about their expectations. A fledgling manager is not going to be as effective as someone who has been on the job for five years, and is likely to err in predictable ways. In the first months, she is trying to figure out what her job really is and to cope with its conflicts and **ambiguities.** As she comes to understand the full range of her new position's demands, she can find herself revisiting unsettling questions: Do I really want to be a manager? Do I have what it takes?

In the second part of that first year, she is struggling to learn the fundamentals of exercising authority and managing subordinate performance. At this juncture, bosses need to provide constructive and timely feedback, not simply on what she has been doing but also on how she has been doing it. The bosses most helpful to new managers are those who adopt a stance of supportive autonomy and a problem-solving approach that might include a kind of **Socratic dialogue.** Remember that new managers will not necessarily seek out feedback; bosses often have to initiate contact. They should also encourage new managers to seek out others, to develop a resource base consisting of previous bosses, current and former peers, and the human-resources department.

It isn't always easy to adopt a developmental point of view when working with a new manager, particularly now, when the pressure for short-term results is intense. Still, the benefits of investing in a new manager have been clearly documented. First-line management is the level in the organization that generates the most frequent reports of incompetence, burnout and excessive attrition. The human and financial costs are staggering for both those who fail to make the transition and the businesses that promote them. Moreover, the first management assignment is a key developmental experience for the company's future executives. It is the time during which basic management philosophies and styles are profoundly shaped.

Questions on Content

1. List three critical lessons that new managers must master.
2. Explain the analogy of an orchestral conductor and a manager.
3. List the three sets of paradoxical tensions that a new manager must juggle.
4. How do managers learn to delegate?
5. Explain the statement: "A new manager, by definition, is a work-in-progress."

Vocabulary

Define the following words as they are used in the article (the words are printed in **boldface** type):

1. credibility
2. rookie
3. autocratic
4. paradoxical
5. ambiguities
6. Socratic dialogue

Suggestions for Response and Reaction

1. Following the guidelines for setting up a classification system (see pages 133–138), explain the categories of the hardest lessons for first-time managers.
2. Compare and contrast the role of a manager with the role of a subordinate.
3. Would you rather be a manager or a subordinate? Explain the reasons for your answer.

Quality

JOHN GALSWORTHY

This short story portrays the unrelenting struggle of the craftsman who takes pride in the quality of his work but who is slowly starved to death by mass production.

I knew him from the days of my extreme youth, because he made my father's boots; inhabiting with his elder brother two little shops let into one, in a small by-street—now no more, but then most fashionably placed in the West End.

That tenement had a certain quiet distinction; there was no sign upon its face that he made for any of the Royal Family—merely his own German name of Gessler Brothers; and in the window a few pairs of boots. I remember that it always troubled me to account for those unvarying boots in the window, for he made only what was ordered, reaching nothing down, and it seemed so inconceivable that what he made could ever have failed to fit. Had he bought them to put there? That, too, seemed inconceivable. He would never have tolerated in his house leather on which he had not worked himself. Besides, they were too beautiful—the air of pumps, so inexpressibly slim, the patent leathers with cloth tops, making water come into one's mouth, the tall brown riding-boots with marvellous sooty glow, as if, though new, they had been worn a hundred years. Those pairs could only have been made by one who saw before him the Soul of Boot—so truly were they **prototypes** incarnating the very spirit of all footgear. These thoughts, of course, came to me later, though even when I was promoted to him, at the age of perhaps fourteen, some inkling haunted me of the dignity of himself and brother. For to make boots—such boots as he made—seemed to me then, and still seems to me, mysterious and wonderful.

I remember well my shy remark, one day, while stretching out to him my youthful foot:

"Isn't it awfully hard to do, Mr. Gessler?"

And his answer, given with a sudden smile from out of the **sardonic** redness of his beard: "Id is an Ardt!"

Himself, he was a little as if made from leather, with his yellow crinkly face, and crinkly reddish hair and beard, and neat folds slanting down his cheeks to the corners of his mouth, and his guttural and one-toned voice; for leather is a sardonic substance, and stiff and slow of purpose. And that was the character of his face, save that his eyes, which were grey-blue, had in them the simple gravity of one secretly possessed by the Ideal. His elder brother was so very like him—though watery, paler in every way, with a great industry—that sometimes in early days I was not quite sure of him until the interview was over. Then I knew it was he, if the words, "I will ask my brudder," had not been spoken; and that, if they had, it was his elder brother.

When one grew old and wild and ran up bills, one somehow never ran them up with Gessler Brothers. It would not have seemed becoming to go in there and stretch out one's foot to that blue iron-spectacled glance, owing him for more than—say—two pairs, just the comfortable reassurance that one was still his client.

For it was not possible to go to him very often—his boots lasted terribly, having something beyond the temporary—some, as it were, essence of boot stitched into them.

One went in, not as into most shops, in the mood of: "Please serve me, and let me go!" but restfully, as one enters a church; and, sitting on the single wooden chair, waited—for there was never anybody there. Soon, over the top edge of that sort of well—rather dark, and smelling soothingly of leather—which formed the shop, there would be seen his face, or that of his elder brother, peering down. A guttural sound, and tip-tap of bast slippers beating the narrow wooden stairs, and he would stand before one without coat, a little bent, in leather apron, with sleeves turned back, blinking—as if awakened from some dream of boots, or like an owl surprised in daylight and annoyed at this interruption.

And I would say: "How do you do, Mr. Gessler? Could you make me a pair of Russian leather boots?"

Without a word he would leave me, retiring **whence** he came, or into the other portion of the shop, and I would continue to rest in the wooden chair, inhaling the incense of his trade. Soon he would come back, holding in his thin, veined hand a piece of gold-brown leather. With eyes fixed on it, he would remark: "What a beadiful biece!" When I, too, had admired it, he would speak again. "When do you wand dem?" And I would answer: "Oh! As soon as you conveniently can." And he would say: "To-morrow fordnighd?" Or if he were his elder brother: "I will ask my brudder!"

Then I would murmur: "Thank you! Good-morning, Mr. Gessler." "Goot-morning!" he would reply, still looking at the leather in his hand. And as I moved to the door, I would hear the tip-tap of his bast slippers restoring him, up the stairs, to his dream of boots. But if it were some kind of footgear that he had not yet made me, then indeed he would observe ceremony—**divesting** me of my boot and holding it long in his hand, looking at it with eyes at once critical and loving, as if recalling the glow with which he had created it, and rebuking the way in which one had disorganized the masterpiece. Then, placing my foot on a piece of paper, he would two or three times tickle the outer edges with a pencil and pass his nervous fingers over my toes, feeling himself into the heart of my requirements.

I cannot forget that day on which I had occasion to say to him: "Mr. Gessler, that last pair of town walking-boots creaked, you know."

He looked at me for a time without replying, as if expecting me to withdraw or qualify the statement, then said:

"Id shouldn't 'ave greaked."

"It did, I'm afraid."

"You godden wed before dey found demselves?"

"I don't think so."

At that he lowered his eyes, as if hunting for memory of those boots, and I felt sorry I had mentioned this grave thing.

"Zend dem back!" he said; "I will look at dem."

A feeling of compassion for my creaking boots surged up in me, so well could I imagine the sorrowful long curiosity of regard which he would bend on them.

"Zome boods," he said slowly, "are bad from birdt. If I can do noding wid dem, I dake dem off your bill."

Once (only once) I went absent-mindedly into his shop in a pair of boots bought in an emergency at some large firm. He took my order without showing me any leather, and I could feel his eyes penetrating the inferior **integument** of my foot. At last he said:

"Dose are nod my boods."

The tone was not one of anger, nor of sorrow; not even of contempt, but there was in it something quiet that froze the blood. He put his hand down and pressed a finger on the place where the left boot, endeavoring to be fashionable, was not quite comfortable.

"Id 'urds you dere," he said. "Dose big virms 'ave no self-respect. Drash!" And then, as if something had given way within him, he spoke long and bitterly. It was the only time I ever heard him discuss the conditions and hardships of his trade.

"Dey get id all," he said, "dey get id by adverdisement, nod by work. Dey dake it away from us, who lofe our boods. Id gomes to this—bresently I haf no work. Every year id gets less—you will see." And looking at his lined face I saw things I had never noticed before, bitter things and bitter struggle—and what a lot of grey hairs there seemed suddenly in his red beard!

As best I could, I explained the circumstances of the purchase of those ill-omened boots. But his face and voice made so deep an impression that during the next few minutes I ordered many pairs. **Nemesis** fell! They lasted more terribly than ever. And I was not able conscientiously to go to him for nearly two years.

When at last I went I was surprised to find that outside one of the two little windows of his shop another name was painted, also that of a boot-maker—making, of course, for the Royal Family. The old familiar boots, no longer in dignified isolation, were huddled in the single window. Inside, the now contracted well of the one little shop was more scented than ever. And it was longer than usual, too, before a face peered down, and the tip-tap of the bast slippers began. At last he stood before me, and gazing through those rusty iron spectacles, said:

"Mr—, isn'd it?!"

"Ah, Mr. Gessler," I stammered, "but your boots are really *too* good, you know! See, these are quite decent still!" And I stretched out to him my foot. He looked at it.

"Yes," he said, "beople do nod wand good boods, id seems."

To get away from his reproachful eyes and voice I hastily remarked: "What have you done to your shop?"

He answered quietly: "Id was too exbensif. Do you wand some boods?"

I ordered three pairs, though I had only wanted two, and quickly left. I had, I do not know quite what feeling of being part, in his mind, of a conspiracy against him; or not perhaps so much against him as against his idea of boot. One does not, I suppose, care to feel like that; for it was again many months before my next visit to his shop, paid, I remember, with the feeling: "Oh! well, I can't leave the old boy—so here goes! Perhaps it'll be his elder brother!"

For his elder brother, I knew, had not character enough to reproach me, even dumbly.

And, to my relief, in the shop there did appear to be his elder brother, handling a piece of leather.

"Well, Mr. Gessler," I said, "how are you?"

He came close, and peered at me.

"I am breddy well," he said slowly, "but my elder brudder is dead."

And I saw that it was indeed himself—but how aged and wan! And never before had I heard him mention his brother. Much shocked, I murmured: "Oh! I am sorry!"

"Yes," he answered. "He was a good man, he made a good bood; but he is dead." And he touched the top of his head, where the hair had suddenly gone as thin as it had been on that of his poor brother, to indicate, I suppose, the cause of death. "He could nod ged over closing de oder shop. Do you wand any boods?" And he held up the leather in his hand: "Id's a beaudiful biece."

I ordered several pairs. It was very long before they came out—but they were better than ever. One simply could not wear them out. And soon after that I went abroad.

It was over a year before I was again in London. And the first shop I went to was my old friend's. I had left a man of sixty; I came back to one of seventy-five, pinched and worn and **tremulous,** who genuinely, this time, did not at first know me.

"Oh! Mr. Gessler," I said, sick at heart, "how splendid your boots are! See, I've been wearing this pair nearly all the time I've been abroad; and they're not half worn out, are they?"

He looked at my boots—a pair of Russian leather, and his face seemed to regain steadiness. Putting his hand on my instep, he said:

"Do they vid you here? I 'ad drouble wid dat bair, I remember."

I assured him that they had fitted beautifully.

"Do you wand any boods?" he said. "I can made dem quickly; id is a slack dime."

I answered: "Please, please! I want boots all round—every kind!"

"I will make a vresh model. Your food must be bigger." And with utter slowness, he traced round my foot, and felt my toes, only once looking up to say: "Did I dell you my brudder was dead?"

To watch him was painful, so feeble had he grown; I was glad to get away.

I had given those boots up, when one evening they came. Opening the parcel, I set the four pairs out in a row. Then one by one I tried them on. There was no doubt about it. In shape and fit, in finish and quality of leather, they were the best he had ever made me. And in the mouth of one of the two walking-boots I found his bill. The amount was the same as usual, but it gave me quite a shock. He had never before sent it in till quarter day. I flew downstairs and wrote a cheque, and posted it at once with my own hand.

A week later, passing the little street, I thought I would go in and tell him how splendidly the new boots fitted. But when I came to where his shop had been, his name was gone. Still there, in the window, were the slim pumps, the patent leathers with cloth tops, the sooty riding boots.

I went in, very much disturbed. In the two little shops—again made into one— was a young man with an English face.

"Mr. Gessler in?" I said.

He gave me a strange, **ingratiating** look.

"No, sir," he said, "no. But we can attend to anything with pleasure. We've taken the shop over. You've seen our name, no doubt, next door. We make for some very good people."

"Yes, yes," I said; "but Mr. Gessler?"

"Oh!" he answered; "dead."

"Dead!! But I only received these boots from him last Wednesday week."

"Ah!" he said; "a shockin' go. Poor old man starved 'imself."

"Good God!"

"Slow starvation, the doctor called it! You see he wanted to work in such a way! Would keep the shop on; wouldn't have a soul touch his boots except himself. When he got an order, it took him such a time. People won't wait. He lost everybody. And there he'd sit, goin' on and on—I will say that for him—not a man in London made a better boot. But look at the competition! He never advertised! Would 'ave the best leather, too, and do it all 'imself. Well, there it is. What could you expect with his ideas?"

"But starvation—!"

"That may be a bit flowery, as the sayin' is—but I know myself he was sitting' over his boots day and night, to the very last. You see I used to watch him. Never gave 'imself time to eat; never had a penny in the house. All went in rent and leather. How he lived so long I don't know. He regular let his fire go out. He was a character. But he made good boots."

"Yes," I said. "He made good boots."

Questions on Content

1. How does Galsworthy describe the central character?
2. Explain the procedure for ordering a pair of boots from Mr. Gessler.
3. What was Gessler's attitude toward boots? Point out lines in the story to support your answer.

Vocabulary

Define the following words as they are used in the article (the words are printed in **boldface** type):

1. prototypes
2. sardonic
3. whence
4. divesting
5. integument
6. nemesis
7. tremulous
8. ingratiating

Suggestions for Response and Reaction

1. What *really* caused Mr. Gessler's death?
2. Evaluate the effectiveness of the story's ending.
3. a. As a skilled worker or technician, explain one advantage in working under the old system of individual craftsmanship and one advantage in working under the system of mass production.
   b. What is a disadvantage of working in each system?
4. As a skilled worker in a technical field, defend or attack the assertion that mass production has eliminated personal pride in craftsmanship.
5. Following the principles for cause-effect explanation (see pages 177–183), show how the general public has benefited and suffered from the shift from one production system to the other.
6. Write a documented paper (see pages 394–409) on some aspect of mass production.

Excerpt from *Zen and the Art of Motorcycle Maintenance*

ROBERT M. PIRSIG

The narrator, in explaining the views of his friends John and Sylvia and their dislike for maintaining their own motorcycles, gets at the larger issues of one's relationship to technology.

Disharmony I suppose is common enough in any marriage, but in John and Sylvia's case it seems more tragic. To me, anyway.

It's not a personality clash between them; it's something else, for which neither is to blame, but for which neither has any solution, and for which I'm not sure I have any solution either, just ideas.

The ideas began with what seemed to be a minor difference of opinion between John and me on a matter of small importance: how much one should maintain one's own motorcycle. It seems natural and normal to me to make use of the small tool kits and instruction booklets supplied with each machine, and keep it tuned and adjusted myself. John **demurs.** He prefers to let a competent mechanic take care of these things so that they are done right. Neither viewpoint is unusual, and this minor difference would never have become magnified if we didn't spend so much time riding together and sitting in country roadhouses drinking beer and talking about whatever comes to mind. What comes to mind, usually, is whatever we've been thinking about in the half hour or forty-five minutes since we last talked to each other. When it's roads or weather or people or old memories or what's in the newspapers, the conversation just naturally builds pleasantly. But whenever the performance of the machine has been on my mind and gets into the conversation, the building stops. The conversation no longer moves forward. There is a silence and a break in the continuity. It is as though two old friends, a Catholic and a Protestant, were sitting drinking beer, enjoying life, and the subject of birth control somehow came up. Big freeze-out.

And, of course, when you discover something like that it's like discovering a tooth with a missing filling. You can never leave it alone. You have to probe it, work around it, push on it, think about it, not because it's enjoyable but because it's on your mind and it won't get off your mind. And the more I probe and push on this subject of cycle maintenance the more irritated he gets, and of course that makes me want to probe and push all the more. Not deliberately to irritate him but because the irritation seems **symptomatic** of something deeper, something under the surface that isn't immediately apparent.

When you're talking birth control, what blocks it and freezes it out is that it's not a matter of more or fewer babies being argued. That's just on the surface. What's underneath is a conflict of faith, of faith in empirical social planning versus faith in the authority of God as revealed by the teaching of the Catholic Church. You can prove the practicality of planned parenthood till you get tired of listening to yourself and it's going to go nowhere because your **antagonist** isn't buying the assumption that anything socially practical is good per se. Goodness for him has other sources which he values as much as or more than social practicality.

So it is with John. I could preach the practical value and worth of motorcycle maintenance till I'm hoarse and it would make not a dent in him. After two sentences on the subject his eyes go completely glassy and he changes the conversation or just looks away. He doesn't want to hear about it.

Sylvia is completely with him on this one. In fact she is even more emphatic. "It's just a whole other thing," she says, when in a thoughtful mood. "Like garbage," she says, when not. They want *not* to understand it. Not to *hear* about it. And the more I try to fathom what makes me enjoy mechanical work and them hate it so, the more elusive it becomes. The ultimate cause of this originally minor difference of opinion appears to run way, way deep.

Inability on their part is ruled out immediately. They are both plenty bright enough. Either one of them could learn to tune a motorcycle in an hour and a half if they put their minds and energy to it, and the saving in money and worry and delay would repay them over and over again for their effort. And they *know* that. Or maybe they don't. I don't know. I never confront them with the question. It's better to just get along.

But I remember once, outside a bar in Savage, Minnesota, on a really scorching day when I just about let loose. We'd been in the bar for about an hour and we came out and the machines were so hot you could hardly get on them. I'm started and ready to go and there's John pumping away on the kick starter. I smell gas like we're next to a refinery and tell him so, thinking this is enough to let him know his engine's flooded.

"Yeah, I smell it too," he says and keeps on pumping. And he pumps and pumps and pumps and pumps and *I* don't know what more to say. Finally, he's really winded and sweat's running down all over his face and he can't pump anymore, and so I suggest taking out the plugs to dry them off and air out the cylinders while we go back for another beer.

"Oh my God no!" He doesn't want to get into all that stuff.

"All what stuff?"

"Oh, getting out the tools and all that stuff. There's no reason why it shouldn't start. It's a brand-new machine and I'm following the instructions perfectly. See, it's right on full choke like they say."

"Full *choke!*"

"That's what the instructions say."

"That's for when it's *cold!*"

"Well, we've been in there for a half an hour at least," he says.

It kind of shakes me up. "This is a hot day, John," I say. "And they take longer than that to cool off even on a freezing day."

He scratches his head. "Well, why don't they tell you that in the instructions?" He opens the choke and on the second kick it starts. "I guess that was it," he says cheerfully.

And the very next day we were out near the same area and it happened again. This time I was determined not to say a word, and when my wife urged me to go over and help him I shook my head. I told her that until he had a real felt need he was just going to resent help, so we went over and sat in the shade and waited.

I noticed he was being superpolite to Sylvia while he pumped away, meaning he was furious, and she was looking over with a kind of "Ye gods!" look. If he had

Robert M. Pirsig
Excerpt from
*Zen and the Art
of Motorcycle
Maintenance*

asked any single question I would have been over in a second to diagnose it, but he wouldn't. It must have been fifteen minutes before he got it started.

Later we were drinking beer again over at Lake Minnetonka and everybody was talking around the table, but he was silent and I could see he was really tied up in knots inside. After all that time. Probably to get them untied he finally said, "You know . . . when it doesn't start like that it just . . . really turns me into a *monster* inside. I just get **paranoic** about it." This seemed to loosen him up, and he added, "They just had this *one* motorcycle, see? This *lemon*. And they didn't know what to do with it, whether to send it back to the factory or sell it for scrap or what . . . and then at the last moment they saw *me* coming. With eighteen hundred bucks in my pocket. And they knew their problems were over."

In a kind of singsong voice I repeated the plea for tuning and he tried hard to listen. He really tries hard sometimes. But then the block came again and he was off to the bar for another round for all of us and the subject was closed.

He is not stubborn, not narrow-minded, not lazy, not stupid. There was just no easy explanation. So it was left up in the air, a kind of mystery, that one gives up on because there is no sense in just going round and round and round looking for an answer that's not there.

It occurred to me that maybe I was the odd one on the subject, but that was disposed of too. Most touring cyclists know how to keep their machines tuned. Car owners usually won't touch the engine, but every town of any size at all has a garage with expensive lifts, special tools and diagnostic equipment that the average owner can't afford. And a car engine is more complex and inaccessible than a cycle engine so there's more sense to this. But for John's cycle, a BMW R60, I'll bet there's not a mechanic between here and Salt Lake City. If his points or plugs burn out, he's done for. I *know* he doesn't have a set of spare points with him. He doesn't know what points are. If it quits on him in western South Dakota or Montana I don't know what he's going to do. Sell it to the Indians maybe. Right now I know what he's doing. He's carefully avoiding giving any thought whatsoever to the subject. The BMW is famous for not giving mechanical problems on the road and that's what he's counting on.

I might have thought this was just a peculiar attitude of theirs about motorcycles but discovered later that it extended to other things. . . . Waiting for them to get going one morning in their kitchen I noticed the sink faucet was dripping and remembered that it was dripping the last time I was there before and that in fact it had been dripping as long as I could remember. I commented on it and John said he had tried to fix it with a new faucet washer but it hadn't worked. That was all he said. The **presumption** left was that that was the end of the matter. If you try to fix a faucet and your fixing doesn't work, then it's just your lot to live with a dripping faucet.

This made me wonder to myself if it got on their nerves, this drip-drip-drip, week in, week out, year in, year out, but I could not notice any irritation or concern about it on their part, and so concluded they just aren't bothered by things like dripping faucets. Some people aren't.

What it was that changed this conclusion, I don't remember . . . some intuition, some insight one day, perhaps it was a subtle change in Sylvia's mood whenever the dripping was particularly loud and she was trying to talk. She has a very soft voice. And one day when she was trying to talk above the dripping and the kids came in and interrupted her, she lost her temper at them. It seemed that her anger

529

Robert M. Pirsig
Excerpt from
*Zen and the Art
of Motorcycle
Maintenance*

at the kids would not have been nearly as great if the faucet hadn't also been dripping when she was trying to talk. It was the combined dripping and loud kids that blew her up. What struck me hard then was that she was *not* blaming the faucet, and that she was *deliberately* not blaming the faucet. She wasn't ignoring the faucet at all! She was *suppressing* anger at the faucet and that goddamned dripping faucet was just about *killing* her! But she could not admit the importance of this for some reason.

Why suppress anger at a dripping faucet? I wondered.

Then that patched in with the motorcycle maintenance and one of those light bulbs went on over my head and I thought Ahhhhhhhh!

It's not the motorcycle maintenance, not the faucet. It's all of technology they can't take. And then all sorts of things started tumbling into place and I knew that was it. Sylvia's irritation at a friend who thought computer programming was "creative." All their drawings and paintings and photographs without a technological thing in them. Of course she's not going to get mad at that faucet, I thought. You always suppress momentary anger at something you deeply and permanently hate. Of course John signs off every time the subject of cycle repair comes up, even when it is obvious he is suffering for it. That's technology. And sure, of course, obviously. It's so simple when you see it. To get away from technology out into the country in the fresh air and sunshine is why they are on the motorcycle in the first place. For me to bring it back to them just at the point and place where they think they have finally escaped it just frosts both of them, tremendously. That's why the conversation always breaks and freezes when the subject comes up.

Other things fit in too. They talk once in a while in as few pained words as possible about "it" or "it all" as in the sentence, "There is just no escape from it." And if I asked, "From what?" the answer might be "The whole thing," or "The whole organized bit," or even "The system." Sylvia once said defensively, "Well, *you* know how to *cope* with it," which puffed me up so much at the time I was embarassed to ask what "it" was and so remained somewhat puzzled. I thought it was something more mysterious than technology. But now I see that the "it" was mainly, if not entirely, technology. But, that doesn't sound right either. The "it" is a kind of force that gives rise to technology, something undefined, but inhuman, mechanical, lifeless, a blind monster, a death force. Something hideous they are running from but know they can never escape. I'm putting it way too heavily here but in a less emphatic and less defined way this is what it is. Somewhere there are people who understand it and run it but those are technologists, and they speak an inhuman language when describing what they do. It's all parts and relationships of unheard-of things that never make any sense no matter how often you hear about them. And their things, their monster keeps eating up land and polluting their air and lakes, and there is no way to strike back at it, and hardly any way to escape it.

That attitude is not hard to come to. You go through a heavy industrial area of a large city and there it is, the technology. In front of it are high barbed wire fences, locked gates, signs saying No Trespassing, and beyond, through sooty air, you see ugly strange shapes of metal and brick whose purpose is unknown, and whose masters you will never see. What it's for you don't know, and why it's there, there's no one to tell, and so all you can feel is alienated, estranged, as though you didn't belong there. Who owns and understands this doesn't want you around. All this technology has somehow made you a stranger in your own land. Its very shape and appearance and mysteriousness say, "Get out." You know there's an explanation for all this somewhere and what it's doing undoubtedly serves mankind in some indirect way, but that isn't what you see. What you see is the No Trespassing, Keep Out signs and not anything serving people but little people, like ants, serving these strange, incomprehensible shapes. And you think, even if I were a part of this, even if I were not a stranger, I would be just another ant serving the shapes. So the final feeling is hostile, and I think that's ultimately what's involved with this otherwise unexplainable attitude of John and Sylvia. Anything to do with valves and shafts and wrenches is a part of *that* dehumanized world, and they would rather not think about it. They don't want to get into it.

If this is so, they are not alone. There is no question that they have been following their natural feelings in this and not trying to imitate anyone. But many others are also following their natural feelings and not trying to imitate anyone and the natural feelings of very many people are similar on this matter; so that when you look at them collectively, as journalists do, you get the illusion of a mass movement, an **antitechnological** mass movement, an entire political antitechnological left emerging, looming up from apparently nowhere, saying, "Stop the technology. Have it somewhere else. Don't have it here." It is still restrained by a thin web of logic that points out that without the factories there are no jobs or standard of living. But there are human forces stronger than logic. There always have been, and if they become strong enough in their hatred of technology that web can break.

Clichés and stereotypes such as "beatnik" or "hippie" have been invented for the antitechnologists, the antisystem people, and will continue to be. But one does not convert individuals into mass people with the simple coining of a mass term. John and Sylvia are not mass people and neither are most of the others going their way. It is against being a mass person that they seem to be revolting. And they feel that technology has got a lot to do with the forces that are trying to turn them into mass people and they don't like it. So far it's still mostly a passive resistance, flight into the rural areas when they are possible and things like that, but it doesn't always have to be this passive.

I disagree with them about cycle maintenance, but not because I am out of sympathy with their feelings about technology. I just think that their flight from and hatred of technology is self-defeating. The Buddha, the Godhead, resides quite as comfortably in the circuits of a digital computer or the gears of a cycle transmission as he does at the top of a mountain or in the petals of a flower. To think otherwise is to demean the Buddha—which is to demean oneself.

Questions on Content

1. What is the narrator's attitude toward a person's being able to maintain his or her own motorcycle?
2. What is John and Sylvia's attitude toward maintaining their own motorcycles?
3. Why do some people become antitechnological?

Vocabulary

Define the following words as they are used in the article (the words are printed in **boldface** type):

1. demurs
2. symptomatic
3. antagonist
4. paranoic
5. presumption
6. antitechnological

Suggestions for Response and Reaction

1. Are you more like the narrator or like John and Sylvia; that is, if you had a motorcycle, would you prefer maintaining it yourself or taking it to a cycle shop? Or if you had a leaky faucet, would you repair it or put up with the drip? (Other questions: Do you change the oil in your car or take the car to a service station? If mowing the lawn and the mower breaks down, do you attempt to repair the mower yourself or do you take it to a shop for repair?)
2. Write an extended definition (see pages 112–118) of *technology*.
3. Do you think that computer programming can be "creative"? Explain your answer.
4. What is the relationship between the title *Zen and the Art of Motorcycle Maintenance* and the reference to Buddha in the last paragraph?
5. Do you agree or disagree with this statement from the last paragraph: "The Buddha, the Godhead, resides quite as comfortably in the circuits of a digital computer or the gears of a cycle transmission as he does at the top of a mountain or in the petals of a flower"? Explain your answer.
6. Reread the last paragraph of the selection. Think about it for a while. Then describe an enjoyable activity in which you can "lose" yourself to the point that you have a new insight into life. (Examples: Pride in a newly washed and waxed automobile, satisfaction in a more efficient rearrangement of furniture, the contentment of knowing that you gave a patient or client your complete attention.) In your description, explain the situation, the activity, the outcome, your overall assessment, and the impact of your thinking and feelings.

The Science of Deduction

SIR ARTHUR CONAN DOYLE

Sherlock Holmes, the world's only unofficial consulting detective, illustrates his powers of observation, deduction, and knowledge.

"I abhor the dull routine of existence. I crave for mental **exaltation.** That is why I have chosen my own particular profession—or rather created it, for I am the only one in the world."

"The only unofficial detective?" I said, raising my eyebrows.

"The only unofficial consulting detective," [Sherlock Holmes] answered. "I am the last and highest court of appeal in detection. When Gregson or Lestrade or Athelney Jones are out of their depths—which, by the way, is their normal state—the matter is laid before me. I examine the data, as an expert, and pronounce a specialist's opinion. I claim no credit in such cases. My name figures in no newspaper. The work itself, the pleasure of finding a field for my peculiar powers, is my highest reward. But you have yourself had some experience of my methods of work in the Jefferson Hope case."

"Yes, indeed," said I, cordially. "I was never so struck by anything in my life. I even embodied it in a small brochure with the somewhat fantastic title of 'A Study in Scarlet.'"

He shook his head sadly. "I glanced over it," said he. "Honestly, [Dr. Watson,] I cannot congratulate you upon it. Detection is, or ought to be, an exact science, and should be treated in the same cold and unemotional manner. You have attempted to tinge it with romanticism, which produces much the same effect as if you worked a love-story or an elopement into the fifth proposition of Euclid."

"But the romance was there," I remonstrated. "I could not tamper with the facts."

"Some facts should be suppressed, or at least a just sense of proportion should be observed in treating them. The only point in the case which deserved mention was the curious analytical reasoning from effects to causes by which I succeeded in unravelling it."

I was annoyed at this criticism of a work which had been specially designed to please him. I confess, too, that I was irritated by the egotism which seemed to demand that every line of my pamphlet should be devoted to his own special doings. More than once during the years that I had lived with him in Baker Street I had observed that a small vanity underlay my companion's quiet and **didactic** manner. I made no remark, however, but sat nursing my wounded leg. I had a Jezail bullet through it sometime before, and, though it did not prevent me from walking, it ached wearily at every change of the weather.

"My practice has extended recently to the Continent," said Holmes, after a while, filling up his old brier-root pipe. "I was consulted last week by François Le Villard, who, as you probably know, has come rather to the front lately in the French detective service. He has all the Celtic power of quick intuition, but he is deficient in the wide range of exact knowledge which is essential to the higher developments of his art. The case was concerned with a will, and possessed some features of interest. I was able to refer him to two parallel cases, the one at Riga in 1857, and the other at St. Louis in 1871, which have suggested to him the true solution. Here is the letter which I had this morning acknowledging my assistance." He tossed over, as he spoke, a crumpled sheet of foreign note-paper. I glanced my eyes down it, catching a profusion of notes of admiration, with stray "magnifiques," "coup-de-maîtres," and "tours-de-force," all testifying to the ardent admiration of the Frenchman.

"He speaks as a pupil to his master," said I.

"Oh, he rates my assistance too highly," said Sherlock Holmes, lightly. "He has considerable gifts himself. He possesses two out of the three qualities necessary for the ideal detective. He has the power of observation and that of deduction. He is only wanting in knowledge; and that may come in time. He is now translating my small works into French."

"Your works?"

"Oh, didn't you know?" he cried, laughing. "Yes, I have been guilty of several monographs. They are all upon technical subjects. Here, for example, is one 'Upon the Distinction between the Ashes of the Various Tobaccos.' In it I enumerate a hundred and forty forms of cigar-, cigarette-, and pipe-tobacco, with colored plates illustrate the difference in the ash. It is a point which is continually turning up in criminal trials, and which is sometimes of supreme importance as a clue. If you can say definitely, for example, that some murder has been done by a man who was smoking an Indian lunkah, it obviously narrows your field of search. To the trained eye there is as much difference between the black ash of a Trichinopoly and the white fluff of bird's-eye as there is between a cabbage and a potato."

"You have an extraordinary genius for **minutiae,**" I remarked.

"I appreciate their importance. Here is my monograph upon the tracing of footsteps, with some remarks upon the uses of plaster of Paris as a preserver of impresses. Here, too, is a curious little work upon the influence of a trade upon the form of the hand, with lithotypes of the hands of slaters, sailors, cork-cutters, compositors, weavers, and diamond-polishers. That is a matter of great practical interest to the scientific detective—especially in cases of unclaimed bodies, or in discovering the antecedents of criminals. But I weary you with my hobby."

"Not at all," I answered earnestly. "It is of the greatest interest to me, especially since I have had the opportunity of observing your practical application of it. But you spoke just now of observation and deduction. Surely the one to some extent implies the other."

"Why, hardly," he answered, leaning back luxuriously in his arm-chair, and sending up thick blue wreaths from his pipe. "For example, observation shows me that you have been to the Wigmore Street Post Office this morning, but deduction lets me know that when there you despatched a telegram."

"Right!" said I. "Right on both points! But I confess that I don't see how you arrived at it. It was a sudden impulse upon my part, and I have mentioned it to no one."

"It is simplicity itself," he remarked, chuckling at my surprise—"so absurdly simple that an explanation is **superfluous;** and yet it may serve to define the limits of observation and of deduction. Observation tells me that you have a little reddish mold adhering to your instep. Just opposite the Seymour Street Office they have taken up the pavement and thrown up some earth which lies in such a way that it is difficult to avoid treading in it in entering. The earth is of this peculiar reddish tint which is found, as far as I know, nowhere else in the neighborhood. So much is observation. The rest is deduction."

"How, then, did you deduce the telegram?"

"Why, of course I knew that you had not written a letter, since I sat opposite to you all morning. I see also in your open desk there that you have a sheet of stamps and a thick bundle of post-cards. What could you go into the post-office for, then, but to send a wire? Eliminate all other factors, and the one which remains must be the truth."

"In this case it certainly is so," I replied, after a little thought. "The thing, however, is, as you say, of the simplest. Would you think me **impertinent** if I were to put your theories to a more severe test?"

"On the contrary," he answered. . . . . "I should be delighted to look into any problem which you might submit to me."

"I have heard you say that it is difficult for a man to have any object in daily use without leaving the impress of his individuality upon it in such a way that a trained observer might read it. Now, I have here a watch which has recently come into my possession. Would you have the kindness to let me have an opinion upon the character or habits of the late owner?"

I handed him over the watch with some slight feeling of amusement in my heart, for the test was, as I thought, an impossible one, and I intended it as a lesson against the somewhat dogmatic tone which he occasionally assumed. He balanced the watch in his hand, gazed hard at the dial, opened the back, and examined the works, first with his naked eyes and then with a powerful convex lens. I could hardly keep from smiling at his crestfallen face when he finally snapped the case to and handed it back.

"There are hardly any data," he remarked. "The watch has been recently cleaned, which robs me of my most suggestive facts."

"You are right," I answered. "It was cleaned before being sent to me." In my heart I accused my companion of putting forward a most lame and impotent excuse to cover his failure. What data could he expect from an uncleaned watch?

"Though unsatisfactory, my research has not been entirely barren," he observed, staring up at the ceiling with dreamy lack-lustre eyes. "Subject to your correction, I should judge that the watch belonged to your elder brother, who inherited it from your father."

"That you gather, no doubt, from the H. W. upon the back?"

"Quite so. The W. suggests your own name. The date of the watch is nearly fifty years back, and the initials are as old as the watch: so it was made for the last generation. Jewelry usually descends to the eldest son, and he is most likely to have the same name as the father. Your father has, if I remember right, been dead many years. It has, therefore, been in the hands of your eldest brother."

"Right, so far," said I. "Anything else?"

"He was a man of untidy habits,—very untidy and careless. He was left with good prospects, but he threw away his chances, lived for some time in poverty with occasional short intervals of prosperity, and finally, taking to drink, he died. That is all I can gather."

I sprang from my chair and limped impatiently about the room with considerable bitterness in my heart.

"This is unworthy of you, Holmes," I said. "I could not have believed that you would have descended to this. You have made inquiries into the history of my unhappy brother, and you now pretend to deduce this knowledge in some fanciful way. You cannot expect me to believe that you have read all this from his old watch! It is unkind, and, to speak plainly, has a touch of **charlatanism** in it."

"My dear doctor," said he, kindly, "pray accept my apologies. Viewing the matter as an abstract problem, I had forgotten how personal and painful a thing it might be to you. I assure you, however, that I never even knew that you had a brother until you handed me the watch."

"Then how in the name of all that is wonderful did you get these facts? They are absolutely correct in every particular."

"Ah, that is good luck. I could only say what was the balance of probability. I did not at all expect to be so accurate."

"But it was not mere guess-work?"

"No, no: I never guess. It is a shocking habit,—destructive to the logical faculty. What seems strange to you is only so because you do not follow my train of thought or observe the small facts upon which large inference may depend. For example, I began by stating that your brother was careless. When you observe the lower part of that watch-case you notice that it is not only dented in two places, but it is cut and marked all over from the habit of keeping other hard objects, such as coins or keys, in the same pocket. Surely it is of no great feat to assume that a man who treats a fifty-guinea watch so cavalierly must be a careless man. Neither is it a very far-fetched inference that a man who inherits one article of such value is pretty well provided for in other respects."

I nodded, to show that I followed his reasoning.

"It is very customary for pawnbrokers in England, when they take a watch, to scratch the number of the ticket with a pin-point upon the inside of the case. It is more handy than a label, as there is no risk of the number being lost or transposed. There are no less than four such numbers visible to my lens on the inside of this case. Inference,—that your brother was often at low water. Secondary inference,—that he had occasional bursts of prosperity, or he could not have redeemed the pledge. Finally, I ask you to look at the inner plate, which contains the key-hole. Look at the thousands of scratches all round the hole—marks where the key has slipped. What sober man's key could have scored those grooves? But you will never see a drunkard's watch without them. He winds it at night, and he leaves these traces of his unsteady hand. Where is the mystery in all this?"

"It is as clear as daylight," I answered. "I regret the injustice which I did you. I should have had more faith in your marvellous faculty."

Questions on Content

1. According to Sherlock Holmes, what are the three qualities necessary for the ideal detective?
2. What does Sherlock Holmes observe and deduce concerning Dr. Watson's watch?

Vocabulary

Define the following words as they are used in the article (the words are printed in
boldface type):

1. exaltation
2. didactic
3. minutiae

4. superfluous
5. impertinent
6. charlatanism

Suggestions for Response and Reaction

1. Compare and contrast (see pages 200–212) *observation* and *deduction.*
2. Do you think it is really possible for a person to develop his or her powers of
   observation, deduction, and knowledge to the extent of those of Sherlock
   Holmes? Explain your answer.
3. What is the relationship between analysis through effect-cause (see pages
   177–183) and detective work?
4. In what occupations, professions, or types of work (other than that of being a
   detective) are the powers of observation, deduction, and knowledge essential?
   Helpful?
5. Select an object about which you do not have prior knowledge and try out
   your powers of observation and deduction. Suggested objects: a pair of shoes,
   a billfold, a suitcase, a notebook, an automobile, a belt. To record your findings
   and inferences, divide a sheet of paper into two columns, one headed *Observations* and the other, *Deductions.*

Are You Alive?

STUART CHASE

Chase writes interestingly and convincingly on a fundamental question—the enjoyment of life. By a simple process of analysis and classification, he divides life into two categories, living and existing. He then proceeds to show specifically the differences between the two conditions by labeling accordingly phases of his own life.

I have often been perplexed by people who talk about "life." They tend to strike an attitude, draw a deep breath, and wave their hands—triumphantly but somehow vaguely. Americans, they tell me, do not know how to live, but the French—ah, the French!—or the Hungarians, or the Poles, or the Patagonians. When I ask them what they mean by life they look at me with pitying silence for the most part, so that their long thin eyes haunt me. Sometimes they say: "Life is the next emotion. . . . Life is—well, you know what Shaw says . . . Life is a continual becoming." Unfortunately this is all Babu to me. These people do not advance me an inch in my quest of the definition of life.

Edgar Lee Masters is more helpful. In *Spoon River* he talks of a woman who dies at the age of ninety-six. She speaks of her hard yet happy existence on the farm, in the woods by the river gathering herbs, among the neighbors. "It takes life to love life," she says in a ringing climax. This is **cryptic,** but I get a flood of meaning from it.

What does it mean to be alive, to live intensely? What do social prophets mean when they promise a new order of life? Obviously they cannot mean a new quality of life never before enjoyed by anyone, but rather an extension of vitality for the masses of mankind in those qualities of "life" which have hitherto been enjoyed only by a few individuals normally, or by large numbers of individuals rarely.

What is it which is enjoyed, and how is it to be shared more extensively? Behind the phrase and the gesture, can we catch the gleam of **verihood** and hold life on a point for a moment while we examine it?

Initially, we must differentiate between "life" in the sense of not being physically dead, and "life" in the sense of awareness and vitality. Obviously everybody who is not dead is alive—"lives" in a sense. And all experience of undead people is a part of life in its broadest interpretation. But there seems to be an ascending scale of values in life, and somewhere in this scale there is a line—probably a blurred one—below which one more or less "exists," and above which one more or less "lives."

Second, we must examine the distinction that is drawn between physical and mental living. While this distinction is of no consequence so far as the joy of living is concerned, it is of profound consequence so far as the techniques of progress are concerned. If all well-being proceeds from a state of mind with little reference to physical environment, it is evident that all people have to do is to create well-being mentally—to think "life." If well-being on the other hand proceeds primarily from

Reprinted by permission of Stuart Chase.

physical causes, it is necessary to clean up slums, produce more usable goods and distribute them more widely, broaden the arts, look into **eugenics,** establish comprehensive systems of sanitation, release the creative impulse, and what not. Here we collide with a problem which is destined to agitate the race for centuries to come. Probably both approaches have their place, but I for one would stress the engineer against the **metaphysician.**

What, concretely, is this "awareness," this "well-being"? These words are shorthand words, meaningless in themselves unless we know the longhand facts for which they are a symbol. I want in a rather personal way to tell you the facts as I have found them. I want to tell you when I think I live in contradistinction to when I think I "exist." I want to make life very definite in terms of my own experience, for in matters of this nature about the only source of data one has is oneself. I do not know what life means to other people—I can only guess—but I do know what it means to me, and I have worked out a method of measuring it.

I get out of bed in the morning, bathe, dress, fuss over a necktie, hurry down to breakfast, gulp coffee and headlines, demand to know where my raincoat is, start for the office—and so forth. These are the crude data. Take the days as they come, put a plus beside the living hours and a minus beside the dead ones; find out just what makes the live ones live and the dead ones die. Can we catch the verihood of life in such an analysis? The poet will say not, but I am an accountant and only write poetry out of hours.

My hours show a classification of eleven states of being in which I feel I am alive, and five states in which I feel I only exist. These are major states, needless to say. In addition I find scores of sub-states which come from Heaven knows where and are too obscure for me to analyze. The eleven "plus" reactions are these:

I seem to live when I am creating something—writing this article for instance; making a sketch, working on an economic theory, building a bookshelf, making a speech.

Art certainly vitalizes me. A good novel—"The Growth of the Soil," for instance—some poems, some pictures, operas (not concerts), many beautiful buildings and particularly bridges affect me as though I took the artist's blood into my own veins. But I do not only have to be exposed to get this reaction. The operation is more subtle, for there are times when a curtain falls over my perceptions which no artist can penetrate.

In spite of those absurd Germans who used to perspire over Alpine passes shouting "Co-lo-sall!," I admit that mountains and the sea and stars and things—all the old subjects of a thousand poets—renew life in me. As in the case of art, the process is not automatic—I hate the sea sometimes—but by and large the chemistry works, and I feel the line of existence below me when I see these things.

Love, underneath its middle-class manners, and its frequent hypocrisies, is life, vital and intense. Very real to me also is the love one bears one's friends.

I feel very much alive in the presence of a genuine sorrow.

I live when I am in the presence of danger—rock-climbing, or being shadowed by an agent of the Department of Justice.

I live when I play—preferably out-of-doors. Such things as diving, swimming, skating, skiing, dancing, sometimes driving a motor, sometimes walking. . . .

One lives when one takes food after genuine hunger, or when burying one's lips in a cool mountain spring after a long climb.

One lives when one sleeps. A sound, healthy sleep after a day spent out-of-doors gives one the feeling of a silent, whirring dynamo. In vivid dreams I am convinced one lives.

I live when I laugh—spontaneously and heartily.

These are the eleven states of well-being which I have checked off from the hours of my daily life. Observe they only represent causes for states of being which I accept as "life." The definition of what state of being is in itself—the physical and mental chemistry of it, the feeling of expansion, of power, of happiness, or whatever it may be that it is—cannot be attempted at this point. The Indian says: "Those who tell do not know. Those who know do not tell." I deal here only in definite states of being which I recognize by some obscure but infallible sign as "life." *Why* I know it, or what it is, intrinsically, cannot be told.

In contradistinction to "living," I find five main states of "existence," as follows:

I exist when I am doing drudgery of any kind—adding up figures, washing dishes, writing formal reports, answering most letters, attending to money matters, reading newspapers, shaving, dressing, riding on street cars, or up and down in elevators, buying things.

I exist when attending the average social function. The whole scheme of middle-class manners bores me—a tea, a dinner, a call on one's relatives, listening to dull people talk, being polite, discussing the weather.

Eating, drinking, or sleeping when one is already replete, when one's senses are dulled, are states of existence, not life. For the most part, I exist when I am ill, but occasionally pain gives me a **lucidity** of thought which is near to life.

Old scenes, old monotonous things—city walls, too familiar streets, houses, rooms, furniture, clothes—drive one to the existence level. Even a scene that is beautiful to fresh eyes may grow intolerably dull. Sheer ugliness, such as one sees in the stockyards or in a city slum, depresses me intensely.

I retreat from life when I become angry. I feel all my handholds slipping. It is as though I were a deep-sea diver in a bell of compressed air. I exist through rows and misunderstandings and in the blind alleys of "getting even."

So in a general way I locate my line and set life off from existence. It must be admitted of course that "living" is often a mental state quite independent of physical environment or occupation. One may feel—in springtime for instance—suddenly alive in old, monotonous surroundings. Then even dressing and dishwashing become eventful and one sings as one shaves. But these outbursts are on the whole abnormal. By and large there seems to be a definite cause for living and a definite cause for existing. So it is with me at any rate. I am not at the mercy of a blind fate in this respect. I believe that I could deliberately "live" twice as much—in hours—as I do now, if only I could come out from under the chains of an artificial necessity—largely economic—which bind me.

I have indeed made some estimate of the actual time I have spent above and below the "existence" line. For instance, let me analyze a week I spent in the city during the past fall—an ordinary busy week of one who works at a desk. Of the 168 hours contained therein my notes show that I only "lived" about forty of them, or 25 per cent of the total time. This allowed for some creative work, a Sunday's hike, some genuine hunger, some healthy sleep, a little stimulating reading, two acts of a play, part of a moving picture, and eight hours of interesting discussion with various friends, including one informal talk on the technique of industrial administration which stimulated me profoundly.

Twenty-five per cent is not a very high life ratio, although if I may be permitted to guess, I guess it is considerably over the average of that of my fellow Americans. It is extremely doubtful if my yearly average is any higher than that which obtained in this representative week. Some weeks—in vacation time, or when the proportion of creative work is high—may be individually better, but the never-ceasing grind of income-getting, and the never-ceasing obscenity of city dwelling, undermine too ruthlessly the hope of a higher average ratio.

I can conceive of a better ratio in another social order. In such a society I do not see why one should not have the opportunity to do creative work at least three hours a day—allowing two hours for the inevitable drudgery—nor why I should not be out-of-doors playing for two or three hours a day; nor why I should not have a great deal more opportunity than I do now to hear good music, and go to good plays, and see good pictures; while reading, discussing, being with one's friends, sound sleep, and the salt of danger should all have an increased share of my time. I do not see why I should not laugh a lot more in that society. Adding these possibilities up—I will not bother you with the mathematical details—the total shows some one hundred hours, or a life ratio of 60 per cent as against the 25 per cent which is now my portion.

If this is true of me, it may be true of you, of many others perhaps. It may be that the states of being which release life in me release it in most human beings. But this I know and to this I have made up my mind: my salvation is too closely bound up with that of all mankind to hope for any great personal advance. In the last analysis—despite much beating against the bars—my ratio of living can only grow with that of the mass of my fellow-men.

Questions on Content

1. Chase divides life into two categories. Name them.
2. In examining the distinction between physical and mental living, which approach does the author favor? Write the sentence in the essay that tells you so.
3. List five of the eleven states of being in which the author feels he is alive. List five states of existence for him.

Vocabulary

Define the following words as they are used in the article (the words are printed in **boldface** type):

1. cryptic
2. verihood
3. eugenics
4. metaphysician
5. lucidity

Suggestions for Response and Reaction

1. a. List and explain at least three states of being that occur during your average work (school) day in which you are truly alive in the author's sense of the term.
   b. List at least three states of existing. Be very specific in describing times or activities.
   c. Present the material in *a* and *b* in an organized paper or speech, using the Chase essay as a model.
2. Excluding hours asleep, for an average work (school) day, what percentage of your time is spent in living and what in existing? Following the steps in cause/effect reasoning (see pages 177–183), explain why you do or do not believe you could increase the percentage of time living. Construct a pie chart (see pages 470–471) to accompany your explanation.
3. For an ideal holiday or weekend day (just one day), list and explain five states of living.
4. Write an informative summary (see pages 224–233) of this selection.

Clear Writing Means Clear Thinking Means . . .

MARVIN H. SWIFT

As a professor of communication at the General Motors Institute, Marvin H. Swift analyzes the way in which a manager reworks and rethinks a memo of minor importance, to point up a constant management challenge of major importance—the clear and accurate expression of a well-focused message.

If you are a manager, you constantly face the problem of putting words on paper. If you are like most managers, this is not the sort of problem you enjoy. It is hard to do, and time consuming; and the task is doubly difficult when, as is usually the case, your words must be designed to change the behavior of others in the organization.

But the chore is there and must be done. How? Let's take a specific case.

Let's suppose that everyone at *X* Corporation, from the janitor on up to the chairman of the board, is using the office copiers for personal matters; income tax forms, church programs, children's term papers, and God knows what else are being duplicated by the gross. This minor piracy costs the company a pretty penny, both directly and in employee time, and the general manager—let's call him Sam Edwards—decides the time has come to lower the boom.

Sam lets fly by dictating the following memo to his secretary:

[Draft 1]

To: All Employees
From: Samuel Edwards, General Manager
Subject: Abuse of Copiers

It has recently been brought to my attention that many of the people who are employed by this company have taken advantage of their positions by availing themselves of the copiers. More specifically, these machines are being used for other than company business.

Obviously, such practice is contrary to company policy and must cease and desist immediately. I wish therefore to inform all concerned—those who have abused policy or will be abusing it—that their behavior cannot and will not be tolerated. Accordingly, anyone in the future who is unable to control himself will have his employment terminated.

If there are any questions about company policy, please feel free to contact this office.

Now the memo is on his desk for his signature. He looks it over; and the more he looks the worse it reads. In fact, it's lousy. So he revises it three times, until it finally is in the form that follows:

543

Marvin H Swift
Clear Writing
Means Clear
Thinking
Means . . .

[Final Version]

To: All Employees
From: Samuel Edwards, General Manager
Subject: Use of Copiers

We are revamping our policy on the use of copiers for personal matters. In the past we have not encouraged personnel to use them for such purposes because of the costs involved. But we also recognize, perhaps belatedly, that we can solve the problem if each of us pays for what he takes.

We are therefore putting these copiers on a pay-as-you-go basis. The details are simple enough. . . .

Samuel Edwards

This time Sam thinks the memo looks good, and it *is* good. Not only is the writing much improved but the problem should now be solved. He therefore signs the memo, turns it over to his secretary for distribution, and goes back to other things.

From Verbiage to Intent

I can only speculate on what occurs in a writer's mind as he moves from a poor draft to a good revision, but it is clear that Sam went through several specific steps, mentally as well as physically, before he had created his end product.

- He eliminated wordiness.
- He modulated the tone of the memo.
- He revised the policy it stated.

Let's retract his thinking through each of these processes.

Eliminating Wordiness

Sam's basic message is that employees are not to use the copiers for their own affairs at company expense. As he looks over his first draft, however, it seems so long that this simple message has become diffused. With the idea of trimming the memo down, he takes another look at his first paragraph.

It has recently been brought to my attention that many of the people who are employed by this company have taken advantage of their positions by availing themselves of the copiers. More specifically, these machines are being used for other than company business.

He edits it like this:

Item: "recently"
Comment to himself: Of course; else why write about the problem? So delete the word.
Item: "It has been brought to my attention"
Comment: Naturally. Delete it.
Item: "the people who are employed by this company"
Comment: Assumed. Why not just "employees"?
Item: "by availing themselves" and "for other than company business"
Comment: Since the second sentence repeats the first, why not **coalesce?**

And he comes up with this:

Employees have been using the copiers for personal matters.

He proceeds to the second paragraph. More confident of himself, he moves in broader swoops, so that the deletion process looks like this:

Obviously, such practice is contrary to company policy ~~and must cease and desist immediately. I wish therefore to inform all concerned—those who have abused policy or will be abusing it—that their behavior cannot and will not be tolerated. Accordingly, anyone in the future who is unable to control himself will have his employment terminated.~~

The final paragraph, apart from "company policy" and "feel free," looks all right, so the total memo now reads as follows:

[Draft 2]

To: All Employees
From: Samuel Edwards, General Manager
Subject: Abuse of Copiers

Employes have been using the copiers for personal matters.
Obviously, such practice is contrary to company policy and
will result in dismissal.

If there are any questions, please contact this office.

Sam now examines his efforts by putting these questions to himself:

Question: Is the memo free of deadwood?
Answer: Very much so. In fact, it's good, tight prose.
Question: Is the policy stated?
Answer: Yes—sharp and clear.
Question: Will the memo achieve its intended purpose?
Answer: Yes. But it sounds foolish.
Question: Why?
Answer: The wording is too harsh; I'm not going to fire anybody over this.
Question: How should I tone the thing down?

To answer this question, Sam takes another look at the memo:

Correcting the Tone

545

Marvin H Swift
Clear Writing
Means Clear
Thinking
Means . . .

What strikes his eye as he looks it over? Perhaps these three words:

- Abuse . . .
- Obviously . . .
- . . . dismissal . . .

The first one is easy enough to correct: he substitutes "use" for "abuse." But "obviously" poses a problem and calls for reflection. If the policy is obvious, why are the copiers being used? Is it that people are outrightly dishonest? Probably not. But that implies the policy isn't obvious; and whose fault is this? Who neglected to clarify policy? And why "dismissal" for something never publicized?

These questions impel him to revise the memo once again:

[Draft 3]

To: All Employees
From: Samuel Edwards, General Manager
Subject: Use of Copiers

Copiers are not to be used for personal matters. If there
are any questions, please contact this office.

Revising the Policy Itself

The memo now seems courteous enough—at least it is not discourteous—but it is just a blank, perhaps overly simple, statement of policy. Has he really thought through the policy itself?

Reflecting on this, Sam realizes that some people will continue to use the copiers for personal business anyhow. If he seriously intends to enforce the basic policy (first sentence), he will have to police the equipment, and that raises the question of costs all over again.

Also, the memo states that he will maintain an open-door policy (second sentence)—and surely there will be some, probably a good many, who will stroll in and offer to pay for what they use. His secretary has enough to do without keeping track of affairs of that kind.

Finally, the first and second sentences are at odds with each other. The first says that personal copying is out, and the second implies that it can be arranged.

The facts of organizational life thus force Sam to clarify in his own mind exactly what his position on the use of copiers is going to be. As he sees the problem now, what he really wants to do is put the copiers on a pay-as-you-go basis. After making that decision, he begins anew:

```
                          [Final Draft]

To: All Employees
From: Samuel Edwards, General Manager
Subject: Use of copiers

We are revamping our policy on the use of copiers for personal
use....
```

This is the draft that goes into distribution and now allows him to turn his attention to other problems.

The Chicken or the Egg?

What are we to make of all this? It seems a rather lengthy and **tedious** report of what, after all, is a routine writing task created by a problem of minor importance. In making this kind of analysis, have I simply labored the obvious?

To answer this question, let's drop back to the original draft. If you read it over, you will see that Sam began with this kind of thinking:

- "The employees are taking advantage of the company."
- "I'm a nice guy, but now I'm going to play Dutch uncle." . . . "I'll write them a memo that tells them to shape up or ship out."

In his final version, however, his thinking is quite different:

- "Actually, the employees are pretty mature, responsible people. They're capable of understanding a problem."
- "Company policy itself has never been crystallized. In fact, this is the first memo on the subject."
- "I don't want to overdo this thing—any employee can make an error in judgment."
- . . . I'll set a reasonable policy and write a memo that explains how it ought to operate."

Sam obviously gained a lot of ground between the first draft and the final version, and this implies two things. First, if a manager is to write effectively, he needs to isolate and define, as fully as possible, all the critical variables in the writing process and **scrutinize** what he writes for its clarity, simplicity, tone, and the rest. Second, after he has clarified his thoughts on paper, he may find that what he has written is not what has to be said. In this sense, writing is feedback and a way for the manager to discover himself. What are his real attitudes toward the **amorphous,** undifferentiated gray mass of employees "out there"? Writing is a way of finding out. By objectifying his thoughts in the medium of language, he gets a chance to see what is going on in his mind.

In other words, *if the manager writes well, he will think well*. Equally, the more clearly he has thought out his message before he starts to dictate, the more likely he is to get it right on paper the first time round. In other words, *if he thinks well, he will write well.*

Hence we have a chicken-and-the-egg situation: writing and thinking go hand in hand; and when one is good, the other is likely to be good.

547

Marvin H Swift
Clear Writing
Means Clear
Thinking
Means . . .

Revision Sharpens Thinking

More particularly, rewriting is the key to improved thinking. It demands a real open-mindedness and objectivity. It demands a willingness to cull **verbiage** so that ideas stand out clearly. And it demands a willingness to meet logical contradictions head on and trace them to the premises that have created them. In short, it forces a writer to get up his courage and expose his thinking process to his own intelligence.

Obviously, revising is hard work. It demands that you put yourself through the wringer, intellectually and emotionally, to squeeze out the best you can offer. Is it worth the effort? Yes, it is—if you believe you have a responsibility to think and communicate effectively.

Questions on Content

1. List the four stages that Manager Sam went through in revising the memo.
2. Explain the chicken-and-egg analogy in the article as it relates to writing and thinking.
3. What conclusion does Swift reach about the worth of revision?

Vocabulary

Define the following words as they are used in the article (the words are printed in **boldface** type):

1. coalesce
2. tedious
3. scrutinize
4. amorphous
5. verbiage

Suggestions for Response and Reaction

1. From your own experience, describe a situation in which a piece of correspondence written or received by you caused confusion, embarrassment, or anger. This could have taken place in a school, job, or social situation.
2. Explain why the tone of a business letter or a memo is significant.
3. Write an interoffice memorandum (see pages 251–253) to call attention to and suggest an alternate plan to correct an undesirable situation.
4. Define (see pages 109–118) *revision* and *draft.* Also, consult handbooks and other composition texts. Then make your own list of suggestions to follow in revising your writing through several drafts.

A Technical Writer Considers His Latest Love Object

CHICK WALLACE

A poet, professor, and technical writer asks probing questions about the nature of technical writing and of love.

1. How can "X" be described?
2. What is the essential function of "X"?
3. What are the components of "X"?
4. How is "X" made or done?
5. How should "X" be made or done?
6. What are the types of "X"?
7. How does "X" compare to "Y"?
8. What are the causes of "X"?
9. What are the consequences of "X"?
10. What kind of representative (*enter* **species** *here*) is "X"?
11. What is my memory of "X"?
12. What is the **efficiency ratio** of the performance of "X"?
13. Is "X" profitable in the long or short term?
14. What is my personal response to "X"?
15. What is the present status of "X"?
16. How should "X" be interpreted?
17. What is the present value of "X"?
18. What case can be made for and against "X"?
19. Conclusion: Hello, darling. I've been thinking. . . .
20. Results: Wedding scheduled 4:00 p.m. 6/27. Take the weekend off.

Questions on Content

549

Chick Wallace
A Technical
Writer Considers
His Latest Love
Object

1. What are the kinds of questions a technical writer asks?
2. List the specific questions in the poem that precipitate the shift from objective to personal considerations.
3. In question 10, what do the words in parentheses mean?

Vocabulary

Define the following words as they are used in the poem (the words are printed in **boldface** type).

1. species
2. efficiency ratio

Suggestions for Response and Reaction

1. In the title of the poem, what is the attitude suggested by the words in the phrase "Latest Love Object"?
2. How is partition (see pages 158–160) in technical writing used in listing the questions in the poem?
3. If you were considering marriage, what questions would you ask? How do your questions differ from those in the poem?
4. What is implied in the closing sentence "Take the weekend off"?
5. In groups of three or four students, compare and contrast (see pages 200–212) the qualities of technical writing in the section Technical Communication: Getting Started (see pages 2–19) with the qualities of technical writing presented in the poem.

Blue Heron

CHICK WALLACE

A poet ponders the problems of environmental pollution for human beings and animals.

Blue Heron

Blue heron, is your corner grocery store back in the
   marsh
So full of man and man's poisonous **additives** to
Your daily fare of crab and minnow, now so harsh
That you are driven to seek your **sustenance** new
In unadapted spaces, like this snow-fluff tide
Receding on my ocean shore
Where this season of the year
Not even gulls and willets stop here for long for tea,
Knowing nothing's here?

You won't let me close enough to say it:
Heron blue, come home with me, and you
And I and Dog may share my bowl of deathly steaming
   stew,
Spiced by Man's poisons and **debris**—just like your
   native **slew.**

Questions on Content

1. What comparisons does the poet use to emphasize the widespread problem of pollution?
2. What does the poet mean by "unadapted spaces."
3. What does the poet imply about the future of the blue heron? Of human beings?

Vocabulary

Define the following words as they are used in the article (the words are printed in **boldface** type).

1. additives
2. sustenance
3. debris
4. slew

Suggestions for Response and Reaction

1. Evaluate the effectiveness of the comparison between the heron's environment and the poet's and dog's environment.
2. Classify (see pages 131–149) types of environmental pollution.
3. Have you noticed evidence of environmental pollution? Have you read about environmental pollution? As a class discuss some of the problems and possible solutions.
4. Working in groups, research and identify one law designed to control environmental pollution. Make a collective list of the laws.
5. Write an extended definition (see pages 112–118) of *pollution*.
6. Using cause and effect analysis (see pages 177–183), write a paper in which you identify ways persons pollute wildlife habitats and the effects such polluting has on wildlife.
7. Identify an environmental concern in your town or state. Write a letter (see pages 254–262) to the local paper suggesting a way or ways an individual can address this concern.
8. Write an observation report (see pages 320–333) in which you describe a local, state, or national environmental problem.
   a. Write the report for publication in a magazine for middle-school students.
   b. Write the report for oral presentation (see pages 441–448) to a local civic organization whose members want to be involved in activities to solve environmental problems.

The Effects of Perceived Justice on Complainants' Negative Word-of-Mouth Behavior and Repatronage Intentions

JEFFREY G. BLODGETT, DONALD H. GRANBOIS, AND ROCKNEY G. WALTERS

When a customer is dissatisfied, the wise retailer will make every effort to remedy the problem quickly and courteously. In this formal report of their research, the authors include an executive summary focusing on implications for management.

Executive Summary

Previous research has found that dissatisfied consumers choose to seek redress (i.e., request a refund, an exchange, or repair, etc.), engage in negative word-of-mouth behavior (i.e., tell friends and family about their dissatisfaction), and exit (i.e., vow never to repatronize the retailer) based upon the perceived likelihood of successful redress, their attitude toward seeking redress, the level of importance they attach to the product, and whether they perceive the problem to be stable (i.e., whether similar problems are expected to occur in the future) or to have been controllable (i.e., whether the retailer could have prevented the problem). Although much progress has been made in our understanding of complaining behavior, current models are limited in that they view complaining behavior as a static phenomenon, without taking into account the outcome of the redress seeking episode. In contrast, our model explicitly recognizes that consumer complaining behavior is a dynamic process, and that once a dissatisfied consumer seeks redress other complaining behaviors (i.e., negative word-of-mouth behavior and exit, or repatronage intentions) are dependent upon the consumer's post-complaint perception of justice. The purpose of this study, then, was to develop and test a model of consumer complaining behavior, in which complainants' negative word-of-mouth behavior and their repatronage intentions are dependent, in large part, upon their perceptions of justice, and to a lesser degree upon the perceived likelihood of successful redress, their attitudes toward seeking redress, the level of product importance, and whether the problem was perceived to be stable or to have been controllable. The model was tested using data from a large sample of dissatisfied consumers who sought redress from a number of different retailers (e.g., department stores, discount stores, specialty stores) concerning a variety of soft (e.g., apparel) and hard (e.g., small appliances) goods at different price levels.

Reprinted by permission from *Journal of Retailing* 69 (Winter 1993): 367–428. Ellipses (three dots) indicate where original material has been omitted.

As hypothesized, perceived justice was the main determinant of complainants' negative word-of-mouth behavior and their repatronage intentions. The significant role that perceived justice plays in consumer complaining behavior suggests that dissatisfied consumers are quite willing to give the retailer another chance *if* the retailer stands behind the product and guarantees customer satisfaction. As long as the retailer ensures satisfaction most complainants will not engage in negative word-of-mouth behavior or exit. Rather, because they perceive the retailer as being fair and just, these complainants may actually become more loyal customers! However, if the retailer does not stand behind the product and ensure customer satisfaction, complainants are likely to react by telling several friends about their dissatisfying experience and to vow never to repatronize the offending retailer.

Other key findings are that dissatisfied consumers who initially perceived little likelihood of successful redress or who were dissatisfied with products they felt were important were more likely to engage in negative word-of-mouth behavior. These consumers probably were very frustrated with the retailer and felt a need to "get it off their chest" by telling friends and relatives. These findings illustrate the importance of the customer service/customer satisfaction concept. Firms that adopt this philosophy and communicate it to their customers are less likely to be the subject of negative word-of-mouth communications when problems do occur because their customers are more confident that their problems will be resolved satisfactorily. Dissatisfied customers who perceive a high likelihood of success, and who feel that their problems are equally as important to the retailer, most likely will first seek redress and will not engage in negative word-of-mouth unless they are dissatisfied with the retailer's response to their complaint.

Another important finding is that dissatisfied consumers who perceived the problem to be stable or controllable were less likely to repatronize the offending retailer. Consumers who perceived the problem to be stable probably wished to avoid that retailer in the future, while consumers who perceived that the retailer could have prevented the problem probably were angry and may have vowed to "get even" by never shopping there again. This finding should be of particular concern to retailers. In order to retain these customers' business it is important that the retailer apologize (Bies and Shapiro 1987) and take responsibility for any problems that may have occurred. The retailer should also thank the customer for bringing the problem to the retailer's attention, and should make a commitment to the customer to do better in the future. Complainants who perceive that the retailer is sincerely concerned, and is genuinely committed to improve, may be more likely to give the retailer another chance.

Because of the impact that perceived justice has on retail sales and profits, remedying consumer complaints should be thought of as a key marketing variable. It is much easier to keep current customers satisifed than it is to attract new customers, especially given slow growth in retail sales. A widely promoted and successfully executed policy of "satisfaction guaranteed" gives customers greater confidence that any problems they might encounter will be remedied in a timely and courteous manner. Because of this confidence, dissatisfied consumers are more likely to first seek redress, thus giving the retailer a chance to remedy the problem. As a result of the retailer standing behind the product and ensuring customer satisfaction, many complainants will henceforth become loyal customers, and will spend an even greater percentage of their sales dollars at that particular retail store.

553

Jeffrey G. Blodgett, Donald H. Granbois, and Rockney G. Walters
The Effects of Perceived Justice on Complainants' Negative Word-of Mouth Behavior and Repatronage Intentions

The Effects of Perceived Justice on Complainants' Negative Word-of-Mouth Behavior and Repatronage Intentions

Previous research has found that dissatisfied consumers choose to seek redress, engage in negative word-of-mouth behavior, and exit (i.e., vow never to repatronize the retailer) based upon the perceived likelihood of successful redress, their attitude toward complaining, the level of importance they attach to the defective product, and whether they perceive the problem to be stable or to have been controllable. The authors extend previous research by modeling consumer complaining behavior as a complex, dynamic process, hypothesizing that once a consumer seeks redress, negative word-of-mouth behavior and repatronage intentions are dependent (primarily) upon the consumer's post-complaint perception of justice. As hypothesized, perceived justice was found to be the main determinant of complainants' negative word-of-mouth behavior and their repatronage intentions, and was found to mediate the effects of likelihood of success, attitude toward complaining, product importance, and stability and controllability on complaining behavior. The model fit the data very well, explaining 49.1 percent of the variance of negative word-of-mouth and 68.5 percent of the variance of repatronage intentions. These findings point to the importance of customer service/customer satisfaction, especially since the cost of keeping a current customer satisfied is much less than the cost of attracting a new customer.

Introduction

The basic premise of the marketing concept is that marketers should strive to create customer satisfaction. Full implementation of this concept requires that marketers should also strive to remedy customer *dissatisfaction*. While satisfaction is assumed to lead to brand loyalty, goodwill, and repeat sales, dissatisfaction with products oftentimes leads to *redress seeking behavior* (i.e., a request for a refund, exchange, or repair, etc.). When a classified consumer seeks **redress** the retailer is given the opportunity to remedy the situation. **Complainants** who feel that justice has been served are likely to repatronize that retailer (and may even become more loyal customers), whereas complainants who perceive a lack of justice are likely to engage in *negative word-of-mouth behavior* (i.e., complain about the **retailer** to family and friends) and to *exit* (i.e., vow never to patronize that particular retail store again). In fact, one study found that dissatisfied consumers, on average, told nine others about their negative experience, and that some businesses may lose ten to fifteen percent of their annual volume each year because of poor service (Technical Assistance Research Programs, or TARP 1981). Considering that it costs five times as much to attract a new customer as it does to retain an old one (Desatnick 1988), it is essential that retailers pay attention to, and resolve, customer complaints.

The study of consumer complaining behavior has progressed steadily throughout the years (see Day and Landon 1976; Day and Bodur 1978; Day and Ash 1979; Day, Grabicke, Schaetzle, and Staubach 1981; Gilly and Gelb 1982; Bearden and Teel 1983; Richins 1983a, 1983b, 1985, 1987; Bearden and Mason 1984; Folkes 1984; Singh 1990). The focus of recent research has been to explain which particular type of complaint behavior—redress seeking, negative word-of-mouth, or exit—a dissatisfied consumer might choose (Singh 1990). Briefly, previous researchers have found that dissatisfied consumers choose to seek redress, engage in negative word-of-mouth behavior, or exit based upon the perceived *likelihood of successful redress* (Singh 1990; Richins 1987, 1985, 1983a; Granbois, Summers, and Frazier 1979; Day and Landon 1976), their *attitude toward complaining* (Richins 1980, 1982, 1983b, 1987; Bearden and Mason 1984), the level of *product importance* (Richins 1985), and whether they perceive the problem to be *stable* or to have been *controllable* (Folkes 1984). Consumers who perceive a high likelihood of successful redress, who have a favorable attitude toward complaining (i.e., toward seeking redress), or who are dissatisfied with a product that they feel is important (or *worthwhile*, see Singh 1990) are more likely to seek redress, whereas those consumers who perceive little likelihood of success, whose attitude is such that they are not predisposed toward seeking redress, or who perceive the problem to be stable or to have been controllable are more likely to engage in negative word-of-mouth behavior and to exit.

Although much progress has been made in our understanding of consumer complaining behavior, current models are limited in that they treat complaining behavior as a stasis phenomenon. In contrast, the proposed model explicitly recognizes that consumer complaining behavior is a dynamic process, and that *once a consumer seeks redress* other complaining behaviors (e.g., negative word-of-mouth and exit) are dependent primarily upon the consumer's post-complaint perception of justice. Though other authors have implicitly recognized the important role perceived justice plays in the complaining process (see Singh 1990; Richins 1987; Day et al. 1981; Day and Landon 1976; Hirschman 1970), little published research to date has explicitly examined the impact of perceived justice on complaining behaviors such as negative word-of-mouth and exit (or, more accurately, *repatronage intentions*). Research that has been conducted suggests that perceived justice is indeed a major determinant of complainants' repatronage intentions and negative word-of-mouth behavior (Tax and Chandrashekaran 1992; Goodwin and Ross 1989; Gilly 1987; Gilly and Gelb 1982; TARP 1981).

The purpose of the present study is to develop and test a model of consumer complaining behavior in which complainants' negative word-of-mouth behavior and repatronage intentions are dependent, in large part, upon their perceptions of justice. More specifically, the purpose is to assess the effects of perceived justice on complainants' negative word-of-mouth behavior and repatronage intentions in relation to other, previously identifed determinants of complaining behavior (specifically, the perceived likelihood of successful redress, attitude toward complaining, product importance, and stability and controllability attributions). Our model depicts complaining behavior as a complex, dynamic process in which complainants' perceptions of justice are in turn influenced by the perceived likelihood of successful redress, their attitudes toward complaining, the level of importance they attach to the defective product, and whether they perceive the problem to be stable or to have been controllable. The model is calibrated using data from a large sample of dissatisfied consumers who sought redress from a number of different retailers (e.g., department stores, discount stores, specialty stores) concerning a variety of soft (e.g., apparel) and hard (e.g., small appliances) goods at different price levels. We test the model using a structural equations modeling approach, thus permitting the identification of both direct and indirect causal effects.

555

Jeffrey G. Blodgett, Donald H. Granbois, and Rockney G. Walters
The Effects of Perceived Justice on Complainants' Negative Word-of Mouth Behavior and Repatronage Intentions

This study had important implications for researchers studying consumer complaining behavior. Including perceived justice into models of consumer complaining behavior provides both researchers, and managers, with greater insights into *why* consumers engage in negative word-of-mouth behavior and exit. Modeling complaining behavior as a complex, dynamic process strengthens the model building efforts of marketing scholars working in the areas of complaining behavior and retail management. The model is also of practical importance; linking complainants' levels of perceived justice to their repatronage intentions and negative word-of-mouth behavior facilitates managerial efforts to measure the effectiveness of their complaint handling policies and procedures in terms of retail sales and profits. Remedying consumer complaints can then be looked upon as a key marketing variable, with an expected return, just like advertising and promotion. An understanding of the benefits (and costs) of this key marketing variable could motivate retail managers to develop better complaint handling policies and procedures to build customer satisfaction.

Theoretical Development and Hypotheses

Theoretical Development

There is no single theory of consumer complaining behavior; rather, the study of complaining behavior is based upon several different theories from various fields of study. The confirmation/disconfirmation paradigm (Oliver 1980) and Hirschman's (1970) theory of exit, voice, and loyalty provide the basic framework for the study of complaining behavior, while conceptualization by Day (1984) and Day, Grabicke, Schaetzle, and Staubach (1981) provides important insights into the role of dissatisfaction. Attribution theory (Folkes 1984) and equity theory (Adams 1965) also have important implications for the study of consumer complaining behavior. This section will briefly discuss and apply these various theories (and conceptualization) to the study of consumer complaining behavior, in the hopes of creating a more unified theory, or framework, of complaining behavior.

• • •

Proposed Model

The model to be tested hypothesizes that complainants' negative word-of-mouth behavior and repatronage intentions are dependent, in part, upon their perceptions of the likelihood of successful redress, their attitudes toward complaining, the level of importance they attach to the defective product, and whether or not they perceive the problem to be stable or to have been controllable. One of the key contributions of our study is that we extend previous research by hypothesizing that—*once a consumer seeks redress*—negative word-of-mouth behavior and repatronage intentions are dependent, in large part, upon the complainants' ensuing level of perceived justice. In addition to having direct effects, we also hypothesize that likelihood of success, attitude toward complaining, product importance, and stability and controllability *indirectly* affect complainants' negative word-of-mouth behavior and repatronage intentions by influencing their perceptions of justice.

Hypotheses

557

Jeffrey G.
Blodgett,
Donald H.
Granbois, and
Rockney G.
Walters
The Effects of
Perceived
Justice on
Complainants'
Negative
Word-of Mouth
Behavior and
Repatronage
Intentions

Within the model a total of 14 **hypotheses** are tested. Eight of these are new hypotheses, whereas six have been examined in previous studies, but in a slightly different context (i.e., this study examines the complaining behavior of only those dissatisfied consumers who actually sought redress from the retailer).

• • •

Research Methodology

Unit of Analysis

Since the primary purpose of this study was to examine the effect of perceived justice on complainants' negative word-of-mouth behavior and repatronage intentions *this study included only those dissatisifed consumers who sought redress from the retailer.* (Dissatisfied consumers who did not seek redress would not have experienced a post-complaint feeling of justice, or a lack thereof; therefore, they were excluded from this study.) Respondents were asked to report on their most recent dissatisfying experience (that occurred within the last twelve months) regarding a product purchased at a retail store. The instance of dissatisfaction was to be one in which the consumer was truly dissatisfied with the product (i.e., because the product did not perform to expectations) rather than an instance in which the consumer returned a product because it was the wrong size, color, etc.

Method

A total of 201 useable surveys were collected. The data were collected via a self-report questionnaire administered to staff employees at two large universities (a midwestern university and a university in the mid-south) and to households in a large midwestern city. Given Anderson and Gerbing's (1988) recommendation of a minimum sample size of 150 when testing a structural model via LISREL, a sample size of 201 appears to be adequate. Based on a more stringent requirement of an n of 5 for each item, plus an additional n of 5 for each structural parameter to be estimated, 175 subjects were needed to obtain meaningful parameter estimates; our sample size exceeded this level.

Descriptive Statistics

Respondent Demographics

Of the 201 respondents, 83.6% were female and 16.4% were male. Ninety-four percent were Caucasian and six percent were African-American, Asian, or Native American. Approximately 11% were between the ages of 18–24, 27% were between the ages of 25–34, 37% were in the 35–44 age group, 21% were in the 45–65 age group, and 3% were age 65 or older. Forty-eight percent reported earning household income between $20,000 and $34,999, 26% earned between $35,000 and $44,999, 22% earned between $45,000 and $64,999, and 3.5% earned in excess of $65,000. Approximately 25% reported that their highest educational level was high school graduate, while 40% had attended some college, 22% were college graduates, and

13% had done graduate work. Approximately 14% held professional and 10% held white collar jobs, while 69% worked in clerical positions, and 7% worked in skilled, blue collar, or other occupations. Ninety percent reported buying the product for themselves or their family, while 10% purchased the product as a gift.

Types of Products and Prior Purchases

A wide cross-section of retailers were represented in the study, including mass merchants, department stores, discount stores, specialty stores, variety stores, and superstores. Complaints arose over a variety of products, including clothing (sweaters, dresses, jackets, etc.), small appliances (answering machines, irons, ceiling fans, etc.), shoes (basketball, jogging), jewelry (watches and rings), electronic items (vacuum cleaners, cameras, etc.), and other miscellaneous items (purses, playpens, toys, hamster cages, etc.). The average cost of the focal product was $90.72 (s.d. = $157.92). Approximately 93% of the respondents had made a purchase at the local retail store prior to their dissatisfying purchase experience, while 7% were first-time customers. Respondents, on average, purchased $270 (s.d. = $361) worth of merchandise at the focal store within the last six months. This figure is in line with research by Sears, which found that its credit card holders spent, on average, $500 per year at its stores (DeMott and Nash 1984).

Negative Word-of-Mouth Behavior

Approximately 75% of all respondents reported that they engaged in negative word-of-mouth behavior. Respondents, on average, told 4.88 (s.d. = 6.11) people about their dissatisfying experience. This figure is in line with previous research by Richins (1983c, 1987), who found that dissatisfied consumers, on average, told approximately five other people about their dissatisfaction.

Measures

Scale Development

After first specifying the domain of each construct, ad hoc scales were then developed. As recommended by Churchill (1979), Schwab (1986), and Nunnally (1978), multiple items were developed to measure each construct.

• • •

Results

Anderson and Gerbing (1988) recommend a two-stage approach to causal modeling, in which the measurement model is first confirmed and then the structural model is built. The measurement model provides an assessment of **discriminant** and **convergent** validity. Given that the measurement model provides an acceptable fit to the data, the structural model then provides an "assessment of nomological validity" (Anderson and Gerbing 1988, p. 411). This two-stage approach was followed using LISREL VII (Joreskog and Sorbom 1990) with maximum likelihood estimation.

• • •

Discussion and Implications

Jeffrey G. Blodgett, Donald H. Granbois, and Rockney G. Walters
The Effects of Perceived Justice on Complainants' Negative Word-of Mouth Behavior and Repatronage Intentions

Previous research has modeled complaining behavior as a static phenomenon, without taking into account the outcome of the redress seeking episode. The results of this study show that consumer complaining behavior is actually a dynamic process, and that *once a consumer seeks redress* negative word-of-mouth behavior and repatronage intentions are then dependent (primarily) upon the complainants' perception of justice. The significant role that perceived justice plays in consumer complaining behavior suggests that dissatisfied consumers are quite willing to give the retailer another chance *if* the retailer stands behind the product and guarantees customer satisfaction. As long as the retailer ensures satisfaction most complainants will not engage in negative word-of-mouth behavior or exit. Rather, because they perceive the retailer as being fair and just, these complainants may actually become more loyal customers! However, if the retailer does not stand behind the product and ensure customer satisfaction, complainants are likely to react by telling several friends about their dissatisfying experience and by vowing never to repatronize the offending retailer.

Other key findings are that dissatisfied consumers who perceived little likelihood of successful redress, or who were dissatisfied with products they felt were important, were more likely to engage in negative word-of-mouth behavior. These consumers probably were very frustrated with the retailer and felt a need to "get it off their chest" by telling friends and relatives. These findings illustrate the importance of the customer service/customer satisfaction concept. Firms that adopt this philosophy and communicate it to their customers are less likely to be the subject of negative word-of-mouth communications when problems do occur because their customers are more confident that their problems will be resolved satisfactorily. Dissatisfied customers who perceive a high likelihood of success, and who feel that their problems are equally as important to the retailer, most likely will first seek redress and will not engage in negative word-of-mouth behavior unless they are subsequently dissatisfied with the retailer's response to their complaint.

Another important finding is that dissatisfied consumers who perceived the problem to be stable or controllable were less likely to repatronize the offending retailer. Consumers who perceived the problem to be stable probably wished to avoid that retailer in the future, while consumers who perceived that the retailer could have prevented the problem probably were angry and may have vowed to "get even" by never shoping there again. This finding should be of particular concern to retailers. In order to retain these customers' business, it is important that the retailer apologize (Bies and Shapiro 1987) and take responsibility for any problems that may have occurred. The retailer should also thank the customer for bringing the problem to the retailer's attention, and should make a commitment to the customer to do better in the future. Complainants who perceive that the retailer is sincerely concerned, and is genuinely committed to improve, may be more likely to give the retailer another chance.

Other key findings are that likelihood of success, attitude toward complaining, product importance, and stability/controllability indirectly affected complaining behavior by influencing complainants' perceptions of justice. Consumers who perceived little likelihood of success, who had unfavorable attitudes toward complaining, who were dissatisfied with products they felt were important, or who perceived the cause to be stable or controllable were less likely to perceive that justice had been served,

TABLE 1
Means, Standard Deviations, Cronbach Alpha's, and List of Items for Each Construct[1]

	Mean	S.D.	Alpha
Likelihood of Success:	5.46	1.31	.76
—This store encourages its customers to return items that they are not satisfied with.			
—When I first bought this product, this store had a reputation for "Satisfaction guaranteed, or your money back!"			
—When this problem first occurred, I was confident that the store would let me exchange the product, give me a refund, or would repair the product.			
Attitude Toward Complaining*:	4.66	1.35	.73
—If a defective product is inexpensive, I usually keep it rather than ask the retailer for a refund, or an exchange.			
—I am usually reluctant to complain to a store regardless of how bad a product is.			
—In general, I am *more* likely to return an unsatisfactory product than most people I know.			
Product Importance:	4.65	1.46	.81
—I depend upon this product a great deal.			
—This product means a lot to me.			
—Compared to most products I buy, this was a fairly important product.			
Stability/Controllability:[2]	3.28	1.23	.72
—This type of thing probably happens all the time at this store.			
—This store hardly ever makes mistakes.			
—The retailer could have taken steps to prevent this problem from occurring.			
—If the retailer had just paid more attention to what it was doing, the problem never would have happened in the first place.			
Perceived Justice:	4.98	2.03	.93
—I was very *dissatisfied* with the store's response to my complaint!			
—Overall, I think that the store treated me fairly regarding my complaint.			
—When I complained to the retailer about this product, I got pretty much what I asked for (regarding a refund or exchange, etc.).			
Repatronage Intentions:	4.95	1.92	.87
—Knowing what I do now, if I had to do it all over again, I would not shop at this store—for *this* type of product.			
—Because of what happened, I will never shop at this store again— for *any* kind of product.			
—I would recommend to a friend that he/she shop at this store.			
Negative Word-of-Mouth:[3]	4.88	6.11	—
—How many friends or relatives (relatives not living at home) did you tell?			

[1] Seven point (1 = strongly disagree, 7 = strongly agree) scales were employed.
[2] Stability/controllability is the sum score of two multiplicative terms, and is measured on a scale of 1–7.
[3] This is the actual number of friends and relatives. For causal modeling purposes, negative word-of-mouth was subsequently rescaled on a 7-category, ordinal scale.
*The two items that were dropped are:
I feel uncomfortable when I have to return a defective product to a store. (Attitude Toward Complaining)
I would recommend to a friend that he/she shop at this store. (Repatronage Intentions)

and hence were more likely to engage in negative word-of-mouth behavior and less likely to repatronize the offending retailer. Interestingly, when their indirect effects are taken into consideration one finds that likelihood of success and product importance substantially impacted on complainants' negative word-of-mouth behavior (with total effects of −.367 and .419, respectively), and that stability/controllability had a major impact on complainants' repatronage intentions (with a total effect of −.446). These findings underscore the complex nature of complaining behavior. In order to better uncover complex relationships such as these, another key implication is that complaining behavior should be examined within the context of a structural equations model. Because structural models can uncover both direct and indirect (and reciprocal; Long 1983) relationships among variables, we recommend that future researchers continue this approach.

Five hypotheses were not supported. Neither likelihood of success, attitude toward complaining, nor product importance had a significant direct effect on complainants' repatronage intentions, whereas neither attitude toward complaining nor stability/controllability had a significant direct effect on complainants' negative word-of-mouth behavior. It should be noted that these findings are due to the strength of perceived justice as a **mediator** between these variables and complaining behavior. Although previous research has found that likelihood of success, attitude toward complaining, and product importance play a critical role in determining whether a dissatisfied consumer will *first* seek redress, engage in negative word-of-mouth behavior, or exit (Singh 1990; Richins 1987, 1985, 1983a; Bearden and Mason 1984), this study shows that *once* a dissatisfied consumer seeks redress, negative word-of-mouth behavior and repatronage intentions are then dependent (primarily) upon the complainants' perceptions of justice. On the whole, the entire set of findings indicates that failure to include perceived justice in models of consumer complaining behavior can result in misspecified models and unreliable findings.

Limitations

This study is subject to several limitations. First, the sampling frame included only those dissatisfied consumers who complained to the retailer (i.e., sought redress); dissatisfied consumers who did not seek redress were excluded from this study. Consequently, this study does not provide a complete picture of the complaining behavior process. In order to better understand the roles that variables such as likelihood of successful redress, attitude toward complaining, product importance, and stability/controllability play in determining whether dissatisfied consumers will seek redress, engage in negative word-of-mouth behavior, or vow never to repatronize the retailer (i.e., exit), future research should examine how these relationships differ between those dissatisfied consumers who *sought* redress and those dissatisfied consumers who *did not seek* redress. For example, one might find that the relationship between likelihood of success and repatronage intentions is stronger for that group of dissatisfied consumers who did not seek redress than it is for that group of dissatisfied consumers who did seek redress. Secondly, because the current sample was comprised mainly of white, middle-class consumers living in the midwest or mid-south, and was limited to products purchased at retail stores, the generalizability (Cook and Campbell 1979) of our model is somewhat limited. An interesting project would be to test the model using data from a broader mix of respondents

561

Jeffrey G. Blodgett, Donald H. Granbois, and Rockney G. Walters
The Effects of Perceived Justice on Complainants' Negative Word-of Mouth Behavior and Repatronage Intentions

(e.g., different ethnic groups, such as Hispanics, Asians, etc.) and to extend it to other purchase settings (e.g., automobiles, major appliances, direct mail purchases, and services industries such as banking, airlines, and hotels, etc.). The lack of a validation sample is also a limitation of this study. A validation sample would provide strong evidence as to the validity of the model (Cook and Campbell 1979); however, the size of our sample ($n = 201$) precluded us from validating our findings. Finally, it appears that key explanatory variables were not included in the model. Although we were able to explain 68.5% of the variance of repatronage intentions, we were able to explain less than half (49.1%) of the variance of negative word-of-mouth. Clearly, other variables, such as *negative affect* (or emotion, Westbrook 1987), are missing from the model. Future research should expand the model to include other possible determinants of complaining behavior.

Summary

Because of the impact that perceived justice has on retail profits, remedying consumer complaints should be thought of as a key marketing variable. It is much easier to keep current customers satisfied than it is to attract new customers, especially given slow growth in retail sales. A widely promoted and successfully executed policy of "satisfaction guaranteed" gives customers greater confidence that any problems they might encounter will be remedied in a timely and courteous manner. Because of this confidence, dissatisfied consumers are more likely to first seek redress, thus giving the retailer a chance to remedy the problem. As a result of the retailer standing behind the product and ensuring customer satisfaction, many complainants will henceforth become loyal customers, and will spend an even greater percentage of their sales dollars at that particular retailer.

· · ·

Questions on Content

1. How much more does it cost a retailer to attract a new customer than it costs to retain an old one?
2. List three types of complaint behavior.
3. Approximately how many other people do dissatisfied consumers tell about their dissatisfaction?
4. How can retailers avoid negative word-of-mouth behavior or exit?
5. List the major findings of this study.
6. Name three limitations of the study.

Vocabulary

Define the following words as they are used in the report (the words are printed in **boldface** type):

1. redress
2. complainants
3. retailer
4. hypotheses
5. discriminant
6. convergent
7. mediator

Suggestions for Response and Reaction

1. Define (see pages 109–118) "perceived justice" as the term is used in the report.
2. Describe an experience when you bought a product you found unsatisfactory and explain how you sought redress from the retailer (see complaint letter, pages 268–271).
3. Using data from the respondent demographics section under "Descriptive Statistics," construct a pie chart (see pages 470–471) or a bar chart (see pages 471–473) showing the household incomes, the age groups, or the education level of the respondents in the study.
4. Compare and contrast (see pages 200–212) the executive summary, the abstract, and the closing summary of the report. What purpose does each serve? (See Chapter 6 The Summary, especially informative summary, pages 224–233.)
5. What are the qualities that make this a formal report (see pages 309–311)?
6. In the first two paragraphs of the Introduction, explain why and how the authors document their statements (see Documentation, pages 404–409).

Excerpt from "Non-Medical Lasers: Everyday Life in a New Light"

REBECCA D. WILLIAMS

Williams gives a definition of a laser and a brief, simple explanation of how a laser works. She then focuses on ways lasers affect our daily lives.

Light is old as the stars, but … laser light is a relative newcomer under the sun.

Developed just over 30 years ago, lasers have revolutionized industry, medicine and science. They play our compact discs. They can carry our phone messages. They entertain us with light shows in the night sky. They are powerful enough to cut through an inch of steel and precise enough to drill 200 holes on the head of a pin.

Perhaps lasers are best known for their feats in medicine. Doctors use them to vaporize or mend tissue with minimal scarring or swelling.

Lesser known, however, are the hundreds of ways in which non-medical lasers affect our lives every day. Like medical lasers, these are regulated by FDA for safety. Non-medical lasers are in the grocery store, the office, our schools, and even our homes.

"They've reached that point where they're everywhere. You can't get away from them," says Peter Baker, executive director of the Laser Institute of America in Orlando, Fla.

"The old saying was that lasers were a solution looking for a problem," Baker adds. "In the early years, they tended to be a bit exotic and unreliable. You had to have a Ph.D. to know how to keep them tweaked up and reliable."

Today, however, lasers are dependable enough to be useful in a wealth of technologies. They're not just for Ph.D.s anymore

What Is a Laser?

Laser is an **acronym** for "light amplification by stimulated emission of radiation." Radiation, in this case, is another word for **electromagnetic energy,** which includes light.

Laser light has several properties that make it different from regular light. First, it is often collimated, which means it travels in a narrow beam for long distances, rather than going off in many directions as regular light does.

Laser life is also coherent, which means that the light waves stay synchronized over long distances. And it is monochromatic, of one color. Some laser beams are invisible, producing light in the infrared or ultraviolet wavelengths.

A laser can produce a short burst of light or a continuous beam. Because it can focus narrowly, laser light can be much more intense than regular light, especially in bursts. Laser beams range in power from a few microwatts to several billion watts in short bursts.

Reprinted by permission from *FDA Consumer* (July–August 1991), pp. 32–35.

How a Laser Works

Rebecca D. Williams
Excerpt from
"Non-Medical Lasers: Everyday Life in a New Light"

To understand how a laser works, first you have to understand something about light.

Light is the visible form of electromagnetic energy. Some electromagnetic energy cannot be seen, but it is everywhere in the universe. It is made of tiny energy packets, called photons, traveling in a stream of wave patterns, called electromagnetic waves.

All objects emit electromagnetic energy in varying wavelengths and directions. Photons are emitted when atoms spontaneously change from one energy state to another. We can feel some of them as heat, and we can see others as light. But even when we can't see or feel them, photons are moving all around us in different directions, getting absorbed by atoms and released again.

In a laser, however, the photons are emitted in the same general wavelength and direction.

It was Albert Einstein who first conceived of laser light. In 1917, he theorized that atoms could be stimulated to release energy with a specific wavelength and direction. But it wasn't until 1960 that Theodore Maiman, an American scientist, put together the first successful laser.

Like Maiman's invention, many lasers today contain three major parts, as shown in the diagram. First, there's a cylinder-shaped rod called the lasing medium. Maiman used a synthetic crystal ruby, but today's lasing mediums are often glass tubes filled with liquid or gases such as helium, neon, argon, and carbon dioxide.

Second, a laser has a power supply that excites the atoms in the lasing medium with energy. The power supply charges the atoms to their capacity, eventually causing them to release photons in one wavelength and direction.

The third part of a laser is a set of mirrors, one on each end of the lasing tube. As the atoms release photons, the mirrors reflect them back and forth. With each pass through the lasing medium, the energy is amplified. One of the mirrors reflects only part of the light, so some of the photons escape through the end of the tube as a narrow laser beam.

This simplified diagram of a common laser shows its three major parts: a lasing medium (solid, liquid or gas), two mirrors at either end, and a power supply.

The power supply provides the energy to charge the atoms of the lasing medium, so that they emit photons. The photons bounce back and forth between the mirrors and with each pass, the energy is amplified. Some of it escapes through the partially reflecting mirror at one end as a laser beam.

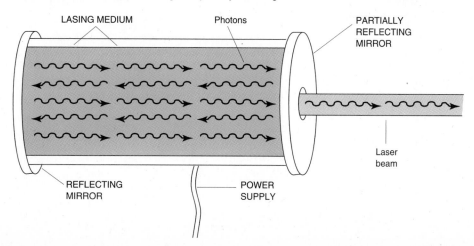

How FDA Monitors Laser Safety

FDA regulates all kinds of lasers, from the ones used in surgery, to those found in supermarket checkout scanners. The agency has authority to regulate them under the Radiation Control for Health and Safety Act and the Medical Device Amendments to the federal Food, Drug, and Cosmetic Act.

In monitoring laser safety, FDA recognizes four major classes and two subclasses of lasers, ranging from those that pose no known hazard to those that pose serious danger if used improperly. The higher the class, the more powerful the laser. Depending on the strength of the laser, FDA requires a variety of safety features such as safety locks, emission indicators, and switches that automatically turn off the laser in certain circumstances.

FDA requires that most lasers bear warning labels about radiation and other hazards, and all must display a certification label stating that the laser complies with FDA safety regulations.

FDA inspectors check 220 of the approximately 1,000 laser manufacturers in the United States each year, according to Jerome Dennis of the division of standards enforcement in FDA's Center for Devices and Radiological Health. FDA repeatedly inspects plants that have frequent violations. The center has a laser laboratory in Rockville, Md., to provide technical support to the inspectors in the field.

FDA also performs 110 tests on lasers in use each year, such as ones operating in factories and supermarkets. Producers of laser light shows, for example, are required to tell the FDA where they are planning a display, so that the agency can inspect it if possible. "We do try to pop in on them unannounced and unexpected," says Dennis.

The Center for Devices and Radiological Health keeps track of laser injuries through several voluntary reporting programs. There have been serious eye injuries, skin burns, and electrocutions reported to FDA over the years. Most of the injuries, however, were the result of human error and misuse, according to Richard Felten, a regulatory review scientist at the center. Either the laser users didn't wear protective eyewear, or they wore the wrong kind for that particular laser, or they didn't heed the warning labels on the device.

Eye protection is not a foolproof defense against laser injury and should only be used as a backup to other safety measures, Dennis advises. No one should look directly into a laser beam even while wearing a pair of goggles.

"From the information we have, it appears that many reported eye injuries and skin burns occur because people are doing something incorrect," says Felten. "However, when the user follows the recommended safety procedures, lasers can be used in a safe manner."

—*R.D.W.*

There are a number of variations to lasers. Chemical lasers, for example, get their energy from chemical reactions instead of electricity. Lasers filled with liquid dye use complex molecules to produce more than one wavelength. And semiconductor diode lasers, some of which are smaller than a grain of salt, are made from circuitry similar to the light-emitting diodes in digital watches and calculators. These lasers emit tiny beams of coherent light.

Today, laser production is a billion-dollar industry worldwide, according to *Laser Focus World* magazine. Manufacturers, who began by replacing one or two drills with lasers, now design the whole of their manufacturing techniques around them. Some industries have even redesigned their products to be more accessible to laser manufacturing.

Communications and Information

By far, the most common lasers are those that transmit and store information. These are the lasers inside compact disc players, laser printers, supermarket scanners, and fiber-optic cables used in telephone lines. Most of the time, these technologies use tiny semiconductor lasers.

Telephone companies convert sound to electrical pulses, then convert the pulses to laser light that can travel through fiber-optic cables. This is more efficient than traditional phone technology, which sends calls over copper wires. With a laser that pulses very quickly, one glass fiber the width of a human hair can transmit the calls of more than 20,000 wires.

Lasers can also record and reproduce music on compact discs. The sound is recorded by a laser beam that burns a pattern of dots into a disc. A tiny semiconductor laser in a compact disc player reads the dots and converts them back to sound. Since only light touches the disc, a CD player gives very clean, clear sound, without the background noise from scratches and dust that you hear on records.

Compact discs are quickly replacing other kinds of recordings, and not just for home stereos. Libraries use them to store databases, encyclopedias, and other information to access by computer.

What if you need to print out that information? A laser printer can do it quickly in type that looks professionally printed. Inside the printer, a scanning laser moves across a light-sensitive drum. The drum attracts ink where the laser has hit it and transfers the ink to the paper as it rolls by.

A scanning laser can read as well as write. Many grocery and department stores tag their merchandise with Universal Product Codes, labels with striped patterns. Low-powered lasers can "read" the stripes and signal the store's computer to deduct the product from inventory and ring up the price.

Industry and the Military

Factories use powerful lasers to manufacture a wide range of products. Lasers can drill or weld metal, cut holes in the tips of baby bottles, or solder tiny circuits for electronic parts.

They can strip wires, etch trademarks, drill tiny holes for watch jewels, and cut out cloth for dozens of suits at once.

Although armies don't vaporize their enemies with lasers like they do in science fiction movies, the military has found lots of other uses for lasers.

Lasers guided the "smart bombs" used during the Persian Gulf War, for example. Lasers transmit communications on the battlefield, and in training, the Army uses laser "guns" to simulate the firing of real ammunition.

Arts and Entertainment

Dazzling laser light shows are standard features at many amusement parks, rock concerts, and planetarium shows, and they are monitoried for safety by FDA inspectors. . . . Programmed to move to the music, the laser beams dance and flash across a suspended screen, the night clouds, and under careful safety precautions, even the audience.

And in photography and art, lasers can produce three-dimensional photographs called **holograms.** You may have seen them on your credit cards—banks use holographic emblems to protect against counterfeiting. Scientists say someday holograms will give us three-dimensional pictures of everything from molecules to city streets. Perhaps we'll even watch three-dimensional television on table tops.

Measuring Up

Because of their precision, lasers have offered a new way of taking measurements.

With lasers, scientists can calculate the distance to the moon more accurately than ever before. In 1969, astronauts placed an object on the lunar surface that can reflect a laser beam back to its precise origin on Earth. Using a 4-billion-watt pulse laser, scientists now measure the distance to the moon to within 6 centimeters, or 2.4 inches.

Lasers can also take small measurements, such as the vibration of atoms and molecules, the frequency of light, and the amount of trace pollutants in air and water.

Lasers for the Future

Lasers have the potential to generate clean, powerful energy from the world's most abundant natural resource—sea water—using the same process that fuels the sun and stars. Called **fusion,** the process involves compressing atoms together to release energy.

Fusion on the sun is produced by tremendous gravity forcing atoms together. On Earth, scientists can briefly produce fusion with lasers. At the Lawrence Livermore National Laboratory in Livermore, Calif., scientists have developed a powerful laser called NOVA. In a fraction of a second, this laser can heat a tiny pellet of frozen hydrogen several million degrees. The heating is so fast that the atoms don't have a chance to escape as vapor. Instead, they compress and implode. For a brief instance, fusion results. A mini-star is born.

So far, however, the laser is not powerful enough to produce sustained fusion, and the technology is still decades away.

Scientists already know how to split an atom to produce energy. Called **fission,** the reaction is used in nuclear power plants and atomic bombs. Unlike fission, laser-generated fusion would produce few hazardous wastes—some scientists say none at all.

As the laser industry grows, scientists continue to discover new applications for this special light. Some speculate the next generation of scientists will use lasers to explore other star systems or fuel spaceships far into the galaxy. . . .

Questions on Content

1. List ten non-medical uses for lasers.
2. List the properties that make laser light different from regular light.
3. Name two early experimenters with laser light.
4. Name the three major parts of a laser and tell what each does.

Vocabulary

Define the following words as they are used in the article (the words are printed in **boldface** type):

1. acronym
2. electromagnetic energy
3. holograms

4. fusion
5. fission

Suggestions for Response and Reaction

1. Write a paragraph explaining how a compact disc reproduces music. Review giving a process explanation (see pages 49–60).
2. Show how the description of a laser and the diagram are representative of many types of lasers (see pages 77–83).
3. Set up a classification system (see pages 131–149) for non-medical laser uses, using the information supplied in the article. Show the system in an outline form (see pages 139–140).
4. Explain in a paragraph how the FDA uses classification (see pages 131–149) to monitor laser safety.
5. Write a documented report (see pages 394–409) on using lasers in fusion and fission. Include a definition of each and focus the report on the uses, advantages, and disadvantages of both fusion and fission.
6. Evaluate the use of headings (see pages 9–11 and 311–312), the box (pages 6–7), and the diagram (see pages 480–482) in the article.

"Journeying Through the Cosmos" from *The Shores of the Cosmic Ocean*

CARL SAGAN

In poetic language, Sagan talks of the magnitude and the mystery of the Cosmos.

The **Cosmos** is all that is or ever was or ever will be. Our feeblest contemplations of the Cosmos stir us—there is a tingling in the spine, a catch in the voice, a faint sensation, as if a distant memory, of falling from a height. We know we are approaching the greatest of mysteries.

The size and age of the Cosmos are beyond ordinary human understanding. Lost somewhere between immensity and eternity is our tiny planetary home. In a cosmic perspective, most human concerns seem insignificant, even petty. And yet our species is young and curious and brave and shows much promise. In the last few **millennia** we have made the most astonishing and unexpected discoveries about the Cosmos and our place within it, explorations that are exhilarating to consider. They remind us that humans have evolved to wonder, that understanding is a joy, that knowledge is prerequisite to survival. I believe our future depends on how well we know this Cosmos in which we float like a mote of dust in the morning sky.

Those explorations require skepticism and imagination both. Imagination will often carry us to worlds that never were. But without it, we go nowhere. Skepticism enables us to distinguish fancy from fact, to test our speculations. The Cosmos is rich beyond measure—in elegant facts, in exquisite interrelationships, in the subtle machinery of awe.

The surface of the Earth is the shore of the cosmic ocean. From it we have learned most of what we know. Recently, we have waded a little out to sea, enough to dampen our toes or, at most, wet our ankles. The water seems inviting. The ocean calls. Some part of our being knows this is from where we came. We long to return. These aspirations are not, I think, irreverent, although they may trouble whatever gods may be.

The dimensions of the Cosmos are so large that using familiar units of distance, such as meters or miles, chosen for their utility on Earth, would make little sense. Instead, we measure distance with the speed of light. In one second a beam of light travels 186,000 miles, nearly 300,000 kilometers or seven times around the Earth. In eight minutes it will travel from the Sun to the Earth. We can say the Sun is eight light-minutes away. In a year, it crosses nearly ten trillion kilometers, about six trillion miles of intervening space. It measures not time but distances—enormous distances.

The Earth is a place. It is by no means the only place. It is not even a typical place. No planet or star or galaxy can be typical, because the Cosmos is mostly empty. The only typical place is within the vast, cold, universal vacuum, the ever-lasting night of intergalactic space, a place so strange and desolate that, by comparison, planets and stars and galaxies seem achingly rare and lovely. If we were randomly inserted into the Cosmos, the chance that we would find ourselves on or near a planet would be less than one in a billion trillion trillion* (10^{33}, a one followed by 33 zeroes). In everyday life such odds are called compelling. Worlds are precious.

571

Carl Sagan
Jorneying
Through the
Cosmos from
"The Shores of
the Cosmic
Ocean"

Galaxies

From an intergalactic vantage point we would see, strewn like sea froth on the waves of space, innumerable faint, wispy tendrils of light. These are the galaxies. Some are solitary wanderers: most inhabit communal clusters, huddling together, drifting endlessly in the great cosmic dark. Before us is the Cosmos on the grandest scale we know. We are in the realm of the nebulae, eight billion light-years from Earth, halfway to the edge of the known universe.

A **galaxy** is composed of gas and dust and stars—billions upon billions of stars. Every star may be a sun to someone. Within a galaxy are stars and worlds and, it may be, a proliferation of living things and intelligent beings and spacefaring civilizations. But from afar, a galaxy reminds me more of a collection of lovely found objects—seashells, perhaps, or corals, the productions of Nature laboring for aeons in the cosmic ocean.

There are some hundred billion (10^{11}) galaxies, each with, on the average, a hundred billion stars. In all the galaxies, there are perhaps as many planets as stars, $10^{11} \times 10^{11} = 10^{22}$, ten billion trillion. In the face of such overpowering numbers, what is the likelihood that only one ordinary star, the Sun, is accompanied by an inhabited planet? Why should we, tucked away in some forgotten corner of the Cosmos, be so fortunate? To me, it seems far more likely that the universe is brimming over with life. But we humans do not yet know. We are just beginning our explorations. From eight billion light-years away we are hard pressed to find even the cluster in which our Milky Way Galaxy is embedded, much less the Sun or the Earth. The only planet we are sure is inhabited is a tiny speck of rock and metal, shining feebly by reflected sunlight, and at this distance utterly lost.

But presently our journey takes us to what astronomers on Earth like to call the Local Group of galaxies. Several million light-years across, it is composed of some twenty constituent galaxies. It is a sparse and obscure and unpretentious cluster. One of these galaxies is M31, seen from the Earth in the constellation Andromeda. Like other spiral galaxies, it is a huge pinwheel of stars, gas and dust. M31 has two small satellites, dwarf elliptical galaxies bound to it by gravity, by the identical law of physics that tends to keep me in my chair. The laws of nature are the same throughout the Cosmos. We are now two million light-years from home.

Beyond M31 is another, very similar galaxy, our own, its spiral arms turning slowly, once every quarter billion years. Now, forty thousand light-years from home, we find ourselves falling toward the massive center of the Milky Way. But if we wish to find the Earth, we must redirect our course to the remote outskirts of the Galaxy, to an obscure locale near the edge of a distant spiral arm.

*We use the American scientific convention for large numbers, one billion = 1,000,000,000 = 10^9; one trillion = 1,000,000,000,000 = 10^{12}, etc. The exponent counts the number of zeroes after the one.

Stars

Our overwhelming impression, even between the spiral arms, is of stars streaming by us—a vast array of exquisitely self-luminous stars. Some as flimsy as a soap bubble and so large that they could contain ten thousand Suns or a trillion Earths; others the size of a small town and a hundred trillion times denser than lead. Some stars are solitary, like the Sun. Most have companions. Systems are commonly double, two stars orbiting one another. But there is a continuous gradation from triple systems through loose clusters of a few dozen stars to the great globular clusters, resplendent with a million suns. Some double stars are so close that they touch, and starstuff flows between them. Most are as separated as Jupiter is from the Sun. Some stars, the supernovas, are as bright as the entire galaxy that contains them; others, the black holes, are invisible from a few kilometers away. Some shine with a constant brightness; others flicker uncertainly or blink with an unfaltering rhythm. Some rotate in stately elegance; others spin so feverishly that they distort themselves to oblateness. Most shine mainly in visible and infrared light; others are also brilliant sources of X-rays or radio waves. Blue stars are hot and young; yellow stars, conventional and middle-aged; red stars, often elderly and dying; and small white or black stars are in the final throes of death. The Milky Way contains some 400 billion stars of all sorts moving with a complex and orderly grace. Of all the stars, the inhabitants of earth know close-up, so far, but one.

Each star system is an island in space, quarantined from its neighbors by the light-years. I can imagine creatures evolving into glimmerings of knowledge on innumerable worlds, every one of them assuming at first their puny planet and paltry few suns to be all that is. We grow up in isolation. Only slowly do we teach ourselves the Cosmos.

Some stars may be surrounded by millions of lifeless and rocky worldlets, planetary systems frozen at some early stage in their evolution. Perhaps many stars have planetary systems rather like our own: at the periphery, great gaseous ringed planets and icy moons, and nearer to the center, small, warm, blue-white, cloud-covered worlds. On some, intelligent life may have evolved, reworking the planetary surface in some massive engineering enterprise. These are our brothers and sisters in the Cosmos. Are they very different from us? What is their form, biochemistry, neurobiology, history, politics, science, technology, art, music, religion, philosophy? Perhaps some day we will know them.

We have now reached our own backyard, a light-year from Earth. Surrounding our Sun is a spherical swarm of giant snowballs composed of ice and rock and organic molecules: the cometary nuclei. Every now and then a passing star gives a tiny gravitational tug, and one of them obligingly careens into the inner solar system. There the Sun heats it, the ice is vaporized and a lovely cometary tail develops.

Planets

We approach the planets of our system, largish worlds, captives of the Sun, gravitationally constrained to follow nearly circular orbits, heated mainly by sunlight. Pluto, covered with methane ice and accompanied by its solitary giant moon Charon, is illuminated by a distant Sun, which appears as no more than a bright point of light in a pitch-black sky. The giant gas worlds, Neptune, Uranus, Saturn—the jewel of the solar system—and Jupiter all have an entourage of icy moons. Interior to the region of gassy planets and orbiting icebergs are the warm, rocky provinces of the inner solar system. There is, for example, red planet Mars, with soaring volcanoes, great rift valleys, enormous planet-wide sandstorms, and just possibly, some simple forms of life. All the planets orbit the Sun, the nearest star, an inferno of hydrogen and helium gas engaged in thermonuclear reactions, flooding the solar system with light.

573

Carl Sagan
Jorneying
Through the
Cosmos from
"The Shores of
the Cosmic
Ocean"

Planet Earth

Finally, at the end of all our wanderings, we return to our tiny, fragile, blue-white world, lost in a cosmic ocean vast beyond our most courageous imaginings. It is a world among an immensity of others. It may be significant only for us. The Earth is our home, our parent. Our kind of life arose and evolved here. The human species is coming of age here. It is on this world that we developed our passion for exploring the Cosmos, and it is here that we are, in some pain and with no guarantees, working out our destiny.

Welcome to the planet Earth—a place of blue nitrogen skies, oceans of liquid water, cool forests and soft meadows, a world positively rippling with life. In the cosmic perspective it is, as I have said, poignantly beautiful and rare; but it is also, for the moment, unique. In all our journeying through space and time, it is, so far, the only world on which we know with certainty that the matter of Cosmos has become alive and aware. There must be many such worlds scattered through space, but our search for them begins here, with the accumulated wisdom of the men and women of our species, garnered at great cost over a million years. We are privileged to live among brilliant and passionately **inquisitive** people, and a time when the search for knowledge is generally prized. Human beings, born ultimately of the stars and now for a while inhabiting a world called Earth, have begun their long voyage home.

Ancient Discovery That the Earth Is Round

The discovery that the Earth is a *little* world was made, as so many important human discoveries were, in the ancient Near East, in a time some humans called the third century B.C., in the greatest metropolis of the age, the Egyptian city of Alexandria. Here there lived a man named Eratosthenes. One of his envious contemporaries called him "Beta," the second letter of the Greek alphabet, because, he said, Eratosthenes was second best in the world in everything. But it seems clear that in almost everything Erastosthenes was "Alpha." He was an astronomer, historian, geographer, philosopher, poet, theater critic and mathematician. The titles of the books he wrote range from *Astronomy* to *On Freedom from Pain*. He was also the director of the great library of Alexandria, where one day he read in a papyrus book that in the southern frontier outpost of Syene, near the first cataract of the Nile, at noon on June 21 vertical sticks cast no shadows. On the summer solstice, the longest day of the year, as the hours crept toward midday, the shadows of temple columns grew shorter. At noon, they were gone. The reflection of the Sun could then be seen in the water at the bottom of a deep well. The Sun was directly overhead.

It was an observation that someone else might easily have ignored. Sticks, shadows, reflections in wells, the position of the Sun—of what possible importance could such simple everyday matters be? But Erastosthenes was a scientist, and his musings on these commonplaces changed the world; in a way, they made the world. Erastosthenes had the presence of mind to do an experiment, actually to observe whether in Alexandria vertical sticks cast shadows near noon on June 21. And, he discovered, sticks do.

Erastosthenes asked himself how, at the same moment, a stick in Syene could cast no shadow and a stick in Alexandria, far to the north, could cast a pronounced shadow. Consider a map of ancient Egypt with two vertical sticks of equal length, one stuck in Alexandria, the other in Syene. Suppose that, at a certain moment, each stick casts no shadow at all. This is perfectly easy to understand—provided the Earth is flat. The Sun would then be directly overhead. If the two sticks cast shadows of equal length, that also would make sense on a flat Earth: the Sun's rays would then be inclined at the same angle to the two sticks. But how could it be that at the same instant there was no shadow at Syene and a substantial shadow at Alexandria?

The only possible answer, he saw, was that the surface of the Earth is curved. Not only that: the greater the curvature, the greater the difference in the shadow lengths. The Sun is so far away that its rays are parallel when they reach the Earth. Sticks placed at different angles to the Sun's rays cast shadows of different lengths. For the observed difference in the shadow lengths, the distance between Alexandria and Syene had to be about seven degrees along the surface of the Earth; that is, if you imagine the sticks extending down to the center of the Earth, they would there intersect at an angle of seven degrees. Seven degrees is something like one-fiftieth of three hundred and sixty degrees, the full circumference of the Earth. Erastosthenes knew that the distance between Alexandria and Syene was approximately 800 kilometers, because he hired a man to pace it out. Eight hundred kilometers times 50 is 40,000 kilometers: so that must be the circumference of the Earth.*

This is the right answer. Eratosthenes' only tools were sticks, eyes, feet and brains, plus a taste for experiment. With them he deduced the circumference of the Earth with an error of only a few percent, a remarkable achievement for 2,200 years ago. He was the first person accurately to measure the size of a planet.

The Mediterranean world at that time was famous for seafaring. Alexandria was the greatest seaport on the planet. Once you knew the Earth to be a sphere of modest diameter, would you not be tempted to make voyages of exploration, to seek out undiscovered lands, perhaps even to attempt to sail around the planet? Four hundred years before Erastosthenes, Africa had been circumnavigated by a Phoenician fleet in the employ of the Egyptian Pharaoh Necho. They set sail, probably in frail open boats, from the Red Sea, turned down the east coast of Africa up into the Atlantic, returning through the Mediterranean. This epic journey took three years, about as long as a modern Voyager spacecraft takes to fly from Earth to Saturn.

After Eratosthenes' discovery, many great voyages were attempted by brave and venturesome sailors. Their ships were tiny. They had only rudimentary navigational

---

*Or, if you like to measure things in miles, the distance between Alexandria and Syene is about 500 miles, and 500 miles × 50 = 25,000 miles.

575

Carl Sagan
Jorneying
Through the
Cosmos from
"The Shores of
the Cosmic
Ocean"

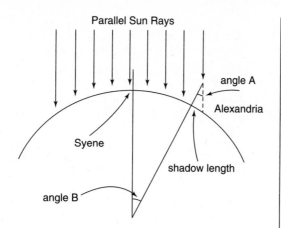

From the shadow length in Alexandria, the angle A can be measured. But from simple geometry ("if two parallel straight lines are transected by a third line, the alternate interior angles are equal"), angle B equals angle A. So by measuring the shadow length in Alexandria, Eratosthenes concluded that Syene was A = B = 7° away on the circumference of the Earth.

instruments. They used dead reckoning and followed coastlines as far as they could. In an unknown ocean they could determine their latitude, but not their longitude, by observing, night after night, the position of the constellations with respect to the horizon. The familiar constellations must have been reassuring in the midst of an unexplored ocean. The stars are the friends of explorers, then with seagoing ships on Earth and now with spacefaring ships in the sky. After Erastosthenes, some may have tried, but not until the time of Magellan did anyone succeed in circumnavigating the Earth. What tales of daring and adventure must earlier have been recounted as sailors and navigators, practical men of the world, gambled their lives on the mathematics of a scientist from Alexandria?

Questions of Content

1. What is a light-year?
2. If the Earth were randomly placed into the cosmos, what are the odds of the Earth coming near another planet?
3. Name two categories of galaxies.
4. Classify stars according to their color.
5. Who was Eratosthenes? Explain how he deduced that the world is curved, not flat. How did he figure the circumference of the Earth?
6. What natural occurrence enabled early voyagers to determine latitude?
7. Who first circumnavigated the Earth?

Vocabulary

Define the following words as they are used in the article (the words are printed in **boldface** type):

1. Cosmos
2. millennia
3. galaxy
4. inquisitive

Suggestions for Response and Reaction

1. Write an extended definition (see pages 112–118) of *cosmos* or of *galaxy*.
2. Sagan says exploration of the cosmos has required *skepticism* and *imagination*. Write a comparison (see pages 200–212) of the two words.
3. Write a documented report (see pages 394–409) on a particular aspect of stars.

"Work" from
The Book of Virtues

EDITED, WITH COMMENTARY, BY WILLIAM J. BENNETT

William Bennett's commentary ponders the many facets of the meaning of work as it relates to living.

"What are you going to be when you grow up?" is a question about **work.** What is your work in the world going to be? What will be your works? These are not fundamentally questions about jobs and pay, but questions about life. Work is applied effort; it is whatever we put ourselves into, whatever we expend our energy on for the sake of accomplishing or achieving something. Work in this fundamental sense is not what we do *for* a living but what we do *with* our living.

Parents and teachers both *work* at the upbringing of children, but only teachers receive paychecks for it. The housework of parents is real work, though it brings in no revenue. The schoolwork, homework, and teamwork of children are all real work, though the payoff is not in dollars. A child's household chores may be accompanied *by* an allowance, but they are not done *for* an allowance. They are done because they need to be done.

The opposite of work is not leisure or play or having fun but idleness—not *investing* ourselves in anything. Even sleeping can be a form of investment if it is done for the sake of future activity. But sleep, like amusement, can also be a form of escape—oblivion sought for its own sake rather than for the sake of renewal. It can be a waste of time. Leisure activity or play or having fun, on the other hand, can involve genuine investment of the self and not be a waste of time at all.

We want our children to **flourish,** to live well and fare well—to be happy. Happiness, as Aristotle long ago pointed out, resides in activity, both physical and mental. It resides in doing things that one can take pride in doing well, and hence that one can *enjoy* doing. It is a great mistake to identify enjoyment with mere amusement or relaxing or being entertained. Life's greatest joys are not what one does *apart from* the work of one's life, but *with* the work of one's life. Those who have missed the joy of work, of a job well done, have missed something very important. This applies to our children, too. When we want our children to be happy, we want them to enjoy *life.* We want them to find and enjoy their work in the world.

How do we help prepare our children for lives like that? Once again, the keys are practice and example: practice in *doing* various things that require a level of effort and engagement **compatible** with some personal investment in the activity, and the examples of our own lives.

The first step in doing things is *learning how* to do them. (And learning how to turn on the television doesn't count—though learning how to turn it *off* might.) Good habits of personal hygiene and helping with meals or bed-making or laundry or caring for pets or any other such household chores all require learning. All can be done well or poorly. All can be done cheerfully and with pride, or grudgingly and with distaste. And which way we do them is really up to us. It is a matter of choice. That is perhaps the greatest insight that the ancient **Stoics** championed for humanity. There are no menial jobs, only menial attitudes. And our attitudes are up to us.

Parents show their children how to enjoy doing the things that have to be done by working with them, by encouraging and appreciating their efforts, and by the witness of their own cheerful and **conscientious** example. And since the possibilities for happy and productive lives are largely opened up for youth by the quality and extent of their education, parents who work most effectively at providing their offspring with what it takes to lead flourishing lives take education very seriously.

Work is effort applied toward some end. The most satisfying work involves directing our efforts toward achieving ends that we ourselves **endorse** as worthy expressions of our talent and character. Volunteer service work, if it is genuinely voluntary and exercises our talents in providing needed service, is typically satisfying in this way. Youth needs experience of this kind of work. It is a good model for our working lives.

Questions on Content

1. Does all work bring a paycheck?
2. What is the opposite of work?
3. Explain Aristotle's definition of happiness.
4. What determines the individual's attitude toward doing things?
5. How do parents show their children how to enjoy doing things that have to be done?

Vocabulary

Define the following words as they are used in the commentary (the words are printed in **boldface** type):

1. work
2. flourish
3. compatible

4. Stoics
5. conscientious
6. endorse

Suggestions for Response and Reaction

1. Write sentence definitions (see pages 109–112) of "work" and "works." Then expand one of the terms to an extended definition (see pages 112–118).
2. Read the short story "Quality," by John Galsworthy, pages 521–525. Write a comparison (see pages 200–212) of "work" as William Bennett discusses the term and "quality" as Galsworthy deals with the term.
3. Explain why and how Bennett uses comparison and contrast (see pages 200–212).

The Active Document: Making Pages Smarter

DAVID D. WEINBERGER

Future documents will think and act for themselves—adding information, changing graphics, and even determining what readers may or may not see.

One of the strengths of paper pages is that, if the paper stock is chemically correct, they can last for hundreds of years unchanged. One of the weaknesses of paper pages is that, while our information needs vary from person to person, our pages don't change.

But imagine a page that changes depending on who is reading it, delivering the appropriate information to each person. Imagine working on a report that can correct your mistakes and send itself via computer to the people who are on a list to review it. Imagine a document that updates itself—text and graphics—as the information it is communicating is altered.

Such technology builds upon the current generation of electronic publishing software, which revolutionized information processing by enabling users to create and revise text and add photographs, illustrations, charts, tables, etc. Now we're seeing the beginning of a second revolution—in which computer software improves the quality of what *doesn't* show up on the page.

Pages generally display information in text and graphic form. But there's more to pages than what they show, including the earlier drafts, the identities of the people who approved them or have the authorization to modify them or see them, the sources of the information on the page, and what's supposed to happen next in the creation-revision-review-distribution process.

With "intelligent" software technology, pages will carry all this information with them. "Active documents" such as these will have the intelligence to access information, evaluate it, and act on it. In fact, documents can become so intelligent that they can correct their authors.

There Will Always Be Pages—But Not Necessarily Paper

The page is and will continue to be our most highly evolved way of giving and getting access to information. While in some situations we're seeing *paper*-less pages increasing in importance (the computer page is a primary example), *page*-less communication will remain relatively rare.

The secret of the success of the page-based model is that page-based documents are structured in visible and expected ways. For example, *The Futurist* is structured

Reprinted, with permission, from *The Futurist,* July–August 1991, published by the World Future Society, 7910 Woodmont Avenue, Suite 450, Bethesda, Maryland 20814.

into different sections. Each section has a title, text, subheads, and perhaps some bulleted lists. Readers navigate the magazine by means of these structures.

The structure of a page-based document is made visible by cues such as the variations in the typeface and sometimes the position on the page (for example, article titles are usually—but not always—at the top of the page).

Because page-based documents have a structure, they can provide formal navigational guides such as a table of contents, index, list of illustrations, etc. You can also turn the pages and easily skip by those you don't want to look at.

It is the versatility of page-based documents in showing information and giving us access to it that ensures its continuing **dominance**—whether on paper or not.

The Invisible Context

A page, by itself, lacks **context.** Computer technology in the future is going to help us navigate not just the page but its invisible context as well.

Think of what's printed on a page as a flat, two-dimensional surface. A page's invisible "depth" consists of information such as where the information came from, what the information is related to (e.g., paragraph "A" describes a summer course in the history department), the changes that have been made since the page was first prepared, by whom and when, and who is authorized to read it.

It can be extremely useful for a document to be able to carry this sort of information with it. But information in general is only useful insofar as it is the basis of action. The same is true for this contextual information.

The major breakthrough in page technology in the 1990s will be the ability of documents to carry embedded contextual information and—more importantly—to act on that information. Here are some examples of actions an intelligent document might take:

- Automatically get the latest version of information whenever the document is opened and update its contents.
- Automatically assemble versions of the document containing only the relevant information for various users.
- Show previous versions of the document and indicate who has made the changes.
- Present versions tailored to particular user-skill levels or authorized levels of access to information.
- Automatically send the document to the next reviewer and format it to that person's preference.

Active documents will do all of this, plus much more. There are four main ways in which active documents can help, all closely related. They can automate the accessing and transforming of information, deliver customized information to each user, ease the workflow, and provide tools tailored for individual needs.

Gathering and Transforming Information

The information on pages generally comes from other sources (except for fiction). Active documents can automate the process of gathering the information and transforming it into effective text and graphics.

For example, you may have to do a weekly report on regional sales. Your electronic publishing system knows this, so every Friday it automatically creates a new weekly report form. It automatically accesses databases around the country and gathers information about each of the regions. Since active documents can evaluate information as well as access it, the document might notice that northeastern regional sales are more than 20% above southeastern sales and that, while midwestern sales are down, sales per salesperson in that area are up.

Because active documents can act on that information, the document could automatically transform this information into charts and insert them into your weekly report, draw up a map with each region's sales figures represented by various shades of blue, and print out certificates of merit for each of the midwestern sales reps who is over quota. Exactly what actions are initiated by the events depends entirely on how you've set up your active documents.

This sort of capability exceeds standard electronic publishing capabilities in several ways. Active documents can go beyond including the contents of already-existing files to the gathering and generating of new data directly relevant to a particular user's needs. Active documents also can evaluate information and act on it. And they can be set to act without human intervention. For example, the publishing software can be instructed to create a new weekly report based on the day of the week. The report could be set up so that it would automatically query a database for more information based on the information it has already received. This further automates the information-gathering process—and gives active documents the appearance of autonomy.

581

David D.
Weinberger
The Active
Document:
Making Pages
Smarter

How Active Documents Work

Active documents are computer-based electronic documents that can be created and edited (text and graphics) using normal electronic publishing tools.

But under their electronic skin, they have programs attached to the various parts of the document. These programs can do anything that users can do by hand with publishing software (such as creating graphics, cutting and pasting contents, creating indexes, etc.), and they can also handle a range of capabilities normally handled by programming language (such as mathematics and other forms of data crunching).

The programs are activated by a wide variety of devices. For example, making a choice on a menu might cause a program to run. So might simply pointing at a graphic within the document with a computer mouse. Other devices that could activate a program include a certain amount of time passing, a document becoming more than 10 pages long, a user making more than five spelling errors, a file elsewhere on a computer network being updated, etc.

Because active documents use "object-oriented" programming, the programs that are attached to the document travel with the objects. So if you electronically cut a graphic from one document and paste it onto another, the graphic's intelligence goes with it. This means users can build libraries of programmed objects and then cut and paste them together to make new active documents.

—David D. Weinberger

Customizing Information

The problem these days clearly isn't getting information; it's limiting the information we receive to the information we need. An active document can change itself, depending on who is viewing it, so that each person who sees it gets precisely the information he or she needs.

For example, when I do my weekly report, the active document queries the corporate database for information relevant to my job. When I complete the document and send it electronically to my manager, she can have the document query the database for the information *she* needs.

In environments where security is important, active documents can hide or show various contents depending on the authorizations (or skill levels) of the various users. These levels can be set by having the document look up the user's name on a list, by having the user enter a password, or through any other method that makes sense.

Comments and **annotations** are another form of information that needs to be directed to particular people. Active documents can be set up to show or hide the appropriate comments. An active document can even be connected to a "voice

583

David D.
Weinberger
The Active
Document:
Making Pages
Smarter

annotation" system—much like a phone-mail message—so that when a particular spot in the document is reached a comment on that point can be heard as well as read.

As the flood of information we encounter grows, the ability to get only the information we need and to screen out the rest will become increasingly valuable. Active documents will help automate that process.

Documents are almost always "in process." Even after the last person has signed off on them and they've been cataloged and shelved, there will come a time when someone will want to reuse the information in it. That's why the document was cataloged and shelved in the first place. Active documents can help take themselves through the work process.

For example, suppose you're working on a corporate-policy bulletin that will be printed in two columns. But you like working in a single-column format. When you finish your draft, the document can be programmed to know that the next person who needs to see it is your editor, who likes to work in full "what-you-see-is-what-you-get" (WYSIWYG) mode. So the document sends itself to your editor, reformats itself to meet the editor's preferences, and even displays the comments of those who have already worked on the document.

Work patterns can be much more complex than that, and active documents can accommodate them. For example, you might have a document that knows not to allow itself to be printed until the company lawyer has approved it. And once that approval has been given, it might know not to let anyone make any alterations. A document might know to display the word "Draft" at its beginning until it has been approved by everyone on the review list. And once everyone has signed off on it, it might automatically distribute itself electronically around the company.

Active documents can do this because they can be programmed with the intelligence to know who needs to see them next and what information and format is appropriate at each stop along the way.

Tailored Tools

Authors in particular environments need specialized tools that may be of practically no help to the vast majority of users. Librarians need access to the details of the Dewey decimal system, research chemists need access to the periodic table of elements, and so forth. Reprogramming software to come up with job-specific versions remains very expensive, but active-document technology lowers the costs dramatically by allowing users to reprogram their documents rather than their software. The tools that people need for working on a particular document can be built into the document itself.

The active document will be useful not just for particular industries, but also for the documents that almost everyone uses. For example, the common expense report could do the math automatically, like a spreadsheet, or do currency conversions based on information from on-line services. A report could know how to do its own bibliography. An address list could figure out ZIP codes based on the city and street information. Active documents—rather than the software on which they run—could be instructed to do these sorts of things.

Active documents combine two of the most important computer technologies: information management systems, which give people access to the data they need, and electronic and desktop publishing systems, which allow people to display data effectively. Active documents will integrate the power of information management with the visual impact of computer-aided publishing systems.

One might say that a new page in communications is about to be written.

Questions on Content

1. Explain what the author means by a page's invisible depth.
2. Name three examples of actions an intelligent document might take.
3. List four main ways the active document will enhance readability and information.
4. Explain the statement: "Documents are almost always 'in process'."
5. What two computer technologies combine to make the active document possible?

Vocabulary

Define the following words as they are used in the article (the words are printed in **boldface** type):

1. dominance
2. context
3. annotations

Suggestions for Response and Reaction

1. For the boxed information, "How Active Documents Work," add headings to the second, third, and fourth paragraphs (see Explanation of a Process, pages 49–68).
2. Write a paragraph definition (see pages 112–118) of the active, or intelligent, document.
3. In a paragraph explain how active documents exceed the capabilities of standard electronic publishing capabilities.
4. Write an evaluative summary (see pages 233–236) of this article.
5. You are district sales manager for a sporting goods company. Write a letter to your supervisor to convince him or her (see pages 254–262) that converting to an active document capability would benefit the company.
6. Evaluate the use of headings (see pages 9–11 and 311–312) in the article.

The Power of Mistakes

STEPHEN NACHMANOVITCH

The resourceful person can turn mistakes into fortunate discoveries.

Do not fear mistakes. There are none.
MILES DAVIS

Poetry often enters through the window of irrelevance.
M. C. RICHARDS

We all know how pearls are made. When a grade of grit accidentally slips into an oyster's shell, the oyster **encysts** it, secreting more and more of a thick, smooth mucus that hardens in microscopic layer after layer over the foreign irritation until it becomes a perfectly smooth, round, hard, shiny thing of beauty. The oyster thereby transforms both the grit and itself into something new, transforming the intrusion of error or otherness into its system, completing the **gestalt** according to its own oyster nature.

If the oyster had hands, there would be no pearl. Because the oyster is forced to live with the irritation for an extended period of time, the pearl comes to be.

In school, in the workplace, in learning an art or sport, we are taught to fear, hide, or avoid mistakes. But mistakes are of incalculable value to us. There is first the value of mistakes as the raw material of learning. If we don't make mistakes, we are unlikely to make anything at all. Tom Watson, for many years the head of IBM, said, "Good judgment comes from experience. Experience comes from bad judgment." But more important, mistakes and accidents can be the irritating grains that become pearls; they present us with unforeseen opportunities; they are fresh sources of inspiration in and of themselves. We come to regard our obstacles as ornaments, as opportunities to be exploited and explored.

Seeing and using the power of mistakes does not mean that anything goes. Practice is rooted in self-correction and refinement, working toward clearer and more reliable technique. But when a mistake occurs we can treat it either as an invaluable piece of data about our technique or as a grain of sand around which we can make a pearl.

Freud illuminated the fascinating way in which slips of the tongue reveal unconscious material. The unconscious is the very bread and butter of the artist, so mistakes and slips of all kinds are to be treasured as priceless information from beyond and within.

As our craft and life develop toward greater clarity and deeper individuation, we begin to have an eye for spotting these essential accidents. We can use the mistakes we make, the accidents of fate, and even weaknesses in our own makeup that can be turned to advantage.

Often the process of our artwork is thrown onto a new track by the inherent balkiness of the world. Murphy's law states that if anything can go wrong, it will. Performers experience this daily and hourly. When dealing with instruments, tape records, projectors, computers, sound systems, and theater lights, there are inevitable breakdowns before a performance. A performer can become sick. A valued assistant can quite at the last minute, or lose his girlfriend and become mentally incapacitated. Often it is these very accidents that give rise to the most ingenious solutions, and sometimes to off-the-cuff creativity of the highest order.

Equipment breaks down, it is Sunday night, the stores are all closed, and the audience is arriving in an hour. You are forced to a little *bricolage,* improvising some new and crazy contraption. Then you attain some of your best moments. Ordinary objects or trash suddenly become valuable working materials, and your perceptions of what you need and what you don't need radically shift. Among the things I love so much about performing are those totally unforeseen, impossible calamities. In life, as in a **Zen koan,** we create by shifting our perspective to the point at which interruptions are the answer. The redirection of attention involved in incorporating the accident into the flow of our work frees us to see the interruption fresh, and find the alchemical gold in it.

Once I was preparing for a full evening poetry performance, with multiscreen slide projections and electronic music I had composed on tape for the occasion. But in the course of overrehearsing during the preceding week, I managed to give myself a case of laryngitis, and woke up the morning of the performance with a ruined voice and a high fever. I was ready to cancel, but in the end decided that would be no fun. Instead I dropped my attachment to my music and preempted the sound system for use as a P.A. I saw in an old wicker wheelchair and croaked into a microphone. My soft, spooky, obsessive, guttural voice, amplified, became an instrument of qualities that totally surprised me, releasing me to find a hitherto unsuspected depth in my own poetic line.

A "mistake" on the violin: I have been playing some pattern; 1, 2, 3, 6; 1, 2, 3, 6. Suddenly I make a slip and play 1, 2, 3, 7, 6. It doesn't matter to me at the time whether I've broken a rule or not; what matters is what I do in the next tenth of a second. I can adopt the traditional attitude, treating what I have done as a mistake: don't do it again, hope it doesn't happen again, and in the meantime, feel guilty. Or I can repeat it, amplify it, develop it further until it becomes a new pattern. Or beyond that I can drop neither the old pattern or the new one but discover the unforseen context that includes both of them.

An "accident" on the violin: I am playing outdoors at night, in misty hills. Romantic? Yes. But also humid. The cold and the humidity take all the poop out of the bottom string, which suddenly slackens and goes out of tune. Out of tune with what? Out of tune with my preconceived benchmark of "in tune." Again I can take the same three approaches. I can tune it back up and pretend that nothing happened. This is what politicians call "toughing it out." I can play the flabby string as it, finding the new harmonies and textures it contains. A low, thick string, when it goes flabby, not only becomes lower in pitch, but because of the flabbiness will give to the bow's weight much more easily and will produce (if lightly touched) more breathy and resonant tones than a normal string. I can have a lot of fun down there in the viola's tonal sub-basement. Or I can detune it even further, until it comes into some new and interesting harmonic relation with the other strings (*scordatura*, a technique the old Italian violinists were fond of). Now I have, instantly, a brand-new instrument with a new and different sonic shape.

An "accident" in computer graphics: I am playing with a paint program, which enables me to create visual art on the screen and then store it on disk as data that can be called up later. I intend to call up the art I was working on yesterday, but I hit the wrong key and call up the zip code index of my mailing list. The thousands of zip codes, transformed into a single glowing screen of abstract color and pattern, turn out as a startling and beautiful scene of other-worldly microscopic life. From this **serendipitous** blunder evolves a technique that I use to create dozens of new artworks.

The history of science, as we well know, is liberally peppered with stories of essential discoveries seed by mistakes and accidents: Fleming's discovery of penicillin, thanks to the dust-borne mold that contaminated his petri dish; Roentgen's discovery of X rays, thanks to the careless handling of a photographic plate. Time after time, the quirks and mishaps that one might be tempted to reject as "bad data" are often the best. Many spiritual traditions point up the vitality we gain by reseeing the value of what we may have rejected as insignificant: "The stone which the builders refused," sing the Psalms of David, "has come to be the cornerstone."

The power of mistakes enables us to reframe creative blocks and turn them around. Sometimes the very skin of omission or commission for which we've been kicking ourselves may be the seed of our best work. (In Christianity they speak of this realization as *felix culpa,* the fortunate fall.) The troublesome parts of our work, the parts that are most baffling and frustrating, are in fact the growing edges. We see these opportunities the instant we drop our preconceptions and our self-importance.

Life throws at us innumerable irritations that can be mobilized for pearl making, including all the irritating people who come our way. Occasionally we are stuck with a petty tyrant who makes our life hell. Sometimes these situations, while miserable at the time, cause us to sharpen, focus, and mobilize our inner resources in the most surprising ways. We become, then, no longer victims of circumstance, but able to use circumstance as the vehicle of creativity. This is the well-known principle of **jujitsu,** taking your opponent's blows and using their own energy to deflect them to your advantage. When you fall, you raise yourself by pushing against the spot where you feel.

The Vietnamese Buddhist poet-priest, Thich Nhat Hanh, devised an interesting telephone meditation. The sound of the telephone ringing, and our semiautomatic instinct to jump up and answer it, seem the very opposite of meditation. Ring and reaction bring out the essence of the choppy, nervous character of the way time is lived in our world. He says the first ring as a reminder, in the midst of whatever you were doing, of mindfulness, a reminder of breath, and of your own center. Use the second and third rings to breathe and smile. If the caller wants to talk, he or she will wait for the fourth ring, and you will be ready. What Thich Nhat Hanh is saying here is that mindfulness, practice, and poetry in life are not to be reserved for a time and place where everything is perfect; we can use the very instrument of society's nervous pressures on us to relieve the pressure. Even under the sound of helicopters—and this is a man who buried many children in Vietnam to the roar of helicopters and bombs—he can say, "Listen, listen; this sound brings me back to my true self."

Questions on Content

1. How is the making of a pearl like a mistake or accident?
2. List at least three examples of how the author turned a mistake or accident into a fortunate discovery.
3. What is Thich Nhat Hanh's telephone meditation?

Vocabulary

Define the following words as they are used in the article (the words are printed in **boldface** type):

1. encysts
2. gestalt
3. Zen koan

4. serendipitous
5. jujitsu

Suggestions for Response and Reaction

1. Find three examples of terms that are used and then defined in the article. Why do you think the author used these terms and then defined them rather than using more common terms that would not need defining?
2. Write a personal essay defining a *mistake* as a possible fortunate discovery, using personal experiences as examples.
3. In what ways are effect-cause analyses (see pages 177–183) used in the article? What is the relationship between an effect-cause analysis and turning a mistake or accident into an unforeseen opportunity?

Self-Motivation: Ten Techniques for Activating or Freeing the Spirit

RICHARD L. WEAVER II

A university professor explains to students that self-motivation is most likely to occur when they can successfully deal with the stresses in their life.

Does it feel like you're being torn in all directions? Like you're getting stressed out? Like every teacher thinks his or her class is the *only* one you're taking? Like everything is coming down on you all at once, and you're not sure you can, or even want to, withstand the pressure? Do things feel like they are out of control? At least you know you're normal! Self-motivation is most likely to occur when you can successfully deal with the stresses in your life. Stress depletes both energy and motivation.

Professor Hans Selye, the world's foremost authority on human stress, defines stress as "the body's nonspecific response to any demand placed on it, pleasant or not." We often recognize these nonspecific responses in a rapid pulse, increased blood pressure, frequent illness, unusual susceptibility to infection. Sometimes it appears as brooding, fuming, shouting, or even in increased use of alcohol, tobacco, or drugs. In some it's as severe as ulcers or a heart attack; in others, like one of my daughters (who takes after her dad in this) it's fingernail biting.

Often, what you need in times of stress is an electrical charge to boost your spirits, reinvigorate your emotions, and to reestablish balance in your life! The techniques that activate or free the spirit are precisely those techniques that you can use to eliminate—or at least reduce—the stress in your life. With the stress reduced, you are far more likely to be self-motivated—reinvigorated or charged up! At least you can regain some control.

What you need to know first is that stress is not always negative. Good stress occurs in life situations toward which you feel positively. I know, from my own personal experience, for example, that I chose this kind of work—being a professor—because the high-pressure schedule it demands seemed to be right for me. I *chose* to live under a fair degree of constant stress.

What you need to know second is that stress is individually determined. Meaning lies in us; thus, we are the ones who determine what is stressful. The useful part of this understanding is that if *we* determine it, then *we* can control it. We determine meaning! We are in control!

We all know *why* we stay in stressful situations. We stay in them for the rewards they offer, for approval, money, status, or self-esteem. Because we so often find ourselves in these stressful situations, what we need to do is to decide at various points in our life—*this* being one of them—are the rewards keeping ahead of the costs? Can we see a light at the end of the tunnel that is still worth pursuing?

Reprinted by permission from *Vital Speeches of the Day* (1 Aug 1991) 57: pp. 620–23. Headings have been added.

Often, the rewards or the goals are precisely what keeps us going through life's little crises. Sometimes, of course, life's little crises look like gigantic **catastrophes!** But do you know what? If we didn't have any crises, what would we talk about? In many cases our crises, along with the weather, are the sum and substance of our conversations with others!

But we *can* develop strategies—or use specific techniques—to cope with stressful situations and, thus, re-kindle the spark that ignites self-motivation. These techniques *can* alter the physiological impact that stress has on us, and they can reduce the harm that we do to ourselves and others in these situations.

What I want to talk about are some techniques for activating or freeing the spirit, but I want you to know that there is an underlying theme to my remarks. Self-motivation does not come from outside of us. It can *only* come from inside. *Change must begin from within.* The key, then, to improvement and change is in recognizing and working with our individuality. What works for one person may not work for another. Also, for some people, it may involve the use of a number of techniques; for others just one may work. Some may have stress under control in ways not mentioned here at all. I will mention ten techniques.

1. Recognize your stress signals.

The first technique is to *recognize your own stress signals.* Stress can creep up on you insidiously—sometimes as the result of prolonged anxiety or even as a result of multiple causes. What if you're having troubles at home, then you're having some relationship problems, and maybe some small problems with a roommate and *then* a couple of exams on top of that? Can you feel the pressure begin to mount?

Stress has its signals, but they may not always be as clear as you would wish. Practice monitoring yourself for the signs of unhealthy pressure. Common signs include irritability, sleeplessness (even when sleepy), rapid weight loss or gain, increased smoking or drinking, as mentioned before, little errors (physical or mental "dumb mistakes"), physical tension, nervous tics, or tightness of breath.

Last fall a student in one of my classes came in to talk to me about three weeks after the semester began. She recognized a number of these symptoms, and because she had them before—to the extent of having to see a doctor—she realized that even though she knew she would enjoy the class, that she would get a great deal out of it, and that she *could* handle the workload, that it would be best for her, considering these early symptoms, to drop the class. And she did—*before* they became severe. We have to be wise in reading the signals, and she withdrew when they appeared—especially because she could easily anticipate the signals continuing and increasing in intensity. In this way, she did not exceed her normal, individual stress endurance. She was simply protecting herself!

2. Engage in exercise and practice good nutrition.

The second thing we need to do—and this is especially important for students—*is to engage in exercise and practice good nutrition.* The two, of course, are related. I often get asked what keeps *me* going. This second technique takes on major importance in my life. Normally, my alarm goes off at 5:30 or 6:00 A.M. I get up and run through about 20 minutes of exercises. Then I run 3 miles, it takes me just under 30 minutes. And that starts my day! When I deliver a 9:30 A.M. lecture, I have already been up for four hours!

Don't underestimate the importance of *exercise* to self-motivation. What happens to students is that many of them forget about exercise. Exercising the body keeps the mind sharp!

It used to be that I hated jogging; now, I look forward to it as an opportunity to plan my day, create ideas, and do some serious reflective thinking. You should know that exercise results in release of **endorphins** that have to do with confidence and self-esteem. It can make a *major* difference in your daily life. With more confidence and self-esteem, you will be more motivated.

Recently, I met an 80-year-old man who looked like 60. He told me his alarm goes off at 5. He gets up, exercises, rides a stationary bike, gets on a treadmill, uses a stairmaster, and then swims ten lengths of a regulation-size pool. He said that he doesn't drink or smoke, and he feels like fifty! That's *my* goal! We all have to get old, but *how* we decide to approach getting old makes all the difference. It's not the destination; it's the ride!

And don't forget the *nutrition.* Many students skip meals or only eat when they are hungry—and then, not very well. We all have different patterns; but I can tell you this. If you stay away from junk food, eat regular meals of high quality, and make certain fresh fruits, fresh vegetables, and grains are well represented in your daily diet, you are far less likely to experience severe stress!—and you will be more motivated as well! Good nutrition can play a major role in motivation.

Daily attention to health care can significantly decrease the effect of stress on both body and mind. Many of my campus meetings occur in the cafeteria. I watch students coming in for their first meal of the day and making a cheeseburger, fries, and a Coke their breakfast. This is horrendous! Breakfast is one of the most important meals of the day.

3. Locate the sources of dangerous stress.
There is a third technique, too. If you are a person who stays unmotivated because of high levels of stress in your life, try to *locate the sources of dangerous stress in your life.* Take a look, first, at your personal relationships with family, employer or employees, friends, roommates, lovers, teachers, and strangers. Often, these forces will accumulate and create a negative stress reaction.

Second, take a look at your own personality. Are you making unreasonable demands on yourself? Do you, for example, insist on getting all "A's"? One of my friends who is in seminary became stressed-out because of this demand on himself. When his wife pointed it out as unreasonable, and when he was able to see it as excessive and unnecessary, he was able to relax and better enjoy his classes.

There are other unreasonable demands you make on yourself. Do you like *never* making a mistake? Do you like never failing? You must learn to accept the possibility of failure without losing your sense of self-worth. Failing will not hurt your health; feeling like a failure will! Reinforce your sense of accomplishment, even in the face of failure, by recalling past achievements. Also, you can talk yourself through failures by:

—viewing failure as part of the process of exploring your world

—accepting failure as a normal part of living

—making a note of the lessons to be learned and then moving on

There are other unreasonable demands we make on ourselves, too. Do you like never being late? Do you like never being in the wrong? Often, you can minimize the stress of frustration and discover renewed enthusiasm and motivation by learning when to stop something, when to acknowledge that the aim is no longer worth fighting for, when to accept your own limitations, or when you are just plain being unreasonable with yourself! That's why constant self-monitoring is healthy. That's some of what I do when I jog. It gives me time to make some of these self-assessments.

Remember what we're doing here. We're trying to locate the sources of dangerous stress in our lives. A third way we have to locate dangerous stress, in addition to looking at our relationships and looking at our own personality, is to look at our personal problems. For example, are you a person who carries grudges? Are you a person who dwells on past unpleasant experiences and incidents? What you may not realize is that carrying a grudge can deplete your naturally limited energy. For your own sake, and for the sake of finding renewed motivation, end unpleasant relationships, and don't wallow in feelings of hostility, antagonism, or revenge. These deplete and destroy motivation.

What a lot of people don't realize is that worry is energy. When we worry, we *lose* energy. So when we stop worrying, we will have *increased* energy. I know this is easier said than done; however, if you recognize yourself involved in excessive worry, and you know this has a direct effect on your energy level—it wears you out!—then you can begin to take specific steps to reduce the worry. Awareness is the first step to control!

4. Turn energy into plans and action.

What you need to do, and this is the fourth technique, now, for activating or freeing the spirit, is to turn energy into *plans and action*. Think of stress as a kind of readiness in the extreme. It is an activation without release. So, when we take on the task at hand, however complex or frightening it might seem as we begin it—planning and then doing the necessary work greatly reduces stress. Not only will it get the work done, but it brings about relief and satisfaction as well.

Procrastination robs you of time, power, and freedom! And the consequences can be extreme: tension, conflict, self-criticism and **deprecation,** embarrassment, anxiety, guilt, panic, depression, physical exhaustion, even physical illness! This is *not* a speech on procrastination. But if this is *the* problem in your life that creates stress and robs you of motivation, let me tell you there are many good books on the subject. The response of most procrastinators to such a suggestion is, "Great. I'll read them later!"

Instead of believing that things are out of control and that you're helpless, you have to start by convincing yourself that you are *in* control and the day belongs to you. When you are *in* control, you are writing your own script, and you are being proactive. But being proactive requires plans and actions.

Let me give you some quick time-management techniques here as part of plans and actions—the fourth technique for reducing stress.

—Do your most demanding work when you're fresh. Since reading is often tough (hard to stay awake through), them make sure you do it when you're the freshest.

—Make double use of waiting time, driving time, walking across campus time, or time when you're doing routine tasks: read, write letters, listen to lectures that you've recorded, listen to notes and outlines you've recorded, make schedules, do **isometric** exercises.

—Get up a half-hour earlier to gain some especially productive time.

—Go to bed earlier. Your lack of sleep, alone, robs you of quality time. You can't be self-motivated when you're half asleep!

—Control interruptions. Don't take phone calls when you're in the midst of important studying. Return calls later during your down time. Don't answer the door. Don't go wandering down the hall looking for someone to talk to.

—Organize your work area so you know where everything is, where everything belongs, and keep things in their place.

—Only deal with things once. Do it now! When you procrastinate, you will have this gnawing feeling of being fatigued and always behind. Procrastinating does *NOT* save your time or energy; it drains away *both* time and energy and leaves you with self-doubt and self-delusion. Memorize and repeat this motto: Action TNT—Action Today, Not Tomorrow! We need these little messages to motivate us.

—Finish what you start. Concentrate all your energy and intensity on the successful completion of your current major project. Completion builds energy and confidence.

—Limit your television viewing to mostly enlightening, educational, or special shows. STS—Stop the Soaps!

5. Live in your own right way.

A fifth technique for activating the spirit is to *live in your own right way.* Some of us need speed and intensity. I, for example, would probably wither away and die under the limited stress of a quiet job or an isolated, do-nothing vacation! For me, because I am accustomed to high levels of activity, there is more stress in suddenly slowing down, than in maintaining my usual frantic pace.

But no matter which way is your own right way, we all need *variety.* Variety offers diversions. Moving from one activity to another is more relaxing than complete rest. As a mattter of fact, frustration often will follow swiftly when stimulation and challenge are removed from our lives.

6. Adjust to unchangeable situations.

The sixth technique is to *try to make whatever adjustments you need to make to unchangeable situations.* In other words, accept the givens; focus on your options. Although it makes sense to fight for your highest attainable goal, it does not make sense to waste your limited energy by resisting or fighting the inevitable. You don't have control over most of a course's assignments, requirements, or expectations just like you don't have control over rush-hour traffic! So face it!

But in *every* situation, you have options. That is, you have opportunities to bring to bear your own unique personality. You have ways to be creative. You have room to offer your own insights. No matter the controls and regulations, no matter the rules and restrictions, the creative person will find ways to be creative.

593

Richard L.
Weaver II
Self-Motivation:
Ten Techniques
for Activating or
Freeing the Spirit

Rather than focusing on the givens, accept those. Focus on the options. Find ways to express the real you! Find the places where you have freedom.

7. Set your own goals.

The seventh technique is to *start setting your own goals*. There is almost unbearable stress—and the resulting lack of motivation—when you try to live out the goals set for you by others. Friends, parents, relationship partners, advisers, and counselors all think they have some stock in *your* life. There is no harm in taking advice, but set your own goals. Over and over, I have witnessed renewed invigoration, a refreshed view of life and classes, and a rejuvenated spirit when students suddenly discover a major they really like and want to pursue. The major is now theirs; they have chosen it; they want to pursue it themselves—for themselves—no one else. Suddenly, they're in control!

And make your goals clear, specific, and achievable. I, personally, like doing things. But I like getting things done, too. Accomplishing, producing, and completing are so important, that I am continually setting new goals—but, believe me, they are ones that I can accomplish.

8. Take direct action when needed.

The eighth technique is to *take direct action when called for*. Doing nothing is no substitute for doing what you feel you must. Let me explain. If you are confronted by a stressful situation that you can change, take direct action to eliminate the source of stress—even if it seems harder to do than accepting things as they are. There is no doubt that it is often hard to come right out and deal with a problem:

—like the choice of a major that isn't working out

—a roommate who never picks anything up

—a relationship partner who has become smothering

—a friend who is demanding too much of your time and energy

But when you weigh the problem against the stress it creates, it is even harder to live with the unrelieved stress!

9. Practice relaxation techniques.

The ninth technique is to learn and regularly practice *relaxation techniques*. When you get stressed out sometime, try sitting there for ten or fifteen minutes and just concentrate on your breathing. Notice each breath as it comes in and goes out. It might even help if you count one as you exhale, then as you exhale again, count two, and so on until four. Then start counting all over again. This helps to focus your attention and regularizes your breathing. This has both a relaxing and kind of a cleansing effect.

Another common relaxation technique is to lie on the floor and progressively tense and relax all the muscles in your body. You can start with your feet and gradually work up to your jaw and eye muscles. Create a pleasant, relaxing scene in your imagination. Make it as vivid and detailed as possible. Visualize yourself in the scene, completely relaxed. If you can turn that scene on in your mind several times a day, especially when you feel tense, it will have a calming effect and give you renewed motivation.

595

Richard L.
Weaver II
Self-Motivation:
Ten Techniques
for Activating or
Freeing the Spirit

10. Find a good support group.

The tenth technique for reducing stress and increasing self-motivation is to find yourself a good support group. When you feel necessary to others, when you have earned the goodwill and support of others, when you feel personal satisfaction in helping others—others offer a stabilizing effect that keeps us in touch, needed, and normal. Friends, family, relationship partners, classmates, fraternity brothers or sorority sisters *can* serve as those who keep your life worthwhile.

Several things, now, should be clear to you. First, self-motivation is directly related to stress-reduction. Second, stress is not always negative. Third, we are the makers of our own stress; thus, we *can* control it. Fourth, there are some useful, workable techniques for controlling it. We can create a great deal of distress for ourselves and others through our reluctance to make some all-out commitment to reducing stress and thus increasing our self-motivation. The real heroes are those who doggedly and joyfully work at becoming more loving and self-actualizing on a daily basis. There are no easy triumphs. Self-motivation is most likely to occur when we can successfully deal with the stresses in our lives. Your personal advance hinges on effort—your willingness to put into practice techniques for activating or freeing the spirit.

Questions on Content

1. What are the ten techniques that the speaker gives for activating or freeing the spirit?
2. How can stress be good for a person?
3. What are examples of unreasonable demands that we make on ourselves?
4. What are the quick time management techniques that the speaker gives?

Vocabulary

Define the following words as they are used in the speech (the words are printed in **boldface** type):

1. catastrophe
2. endorphins
3. deprecation
4. isometric

Suggestions for Response and Reaction

1. What can you assume about the content of the speech from reading only the title and the subtitle?
2. In what ways is this speech appropriate for Weaver's audience, college students? To what extent is it appropriate for others?

3. What does the speaker mean when he says that "we are the ones who determine what is stressful"? In what ways do you agree with the speaker?

4. The speaker gives instructions (see pages 27–44) for bringing stress under control. Choose three of the techniques that you think you need to apply to your life, and explain specifically how you can thus achieve a less stressful life.

5. Analyze the content and the organization of the speech by applying, as appropriate, the steps for preparing an oral presentation, pages 441–444.

Part III
Revising and Editing

Part III of *Technical English: Writing, Reading, and Speaking* is a guide to revising and editing. This part—integral to the process of writing—is a digest, a summary, a revisiting of the major principles presented in Part I Forms of Communication. Part III shows you and your peers how to check for an adequate Plan Sheet; how to determine where you need to add information and details to a draft; how to check whether you have sufficiently considered audience, purpose, organization of material, and other rhetorical matters; and how to review and evaluate a paper.

Further, this guide to revising and editing presents conventional standards to help you edit your paper for accuracy and clarity in language usage and in mechanics.

Section 1
The Entire Paper

Introduction

The writing process in *Technical English: Writing, Reading, and Speaking* is based on a three-part approach:

- Plan Sheet fill-in
- Drafting
- Revising

with peer review in each stage.

No two persons approach the process of writing with the same techniques and strategies; in fact one individual may approach one writing project, such as a set of instructions or a resumé, differently from another writing project, such as a proposal or a feasibility report.

Good writing—writing that gets the job done effectively—requires time, thought, and work. Contrary to an unspoken misconception, there is no such thing as waving a magic pen or touching magic keys so that the finished paper appears effortlessly, all at once. Rather, for most people—professional writers and students alike—writing is a struggle. But that struggle becomes manageable through taking the time to let each stage build upon the previous stages.

Lay aside notions that successful writers move directly from blank sheet of paper to finished product. Writing is a complex process of moving back and forth with ideas, reworking the text, and developing the organization.

Part III Revising and Editing: The Entire Paper will help you look back to the material in Part I Forms of Communication.

Plan Sheet

The Plan Sheet provides prompts with fill-in space for getting started.

Filling in the Plan Sheet may take more time than writing a preliminary draft or preparing the "final" paper, but preparing the Plan Sheet fully is time well spent. The Plan Sheet requires you to think through the concepts of audience and purpose; the Plan Sheet helps you to outline (the substance, the body) the paper—with attention to visuals to explain and support major points.

For examples of filled-in Plan Sheets, look back to pages 38–39, 52–53, 61–62, 79–80, 87–88, 113–114, 141–142, 164–165, 185–186, 204–205, 231–232, 322–323, 338–339, 346–347, and 355–356.

Drafting

From Plan Sheet to draft moves from perspective and data to information blocks and paragraphs and the development of visuals. Figure 1 is a draft of "How to Sharpen a Ruling Pen," based on the Plan Sheet on pages 38–39. This draft incorporates notations for revising.

Sharpening a Ruling Pen

A major instrument in making a mechanical drawing is the ruling pen. *(which)* It is used to ink in straight lines and noncircular curves. *(with a T square, a triangle, a curve, or a straightedge)* *(as a guide)* The shape *(blades resembling tweezers)* of the nibs is the most importance aspect of the pen. The nibs must be rounded (elliptical) to *(create)* ~~creat~~ an ink space between the nibs. *(to assure neat, clean lines,)* The ruling pen must be kept sharp and in good condition. It must be sharpened from time to time after extensive use because the nibs wear down.

Add heading: Equipment and materials *o List with visual marker.*

~~To sharpen the ruling pen, you need a~~ 3- or 4-inch sharpening stone (preferably a hard Arkansas knife piece that has been soaked in oil) and a crocus cloth (optional). Sharpening a ruling pen involves *(seven)* ~~six~~ steps: *Add heading: Overview of procedure* close the nibs, hold the sharpening stone in the left hand, hold the ruling pen in the right hand, round the nibs, sharpen the nibs, polish the nibs (optional), and test the pen.

Add heading: Steps

1. Close the nibs. Turn the *(adjustment)* screw, located above the nibs, to the right until the nibs touch. The nibs can *(then)* be sharpened to exactly the same shape and the same *(length)*.

2. Hold the sharpening stone in the left hand in a usable position. With the stone lying across the palm of the left hand, grasp it with the thmb and fingers to give the best control of the stone.

3. Hold the ruling pen in the right hand. Pick up the ruling pen with the thumb and index finger of the right hand as if it were a drawing crayon. The other three fingers should rest lightly on the pen handle. CAUTION: *Place CAUTION in a box.* failure to hold the sharpening stone and ruling pen correctly may result in an injured hand or a ruined ruling pen. *(See Fig.)* *Show hand holding pen correctly.*

4. Round the nibs. To round (actually to make elliptical) the nibs, stroke the pen back and forth on the stone, starting with the pen at a 30-degree angle to the stone and following through to past a 90-degree angle as the line across the stone *(moves)* forward. Usually four to six strokes are needed.

Be sure that both nibs are the same length and shape so that the tips of both nibs will touch the paper as the pen is used. *(See Fig 2.)* *Show correct shape.* ~~When the nibs are satisfactorily rounded, they will be left dull;~~

5. Sharpen the nibs. Open the blades slightly by turning the adjustment screw gently to the left. Sharpen *(only the outside of)* each blade, one at a time.

To sharpen, hold the stone and the ruling pen in the hands as in rounding the nibs; *(see step 3)* the pen should be at a slight angle with the stone. Rub the pen back and forth *(six to eight times)* with a pendulum motion to restore the original shape.

Make this step 6. If desired, ~~the nibs may be~~ polished *(the nibs)* with a crocus cloth *(by rubbing the cloth lightly over)* to remove *(the nibs)* any rough places. ~~It is a good idea to~~ *Make this step 7* Test the ruling pen after sharpening it. If the *(pen is properly sharpened)* ~~job has been done well,~~ the pen ~~is capable of~~ *(will draw sharp)* ~~making clean sharp~~ *(clean)* lines.

Add a small drop of ink between the nibs. Draw a line along the edge of a T square placed in a piece of paper.

Figure 1. A draft of "How to Sharpen a Ruling Pen," pages 40–41.

Revising

For many writers the revising stage is somewhat tedious, for it requires a close application of analytical skills. Nevertheless it is usually this stage that makes the difference in producing a good paper.

Peer review can be very helpful in the process of revising. Reading the papers of colleagues and their reading yours can prompt suggestions for refining the wording, for further development of the topic, for appropriate visuals, for using standard grammar, and so on. Figure 2 shows an example of a peer reviewer's comments on a draft of how to administer an intramuscular injection.

Overall this is a good paper. I am a phlebotonist (person who draws blood for testing purposes).

1. It would help to make a cushion with thumb and forefinger on the area of skin where shot is to be given.
2. Under Equipment & Materials add: medication to be injected.
3. The definition of I.M. injection is clear.
4. The point about knowing the procedure thoroughly was good, so one wouldn't make a mistake.
5. The caution about accidentally sticking yourself with the needle is important. People should be careful when handling needles.
6. Step 2 could add that the nurse should check to see if there are any air bubbles in the syringe.
7. Good visuals.
8. Check spelling in Step 1.
9. Rethink the closing statement—it doesn't fit.

Figure 2. Peer reviewer's comments (how to administer an intramuscular injection)

The Questions for Reviewing on the front endpaper and the Checklist for Revising on the inside back cover point out specific items to consider when revising a paper.

Convenience of a Word Processor

Using a word processor to compose documents offers many advantages.

- You can insert words, phrases, and sentences at any time and at any point in a document.
- You can change or revise any part of the document at any time.
- You can change punctuation or correct grammatical errors.
- You can vary the size of type and select from a choice of fonts and stylistic enhancements such as bold, italics, and underlining. (A word of caution. Think through the visual effect you want to achieve. Do not mix sizes or fonts or choose stylistic enhancements without good reason.)

- You can easily alter spacing and margins. Spacing between lines can vary, such as single or double; material can be centered; margins can be left-justified, right-justified, or right- and left-justified. Text can be arranged in columns.
- You can use software programs that provide grammar, spelling, and stylistic checks.
- You can prepare graphics and other visuals and insert them into a document.
- You can save a document and later update, revise, and edit it.
- You can, through computer networks and on-line services, write, revise, and edit collaborative documents.

Audience and Purpose

All answers to questions about how to write, what to write, what format to use, what page layout and document design elements to include, and so on, relate to two primary considerations: audience and purpose.

Audience

The audience is the reader, the viewer, or the hearer. For the technical writer, the intended audience, that is, the individual or the group for whom a presentation is prepared, influences decisions the writer makes in preparing the presentation. The audience governs decisions such as the details to include, the length of the presentation, the word choice, the sentence structure, the selection of visuals and headings, and the format.

Audiences may be classified at two extremes on a continuum as lay audiences and expert audiences. Lay audiences typically do not know a great deal about the subject. The population at large is a lay (general) audience. The expert audience, on the other hand, is an audience who is advanced in a field, who understands technical information within the field, and who handles theory and practical application with ease. Audiences between the lay audience and the expert audience can be classified as professionals (nonexperts) who are college-level educated and technicians who understand technical information within their field and who handle practical application with ease.

If you as a writer can classify your audience into one of the categories named above, you can better anticipate how well informed the audience is about the subject, how interested the audience is in the presentation, how receptive, how sophisticated the audience is (the extent of vocabulary, the length and complexity of sentences the audience can follow), and how you and the audience are related (peer to peer, expert to novice, supervisor to employee).

Before working on a presentation, find out as much as you can about the intended audience. If you learn more about the audience as the presentation evolves, adjust the presentation accordingly.

Put yourself in place of the audience for the presentation: Could you understand the material as it is presented? How would you react and respond to it?

You want your audience to respond positively to your presentation.

Purpose

The purpose of a presentation is the reason for the presentation. Why are you writing or speaking? What do you the presenter, expect or require from the intended audience?

Typically, the purpose is to inform, to persuade, to initiate action, or to build good will. In communicating instructions and process explanations, descriptions of mechanisms, definitions, analyses through classification and partition, analyses through effect and cause, comparison and contrast, summaries, and reports (Chapters 1–6 and 8 in Part I), the writer's goal is to inform the audience. In analyses through effect and cause, through comparison and contrast, and in reports, the writer's goal may be persuading as well as informing the audience. See also Chapter 10 Oral Communication.

Organization

The organization of a document or a presentation is the way the writer or speaker arranges the content.

When responsible for preparing a presentation, whether written or oral, you first identify the audience and determine the purpose. Who will read or hear the presentation? What background information does the reader or hearer have? What is the response desired from the reader, viewer, or hearer? What action do you expect or require? See pages 3–4, 28–32, 49–56, 75–76, 106–107, 118–120, 135, 160, 184, and 218–219 for further discussion of audience and purpose.

With the needs of the audience and the purpose of the presentation guiding each step, you gather data; you analyze the data—you sort through the data, discarding some perhaps and adding other data; you organize the data; and you prepare the presentation.

As you organize the data, you may find it helpful to think of the presentation as three major blocks of material: the block of introductory material, the block of support material (body), and the block of closing and concluding material. While dividing a presentation into three major blocks of material—introduction, body, and closing—seems simplistic, such a division guides the beginning writer in organizing material into a clear, coherent presentation. In each of the first five chapters of this textbook, you will find a procedure section.

Chapter 1   Procedure for Giving Instructions, pages 42–43
            Procedure for Explaining a Process, pages 67–68
Chapter 2   Procedure for Describing a Representative Model of a Mechanism, pages 97–98
            Procedure for Describing a Specific Model of a Mechanism, page 98
Chapter 3   Procedure for Giving an Extended Definition, page 122
Chapter 4   Procedure for Giving an Analysis Through Classification, page 149
            Procedure for Giving an Analysis Through Partition, page 171
Chapter 5   Procedure for Giving an Analysis Through Effect and Cause, pages 193–194
            Procedure for Giving an Analysis Through Comparison and Contrast, page 212

In each of these procedure sections, an outline with roman numerals I, II, and III divides the organization of material into three major sections; roman numeral I suggests ways to introduce the material, roman numeral II suggests ways to develop the material, and roman numeral III suggests ways to close or end the presentation.

Introduction

The introduction of a written or oral presentation is important because it is what the audience sees or hears first. The introduction serves an obvious purpose: to let the audience know what the presentation is about and to forecast the content. In other words, the introduction clarifies the topic and the focus.

In many kinds of writing, a very brief opening suggests the focus of the presentation. A letter might begin with the following sentence:

> You will be pleased to know that our company has made the merchandise exchanges that you requested.

A crime lab report might begin as follows:

> The Crime Lab has studied carefully all of the evidence found at the crime scene. The lab analysis indicates . . .

The introductory sentence or the last sentence of the introductory section is typically a sentence that gets the audience "into" the body of the document or presentation. The purpose of the sentence is to guide the audience to what comes next. Functional labels for such sentences are lead-in sentence, key sentence, focus sentence, or outline sentence.

The Lead-In Sentence

The lead-in sentence *suggests* the kinds of information to follow. For example, in a report to stockholders on the quarterly decline in dividends, this lead-in sentence appeared.

SAMPLE LEAD-IN SENTENCE

The Board of Directors for Maximus Corporation believes that the quarterly decline in dividends can be attributed to changes in production.

The audience expects the content following this lead-in sentence to identify the changes in production and explain how these changes have reduced dividends.

A report distinguishing characteristic properties of the upper layer of the earth's surface and the underlying rocks had the following introduction.

SAMPLE PARAGRAPH WITH LEAD-IN SENTENCE

The relatively thin upper layer of the earth's crust is called soil. One dictionary definition of soil is "finely divided rock material mixed with decayed vegetable or animal matter, constituting that portion of the surface of the earth in which plants grow, or may grow." This upper layer of the earth's surface, varying in thickness from 6 to 8 inches in the case of humid soils to 10 or 20 feet in the case of arid soils, possesses characteristic properties that distinguish it from the underlying rocks and rock ingredients.

The reader expects to find in the following body of the report an identification of the characteristic properties and an explanation of how these properties distinguish the upper layer of the earth's surface from underlying layers of rocks and rock ingredients.

The Outline Sentence

The outline sentence *states specifically* the major points included in the presentation.

> SAMPLE OUTLINE SENTENCE
>
> We recommend the purchase of new computers and printers to increase efficiency, to improve the appearance of documents, and to expand storage capabilities.

The audience expects the developed document to include specific details supporting each of the reasons for the purchase of new computers. A typical approach is to use these reasons as headings to indicate the organization of the data.

In the following introductory paragraph, the first four sentences lead up to the outline sentence.

> SAMPLE PARAGRAPH WITH OUTLINE SENTENCE
>
> When eating a mushroom, have you ever wondered how it grows into the umbrella shape? Mushrooms are fungi that grow in decaying vegetable matter. The common table mushroom grows in much the same way as other mushrooms. Belonging to a group of mushrooms called the agarics, the common mushroom grows wild, but it can also be cultivated. The five stages of mushroom growth are the spawn, the pinhead, the button, the cap and gills, and the mature mushroom, shaped like an open umbrella.

By concluding this introductory paragraph with the outline sentence, the writer leads the audience to expect an explanation of what happens during each of the five stages.

As a writer you can choose from various techniques to get to the lead-in or outline sentence. Common techniques include a question, a definition, a startling statistic or an unusual fact, background or history, a quotation, an objective.

Review roman numeral section I of the procedure outlines in each of Chapters 1–5 for suggested ways to introduce specific presentations. Look at the sample presentations throughout the textbook; the introductions for the several examples illustrate a variety of ways to introduce a presentation.

The best guideline for writing an introduction is answering this question: How can I write the introduction to make clear to the audience exactly what the presentation will be about (the topic) and what direction the development will take (the focus).

Body

The body of a presentation—the substance—typically is a cluster of paragraphs that supports or develops the subject. The paragraphs develop the subject in a clearly organized, coherent presentation, supporting the focus set forth in the introduction. For instance, if an industry plans to relocate and wants information about available land and cost of the land for a building site, the report's overall topic is the move of the industry to a new location. The subject of the report is locating and purchasing a building site; the focus is identifying available land and the cost of that land. The body of the report includes related data. A memo to sales associates about ways to increase sales identifies the ways and elaborates on how sales associates might follow the suggestions.

The amount and depth of detail in the body of any presentation depends on several factors: the audience and their knowledge of the subject, the purpose of the presentation, and the complexity of the subject.

Review roman numeral section II of the procedure outlines in Chapters 1–5 for suggested content for specific types of presentations. Look at the sample presentations throughout the book and how they illustrate various ways to organize and develop content.

Closing, Conclusion, and Recommendations

The closing of a presentation should be compatible with the audience, the purpose, and the emphasis of the presentation. Some presentations, for instance, logically end with a last developed point. If presenting a set of instructions introduced with the sentence, "Complete these five procedures before filling out the patient's chart," the instructions would obviously be complete when you have explained the fifth procedure.

Other presentations require a closing to signal the reader that the presentation is completed. Such closings reemphasize the main points discussed in the presentation, draw a conclusion about the material, make a recommendation, or make a forecast. A report explaining the primary differences between plants and animals ends with the following summary paragraph.

SAMPLE CLOSING REEMPHASIZING THE MAIN POINTS

The primary differences between plants and animals then are these: animals can move about, animals sense their surroundings, animals live on ready-made foods, and animals do not contain cellulose.

The closing reemphasizes the main points in the report. A lengthy presentation often closes with a summary of the main points.

A presentation discussing the pros and cons of advertising concludes with the following paragraph.

SAMPLE CLOSING STATING A CONCLUSION

Many other good and bad points of advertising could be given. But these are sufficient to suggest the relation of advertising to people's lives, the vast power it controls, and the social responsibilities it carries.

The writer's conclusion here is that the points about advertising sufficiently suggest its relationship to people's lives, its power, and its social responsibilities.

A closing for a report about possible changes in assembly of a product could end with a recommendation for a specific change or a recommendation to investigate additional options.

Review roman numeral III in each of the procedure outlines at the ends of Chapters 1–5 for suggested ways to close specific types of presentations. Review also the sample presentations in the book, noting the variety of closing methods.

No rules tell you when you need a formal closing for a presentation. A good rule of thumb is to review the presentation; if it seems to end abruptly, consider adding a closing, usually one reemphasizing the main points within the presentation or reaching a conclusion based upon the main points.

No rules tell you the best length of a closing for a presentation. A closing could be a single sentence, a paragraph, or even several paragraphs. You as a writer make the decision about the type of closing and the length of the closing.

A word of caution. Do not introduce a new topic in the closing. The closing should always relate to and reinforce the ideas discussed in the body.

Order

Common kinds of order are time, space, general to particular, particular to general, and climax. (See also Chapter 4 Classification and Partition, pages 138–139.) You must decide which order best arranges your selected details.

Time Order

In a presentation using time (chronological) order, arrange the events according to *when* they happen. For example, you might use time order to report your progress in completing course requirements as a computer engineering major. A salesperson uses time order to explain how a sale is initiated and concluded; a nurse uses time order in charting the hospitalization, surgery, and recovery of a patient.

Study the following paragraph; the writer, giving directions for designing a furnace, uses time order.

> The first step in designing a furnace is to measure the fireplace to determine the length of pipe needed.
>
> 1. For depth, start at the front of the fireplace and measure from there to the back wall.
> 2. For height, measure from the floor of the fireplace to the top of the wall.
> 3. For width, measure the back wall along the top part of the fireplace.
>
> Subtract 7 inches from the measurement of floor to top wall so that the furnace can rest 2 inches off the floor and still have 5 inches of clearance between the furnace and the top of the fireplace. Add all three of these measurements to find how much pipe is needed.

Time order is used to develop instructions and process explanations (Chapter 1 in Part I); time order might also be used in business letters (Chapter 7 in Part I) in which the time that events or steps occur is of major importance.

Space Order

In a presentation using spatial order, discuss the items according to *where* they are. For example, in describing the physical layout of a college, you might show certain buildings located to the south of a main building, others to the north, and still others directly to the east. Also, in describing the luxurious features of a house room by room, you might use space order.

In the following paragraph, which explains a shop layout, the details are arranged in space order.

Entering the front door of the machine shop at Westhaven Area Technical School, an industrialist sees a well-organized shop. To her left she sees a group of lathes, turret lathes and engine lathes. In the left back corner she sees a tool room from which students check out materials and tools. Just to the right of the tool room is a cutting saw. In the center of the shop she sees a line of grinding machines. Included are surface grinders, tool-cutter grinders, cylindrical grinders, and centerless grinders. Nearer the center back wall is the heat-treating section where the industrialist sees furnaces used in heat treatment. To her right she sees two other types of machines: the milling machines and the drill presses. Milling machines are of the vertical and the horizontal types; drill presses included are the multispindle and the radial drill.

Space order may also be used to describe a mechanism (Chapter 2 in Part I).

General to Particular (Specific) Order

In a presentation using general to particular order, the most general statement (lead-in sentence, outline sentence) within the presentation, appears near the beginning. Following the general statement is the particular, or the specific, material of the presentation. This material makes up the body and includes details giving additional information, examples, comparisons, contrasts, reasons, and so forth. This is the order most often used by students and with good reason. Placing a general statement near the beginning of the presentation allows you to reread it often as you write and helps you stick to the main idea.

The following paragraph illustrates general to particular order.

Lures for bass fishing should be selected according to the time of day the fishing will take place. For early morning and late afternoon fishing, use top-water baits that make a bubbling or splashing sound. For midday fishing, select deep-running lures and plastic worms. And for night fishing try any kind of bait that makes a lot of noise and has a lot of action. The reason for selecting these different baits is that bass feed on top of shallow water in the early morning, late afternoon, and at night; the only way to get their attention is to use a noisy, top-water lure. During the middle of the day, however, bass swim down to cooler water; then a deep-running lure is best.

Particular (Specific) to General Order

A presentation using particular to general order, of course, reverses the general to particular order. The most general statement, or the topic statement, is near the end of the presentation, and the specific material—reasons, examples, comparisons, details, and the like—precedes and leads up to it, as in the following paragraph.

The human body requires energy for growth and activity; this energy is provided through calories in food. The weight of an adult is determined in general by the balance of the intake of energy in food with the expenditure of energy in activity and growth. When intake and outgo of energy are equal, weight will stay the same. Similarly weight is reduced when the body receives fewer calories from food than it uses, and weight is increased when the body receives more calories than it uses. Weight can be controlled therefore by regulating either the amount of food eaten, or the extent of physical activity, or both.

Climax

In a presentation using the order of climax, arrange the ideas so that each succeeding idea is more important or significant or appealing than the preceding. The sentences and paragraphs build to a climax at the end of the presentation.

People often drink water without knowing for sure that it is safe to drink. They drink water from driven wells into which potentially impure surface water has drained. They drink water from springs without checking around the outlet of the spring to be sure that no surface pollution from human or animal wastes has occurred. Without knowing the rate of flow, the degree of pollution, or the temperature, they drink running water, thinking it safe because of a popular notion that running water purifies itself. These are dangerous acts; polluted water can kill!

Other Orders

Among other orders of development are alphabetical and random listings, least important to most important, least pertinent to most pertinent, least likely to most likely, simply to complex, known to unknown.

Paragraph of Transition

The paragraph of transition connects two paragraphs or two groups of paragraphs. Usually brief (often only one or two sentences), the transitional paragraph signals a marked shift in thought between two sections of a presentation. The transitional paragraph looks back to the preceding material, summarizes the preceding material, or indicates the material to come.

Not all writing, of course, requires transitional paragraphs. Frequently a transitional word, phrase, or sentence is sufficient to connect the sections. You must decide, after considering the purpose and method of development in your presentation, whether it is more effective to show transition between paragraphs by means of the opening and closing sentences of the adjacent paragraphs or by means of a transitional paragraph.

For instance, in a presentation discussing the advantages and disadvantages of living at home while attending college, you might decide that you need a transitional paragraph to connect the "advantages" section with the "disadvantages" section.

> Thus, the student who lives at home while attending college has fewer living expenses, is free from dormitory rules, and can enjoy the conveniences of family living. On the other hand, the student must consider that he or she misses the experience of dormitory life, has less opportunity to meet other students, spends a lot of time and money commuting, and has less opportunity to become independent in making decisions.

The first sentence summarizes the first section of the report; the second sentence looks forward to (forecasts) the next section by suggesting the material that is to come.

The following paragraph separates a discussion of general anesthetics and local anesthetics.

> Not all surgery requires that the patient be unconscious. For many minor operations, only a restricted, or local, area of the body need be made insensible to pain; thus a local anesthetic is administered. The local anesthetic prevents sensations of pain and touch from traveling through the nerves in the drugged area.

In "Clear Only If Known" (see pages 496–498), Edgar Dale uses the following transitional paragraph as he moves from a discussion of why people give poor directions to one of why people do not follow good directions.

But we must not put too much of the blame for inadequate directions on whose who give them. Sometimes the persons who ask for help are also at fault. Communication, we must remember, is a two-way process. (page 498).

Transition (Coherence)

Transition aids in the smooth flow of thought from sentence to sentence and from paragraph to paragraph. It relates thoughts so there is a clear, coherent progression from one thought to the next. Transition is achieved through *transitional words, phrases, and paragraphs.* As the term implies, these are words or phrases—even sentences and groups of sentences—that relate a preceding thought or topic to a succeeding thought or topic. Transition may also be achieved through the use of other techniques.

Transition Through Conjunctions
Conjunctions are familiar transitional words. To show that several items are of equal value, use the word *and:* item 1, item 2, item 3, *and* item 4. *And,* of course, is the conjunction that means "also" or "in addition," and it indicates a relationship of sameness or similarity between items or ideas.

To show difference or contrast between items, use *but.*

Other conjunctions and the relationships they show include: *for* (condition), *nor* (exclusion), *or* (choice), and *yet* (condition).

Transition Through Conventional Words or Phrases
In our language a large group of transitional words and phrases shows relationships between ideas and facts. These words are signs that the ideas and facts are related in a specific way to the preceding ideas and facts. For example, if in reading you find the phrase *in contrast,* you know the following material will be opposite in meaning to the material preceding the phrase.

Listed below are some of the more commonly used words and phrases and the relationship they signal.

Relationship	*Transitional Words/Phrases*
addition	also, and, first, moreover, too, next, furthermore, similarly
contrast	but, however, nevertheless, on the contrary, on the other hand
passage of time	meanwhile, soon, presently, immediately, afterward, at last
illustration	for example, to illustrate, for instance, in other words
comparison	likewise, similarly
conclusion	therefore, consequently, accordingly, as a result, thus
summary	to sum up, in short, in brief, to summarize
concession	although, of course, at the same time, after all

Transition Through Other Techniques
Other techniques used to ensure the smooth flow of sentences and thoughts are pronoun reference, repetition of key terms, continuity of sentence subjects, parallel sentence patterns, or a combination of these. In the following paragraph from Abraham Lincoln's "Gettysburg Address," italics have been added to indicate these transition techniques.

CONTINUITY OF SUBJECT "WE"

REPETITION OF KEY TERMS:
"WAR," "NATION," AND
"BATTLEFIELD"

PARALLEL SENTENCE PATTERNS

PRONOUN REFERENCE

Now *we* are engaged in a great civil *war,* testing whether that *nation,* or any *nation* so conceived and so dedicated, can long endure. *We* are met on a great *battlefield* of that *war.* We have come to dedicate a portion of *that field* as a final resting-place for those who *here* gave their lives that that *nation* might live. It is altogether fitting and proper that *we* should do *this.*

Transition and Continuity

Read the following paragraph.

> The diesel engine has its disadvantages. The higher compresson that makes the diesel more efficient and clean burning necessitates the use of heavier engine components such as pistons, rods, and cylinder heads. Diesels are heavier than gasoline engines of the same displacement. Diesel engines cost slightly more than gasoline engines. Diesel engines are difficult to start in cold weather. "Glow plugs" are available to preheat the combustion chambers and engine block heaters to provide easier starting. A diesel engine has relatively poor acceleration compared to the gasoline engine. It takes the diesel longer to rev up to its peak horsepower. Diesel engines have been noted for their noise, vibration, and smoke. These three characteristics have been reduced to a minimum in recent diesel-powered vehicles.

Although the preceding paragraph contains information, it presents some difficulty in reading and understanding because the sentences are choppy and there are no transitional words or phrases, or any other techniques, to relate one idea to another. Read the revised paragraph, noting the words in boldface.

> The diesel engine has its disadvantages. **For one,** the higher compression that makes the diesel more efficient and clean burning necessitates the use of heavier engine components such as pistons, rods, and cylinder heads; **thus,** diesel engines are heavier than gasoline engines. **Also** diesel engines cost slightly more than gasoline engines. **Another disadvantage** is that diesel engines are difficult to start in cold weather. **However,** glow plugs preheat the combustion chambers and engine block heaters to provide easier starting. **Further,** a diesel engine has relatively poor acceleration compared to the gasoline engine; **thus,** the diesel takes longer to rev up to its peak horsepower. **Finally,** diesel engines are noted for their noise, vibration, and smoke, **but** these characteristics have been reduced to a minimum in recent diesel-powered vehicles.

Notice the words in boldface print; these added words, transitional markers, indicate clearly the relationship of ideas. The paragraph now reads smoothly and is easily understood because there is continuity, a smooth flow of sentences moving from one to another. The stop-go effect characterizing the first writing of the paragraph is eliminated. In your writing, strive for continuity, for a smooth flow of sentences moving clearly from one sentence to another and from paragraph to paragraph. Make clear the relationship of ideas expressed in the sentences and paragraphs.

Section 2

Accuracy and Clarity Through Standard Language Usage

Introduction

English language usage varies and constantly changes. Attempts to classify language according to its usage in different dialects (forms of a language spoken in different geographic areas), social statuses, educational levels or styles (formal and informal) are all incomplete and overlapping. Amid these differences, however, are generally accepted standard practices in business and industry.

Usage does make a difference. Select the usage most likely to get the desired results, recognizing that usages, like people, have different characteristics. For example, you say or write *I have known him a week* rather than *I have knowed him a week* because *I have known him a week* is more acceptable to most people. You base this decision, consciously or unconsciously, on the grammatical expressions of people you respect or of people with whom you want to "fit in," and on the reactions of people to these expressions.

Because judgments are based on knowledge, the following guides are designed to improve your judgment by increasing your knowledge of acceptable grammatical usage and word choice in standard English.

The remainder of this section is a discussion of standard language usage, organized alphabetically.

Adjectives

Adjectives show degrees of comparison in quality, quantity, and manner by affixing *-er* and *-est* to the positive form or by adding *more* and *most* or *less* and *least* to the positive form.

Degrees of Comparison

POSITIVE	tall	beautiful	wise
COMPARATIVE	taller	more beautiful	less wise
SUPERLATIVE	tallest	most beautiful	least wise

The comparative degree is used when speaking of two things, and the superlative when speaking of three or more. One-syllable words generally add *-er* and *-est* while words of two or more syllables generally require *more* and *most.* To show degrees of inferiority, *less* and *least* are added.

- A double comparison, such as *most beautifulest,* should not be used; either *-est* or *most* makes the superlative degree, not both.
- Adjectives that indicate absolute qualities or conditions (such as *dead, round,* or *perfect* cannot be compared. Thus: *dead, more nearly dead, most nearly dead* (not *deader, deadest*).
- Adjective clauses are punctuated according to the purpose they serve. If they are essential to the meaning of the sentence, they are not set off by commas. If they simply give additional information, they are set off from the rest of the sentence. (See comma usage, item 6, page 651.) Adjective clauses beginning with *that* are essential and are not set off (with one exception: when *that* is used as a substitute for *which* to avoid repetition).

Edward Jenner is the man *who experimented with smallpox vaccination.* (essential to meaning)

Bathrooms are often decorated in colors *that suggest water.* (essential to meaning)
My ideas about exploration of space, *which are different from most of my friends' ideas,* were formulated mainly through reading. (gives additional information)

615

Section 2
Accuracy and
Clarity Through
Standard
Language
Usage

Adverbs

- An adjective should not be used when an adverb is needed.

  INCORRECT   The new employee did *good* on her first report. (adverb needed)
  CORRECT   The new employee did *well* on her first report.

  INCORRECT   For a department to operate *smooth* and *efficient,* all employees must cooperate. (adverbs needed)
  CORRECT   For a department to operate *smoothly* and *efficiently,* all employees must cooperate.

- Introductory adverb clauses are generally followed by a comma. Adverb clauses at the end of a sentence are not set off. (See comma usage, item 7, page 651.)

  When the investigation is completed, the company will make its decision.
  The company will make its decision *when the investigation is completed.*

Comma Splice

The comma splice occurs when a comma is used to connect, or splice together, two sentences. The comma splice, a common error in sentence punctuation, is also called the comma fault. (See also the discussion of Run-on, or Fused, Sentence, page 626.)

Mahogany is a tropical American timber tree, its wood turns reddish brown at maturity.

The comma splice in the above sentence may be corrected by several methods.

- *Replacing the comma with a period*

  Mahogany is a tropical American timber tree. Its wood turns reddish brown at maturity.

- *Replacing the comma with a semicolon*

  Mahogany is a tropical American timber tree; its wood turns reddish brown at maturity.

- *Adding a coordinate conjunction, such as "and"*

  Magogany is a tropical American timber tree, and its wood turns reddish brown at maturity.

- *Recasting the sentence*

  Mahogany is a tropical American timber tree whose wood turns reddish brown at maturity.

Conjunctions

A conjunction connects words, phrases, or clauses. Conjunctions may be divided into two general classes: coordinating conjunctions and subordinating conjunctions.

1. Coordinating Conjunctions—Such as *and, but, or*—Connect Words, Phrases, or Clauses of Equal Rank.

I bought nails, brads, staple, *and* screws. (connects nouns in a series)
Nancy or Elaine will serve on the committee. (connects nouns)

- Conjunctive adverbs—such as *however, moreover, therefore, consequently, nevertheless*—are a kind of coordinating conjunction. They link the independent clause in which they occur to the preceding independent clause. The clause they introduce is grammatically independent, but it depends on the preceding clause for complete meaning. Note that a semicolon separates the independent clauses (see semicolon usage, pages 653–654) and that the conjunctive adverb is usually set off with commas (see comma usage, item 10, page 651).

> The film was produced and directed by students; *therefore,* other students should be interested in viewing the film.
>
> Jane had a severe cold; *consequently,* she was unable to participate in the computer tournament.

- Correlative conjunctions—*either . . . or, neither . . . nor, both . . . and, not only . . . but also*—are a kind of coordinating conjunction. Correlative conjunctions are used in pairs to connect words, phrases, and clauses of equal rank.

> The movie was *not only* well produced *but also* beautifully filmed.

- The units joined by coordinate conjunctions should be the same *grammatically*. (See also Coordinate, page 135, and Parallelism in Sentences, pages 621–622.)

CONFUSING	Our store handles two kinds of drills—manual *and* electricity. (adjective and noun)
REVISED	Our store handles two kinds of drills—manual *and* electric. (adjective and adjective)
CONFUSING	I *either* will go today *or* tomorrow. (verb and adverb)
REVISED	I will go *either* today *or* tomorrow. (adverb and adverb)

- The units joined by coordinate conjunctions should be the same *logically*. (See also Coordinate, page 135, and Parallelism in Sentences, pages 621–622.)

CONFUSING	I can't decide whether I want to be a lab technician, a nurse, *or* respiratory therapy. (two people and a field of study)
REVISED	I can't decide whether I want to be a lab technician, a nurse, *or* a respiratory therapist. (three people)
REVISED	I can't decide whether I want to study lab technology, nursing, *or* respiratory therapy. (three fields of study)

2. Subordinate Conjunctions—Such as *when, since, because, although, as, as if*—Introduce Subordinate Clauses.

> *Because the voltmeter was broken,* we could not test the circuit.
>
> The instructor gave credit for class participation *since the primary purpose of the course was to stimulate thought.*

Nonsexist Use of Language

The following section is reprinted from the National Council of Teachers of English pamphlet on nonsexist use of language.*

*Reprinted by permission from the National Council of Teachers of English

This section deals primarily with word choice. Many of the examples are matters of vocabulary; a few are matters of grammatical choice. The vocabulary items are relatively easy to deal with, since the English lexicon has a history of rapid change. Grammar is a more difficult area, and we have chosen to use alternatives that already exist in the language rather than to invent new constructions. In both cases, recommended alternatives have been determined by what is graceful and unobtrusive. The purpose of these changes is to suggest alternative styles.

617

Section 2
Accuracy and
Clarity Through
Standard
Language
Usage

Generic "Man"

1. Since the word *man* has come to refer almost exclusively to adult males, it is sometimes difficult to recognize its generic meaning.

Problems	*Alternatives*
mankind	humanity, human beings, people*
man's achievements	human achievements
the best man for the job	the best person for the job
the common man	the average person, ordinary people
cavemen	cave dwellers, prehistoric people

2. Sometimes the combining form *-woman* is used alongside *-man* in occupational terms and job titles, but we prefer using the same titles for men and women when naming jobs that could be held by both. Note, too, that using the same forms for men and women is a way to avoid using the combining form *-person* as substitute for *-woman* only.

Problems	*Alternatives*
chairman/chairwoman	chair, coordinator (of a committee or department), moderator (of a meeting), presiding officer, head, chairperson
buisnessman/businesswoman	business executive, manager
congressman/congresswoman	congressional representative
policeman/policewoman	police officer
salesman/saleswoman	sales clerk, sales representative, salesperson
fireman	fire fighter
mailman	letter carrier

Generic "He" and "His"

Because there is no one pronoun in English that can be effectively substituted for *he* or *his,* we offer several alternatives. The form *he or she* has been the NCTE house style over the last ten years, on the premise that it is less distracting than *she or he* or *he/she*. There are other choices, however. The one you make will depend on what you are writing.

---

*A one-word substitution for *mankind* isn't always possible, especially in set phrases like *the story of mankind*. Sometimes recasting the sentence altogether may be the best solution.

1. Sometimes it is possible to drop the possessive form *his* altogether or to substitute an article.

Problems	**Alternatives**
The average student is worried about his grades.	The average student is worried about grades.
When the student hands in his paper, read it immediately.	When the student hands in the paper, read it immediately.

2. Often, it makes sense to use the plural instead of the singular.

Problems	**Alternatives**
Give the student his grade right away.	Give the students their grades right away.
Ask the student to hand in his work as soon as he is finished.	Ask students to hand in their work as soon as they are finished.

3. The first or second person can sometimes be substituted for the third person.

Problems	**Alternatives**
As a teacher, he is faced daily with the problem of paperwork.	As teachers, we are faced daily with the problem of paperwork.
When a teacher asks his students for an evaluation, he is putting himself on the spot.	When you ask your students for an evaluation, you are putting yourself on the spot.

4. In some situations, the pronoun *one (one's)* can be substituted for *he (his),* but it should be used sparingly. Notice that the use of *one*—like the use of *we* or *you*—changes the tone of what you are writing.

Problem	**Alternative**
He might well wonder what his response should be.	One might well wonder what one's response should be.

5. A sentence with *he* or *his* can sometimes be recast in the passive voice or another impersonal construction.

Problems	**Alterntives**
Each student should hand in his paper promptly.	Papers should be handed in promptly.
He found such an idea intolerable.	Such an idea was intolerable.

6. When the subject is an indefinite pronoun, the plural form *their* can occasionally be used with it, especially when the referent for the pronoun is clearly understood to be plural.

Problem	**Alternative**
When everyone contributes his own ideas, the discussion will be a success.	When everyone contributes their own ideas, the discussion will be a success.

   But since this usage is transitional, it is usually better to recast the sentence and avoid the indefinite pronoun.

619

Section 2
Accuracy and
Clarity Through
Standard
Language
Usage

Problem	**Alternative**
When everyone contributes his own ideas, the discussion will be a success.	When all the students contribute their own ideas, the discussion will be a success.

7. Finally, sparing use can be made of *he or she* and *his or her*. It is best to restrict this choice to contexts in which the pronouns are not repeated.

Problems	**Alternatives**
Each student will do better if he has a voice in the decision.	Each student will do better if he or she has a voice in the decision.
Each student can select his own topic.	Each student can select his or her own topic.

Sex-Role Stereotyping

Word choices sometimes reflect unfortunate and unconscious assumptions about sex roles—for example, that farmers are always men and elementary school teachers are always women; that men are valued for their accomplishments and women for their physical attributes; or that men are strong and brave while women are weak and timid. We need to examine the assumptions inherent in certain stock phrases and choose nonstereotyped alternatives.

1. Identify men and women in the same way. Diminutive or special forms to name women are usually unnecessary. In most cases, generic terms such as *doctor* or *actor* should be assumed to include both men and women. Only occasionally are alternate forms needed, and in these cases, the alternate form replaces both the masculine and the feminine titles.

Problems	**Alternatives**
stewardess	flight attendant (for both *steward* and *stewardess*)
authoress	author
waitress	server, food server
poetess	poet
coed	student
lady lawyer	lawyer . . . she
male nurse	nurse . . . he

2. Do not represent women as occupying only certain jobs or roles and men as occupying only certain others.

Problems	**Alternatives**
the kindergarten teacher . . . she	*occasionally use* the kindergarten teacher . . . he *or* kindergarten teachers . . . they
the principal . . . he	*occasionally use* the principal . . . she *or* principals . . . they
Have your mother send a snack for the party.	Have a parent send a snack for the party. *occasionally use* Have your father. . . . *or* Have your parents. . . .

NCTE conventiongoers and their wives are invited.	NCTE conventiongoers and their spouses are invited.
Writers become so involved in their work that they neglect their wives and children.	Writers become so involved in their work that they neglect their families.

3. Treat men and women in a parallel manner.

Problems	***Alternatives***
The class interviewed Chief Justice Burger and Mrs. O'Connor.	The class interviewed Warren Burger and Sandra O'Connor.
	or . . . Mr. Burger and Ms. O'Connor
	or . . . Chief Justice Burger and Justice O'Connor.
The reading list included Proust, Joyce, Gide, and Virginia Woolf.	The reading list included Proust, Joyce, Gide, and Woolf.
	or . . . Marcel Proust, James Joyce, André Gide, and Virginia Woolf.
Both Bill Smith, a straight-A sophomore, and Kathy Ryan, a pert junior, won writing awards.	Both sophomore Bill Smith, a straight-A student, and junior Kathy Ryan, editor of the school paper, won writing awards.

4. Seek alternatives to language that patronizes or trivializes women, as well as to language that reinforces stereotyped images of both women and men.

Problems	***Alternatives***
The president of the company hired a gal Friday.	The president of the company hired an assistant.
I'll have my girl do it.	I'll ask my secretary to do it.
Stella is a career woman.	Stella is a professional.
	or Stella is a doctor (architect, etc.).
The ladies on the committee all supported the bill.	The women on the committee all supported the bill.
Pam had lunch with the girls from the office.	Pam had lunch with the women from the office.
This is a man-sized job.	This is a big (huge, enormous) job.
That's just an old wives' tale.	That's just a superstition (superstitious story).
Don't be such an old lady.	Don't be so fussy.

Sexist Language in a Direct Quotation

Quotations cannot be altered, but there are other ways of dealing with this problem.

1. Avoid the quotation altogether if it is not really necessary.
2. Paraphrase the quotation, giving the original author credit for the idea.

3. If the quotation is fairly short, recast it as an indirect quotation, substituting nonsexist words as necessary.

621

Section 2
Accuracy and
Clarity Through
Standard
Language
Usage

Problem

Among the questions asked by the school representatives was the following: "Considering the ideal college graduate, what degree of knowledge would you prefer him to have in each of the curricular areas?"

Alternative

Among the questions asked by the school representatives was one about what degree of knowledge the ideal college graduate should have in each of the curricular areas.

Sample Revised Passage

Substantial revisions or deletions are sometimes necessary when problems overlap or when stereotyped assumptions about men and women so pervade a passage that simple replacement of words is inadequate.

Problem

Each student who entered the classroom to find himself at the mercy of an elitist, Vassar-trained Miss Fidditch, could tell right away that the semester would be a trial. The trend in composition pedagogy toward student-created essays and away from hours of drill on grammatical correctness has meant, at least for him, that he can finally learn to write. But Macrorie, Elbow, and Janet Emig could drive the exasperated teacher of a cute and perky cheerleader type to embrace the impersonal truth of *whom* as direct object rather than fight his way against the undertow of a gush of personal experience. As Somerset Maugham remarked, "Good prose should resemble the conversation of a well-bred man," and both Miss Fidditch and the bearded guru who wants to "get inside your head" must realize it.

Alternative

The trend in composition pedagogy toward student-centered essays, represented by such writers as Ken Macrorie, Peter Elbow, and Janet Emig, has meant that some students are finally learning to write. Yet the movement away from hours of drill on grammatical correctness has brought with it a new problem: in the hands of the inexperienced teacher, student essays can remain little more than unedited piles of personal experiences and emotions.

Parallelism in Sentences

Parallel structure involves getting like ideas into like constructions. A coordinate conjunction, for example, joins ideas that must be stated in the same grammatical form. Other examples: an adjective should be parallel with an adjective, a verb with a verb, an adverb clause with an adverb clause, and an infinitive phrase with an infinitive phrase. Parallel structure in grammar helps to make parallel meaning clear.

> A *computer,* a *table,* and a *filing cabinet* were delivered today. (nouns in parallel structure)
> *Whether you accept the outcome of the experiment* or *whether I accept it* depends on our individual interpretation of the facts. (dependent clauses in parallel structure)

Failure to express all of the ideas in the same grammatical form results in faulty parallelism.

FAULTY	This study should help the new spouse *learn* skills and *to be knowledgeable* about sewing and cooking. (verb and infinitive phrase)
REVISED	This study should help the new spouse *learn* skills and *become* knowledgeable about sewing and cooking. (verbs in parallel structure)
FAULTY	Three qualities of tungsten steel alloys are *strength, ductility,* and *they have to be tough.* (noun, noun, and independent clause)
REVISED	Three qualities of tungsten steel alloys are *strength, ductility,* and *toughness.* (nouns in parallel structure)
FAULTY	The best lighting in a study room can be obtained *if windows are placed in the north side, if tables are placed so that the light comes over the student's left shoulder,* and *by painting the ceiling a very light color.* (dependent clause, dependent clause, phrase)
REVISED	The best lighting in a study room can be obtained *if windows are placed in the north side, if tables are placed so that the light comes over the student's left shoulder,* and *if the ceiling is painted a very light color.* (dependent clauses in parallel structure)

See also Part I, Chapter 4 Analysis Through Classification and Partition especially Coordinate discussion, page 135.

Placement of Modifiers

Modifiers are words, phrases, or clauses, either adjective or adverb, that limit or restrict other words in the sentence. Careless construction and placement of these modifiers may cause problems such as dangling modifiers, dangling elliptical clauses, misplaced modifiers, and squinting modifiers.

Dangling Modifiers

Dangling modifiers or dangling phrases occur when the word the phrase should modify is hidden within the sentence or is missing.

WORD HIDDEN IN SENTENCE	Holding the bat tightly, the ball was hit by the boy. (Implies that the ball was holding the bat tightly)
REVISION	Holding the bat tightly, the boy hit the ball.
WORD MISSING	By placing a thermometer under the tongue for approximately three minutes, a fever can be detected. (Implies that a fever places a thermometer under the tongue)
REVISION	By placing a thermometer under the tongue for approximately three minutes, anyone can tell if a person has a fever.

- Dangling modifiers may be corrected by either rewriting the sentence so that the word modified by the phrase immediately follows the phrase (this word is usually the subject of the sentence) or rewriting the sentence by changing the phrase to a dependent clause.

623

Section 2
Accuracy and
Clarity Through
Standard
Language
Usage

DANGLING MODIFIER	*Driving down Main Street,* the city auditorium came into view.
SENTENCE REWRITTEN TO CLARIFY WORD MODIFIED	Driving down Main Street, I saw the city auditorium.
SENTENCE REWRITTEN WITH A DEPENDENT CLAUSE	As I was driving down Main Street, the city auditorium came into view.

Dangling Elliptical Clauses

In elliptical clauses some words are understood rather than stated. For example, the dependent clause in the following sentence is elliptical: *When measuring the temperature of a conductor, you must use the Celsius scale;* the subject and part of the verb are omitted. The understood subject of an elliptical clause is the same as the subject of the sentence. This is true in the example: *When (you are) measuring the temperature of a conductor, you must use the Celsius scale.* If the understood subject of the elliptical clause is not the same as the subject of the sentence, the clause is a dangling clause.

| DANGLING CLAUSE | *When using an electric saw,* safety glasses should be worn. |
| REVISION | When using an electric saw, wear safety glasses. |

- Dangling elliptical clauses may be corrected by either including within the clause the missing words or rewriting the main sentence so that the stated subject of the sentence and the understood subject of the clause will be the same.

DANGLING CLAUSE	*After changing the starter switch,* the car still would not start.
SENTENCE REVISED; MISSING WORDS INCLUDED	After *the mechanic* changed the starter switch, the car still would not start.
SENTENCE REVISED; UNDERSTOOD SUBJECT AND STATED SUBJECT THE SAME	After changing the starter switch, *the mechanic* still could not start the car.

Misplaced Modifiers

Place modifiers near the word or words modified. If the modifier is correctly placed, there should be no confusion. If the modifier is incorrectly placed, the intended meaning of the sentence may not be clear.

MISPLACED MODIFIER	The machinist placed the work to be machined in the drill press vise *called the workpiece.*
CORRECTLY PLACED MODIFIER	The machinist placed the work to be machined, *called the workpiece,* in the drill press vise.
MISPLACED MODIFIER	Where are the shirts for children *with snaps?*
CORRECTLY PLACED MODIFIER	Where are the children's shirts *with snaps?*

Squinting Modifiers

A modifier should clearly limit or restrict *one* sentence element. If a modifier is so placed within a sentence that it can be taken to limit or restrict either of two elements, the modifier is squinting; that is, the reader cannot tell which way the modifier is looking.

SQUINTING MODIFIER When the student began summer work *for the first time* she was expected to follow orders promptly and exactly. (The modifying phrase could belong to the clause that precedes it or to the clause that follows it.)

Punctuation may solve the problem. The sentence might be written in either of the following ways, depending on the intended meaning.

REVISION When the student began summer work *for the first time,* she was expected to follow orders promptly and exactly.

REVISION When the student began summer work, *for the first time* she was expected to follow orders promptly and exactly.

- Squinting modifiers often are corrected by shifting their position in the sentence. Sentence meaning determines placement of the modifier.

SQUINTING MODIFIER The students were advised *when it was midmorning* the new class schedule would go into effect.

REVISION *When it was midmorning,* the students were advised that the new class schedule would go into effect.

REVISION The students were advised that the new class schedule would go into effect *when it was midmorning.*

- Particularly difficult for some writers is the correct placement of *only, almost,* and *nearly.* These words generally should be placed immediately before the word they modify since changing position of these words within a sentence changes the meaning of the sentence.

In the reorganization of the district the member of congress *nearly* lost a hundred voters. (didn't lose any voters)

In the reorganization of the district the member of congress lost *nearly* a hundred voters. (lost almost one hundred voters)

The customer *only* wanted to buy soldering wire. (emphasizes *wanted*)

The customer wanted to buy *only* soldering wire. (emphasizes *soldering wire*)

Preposition at End of Sentence

- A preposition may end a sentence when any other placement of the preposition would result in a clumsy, unnatural sentence.

UNNATURAL Sex is a topic *about* which many people think.

NATURAL Sex is a topic which many people think *about.*

Pronouns

Section 2
Accuracy and
Clarity Through
Standard
Language
Usage

- *Myself* should not be used in the place of *I* or *me*.

INCORRECT	Mother or *myself* will be present.
CORRECT	Mother or *I* will be present.
INCORRECT	Reserve the conference room for the supervisor and *myself*.
CORRECT	Reserve the conference room for the supervisor and *me*.

- *Themself, theirselves,* and *hisself* should never be used; they are not acceptable forms.

  | CORRECT | The members of the team *themselves* voted not to go. |
  | CORRECT | Henry cut *himself*. |

Pronoun and Antecedent Agreement

Pronouns take the place of or refer to nouns. They provide a good way to econo-mize in writing. For example, writing a paper on Anthony van Leeuwenhoek's pio-neer work in developing microscopes, you could use the pronoun forms *he, his,* and *him* to refer to Leeuwenhoek rather than repeat his name numerous times. Since pronouns take the place of or refer to nouns, there must be number (singular or plural) agreement between the pronoun and its antecedent (the word the pro-noun stands for or refers to).

- A singular antecedent requires a singular pronoun; a plural antecedent requires a plural pronoun.

  | NONAGREEMENT | Gritty ink *erasers* should be avoided because *it* invariably damages the working surface of the paper. |
  | AGREEMENT | Gritty ink *erasers* should be avoided because *they* invariably damage the working surface of the paper. |

- Two or more subjects joined by *and* must be referred to by a plural pronoun.

  The *business manager* and the *accountant* will plan *their* budget.

- Two or more singular subjects joined by *or* or *nor* must be referred to by singular pronouns.

  Laticia or *Margo* will give *her* report first.
  Neither *Conway* nor *Freeman* will give *his* report.

- Partitive nouns, such as *part, half, some, rest,* and *most* may be singular or plural. The number is determined by a phrase following the noun.

  Most of the *students* asked that *their* grades be mailed.
  Most of the *coffee* was stored in *its* own container.

- The indefinite pronouns *each, every, everyone* and *everybody, nobody, either, neither, one, anyone* and *anybody,* and *somebody* are singular and thus have singular pronouns referring to them.

  Each of the sorority members has indicated *she* wishes *her* own room.
  Neither of the machines has *its* motor repaired.

- The indefinite pronouns *both, many, several,* and *few* are plural; these forms require plural pronouns.

    Several realized *their* failure was the result of not studying.
    Both felt that nothing could save *them.*

- Still other indefinite pronouns—*most, some, all, none, any,* and *more*—may be either singular or plural. Usually a phrase following the indefinite pronoun will reveal whether the pronoun is singular or plural in meaning.

    Most of the *patients* were complimentary of *their* nurses.
    Most of the *money* was returned to *its* owner.

Run-On, or Fused, Sentence

The run-on, or fused, sentence occurs when two sentences are written with no punctuation to separate them. (See also Comma Splice, page 615)

> The evaluation of the Automated Drafting System is encouraging we expect to operate the automated system in an efficient and profitable manner.

Obviously the sentence above needs punctuation to make it understandable. Several methods could be used to make it into an acceptable form.

- *Period between "encouraging" and "we"*

    The evaluation of the Automated Drafting System is encouraging. We expect to operate the automated system in an efficient and profitable manner.

- *Semicolon between "encouraging" and "we"*

    The evaluation of the Automated Drafting System is encouraging; we expect to operate the automated system in an efficient and profitable manner.

- *Comma plus "and" between "encouraging" and "we"*

    The evaluation of the Automated Drafting System is encouraging, and we expect to operate the automated system in an efficient and profitable manner.

- *Recasting the sentence*

    Since the evaluation of the Automated Drafting System is encouraging, we expect to operate the automated system in an efficient and profitable manner.

    See also Punctuation, especially usage of the period, the comma, and the semicolon, pages 648–649, 650–653, 653–654.

Sentence Fragments

A group of words containing a subject and a verb and standing alone as an independent group of words is a sentence. If a group of words lacks a subject or a verb or cannot stand alone as an independent group of words, the group of words is called a sentence fragment. Sentence fragments generally occur because:

- A noun (subject) followed by a dependent clause or phrase is written as a sentence. The omitted unit is the verb.

FRAGMENT   The engineer's scale, which is graduated in the decimal system.
FRAGMENT   Colors that are opposite.

To make the sentence fragment into an acceptable sentence, add the verb and any modifiers needed to complete the meaning.

SENTENCE   The engineer's scale, which is graduated in the decimal system, is often called the decimal scale.
SENTENCE   Colors that are opposite on the color circle are used in a complementary color scheme.

- Dependent clauses or phrases are written as complete sentences. These are introductory clauses and phrases that require an independent clause.

FRAGMENT   While some pencil tracings are made from a drawing placed beneath the tracing paper.
FRAGMENT   Another development that has promoted the recognition of management.

To make the fragments into sentences, add the independent clause.

SENTENCE   While some pencil tracings are made from a drawing placed underneath the tracing paper, most drawings today are made directly on pencil tracing paper, cloth, vellum, or bond paper.
SENTENCE   Another development that has promoted the recognition of management is the separation of ownership and management.

Acceptable Fragments

Types of fragments, often called nonsentences, are acceptable in certain situations.

- *Emphasis*

    Open the window. Right now.

- *Transition Between Ideas in a Paragraph or Composition*

    Now to the next point of the argument.

- *Dialogue*

    "How many different meals have been planned for next month?"
    "About ten."

Shift in Focus

The writer may begin a sentence that expresses one thought but, somehow, by the end of the sentence has unintentionally shifted the focus. Consider this sentence:

There are several preparatory steps in spray painting a house, which is easy to mess up if you are not careful.

The focus in the first half of the sentence is on preparatory steps in spray painting a house; the focus in the second half is on how easy it is to mess up when spray painting a house. Somehow, in the middle of the sentence, the writer shifted the focus of the sentence. The result is a sentence that is confusing to the reader.

627

Section 2
Accuracy and
Clarity Through
Standard
Language
Usage

To correct such a sentence, the writer should think through the intended purpose and emphasis of the sentence and should review preceding and subsequent sentences. The poorly written sentence above might be

- divided into two sentences

  There are several preparatory steps in spray painting a house. Failure to follow these steps may result in a messed-up paint job.

- recast as one sentence

  Following several preparatory steps in spray painting a house will help to keep you from messing up.

  To help assure satisfactory results when spray painting a house, follow these preparatory steps.

Shifts in Person, Number, and Gender

Avoid shifting from one person to another, from one number to another, or from one gender to another—such shifting creates confusing, awkward constructions.

SHIFT IN PERSON	When *I* was just an apprentice welder, the boss expected *you* to know all the welding processes. (*I* is first person; *you* is second person)
REVISION	When *I* was just an apprentice welder, the boss expected *me* to know all the welding processes.
SHIFT IN NUMBER	*Each* person in the space center control room watched *his or her* dial anxiously. *They* thought the countdown to "zero" would never come. (*Each* and *his or her* are singular pronouns; *they* is plural)
REVISION	*Each* person in the space center control room watched *his* or *her* dial anxiously and thought the countdown to "zero" would never come.
SHIFT IN GENDER	The mewing of my cat told he *it* was hungry, so I gave *her* some food. (*It* is neuter; *her* is feminine)
REVISION	The mewing of my cat told me *she* was hungry, so I gave *her* some food.

Subject and Verb Agreement

A subject and verb agree in person and number. Person denotes person speaking (first person), person spoken to (second person), and person spoken of (third person). The person used determines the verb form that follows.

Present Tense

FIRST PERSON	I go	we go	I am	we are
SECOND PERSON	you go	you go	you are	you are
THIRD PERSON	he goes	they go	he is	they are
IMPERSONAL	one goes		one is	

- A subject and verb agree in number; that is, a plural subject requires a plural verb and a singular subject requires a singular verb.

The *symbol* for hydrogen *is* H. (singular subject; singular verb)
Comic *strips are* often vignettes of real-life situations. (plural subject; plural verb)

629

Section 2
Accuracy and
Clarity Through
Standard
Language
Usage

- A compound subject requires a plural verb.

  Employees and *employers work* together to formulate policies.

- The pronoun *you* as subject, whether singular or plural in meaning, takes a plural verb.

  You were assigned the duties of staff nurse on fourth floor.
  Since *you are* new technicians, *you are* to be paid weekly.

- In a sentence containing both a positive and a negative subject, the verb agrees with the positive.

  The *employees,* not the *manager, were asked* to give their opinions regarding working conditions. (Positive subject plural; verb plural)

- If two subjects are joined by *or* or *nor,* the verb agrees with the nearer subject.

  Graphs or a *diagram aids* the interpretation of statistical reports. (subject nearer verb singular; verb singular)
  A *graph* or *diagrams aid* the interpretation of this statistical report. (subject nearer verb plural; verb plural)

- A word that is plural in form but names a single object or idea requires a singular verb.

  The *United States has changed* from an agricultural to a technical economy.
  Twenty-five *dollars was offered* for any usable suggestion.
  Six *inches* is the length of the narrow rule.

- The term *the number* generally takes a singular verb; *a number* takes a plural verb.

  The number of movies attended each year by the average Canadian *has decreased.*
  A *number* of modern inventions *are* the product of the accumulation of vast storehouses of smaller, minor discoveries.

- When certain subjects are followed by an *of* phrase, the number (singular or plural) of the subject is determined by the number of the object of the *of.* These certain subjects include fractions, percentages, and the words *all, more, most, some,* and *part.*

  Two-thirds of his *discussion was* irrelevant. (object of *of* phrase, *discussion,* is singular; verb is singular)
  Some of the *problems* in a college environment *require* careful analysis by both faculty and students. (object of *of* phrase, *problems,* is plural; verb is plural)

- Elements that come between the subject and the verb ordinarily do not affect subject-verb agreement.

  One of the numbers *is* difficult to represent because of the great number of zeros necessary.
  Such *factors* as temperature, available food, age of organism, or nature of the suspending medium *influence* the swimming speed of a given organism.

- Occasionally the subject may follow the verb, especially in sentences beginning with the expletives *there* or *here;* however, such word order does not change the subject-verb agreement. (The expletives *there* and *here*, in their usual meaning, are never subjects.)

> There *are* two common temperature *scales:* Fahrenheit and Celsius.
> There *is* a great *deal* of difference in the counseling techniques used by contemporary ministers.
> Here *are* several *types* of film.

- The introductory phrase *It is* or *It was* is always singular, regardless of what follows.

> *It is* these problems that overwhelm me.
> *It was* the officers who met.

- Relative pronouns *(who, whom, whose, which, that)* may require a singular or a plural verb, depending on the antecedent of the relative pronoun.

> She is one of those people *who keep* calm in an emergency. (antecedent of *who* is *people*, a plural form; verb plural)
> The earliest lab *that* is offered is at 11:00. (antecedent of *that* is *lab*, a singular form; verb singular)

Verbs

- Use the verb tense that accurately conveys the sequence of events. Do not needlessly shift from one tense to another. Since verb tenses tell the reader when the action is happening, inconsistent verb tenses confuse the reader.

SHIFTED VERBS (CONFUSING)   As time *passed,* technology *becomes* more complex.
passed—past tense verb
becomes—present tense verb

Since the two indicated actions occur at the same time, the verbs should be in the same tense.

CONSISTENT VERBS   As time *passed,* technology *became* more complex.
CONSISTENT VERBS   As time *passes,* technology *becomes* more complex.

passed }
became } —past tense verbs

passes }
becomes } —present tense verbs

SHIFTED VERBS (CONFUSING)   While I *was installing* appliances last summer, I *had learned* that customers appreciate promptness.
was installing—past tense (progressive) verb
had learned—past perfect tense verb

The verb phrase *was installing* indicates a past action in progress while the verb phrase *had learned* indicates an action completed prior to some stated past time. To indicate that both actions were in the past, the sentence might be written as follows:

CONSISTENT VERBS   While I *was installing* appliances last summer, I *learned* that customers appreciate promptness

was installing }
learned } —past tense verbs

Active and Passive Verb Usage

631

Section 2
Accuracy and
Clarity Through
Standard
Language
Usage

- Verbs have two voices, active and passive.

  1. The active voice indicates action done by the subject. The active voice is forceful and emphatic.

     Roentgen *won* the Nobel Prize for his discovery of X rays.
     Meat slightly marbled with fat *tastes* better.
     The scientist *thinks*.

  2. The passive voice indicates action done to the subject. The recipient of the action receives more emphasis than the doer of the action. A passive voice verb is always at least two words, a form of the verb *to be* and the past participle (third principal part) of the main verb.

     My husband's surgery *was performed* by Dr. Petronia Vanetti.
     Logarithm tables *are found* on page 210.

  See also page 18.

- Choose the voice of the verb that permits the desired emphasis.

  ACTIVE VOICE   The ballistics experts *examined* the results of the tests. (emphasis on *ballistics experts*)
  PASSIVE VOICE   The results of the tests *were examined* by the ballistics experts. (emphasis on *results of the tests*)

  ACTIVE VOICE   Juan *gave* the report. (emphasis on *Juan*)
  PASSIVE VOICE   The report *was given* by Juan. (emphasis on *report*)

- Avoid needlessly shifting from the active to the passive voice.

  NEEDLESS SHIFT IN VOICE (CONFUSING)   Management and labor representatives *discussed* the pay raise, but no decision *was reached*.
  discussed—active voice
  was reached—passive voice

  CONSISTENT VOICE   Management and labor representatives *discussed* the pay raise, but they *reached* no decision.
  discussed
  reached }—active voice

Words Often Confused and Misused

Here is a list of often confused and misused words, with suggestions for their proper use.

a, an   *A* is used before words beginning with a consonant sound; *an* is used before words with a vowel sound. (Remember: Consider sound, not spelling.)

  EXAMPLES   This is *a* banana.   The patient needs *a* unit of blood.
               This is *an* orange.   I'll call in *an* hour.

accept, except   *Accept* means "to take an object or idea offered" or "to agree to something"; *except* means "to leave out" or "excluding."

> EXAMPLES   Please *accept* this gift.
> Everyone *except* Joe may leave.

access, excess   *Access* is a noun meaning "way of approach" or "admittance"; *excess* means "greater amount than required or expected."

> EXAMPLES   The children were not allowed *access* to the laboratory.
> Her income is in *excess* of $50,000.

ad   *Ad* is a shortcut to writing *advertisement*. In formal writing, however, write out the full word. In informal writing and speech, such abbreviated forms as *ad, auto, phone, photo,* and *TV* may be acceptable.

advice, advise   *Advice* is a noun meaning "opinion given," "suggestions"; *advise* is a verb meaning "to suggest," "to recommend."

> EXAMPLES   We accepted the lawyer's *advice.*
> The lawyer *advised* us to drop the charges.

affect, effect   *Affect* as a verb means "to influence" or "to pretend"; as a noun, it is a psychological term meaning "feeling" or "emotion." *Effect* as a verb means "to make something happen"; as a noun, it means "result" or "consequence."

> EXAMPLES   The colors used in a home may *affect* the prospective buyers decision to buy.
> The new technique will *effect* change in the entire procedure.

aggravate, irritate   *Aggravate* means "to make worse or more severe." Avoid using *aggravate* to mean "to irritate" or "to vex," except perhaps in informal writing or speech.

ain't   A contraction for *am not, are not, has not, have not.* This form is regarded as substandard; careful speakers and writers do not use it.

all ready, already   *All ready* means that everyone is prepared or that something is completely prepared; *already* means "completed" or "happened earlier."

> EXAMPLES   We are *all ready* to go.
> It is *already* dark.

all right, alright   Use *all right.* In time, *alright* may become accepted, but for now *all right* is generally preferred.

> EXAMPLE   Your choice is *all right* with me.

all together, altogether   *All together* means "united"; *altogether* means "entirely."

> EXAMPLES   We will meet *all together* at the clubhouse.
> There is *altogether* too much noise in the hospital area.

almost, most   *Almost* is an adverb meaning "nearly"; *most* is an adverb meaning "the greater part of the whole."

> EXAMPLES   The emergency shift *almost* froze.
> The emergency shift worked *most* all night.

a lot, alot, allot   *A lot* is written as two separate words and is a colloquial term meaning "a large amount"; *alot* is a common miswriting for *a lot; allot* is a verb meaning "to give a certain amount."

633

Section 2
Accuracy and
Clarity Through
Standard
Language
Usage

amount, number   *Amount* refers to mass or quantity; *number* refers to items, objects, or ideas that can be counted individually.

EXAMPLES   The *amount* of money for clothing is limited.
            A large *number* of people are enrolled in the class.

and/or   This pairing of coordinate conjunctions indicates appropriate alternatives. Avoid using *and/or* if it misleads or confuses the reader, or if it indicates imprecise thinking. In the sentence, "Jones requests vacation leave for Monday–Wednesday and/or Wednesday–Friday," the request is not clear because the reader does not know whether three days' leave or five days' leave is requested. Use *and* and *or* to mean exactly what you want to say.

EXAMPLES   She is a qualified lecturer *and* consultant.
            For our vacation we will go to London *or* Madrid *or* both.

See also Virgule, page 656.

angel, angle   *Angel* means "a supernatural being"; *angle* means "corner" or "point of view." Be careful not to overuse *angle* meaning "point of view." Use *point of view, aspect,* and the like.

anywheres, somewheres, nowheres   Use *anywhere, somewhere,* or *nowhere.*

as if, like   *As if* is a subordinate conjunction; it should be followed by a subject-verb relationship to form a dependent clause. *Like* is a preposition; in formal writing *like* should be followed by a noun or a pronoun as its object.

EXAMPLES   He reacted to the suggestion *as if* he never heard of it.
            Pines, *like* cedars, do not have leaves.

as regards, in regard to   Avoid these phrases. Use *about* or *concerning.*

average, mean, median   *Average* is the quotient obtained by dividing the sum of the quantities by the number of quantities. For example, for scores of 70, 75, 80, 82, 100, the *average* is 81$\frac{2}{5}$. *Mean* may be the simple average or it may be the value midway between the lowest and the highest quantity (in the scores above, 85). *Median* is the middle number (in the scores above, 80).

balance, remainder   *Balance* as used in banking, accounting, and weighing means "equality between the totals of two sides." *Remainder* means "what is left over."

EXAMPLES   The company's bank *balance* continues to grow.
            Our shift will work overtime the *remainder* of the week.

being that, being as how   Avoid using either of these awkward phrases. Use *since* or *because.*

EXAMPLE   *Because* the bridge is closed, we will have to ride the ferry.

beside, besides   *Beside* means "alongside," "by the side of," or "not part of"; *besides* means "furthermore" or "in addition."

> EXAMPLES   The tree stands *beside* the walk.
> *Besides* the cost there is a handling charge.

bi-, semi-   *Bi-* is a prefix meaning "two" and *semi-* is a prefix meaning "half," or "occurring twice within a period of time."

> EXAMPLES   Production quotas are reviewed *biweekly*. (every two weeks)
> The board of directors meets *semiannually*. (every half year, or twice a year)

brake, break   *Brake* is a noun meaning "an instrument to stop something"; *break* is a verb meaning "to smash," "to cause to fall apart."

> EXAMPLES   The mechanic relined the *brakes* in the truck.
> If the vase is dropped, it will *break*.

can't hardly   This is a double negative; use *can hardly*.

capital, capitol   *Capital* means "major city of a state or nation," "wealth," or, as an adjective, "chief" or "main." *Capitol* means "building that houses the legislature"; when written with a capital "C," it usually means the legislative building in Washington.

> EXAMPLES   Jefferson City is the *capital* of Missouri.
> Our company has a large *capital* investment in preferred stocks.
> The *capitol* is located on Third Avenue.

cite, sight, site   *Cite* is a verb meaning "to refer to"; *sight* is a noun meaning "view" or "spectacle"; *site* is a noun meaning "location."

> EXAMPLES   *Cite* a reference in the text to support your theory.
> Because of poor *sight*, he has to wear glasses.
> This is the building *site* for our new home.

coarse, course   *Coarse* is an adjective meaning "rough," "harsh," or "vulgar"; *course* is a noun meaning "a way," "a direction."

> EXAMPLES   The sandpaper is too *coarse* for this wood.
> The creek follows a winding *course* to the river.

consensus   *Consensus* means "a general agreement of opinion." Therefore, do not write *consensus of opinion*; it is repetitious.

contact   *Contact* is overused as a verb, especially in business and industry. Consider using in its place such exact forms as *write to, telephone, talk with, inform, advise*, or *ask*.

continual, continuous   *Continual* means "often repeated"; *continuous* means "uninterrupted" or "unbroken."

> EXAMPLES   The conference has had *continual* interruptions.
> The rain fell in a *continuous* downpour for an hour.

could of   The correct form is *could have*. This error occurs because of the sound in pronouncing such statements as: We *could have* (could've) completed the work on time.

council, counsel, consul *Council* is a noun meaning "a group of people appointed or elected to serve in an advisory or legislative capacity." *Counsel* as a noun means "advice" or "attorney"; as a verb, it means "to advise." *Consul* is a noun naming the official representing a country in a foreign nation.

635

Section 2
Accuracy and
Clarity Through
Standard
Language
Usage

EXAMPLES   The club has four members on its *council*.
            The *counsel* for the defense advised him to testify.
            The *consul* from Switzerland was invited to our international tea.

criteria, criterion *Criteria* is the plural of *criterion,* meaning "a standard on which a judgment is made." Although *criteria* is the preferred plural, *criterions* is also acceptable.

EXAMPLES   This is the *criterion* that the trainee did not understand.
            On these *criteria,* the proposals will be evaluated.

device, devise *Device* is a noun meaning "a contrivance," "an appliance," "a scheme"; *devise* is a verb meaning "to invent."

EXAMPLES   This *device* will help prevent pollution of our waterways.
            We need to *devise* a safer method for drilling offshore oil wells.

different from, different than Although *different from* is perhaps more common, *different than* is also an acceptable form.

discreet, discrete *Discreet* means "showing good judgment in conduct and especially in speech." *Discrete* means "consisting of distinct, separate, or unconnected elements."

EXAMPLES   The administrative assistant was *discreet* in her remarks.
            Discrete electronic circuitry was the standard until the advent of integrated
            circuits.

dual, duel *Dual* means "double." *Duel* as a noun means a fight or contest between two people; as a verb, it means "to fight."

EXAMPLES   The car has a *dual* exhaust.
            He was shot in a *duel.*

due to Some authorities object to *due to* in adverbial phrases. Acceptable substitutes are *owing to* and *because of.*

each and every Use one or the other. *Each and every* is a wordy way to say *each* or *every.*

EXAMPLE   *Each* person should make a contribution.

except for the fact that Avoiding using this wordy and awkward phrase.

fact, the fact that Use *that.*

EXAMPLE   *That* he was late is indisputable.

field Used too often to refer to an area of knowledge or a subject.

had ought, hadn't ought Avoid using these phrases. Use *ought* or *should.*

EXAMPLE   He *should not* speak so loudly.

hisself Use *himself.*

imply, infer   *Imply* means to "express indirectly." *Infer* means to "arrive at a conclusion by reasoning from evidence."

> EXAMPLES   The supervisor *implied* that additional workers would be laid off.
> I *inferred* from her comments that I would not be one of them.

in case, in case of, in case that   Avoid using this overworked phrase. Use *if*.

in many instances   Wordy. Use *frequently* or *often*.

in my estimation, in my opinion   Wordy. Use *I believe* or *I think*.

irregardless   Though you hear this double negative and see it in print, it should be avoided. Use *regardless*.

its, it's   *Its* is the possessive form; *it's* is the contraction for *it is*. The simplest way to avoid confusing these two forms is to think *it is* when writing *it's*.

> EXAMPLES   The tennis team won *its* match.
> *It's* time for lunch.

lay, lie   *Lay* means "to put down" or "place." Forms are *lay, laid,* and *laying*. It is a transitive verb; thus it denotes action going to an object or to the subject.

> EXAMPLES   *Lay* the books on the table.
> We *laid* the floor tile yesterday.

Lie means "to recline" or "rest." Forms are *lie, lay, lain,* and *lying*. It is an intransitive verb; thus it is never followed by a direct object.

> EXAMPLES   The pearls *lay* in the velvet-lined case.
> *Lying* in the velvet-lined case were the pearls.

lend, loan   *Lend* is a verb. *Loan* is used as a noun or a verb; however, many careful writers use it only as a noun.

loose, lose   *Loose* means "to release," "to set free," "unattached," or "not securely fastened"; *lose* means "to suffer a loss."

> EXAMPLE   We turned the horses *loose* in the pasture, but we locked the gate so that we wouldn't *lose* them.

lots of, a lot of   In writing, use *many, much,* a *large amount*.

might of, ought to of, must of, would of   *Of* should be *have*. See *could of*.

off of   Omit *of*. Use *off*.

on account of   Use *because*.

one and the same   Wordy. Use *the same*.

outside of   Use *besides, except for,* or *other than*.

passed, past   *Passed* identifies an action and is used as a verb; *past* means "earlier" and is used as a modifier.

> EXAMPLES   He *passed* us going 80 miles an hour.
> In the *past,* bills were sent out each month.

personal, personnel   *Personal* is an adjective meaning "private," "pertaining to the person"; *personnel* is a noun meaning "body of persons employed."

637

Section 2
Accuracy and
Clarity Through
Standard
Language
Usage

EXAMPLES   Please do not open my *personal* mail.

He is in charge of hiring new *personnel*.

plain, plane   *Plain* is an adjective meaning "simple," "without decoration"; *plane* is a noun meaning "airplane," "tool," or "type of surface."

EXAMPLES   The *plain* decor of the room created a pleasing effect.

We worked all day checking the engine in the *plane*.

principal, principle   *Principal* means "highest," "main," or "head"; *principle* means "belief," "rule of conduct," or "fundamental truth."

EXAMPLES   The school has a new *principal*.

The refusal to take a bribe was a matter of principle.

proved, proven   Use either form.

put across   Use more exact terms, such as *demonstrate, explain, prove, establish*.

quiet, quite   *Quiet* is an adjective meaning "silent" or "free from noise"; *quite* is an adverb meaning "completely" or "wholly."

EXAMPLES   Please be *quiet* in the library.

It's been *quite* a while since I've seen him.

raise, rise   *Raise* means "to push up." Forms are *raise, raised, raising*. It is a transitive verb; thus it denotes action going to an object or to the subject.

EXAMPLES   *Raise* the window.

The technician *raised* the impedance of the circuit 300 ohms.

Rise means "to go up" or "ascend." Forms are *rise, rose, risen*. It is an intransitive verb and thus it is never followed by a direct object.

EXAMPLES   Prices *rise* for several reasons.

The sun *rose* at 6:09 this morning.

rarely ever, seldom ever   Avoid using these phrases. Use *rarely* or *seldom*.

read where   Use *read that*.

reason is because, reason why   Omit *because* and *why*.

respectfully, respectively   *Respectfully* means "in a respectful manner"; *respectively* means "in the specified order."

EXAMPLES   I *respectfully* explained my objection.

The capitals of Libya, Iceland, and Tasmania are Tripoli, Reykjavik, and Hobart, *respectively*.

sense, since   *Sense* means "ability to understand"; *since* is a preposition meaning "until now," an adverb meaning "from then until now," and a conjunction meaning "because."

EXAMPLES   At least he has a *sense* of humor.

I have been on duty *since* yesterday.

set, sit   *Set* means "to put down" or "place." Its basic form does not change: *set, set, setting*. It is a transitive verb; thus it denotes action going to an object or to the subject.

EXAMPLES   Please *set* the test tubes on the table.
              The electrician *set* the breaker yesterday.

Sit means "to rest in an upright position." Forms are *sit, sat, sitting*. It is an intransitive verb and thus it is never followed by a direct object.

EXAMPLES   Please *sit* here.
              The electrician *sat* down when the job was finished.

state   Use exact terms such as *say, remark, declare, observe*. To *state* means "to declare in a formal statement."

stationary, stationery   *Stationary* is an adjective meaning "fixed"; *stationery* is a noun meaning "paper used in letter writing."

EXAMPLES   The workbench is *stationary*.
              The school's *stationery* is purchased through our firm.

their, there, they're   *Their* is a possessive pronoun; *there* is an adverb of place; *they're* is a contraction for *they are*.

EXAMPLES   *Their* band is in the parade.
              There goes the parade.
              They're in the parade.

this here, that there   Avoid this phrasing. Use *this* or *that*.

EXAMPLE   *This* machine is not working properly.

thusly   Use *thus*.

till, until   Either word may be used.

to, too, two   *To* is a preposition; *too* is an adverb telling "How much?"; *two* is a numeral.

try and   *Try to* is generally preferred.

type, type of   In writing, use *type of*.

used to could   Use *formerly was able* or *used to be able*.

where . . . at   Omit the *at*. Write "Where is the library" (not "Where is the library at?")

who's, whose   *Who's* is the contraction for *who is; whose* is the possessive form of *who*.

EXAMPLES   *Who's* on the telephone?
              Whose coat is this?

-wise   Currently used and overused as an informal suffix in such words as *timewise, safetywise, healthwise*. Better to avoid such usage.

would of   The correct form is *would have*. This error occurs because of the sound in pronouncing such a statement as this: He *would have* (would've) come if he had not been ill.

Section 3

Accuracy and Clarity Through Mechanics

Introduction

Threw the yrs. certin, conventions in the mechanics of written communication; have developed? These convention: Or generally accepted practicen; eze the communication process for when. These conventional usages or followed! the (readers attention) can berightly focused on the c. of the wtg.

Through the years certain conventions in the mechanics of written communication have developed. These conventions, or generally accepted practices, ease the communication process, for when these conventional usages are followed, the reader's attention can be rightly focused on the content of the writing.

Which is easier to follow—the first or the second paragraph? Undoubtedly the second, because it follows generally accepted practices in the mechanics of written English. The second paragraph permits you to concentrate on what is being said; further, this paragraph is thoughtful of you, the reader, in not requiring you to spend a great deal of time in simply figuring out the words and the sentence units before beginning to understand the subject material. The matter of mechanics, in short, is a matter of convention and of courtesy to the reader.

The following are accepted practices concerning abbreviation, capitalization, numbers, plurals of nouns, punctuation, spelling, and symbols. Applying these practices to your writing will help your reader understand what you are trying to communicate.

Abbreviations

Always consult a recent dictionary for forms you are not sure about. Some dictionaries list abbreviations together in a special section; other dictionaries list abbreviated forms as regular entries in the body of the dictionary.

Always acceptable abbreviations

1. Abbreviations generally indicate informality. Nevertheless, there are a few abbreviations always acceptable when used to specify a time or a person, such as A.M. (ante meridien, before noon), BC (before Christ), AD (anno Domini, in the year of our Lord), Ms. or Ms (combined form of Miss and Mrs.), Mrs. (mistress), Mr. (mister), Dr. (doctor).

   Dr. Ann Meyer and Mrs. James Brown will arrive at 7:30 P.M.

Titles following names

2. Certain titles following a person's name may be abbreviated such as Jr. (Junior), Sr. (Senior), MD (Doctor of Medicine), SJ (Society of Jesus).

   Martin Luther King, Jr., was assassinated on April 4, 1968. (The commas before and after "Jr." are optional.)
   George B. Schimmet, MD, signed the report.

Titles preceding names

3. Most titles may be abbreviated when they precede a person's full name, but not when they precede only the last name.

   Lt. James W. Smith or Lieutenant Smith but *not* Lt. Smith
   The Rev. Arthur Bowman or the Reverend Bowman but *not* Rev. Bowman or the Rev. Bowman

4. Abbreviate certain terms only when they are used with a numeral.

A.M., P.M., BC, AD, No. (number), $ (dollars)

(Careful writers place "BC" after the numeral and "AD" before the numeral: 325 BC, AD 597)

ACCEPTABLE   He arrives at 2:30 P.M.
             Julius Caesar was killed in 44 BC.
             The book costs $12.40.

UNACCEPTABLE He arrives this P.M.
             Julius Caesar was killed a few years B.C.
             The book costs several $.

5. To avoid repetition of a term or a title appearing many times in a piece of writing, abbreviate the term or shorten the title. Write out in full the term or title the first time it appears, followed by the abbreviated or shortened form in parentheses.

Many high school students take the American College Test (ACT) in the eleventh grade; however, a number of students prefer to take the ACT in the twelfth grade.

A Portrait of the Artist as a Young Man (Portrait) is an autobiographical novel by James Joyce. *Portrait* shows the struggle of a young man in answering the call of art.

6. Generally, place a period after each abbreviation. However, there are many exceptions:

- The abbreviations of organizations, of governmental divisions, and of educational degrees usually do not require periods (or spacing between the letters of the abbreviation).

  IBM (International Business Machines), DECA (Distributive Education Clubs of America), IRS (Internal Revenue Service), AAS (Associate in Applied Science) degree.

- The U.S. Postal Service abbreviations for states are written without periods (and in all uppercase—capital—letters). For a complete list of these abbreviations, see page 257 in Chapter 7 Memorandums and Letters.

- Roman numerals in sentences and contractions are written without periods.

  Henry VIII didn't let enemies stand in his way.

- Units of measure, with the exceptions of in. (inch) and at. wt. (atomic weight), are written without periods. See 9 below.

  He weighs 186 lb and stands 6 ft tall.

7. Abbreviations of plural terms are written in various ways. Add *s* to some abbreviations to indicate more than one; others do not require the *s*. See 9 below.

ABBREVIATIONS ADDING "S"	Figs. (or Figures) 1 and 2
	20 vols. (volumes)
ABBREVIATIONS WITHOUT "S"	pp. (pages)
	ff. (and following)

Lowercase letters

8. Generally use lowercase (noncapital) letters for abbreviations except for abbreviations of proper nouns.

mpg (miles per gallon)   Btu (British thermal unit)
c.d. (cash discount)   UN (United Nations)

Units of measure as symbols

9. Increasingly, the designations for units of measure are being regarded as symbols rather than as abbreviations. As symbols, the designations have only one form—regardless of whether the meaning is singular or plural—and are written without a period.

100 kph (kilometers per hour)   200 rpm (revolutions per minute)
50 m (meters)   1 T (tablespoon)

Capitalization

There are very few absolute rules concerning capitalization. Many reputable businesses and publishers, for instance, have their own established practices of capitalization. However, the following are basic conventions in capitalization and are followed by most writers.

Sentences

1. Capitalize the first word of a sentence or a group of words understood as a sentence (except a short parenthetical statement within another).

After the party no one offered to help clean up. Not one person.

Quotations

2. Capitalize the first word of a direct quotation.

Melissa replied, "Tomorrow I begin."

Proper nouns

3. Capitalize proper nouns:

People

- Names of people and titles referring to specific persons

Frank Lloyd Wright   Mr. Secretary
Aunt Marian   the Governor

Places

- Places (geographic locations, streets), but not directions

Canada   Golden Gate Bridge
Canal Street   the Smoky Mountains
the South   the Red River

Go three blocks south; then turn west.

Groups

- Nationalities, organizations, institutions, and members of each

Indian   Bear Creek High School
British   League of Women Voters
a Rotarian   International Imports, Inc.

Calendar divisions

- Days of the week, months of the year, and special days, but not seasons of the year

  Monday   Halloween   New Year's Day
  January   spring   summer

Historic occurrences

- Historic events, periods, and documents

  World War II   the Industrial Revolution
  the Magna Carta   Battle of San Juan Hill

Religions

- Religions and religious groups

  Judaism   the United Methodist Church

Deity

- Names of the Deity and personal pronouns referring to the Deity

  God   Son of God
  Creator   His, Him, Thee, Thy, Thine

Bible

- Bible, Scripture, and names of the books of the Bible. These words are not italicized (underlined in handwriting).

  My favorite book in the Bible is Psalms.

Proper noun derivatives

4. Capitalize derivatives of proper nouns when used in their original sense.

   Chinese citizen   *but* china pattern
   Salk vaccine   *but* pasteurized milk

Pronoun I

5. Capitalize the pronoun *I*.

   In that moment of fear, I could not say a word.

Titles of publications

6. Capitalize titles of books, chapters, magazines, newspapers, articles, poems, plays, stories, musical compositions, paintings, motion pictures, and the like. Ordinarily, do not capitalize articles (*a, an, the*), coordinate conjunctions (*and, or, but*), and prepositions (*or, by, in, with*), unless they are the first word of the title. It is acceptable to capitalize prepositions of five or more letters (*against, between*).

   Newsweek (magazine)
   "The Purloined Letter" (short story)
   "Tips on Cutting Firewood" (article in a periodical)
   The Phantom of the Opera (musical)
   Man Against Himself (book)
   Madonna and Child (painting)

Titles with names

7. Capitalize titles immediately preceding or following proper names.

   Juan Perez, Professor of Computer Science
   Dr. Alicia Strumm
   Dale Jaggers, Member of Congress

Substitute names

8. Capitalize words or titles used in place of the names of particular persons. However, names denoting kinship are not

capitalized when immediately preceded by an article or a possessive.

Last week Mother and my grandmother gave a party for Sis.
Jill's dad and her uncle went hunting with Father and Uncle Bob.

Trade names

9. Capitalize trade names.

Dodge trucks   Hershey bars

Certain words with numerals

10. Capitalize the words *Figure, Number, Table,* and the like (whether written out or abbreviated) when used with a numeral.

See Figure 1. See the accompanying figure.
This is Invoice No. 6143. Check the number on the invoice.

School subjects

11. Capitalize school subjects only if derived from proper nouns (such as those naming a language or a nationality) or if followed by a numeral.

English   Spanish
history   History 1113
algebra   Algebra II

Numbers

The problem with numbers is knowing when to use numerals (figures) and when to use words.

Numerals (Figures)

Dates, houses. telephones, ZIPs, specific amounts, math, etc.

1. Use numerals for dates, house numbers, telephone numbers, ZIP codes, specific amounts, mathematical expressions, and the like.

July 30, 1987 *or* 30 July 1987   857–5969
600 Race Street   10:30 P.M.
61 percent   Chapter 12, p. 14

Decimals

2. Use numerals for numbers expressed in decimals. Include a zero before the decimal point in writing fractions with no whole number (integer).

12.006
0.01
0.500 (The zeros following "5" show that accuracy exists to the third decimal place.)

10 and above; three or more words

3. Use numerals for number 10 and above or numbers that require three or more written words.

She sold 12 new cars in $2\frac{1}{2}$ hours.

Several numbers close together

4. Use numerals for several numbers (including fractions) that occur within a sentence or within related sentences

The recipe calls for 3 cups of sugar, $\frac{1}{2}$ teaspoon of salt, 2 sticks of butter, and $\frac{1}{4}$ cup of cocoa.

The report for this week shows that our office received 127 telephone calls, 200 letters, 30 personal visits, and 3 faxes.

Adjacent numbers

5. Use numerals for one of two numbers occurring next to each other.

12 fifty-gallon containers
two hundred 24 × 36 mats

Words

Approximate or indefinite numbers

1. Use words for numbers that are approximate or indefinite.

If I had a million dollars, I'd buy a castle in Ireland.
About five hundred machines were returned because of faulty assembling.

Fractions

2. Use words for fractions.

The veneer is one-eight of an inch thick.
Our club receives three-fourths of the general appropriation.

Below 10

3. Use words for numbers below 10.

There are four quarts in a gallon.

Beginning of sentence

4. Use words for a number or related numbers that begin a sentence.

Fifty cents is a fair entrance fee.
Sixty percent of the freshmen and seventy percent of the sophomores come from this area.

Note: If using words for a number at the beginning of a sentence is awkward, recast the sentence.

UNACCEPTABLE   2,175 freshmen are enrolled this semester.
AWKWARD   Two thousand, one hundred and seventy-five freshmen are enrolled this semester.
ACCEPTABLE   This semester 2,175 freshmen are enrolled.

Adjacent numbers

5. Use words for one of two numbers occurring next to each other.

50 six-cylinder cars   four 3600-pound loads

Repeating a number

6. Except in special instances (such as in order letters or legal documents), it is not necessary to repeat a written-out number by giving the numerals in parentheses.

ACCEPTABLE   The trumpet was invented five years ago.
UNNECESSARY   The trumpet was invented five (5) years ago.

Plurals of Nouns

The English language has been greatly influenced not only by the original English but also by a number of other languages, such as Latin, Greek, and French. Some nouns in the English language, especially those frequently used, continue to retain plural forms from the original language.

1. Most nouns in the English language form their plural by adding *-s* or *-es*. Add *-s* unless the plural adds a syllable when the singular nouns ends in *s, ch* (soft), *sh, x,* and *z*.

pencil	pencils		mass	masses
desk	desks		church	churches
flower	flowers		leash	leashes
boy	boys		fox	foxes
post	posts		buzz	buzzes

2. If a noun ends in *y* preceded by a consonant sound, change the *y* to *i* and add *-es*.

history	histories
penny	pennies

For other nouns ending in *y*, add *-s*.

monkey	monkeys
valley	valleys

3. A few nouns ending in *f* or *fe* change the *f* to *v* and add *-es* to form the plural. These nouns are

calf	calves	life	lives	shelf	shelves
elf	elves	loaf	loaves	thief	thieves
half	halves	lead	leaves	wife	wives
knife	knives	sheaf	sheaves	wolf	wolves

In addition, several nouns may either add *-s* or change the *f* to *v* and add *-es*. These nouns are

beef	beefs (slang for "complaints"), beeves
scarf	scarfs, scarves
staff	staffs (groups of officers), staffs or staves (poles or rods)
wharf	wharfs, wharves

4. Most nouns ending in *o* add *-s* to form the plural. Among the exceptions, which add *-es,* are the following:

echo	echoes	potato	potatoes
hero	heroes	tomato	tomatoes
mosquito	mosquitoes	veto	vetoes

5. Most compound nouns form the plural with a final *-s* or *-es*. A few compounds pluralize by changing the operational part of the compound noun.

handful	handfuls	son-in-law	sons-in-law
go-between	go-betweens	(and other in-law compounds)	
good-by	good-bys	passer-by	passers-by
court-martial	courts-martial	editor in chief	editors in chief

6. A few nouns form the plural by an internal vowel change. These nouns are

foot	feet	mouse	mice
goose	geese	tooth	teeth
louse	lice	woman	women
man	men		

-en plurals

7. A few nouns form the plural by adding *-en* or *-ren*. These nouns are

ox	oxen
child	children
brother	brothers, brethren

Foreign plurals

8. Several hundred English nouns, originally foreign, have two acceptable plural forms: the original form and the conventional American English *-s* or *-es* form.

memorandum	memoranda	memorandums
curriculum	curricula	curriculums
index	indices	indexes
criterion	criteria	criterions

Some foreign nouns always keep their original forms in the plural.

crisis	crises	die	dice
analysis	analyses	alumna	alumnae (feminine)
bacterium	bacteria	alumnus	alumni (masculine)
basis	bases	thesis	theses
ovum	ova	fabliau	fabliaux

Terms being discussed

9. Letters of the alphabet, signs, symbols, and words used as a topic of discussion form the plural by adding the apostrophe and *-s*.

The *i*'s and *e*'s are not clear.
The sentence has too many *and*'s and *but*'s.

Note: Abbreviations and numbers form the plural regularly, that is, by adding *s* or *-es*.

The three *PhDs* were born in the 1950s.
The number has two *sixes*.

Same singular and plural forms

10. Some nouns have the same form in both the singular and the plural. In general, names of fish and of game birds are included in this group.

cod	swine	sheep
trout	cattle	quail
deer	species	corps

Two forms

11. Some nouns have two forms, the singular indicating oneness or a mass, and the plural indicating different individuals or varieties within a group.

a string of fish	four little fishes
a pocketful of money	money (or monies) appropriated by Congress
fresh fruit	fruits from Central America

Only plural forms

12. Some nouns have only plural forms. A noun is considered singular, however, if the meaning is singular.

measles (Measles is a contagious disease.) mumps
economics   news
mathematics   physics
dynamics   molasses

A noun is considered plural if the meaning is plural.

scissors (The scissors are sharp.)
pants

Punctuation

Punctuation is a necessary part of the written language. Readers and writers depend on marks of punctuation to help prevent vagueness by indicating pauses and stops, separating and setting off various sentence elements, indicating questions and exclamations, and emphasizing main points while subordinating less important sentence content.

Punctuation usage is presented here according to marks used primarily at the end of a sentence (period, question mark, and exclamation point), internal marks that set off and separate (comma, semicolon, colon, dash, and virgule), enclosing marks always used in pairs (quotation marks, parentheses, and brackets), and punctuation of individual words and of terms (apostrophe, ellipsis points, hyphen, and italics).

Punctuation Marks Used Primarily at the End of a Sentence

Period (.)

Statement,
command, or
request

1. Use a period at the end of a sentence (and of words understood as a sentence) that makes a statement, gives a command, or makes a request (except a short parenthetical sentence within another).

James Naismith invented the game of basketball. (statement)
Choose a book for me. (command)
No. (understood as a sentence)
Naismith (he was a physical education instructor) wanted to provide indoor exercise and competition for students. (parenthetical sentence with no capital and no period)

See also Sentence Fragments, pages 626–627.

Note: The polite request phrased as a question is usually followed by a period rather than a question mark.

Will you please send me a copy of your latest sale catalog.

Initials and
abbreviations

2. Use a period after initials and most abbreviations.

Dr. H. H. Wright   p. 31
437 mi   No. 7

The abbreviations of organizations, of governmental divisions, and of educational degrees usually omit periods.

BA (Bachelor of Arts) degree
NFL (National Football League)
FBI (Federal Bureau of Investigation)

The U.S. Postal Service abbreviations for states (see page 257 in Chapter 7 Memorandums and Letters for a list of the abbreviations), contractions, parts of names used as a whole, roman numerals in sentences, and units of measure with the exception of in. (inch) and at. wt. (atomic weight) are written without periods. (Designations for units of measure are increasingly being regarded as symbols rather than as abbreviations. See item 9, page 642.)

Hal C. Johnson IV, who lives 50 mi away in Sacramento, CA, won't be present.

See also Abbreviations, item 6, page 641.

Outline

3. Use a period after each number and letter symbol in an outline.

I.
  A.
  B.

For other examples see pages 139–140.

Decimals

4. Use a period to mark decimals.

$10.52
A reading of 1.260 indicates a full charge in a battery; 1.190, a half charge.

Question Mark (?)

Questions

1. Use a question mark at the end of every direct question, including a short parenthetical question within another sentence.

Have the blood tests been completed?
When you return (When will that be?), please bring the reports.

Note: An indirect question is followed by a period.

He asked if John were present.

Note: A polite request is usually followed by a period.

Will you please close the door.

Uncertainty

2. Use a question mark in parentheses to indicate there is some question as to certainty or accuracy.

Chaucer, 1343(?)–1400
The spindle should revolve at a slow (?) speed.

Exclamation Point (!)

Sudden or strong emotion or surprise

1. Use an exclamation point after words, phrases, or sentences (including parenthetical expressions) to show sudden or strong emotion or force, or to mark the writer's surprise.

   What a day!
   The computer (!) made a mistake.

   Note: The exclamation point can be easily overused, thus causing it to lose its force. Avoid using the exclamation point in place of vivid, specific description.

Internal Marks That Set Off and Separate

Comma (,)

Items in a series

1. Use a comma to separate items in a series. The items may be words, phrases, or clauses.

   How much do you spend each month for food, housing, clothing, and transportation?

Compound sentence

2. Use a comma to separate independent clauses joined by a coordinate conjunction (*and, but, or, either, neither, nor,* and sometimes *for, so, yet*). (An independent clause is a group of related words that have a subject and verb and that could stand alone as a sentence.)

   There was a time when homemakers had few interests outside the home, but today they are leaders in local and national affairs.

   Note: Omission of the coordination conjunction results in a comma splice (comma fault), that is, a comma incorrectly splicing together independent clauses. (See page 615.)

   If the clauses are short, the comma may be omitted.

   I aimed and I fired.

   See also Semicolon, item 1, page 653.

Equal adjectives

3. Use a comma to separate two adjectives of equal emphasis and with the same relationship to the noun modified.

   The philanthropist made a generous, unexpected gift to our college.

   Note: If *and* can be substituted for the comma or if the order of the adjectives can be reversed without violating the meaning, the adjectives are of equal rank and a comma is needed.

Misreading

4. Use a comma to prevent misreading.

   Besides Sharon, Ann is the only available organist.
   Ever since, he has gotten to work on time.

Number units

5. Use a comma to separate units in a number of four or more digits (except telephone numbers, ZIP numbers, house numbers, and the like).

2,560,781   7,868 *or* 7868

651

Section 3
Accuracy and
Clarity Through
Mechanics

Note: The comma may be omitted from four-digit numbers.

Nonrestrictive modifiers

6. Use commas to set off nonrestrictive modifiers, that is, modifiers which do not limit or change the basic meaning of the sentence.

The Guggenheim Museum, *designed by Frank Lloyd Wright,* is in New York City. (nonrestrictive modifier; commas needed)

A museum *designed by Frank Lloyd Wright* is in New York City. (restrictive modifier; no comma needed)

Leontyne Price, *who is a world-renowned soprano,* was awarded the Presidential Medal of Freedom. (nonrestrictive modifier; commas needed)

A world-renowned soprano *who was awarded the Presidential Medal of Freedom* is Leontyne Price. (restrictive modifier; no commas needed)

Introductory adverb clause

7. Use a comma to set off an adverb clause at the beginning of a sentence. An adverb clause at the end of a sentence is not set off.

When you complete these requirements, you will be eligible for the award.

If given the proper care and training, a dog can be an affectionate and obedient pet.

A dog can be an affectionate and obedient pet *if given the proper care and training.*

Introductory verbal modifier

8. Use a comma to set off a verbal modifier (participle or infinitive) at the beginning of a sentence.

Experimenting in the laboratory, Sir Alexander Fleming discovered penicillin. (participle)

To understand the continents better, researchers must investigate the oceans. (infinitive)

Appositives

9. Use commas to set off an appositive. (An appositive is a noun or pronoun that follows another noun or pronoun and renames or explains it.)

Joseph Priestly, *a theologian and scientist,* discovered oxygen.
Opium painkillers, *such as heroin and morphine,* are narcotics.

Note: The commas are usually omitted if the appositive is a proper noun or is closely connected with the word it explains.

My friend Mary lives in Phoenix.
The word *occurred* is often misspelled.

Parenthetical expressions, conjunctive adverbs

10. Use commas to set off a parenthetical expressions or a conjunctive adverb.

A doughnut, *for example,* has more calories than an apple.
I was late; *however,* I did not miss the plane.

See also Semicolon, items 2 and 5, page 654.

Address, dates

11. Use commas to set off each item after the first in an address or a date.

> My address will be 1045 Carpenter Street, Columbia, Missouri 65201, after today. (House number and street are considered one item; state and ZIP code are considered one item.)
> Thomas Jefferson died on July 4, 1826, at Monticello. (Month and day are considered one item.)

> *Note:* If the day precedes the month or if the day is not given, omit the commas.

> Abraham Lincoln was assassinated on 14 April 1865.
> In July 1969 humans first landed on the moon.

Quotations

12. Use commas to set off the *he said* (or similar matter) in a direct quotation.

> "I am going," Mary responded.
> "If you need help," he said, "a student assistant is in the library."

Person or thing addressed

13. Use commas to set off the name of the person or thing addressed.

> If you can, Ms. Yater, I would prefer that you attend the meeting.
> My dear car, we are going to have a good time this weekend.

Mild interjections

14. Use commas to set off mild interjections, such as *well, yes, no, oh*.

> Oh, this is satisfactory.
> Yes, I agree that farming is still the nation's single largest industry.

Inverted name

15. Use commas in an inverted name to set off a surname from a given name.

> Adams, Lucius C., is the first name on the list.

Title after a name

16. Use commas to set off a title following a name. (Setting off Junior or Senior, or their abbreviations, following a name is optional.)

> Patton A. Houlihan, President of Irish Imports, is here.
> Gregory McPhail, DDS, and Harvey D. Lott, DVM, were classmates.

Contrasting elements

17. Use commas to set off contrasting elements.

> The harder we work, the sooner we will finish.
> Leif Ericksson, not Columbus, discovered the North American continent.

Elliptical clause

18. Use a comma to indicate understood words in an elliptical clause.

> Tom was elected president; Jill, vice-president.

Inc., Ltd.

19. Use commas to set off the abbreviation for *incorporated* or *limited* from a company name. (Newer companies tend to omit the commas.)

> Drake Enterprises, Inc., is our major competitor.
> I believe that Harrells, Ltd., will answer our request.

Introductory elements	20. Use a comma to introduce a word, phrase, or clause.

His destination was clearly indicated, New York City.
I told myself, you can do this if you really want to.

See also Colon, item 1, page 654.

Before conjunction "for"	21. Use a comma to precede *for* when used as a conjunction, to prevent misreading.

The plant is closed, for the employees are on strike.
The plant has been closed for a week. (No comma; *for* is a preposition, not a conjunction.)

Correspondence	22. Use a comma to follow the salutation and complimentary close in a social letter and usually the complimentary close in a business letter.

Dear Mother,   Sincerely,
Dear Lynne,   Yours truly,

Note: The salutation in a business letter is usually followed by a colon; a comma may be used if the writer and the recipient know each other well. Some newer business letter formats omit all punctuation following the salutation and the complimentary close. For a full discussion, see pages 256 and 258, Chapter 7 Memorandums and Letters.

Tag question	23. Use commas to set off a tag question (such as *will you, won't you, can you*) from the remainder of the sentence.

You will write me, won't you?

Absolute phrase	24. Use commas to set off an absolute phrase. An absolute phrase (also called a nominative absolute) consists usually of a participle phrase plus a subject of the participle and has no grammatical connection with the clause to which it is attached.

I fear the worst, *his health being what it is.*

Semicolon (;)

Independent clauses, no coordinate conjunction	1. Use a semicolon to separate independent clauses not joined by a coordinate conjunction (*and, but, or*).

Germany has a number of well-known universities; several of them have been in existence since the Middle Ages.
There are four principal blood types; the most common are O and A.

Note 1: If a comma is used instead of the needed semicolon, the mispunctuation is called a comma splice, or comma fault (see page 615).

Note 2: If the semicolon is omitted, the result is a run-on, or fused, sentence (see page 626).

Short, emphatic clauses may be separated by commas.

I came, I saw, I conquered.

Independent clauses,
transitional
connective

2. Use a semicolon to separate independent clauses joined by a transitional connective. Transitional connectives include conjunctive adverbs such as *also, however, moreover, nevertheless, then, thus,* and explanatory expressions such as *for example, in fact, on the other hand.*

> We have considered the historical background of the period; thus we can consider its cultural achievements more intelligently.
> During the Renaissance the most famous Humanists were from Italy; for example, Petrarch, Bocaccio, Ficino, and Pico della Mirandola were all of Italian birth.

See also Comma, item 10, page 651.

Certain independent
clauses, coordinate
conjunction

3. Use a semicolon to separate two independent clauses joined by a coordinate conjunction when the clauses contain internal punctuation or when the clauses are long.

> The room needs a rug, new curtains, and a lamp; but my budget permits only the purchase of a lamp.
> Students in an occupational program of study usually have little time for electives; but these students very often want to take courses in the humanities.

Items in series

4. Use a semicolon to separate items in a series containing internal punctuation.

> The three major cities in our itinerary are London, Ontario, Canada; Washington, DC, USA; and Tegucigalpa, Honduras, Central America.
> The new officers are Betty Harrison, president; Corkren Samuels, vice-president; and Tony McBride, secretary.

Examples

5. Use a semicolon to separate an independent clause containing a list of examples from the preceding independent clause when the list is introduced by *that is, for example, for instance,* or a similar expression.

> Many great writers have had to overcome severe physical handicaps; for instance, John Milton, Alexander Pope, and James Joyce were all handicapped.

Colon (:)

List or series

1. Use a colon to introduce a list or series of items. An expression such as *the following, as follows,* or *these* often precedes the list.

> A child learns responsibility in three ways: by example, by instruction, and by experience.
> The principal natural fibers used in the production of textile fabrics include the following: cotton, wool, silk, and linen.

See also Comma, item 20, page 653.

Explanatory
clause

2. Use a colon to introduce a clause that explains, reinforces, or gives an example of a preceding clause or expression.

> Until recently, American industry used the English system of linear measure as standard: the common unit of length was the inch.
>
> "Keep cool: it will be all one a hundred years hence." (Emerson)

Emphatic appositive

3. Use a colon to direct attention to an emphatic appositive.

> We have overlooked the most obvious motive: love.
> That leaves me with one question: When do we start?

Quotation

4. Use a colon to introduce a long or formal quotation.

> My argument is based on George Meredith's words: "The attitudes, gestures, and movements of the human body are laughable in exact proportion as that body reminds us of a machine."

Formal greeting

5. Use a colon to follow a formal greeting (usually in a business letter).

> Dear Ms. Boxeman: Arletha Balih: Greenway, Inc.:

> See also Comma, item 22, page 653.

Relationships

6. Use a colon to indicate relationships such as volume and page, ratio, and time.

> 42:81–90 (volume 42, pages 81–90)   x:y
> Genesis 4:8   3:1
> 2:50 A.M.

Dash (—)

The dash generally indicates emphasis or a sudden break in thought. Often the dash is interchangeable with a less strong punctuation mark. If emphasis is desired, use a dash; if not, use an alternate punctuation mark (usually a comma, colon, or parentheses). There is no spacing before or after the dash within a sentence.

> I want one thing out of this agreement—my money.
> My mother—she is the president of the company—will call this to the attention of the board of directors.
> If we should succeed—God help us!—all mankind will profit.

> *Note:* Often the writer has choices among dashes, parentheses, and commas. Dashes emphasize; parentheses subordinate; and commas imply additional information.

Sudden change

1. Use a dash to mark a sudden break or shift in thought.

> The murderer is—but perhaps I shouldn't spoil the book for you.
> And now to the next point, the causes of—did someone have a question?

Appositive series

2. Use a dash to set off a series of appositives.

> Four major factors—cost, color, fabric, and fit—influence the purchase of a suit of clothes.

Because of these qualities—beauty, durability, portability, divisibility, and uniformity of value—gold and silver have gradually displaced all other substances as material for money.

Summarizing clause

3. Use a dash to separate a summarizing clause from a series.

Tests, a term paper, and class participation—these factors determined the student's grade.

Note: The summarizing clause usually begins with *this, that, these, those,* or *such.*

Emphasis

4. Use a dash to set off material for emphasis.

Flowery phrases—regardless of the intent—have no place in reports.

Sign of omission

5. Use a dash to indicate the omission of letters or words.

The only letters we have in the mystery word are —*a—lm.*

Virgule (Commonly Called "Slash") (/)

Alternative

1. Use a virgule to indicate appropriate alternatives.

Identify/define these persons, occurrences, and terms.

Poetry

2. Use a virgule to separate run-in lines of poetry. For readability, space before and after the virgule.

"Friends, Romans, countrymen, led me your ears. / I come to bury Caesar, not to praise him." (Shakespeare, *Julius Caesar*)

Per

3. Use a virgule to represent *per* in abbreviations.

12 ft/sec   260 mi/hr

Time

4. Use a single virgule to separate divisions of a period of time.

the fiscal year 1997/98

Enclosing Marks Always Used in Pairs

Quotation Marks (" ")

Direct quotations

1. Use quotation marks to enclose every direct quotation.

The American Heritage Dictionary defines dulcimer as "a musical instrument with wire strings of graduated lengths stretched over a sound box, played with two padded hammers or by plucking."
"Marriage is popular," said George Bernard Shaw, "because it combines the maximum of temptation with the maximum of opportunity."

Note: Quotations of more than one paragraph have quotation marks at the beginning of each paragraph and at the end of the last paragraph. Long quotations, however, are usually set off by indentation, eliminating the need for quotation marks.

Titles

2. Use quotation marks to enclose titles of magazine articles, short poems, songs, television and radio shows, and speeches.

Included in this volume of Poe's works are the poem "Annabel Lee," the essay "The Philosophy of Literary Criticism," and the short story "The Black Cat."

Note: Titles of magazines, books, newspapers, long poems, plays, operas and musicals, motion pictures, ships, trains, and aircraft are italicized (underlined in handwriting).

Different usage level

3. Use quotation marks to distinguish words on a different level of usage.

The ambassador and her delegation enjoyed the "good country eating."

Note: Be sparing in placing quotation marks around words used in a special sense, for this practice is annoying to many readers. Rather than apologizing for a word with quotation marks, either choose another word or omit the quotation marks. Avoid using quotation marks for emphasis.

Nicknames

4. Use quotation marks to enclose nicknames.

William J. "Happy" Kiska is our club adviser.

Note: Quotation marks are usually omitted from a nickname that is well known (Babe Ruth, Teddy Roosevelt) or after the first use in a piece of writing.

Single quotation marks

5. Use single quotation marks (' ') for a quotation within a quotation.

"I was puzzled," confided Mary, "when Jim said that he agreed with the writer's words, 'The world wants to be deceived.'"
This writer states, "Of all the Sherlock Holmes stories, 'The Red-Headed League' is the best plotted."

Own title not quoted

6. At the beginning of a piece of writing, do not put your own title in quotation marks (unless the title is a quotation).

How to Sharpen a Drill Bit
The Enduring Popularity of the Song "White Christmas"

With other punctuation marks

7. Use quotation marks properly with other marks of punctuation.

Period or comma

- The closing quotation mark follows the period or comma.

"File all applications before May," the director of personnel cautioned, "if you wish to be considered for summer work."

Colon or semicolon

- The closing quotation mark precedes the colon or semicolon.

I have just finished reading Shirley Ann Grau's "The Black Prince"; the main character is a complex person.

Question mark,
exclamation
point, or dash

- The closing quotation mark precedes the question mark, exclamation point, or dash when these punctuation marks refer to the entire sentence. The closing quotation mark follows the question mark, exclamation point, or dash when these punctuation marks refer only to the quoted material.

  Who said, "I cannot be here tomorrow"?
  Was it you who yelled, "Fire!"?
  Bill asked, "Did you receive the telegram?"

Parentheses ( )

Additional
material

1. Use parentheses to enclose additional material remotely connected with the remainder of the sentence.

   If I can find a job (I hope I won't need a driver's license), I will pay part of my college expenses.
   Ernest Hemingway (1899–1961) won the Nobel Prize in literature.

   Note: The writer may have a choice among using parentheses, dashes, or commas. Dashes emphasize; parentheses subordinate; and commas imply additional information.

Itemizing

2. Use parentheses to enclose numbers of letters that mark items in a list.

   Government surveys indicate that students drop out of school because they (1) dislike school, (2) think it would be more fun to work, and (3) need money for themselves and their families.

"See" references

3. Use parentheses to enclose material within a sentence directing the reader to see other pages, charts, figures, and the like.

   The average life expectancy in the United States is 70 years (see Figure 3).

Capitalization and
punctuation with
parentheses

4. Use capitalization and punctuation properly within parentheses.

Capitalization

- Do not capitalize the first word of a sentence enclosed in parentheses within a sentence.

  The table shows the allowable loads on each beam in kips (a kip is 1000 pounds).

Period

- Omit the period end punctuation in a sentence enclosed in parentheses within a sentence.

  The Old North Church (the name is now Christ Church) is the oldest church building in Boston.

Question mark,
exclamation
point

- Use a needed question mark or exclamation point with matter enclosed in parentheses.

  Pour the footings below the frost line (What is the frost line?) for a stable foundation.

Separate
sentence

- If the matter enclosed in parentheses is a separate sentence, place the end punctuation inside the closing parentheses.

  The regular heating system will be sufficient. (Infrared heaters are available for spotheating.)

Brackets ([ ])

Insertion in
quotation

1. Use brackets for parentheses within parentheses.

   Susan M. Jones (a graduate of Teachers' College [now the University of Southern Mississippi]) was recognized as Alumna of the Year.

2. Use brackets to insert comments or explanations in quotations.

   "Good design [of automobiles] involves efficient operation, sound construction, and pleasing form."
   "It is this decision [*Miller* v. *Adams*] that parents of juvenile offenders will long remember," said the judge.

Sic in quotation

3. Use brackets to enclose the Latin word *sic* ("so," "thus") to indicate strange usage or an error such as misspelling or incorrect grammar in a quotation.

   According to the report, "Sixty drivers had there [sic] licenses revoked."

Documentation

4. Use brackets to indicate missing or unverified information in documentation.

   James D. Amo. *How to Hang Wallpaper.* [Boston: McGuire Institute of Technology.] 1985.

Punctuation of Individual Words and of Terms

Apostrophe (')

Contractions

1. Use an apostrophe to take the place of a letter or letters in a contraction.

   I'm (I am)
   we've (we have)
   Don't (Do not) come until one o'clock (of the clock).

Singular
possessive

2. Use an apostrophe to show the possessive form of singular nouns and indefinite pronouns.

   citizen's responsibility   Rom's car
   someone's book   everybody's concern

   Note: To form the possessive of a singular noun, add the apostrophe + *s.*

   doctor + ' + s = doctor's, as in doctor's advice.
   Keats + ' + s = Keats's, as in Keats's poems.

   Note: The *s* may be omitted in a name ending in *s,* especially if the name has two or more syllables.

   James's (or James') book
   Ms. Tompkins's (or Ms. Thompkins') cat

Note: Personal pronouns *(his, hers, its, theirs, ours)* do not need the apostrophe because they are already possessive in form.

Plural possessive

3. Use an apostrophe to show the possessive form of plural nouns.

   boys' coats   children's coats

   Note: To show plural possessive, first form the plural; if the plural noun ends in *s,* add only an apostrophe. If the plural noun does not end in *s,* add an apostrophe + *s.*

Certain plurals

4. Use an apostrophe to form the plural of letters and words used as words.

   Don't use too many *and's,* and eliminate *I's* from your report.

   Note: The apostrophe is usually omitted in the plurals of abbreviations and numbers.

Ellipsis Points (. . . or . . .)

Omission sign in quotations

1. Use ellipsis points (plural form: ellipses) to indicate that words have been left out of quoted material. Three dots show that words have been omitted at the beginning of a quoted sentence or within a quoted sentence. Four dots show that words have been omitted at the end of a quoted sentence (the fourth dot is the period at the end of the sentence).

   "The average American family spent about $4,000 on food . . . in 1984."
   "The adoption of standard time in North America stems from the railroad's search for a solution to their chaotic schedules. . . . In November, 1883, rail companies agreed to set up zones for each 15 degrees of longitude, with uniform time throughout each zone."

Hesitation in dialogue

2. Use ellipsis points to indicate hesitation, halting speech, or an unfinished sentence in dialogue.

   "If . . . if it's all right . . . I mean . . . I don't want to cause any trouble," the bewildered child stammered.

Hyphen (-)

Word division

1. Use a hyphen to separate parts of a word divided at the end of a line. (Divide words only between syllables.) Careful writers try to avoid dividing a word because divided words may impede readability.

   On April 15, 1912, the *Titanic* sank after colliding with an iceberg.

Compound numbers; fractions

2. Use a hyphen to separate parts of compound numbers and fractions when they are written as words.

   seventy-four people   twenty-two cars
   one-eighth of an inch   one-sixteenth-inch thickness

Compound
adjectives

3. Use a hyphen to separate parts of compound adjectives when they precede the word modified.

an eighteenth-century novelist   40-hour week

Compound nouns

4. Use a hyphen to separate parts of many compound nouns.

brother-in-law   U-turn
kilowatt-hour   proof-of-purchase

Note: Many compound nouns are written as a single word, such as *notebook* and *blueprint.* Others are written as two words without the hyphen, such as *card table* and *steam iron.* If you do not know how to write a word, look it up in a dictionary.

Compound verbs

5. Use a hyphen to separate parts of a compound verb.

brake-test   oven-temper

Numbers or dates

6. Use a hyphen to separate parts of inclusive numbers or dates.

pages 72–76   the years 1988–92

Prefixes

7. Use a hyphen to separate parts of some words whose prefix is separated from the main stem of the word.

ex-president   self-respect
semi-invalid   pre-Renaissance

Note: A good dictionary is the best guide for determining which words are hyphenated.

Italics (Underlining)

Italics *(such as these words)* are used in print; the equivalent in handwriting is underlining.

Titles

1. Italicize (underline) titles of books, magazines, newspapers, long poems, plays, operas and musicals, motion pictures, ships, trains, and aircraft.

At the library yesterday, I checked out the book *Roots,* read this month's *Reader's Digest,* looked at the sports section in the *Daily Register,* and listened to parts of *The Phantom of the Opera.*

Note: Use quotation marks to enclose titles of magazine articles, book chapters, short poems, songs, television and radio shows, and speeches.

Do not italicize or put in quotation marks titles of sacred writings, editions, series, and the like: Bible, Psalms, Anniversary Edition of the Works of Mark Twain.

Terms as such

2. Italicize words, letters, or figures when they are referred to as such.

People often confuse *to* and *too.*
I cannot distinguish between your *a's* and *o's.*

Foreign terms

3. Italicize words and phrases that are considered foreign.

His novel is concerned with the *nouveau riche.*
This item is included gratis. ("Gratis" is no longer considered foreign.)

Emphasis

4. Italicize a word or phrase for special emphasis. If the emphasis is to be effective, however, you must use italics sparingly.

My final word is *no.*

Spelling

Because of the strong influence of other languages, spelling in the English language is somewhat irregular.

Suggestions for Improving Spelling

1. If you have difficulty with spelling, consider several helpful suggestions:
2. *Keep a study list of words misspelled.* Review the list often, dropping words that you have learned to spell and adding new spelling difficulties.
3. *Attempt to master the spelling of these words from the study list.* Use any method that is successful for you. Some students find that writing one or several words on a card and studying them while riding to school or waiting between classes is an effective technique. Some relate the word in some way, such as "There is 'a rat' in sepa*rat*e."
4. *Use a dictionary.* It is the poor speller's best friend. If you have some idea of the correct spelling of a word but are not sure, consult a dictionary. If you have no idea about the correct spelling, get someone to help you find the word in a dictionary. Or look in a dictionary designed for poor spellers, which lists words by their common misspellings and then gives the correct spelling.
5. *Use an electronic spellchecker.* These spellcheckers indicate misspelled words including typos, and typically they offer choices to help you select the appropriate word for a particular usage. Many spellcheckers, however, include few proper nouns or specialized words often confused and misused (see pages 631–638). For instance, a spellchecker likely will not indicate the errors in this sentence: "After eating to much pizza, I got quiet sick—a phenomena for me." (After eating too much pizza, I got quite sick—a phenomenon for me.")
6. *Proofread everything you write.* Look carefully at every word within a piece of writing. If a word does not look right, check its spelling in a dictionary.
7. *Take care in pronouncing words.* Words sometimes are misspelled because of problems in pronouncing or hearing the words. Examples: *prompness* (misspelled) for *promptness, accidently* (misspelled) for *accidentally, sophmore* (misspelled) for *sophomore.* Pronunciation, of course, is not a guide for a small portion of the words in our language (typically words not of English origin). Examples: *pneumonia, potpourri, tsunami, xylophone.*

Although spelling may be difficult, it can be mastered—primarily because many spelling errors are a violation of conventional practices for use of *ie* and *ei* and of spelling changes when affixes are added.

Using *ie* and *ei*

The following jingle sums up most of the guides for correct *ie* and *ei* usage:

> Use *i* before *e*,
> Except after *c*,
> Or when sounded like *a*,
> As in *neighbor* and *weigh*.

- Generally use *ie* when the sound is a long *e* after any letter except *c*.

believe	chief
grief	niece
piece	relieve

- Generally use *ei* after *c*.

deceive	receive	receipt

  Note: An exception occurs when the combination of letters *cie* is sounded *sh;* in such instances *c* is followed by *ie*.

sufficient	efficient	conscience

- Generally use *ei* when the sound is *a*.

neighbor	freight	sleigh
weigh	reign	vein

Spelling Changes When Affixes Are Added

An affix is a letter or syllable added either at the beginning or at the end of a word to change its meaning. The addition of affixes, whether prefix or suffix, often involves spelling changes.

Prefixes

A prefix is a syllable added to the beginning of a word. One prefix may be spelled in several different ways, usually depending on the beginning letter of the base word. For example, *com, con, cor,* and *co* are all spellings of a prefix meaning "together, with." They are used to form such words as *commit, collect,* and *correspond.*

Common Prefixes

Following are common prefixes and illustrations showing how they are added to base words. The meaning of the prefix is in parentheses.

ad (to, toward) In adding the prefix *ad* to a base, the *d* often is changed to the same letter as the beginning letter of the base.

    ad + breviate = abbreviate
    ad + commodate = accommodate

com (together, with) The spelling is *com* unless the base word begins with *l* or *r;* then the spelling is *col* and *cor,* respectively.

com + mit = commit
com + lect = collect
com + respond = correspond

de (down, off, away) This prefix is often incorrectly written *di.* Note the correct spellings of words using this prefix.

describe   desire
despair   destroyed

dis (apart, from, not) The prefix *dis* is usually added unchanged to the base word.

dis + trust = distrust
dis + satisifed = dissatisfied

in (not) The consonant *n* often changes to agree with the beginning letter of the base word.

in + reverent = irreverent
in + legible = illegible

The *n* may change to *m.*

in + partial = impartial
in + mortal = immortal

sub (under) The consonant *b* sometimes changes to agree with the beginning letter of the base word.

sub + marine = submarine
sub + let = sublet
sub + fix = suffix
sub + realistic = surrealistic

un (not) This is added unchanged.

un + able = unable
un + fair = unfair

Suffixes

A suffix is a letter or syllable added to the end of a word. One suffix may be spelled in several different ways, such as *ance* and *ence.* Also, the base word may require a change in form when a suffix is added. Because of these possibilities, adding suffixes often causes spelling difficulties.

Common Suffixes

Learning the following suffixes and the spelling of the exemplary words will improve your vocabulary and spelling. The meaning of the suffix is in parentheses.

able, ible (capable of being) Adding this suffix to a base word, usually a verb or a noun, forms an adjective.

rely—reliable	sense—sensible
consider—considerable	horror—horrible
separate—separable	terror—terrible
read—readable	destruction—destructible
laugh—laughable	reduce—reducible
advise—advisable	digestion—digestible
commend—commendable	comprehension—comprehensible

ance, ence (act, quality, state of) Adding this suffix to a base word, usually a verb, forms a noun.

appear—appearance	exist—existence
resist—resistance	prefer—preference
assist—assistance	insist—insistence
attend—attendance	correspond—correspondence

Other nouns using *ance, ence* include

ignorance	experience
brilliance	intelligence
significance	audience
importance	convenience
abundance	independence
performance	competence
guidance	conscience

ary, ery (related to, connected with) Adding this suffix to base words forms nouns and adjectives.

boundary	gallery
vocabulary	cemetery
dictionary	millinery
library	
customary	

efy, ify (to make, to become) Adding this suffix forms verbs.

liquefy	ratify
stupefy	testify
rarefy	falsify
putrefy	justify
	classify

ize, ise, yze (to cause to be, to become, to make conform with) These suffixes are verb endings all pronounced the same way.

recognize	revise	analyze
familiarize	advertise	paralyze
generalize	exercise	
emphasize	supervise	
realize		
criticize		
modernize		

Also, some nouns end in *ise*.

exercise	enterprise
merchandise	franchise

ly (in a specified manner, like, characteristic of) Adding this suffix to a base noun forms an adjective; adding *ly* to a base adjective forms an adverb. Generally *ly* is added to the base word with no change in spelling.

monthly	surely
heavenly	softly
earthly	annually
randomly	clearly

If the base word ends in *ic,* usually add *ally*.

critically	drastically
basically	automatically

An exception is *public—publicly*.

ous (full of) Adding this suffix to a base noun forms an adjective.

courageous	outrageous	grievous
dangerous	humorous	mischievous
hazardous	advantageous	beauteous
marvelous	adventurous	bounteous

Other suffixes include:

ant (ent, er, or, ian) meaning "one who" or "pertaining to"

ion (tion, ation, ment) meaning "action," "state of," "result"

ish meaning "like a"

less meaning "without"

ship meaning "skill," "state," "quality," "office"

Final Letters

Final letters of words often require change before certain suffixes can be added.

- **Final e.** Generally keep a final silent *e* before a suffix beginning with a consonant, but drop it before a suffix beginning with a vowel.

use	useful	write	writing
love	lovely	hire	hiring

  Exceptions:

true	truly	due	duly	argue	argument

  Note: In adding *ing* to some words ending in *e*, retain the *e* to avoid confusion with another word.

dye	dyeing	die	dying
singe	singeing	sing	singing

- **Final ce and ge.** Retain the *e* when adding *able* to keep the *c* or *g* soft. If the *e* were dropped, the *c* would have a *k* sound in pronunciation and the *g* a hard sound. For example, the word *change* retains the *e* when *able* is added: *changeable.*

- **Final ie.** Before adding *ing,* drop the *e* and change the *i* to *y* to avoid doubling the *i.*

tie	tying	lie	lying

- **Final y.** To add suffixes to words ending in a final *y* preceded by a consonant, change the *y* to *i* before adding the suffix. In words ending in *y* preceded by a vowel, the *y* remains unchanged before the suffix.

try	tries	survey	surveying

- **Final consonants.** Double the final consonant before adding a suffix beginning with a vowel if the word is one syllable or if the word is stressed on the last syllable.

hop	hopping	hopped	occur	occurred	occurring
plan	planning	planned	refer	referred	referring
stop	stopping	stopped	forget	forgotten	forgetting

- In adding suffixes to some words, the stress shifts from the last syllable of the base word to the first syllable. When the stress is on the first syllable, do *not* double the final consonant.

prefer	preference	confer	conference
refer	reference	defer	deference

Ceed, Sede, Cede Words

The base words *ceed, sede,* and *cede* sound the same when they are pronounced. However, they cannot correctly be interchanged in spelling.

- *ceed:* Three words, all verbs, end in *ceed.*

proceed	succeed	exceed

- *sede:* The only word ending in *sede* is *supersede.*

- *cede:* All other words, excluding the four named above, ending in this sound are spelled *cede.*

  recede   secede   accede
  concede   intercede   precede

Symbols

Symbols are used mostly in tables, charts, figures, drawings, diagrams, and the like; they are not used generally within the text of pieces of writing for most readers.

Symbols cannot be discussed as definitely as abbreviations, capitalization, and numbers because no one group of symbols is common to all specialized areas. Most organizations and subject groups—medicine and pharmacy, mathematics, commerce and finance, engineering technologies—have their own symbols and practices for the use of these symbols. A person who is a part of any such group is obligated to learn these symbols and the accepted usage practices. The following are examples of symbols common to these groups:

- Medicine and Pharmacy

  ℞ take (Latin, *recipe*): used at the beginning of a prescription
  ℥ ounce
  ℨ dram
  s write (Latin, *signā*): on prescription indicates directions to be printed on medicine label)

- Mathematics

  + plus   > is greater than
  − minus   = equals
  × times   ∫ integral
  ÷ divide   ∠ angle

- Commerce and Finance

  # number, as in #7, or pounds, as in 50#
  £ pound sterling, as in British currency
  @ *at:* 10 @ 1¢ each

- Engineering Technologies

Specialized fields, such as electronics, hydraulics, welding, and technical drawing, use standard symbols for different areas. These symbols are usually determined by the American Standards Association.

For example, one electricity text has nine pages of symbols used by electricians. There are resistor and capacitor symbols, contact and push-button symbols, motor and generator symbols, architectural plans symbols, transformer symbols, and switch and circuit breaker symbols. Some examples follow:

Contacts-N.O. (normally open)   Ground   Conductor   Squirrel-Cage Induction Motor   Ceiling Outlet

A technical drawing text has eight pages of symbols including topographic symbols; railway engineering symbols; American Standard piping symbols; heating, ventilating, and ductwork symbols; and plumbing symbols. Some examples:

⊙ County seat   ╫ Flanged joint
---- National or state line   Ⓗⓦ Hot water tank
▨ Wood—with the grain   ⓧ Exit outlet

- Nonspecialized Areas

  Common symbols that persons recognize and use:

%	percent	′ and ″	feet and inches
°	degree	$ and ¢	dollars and cents
&	and		

See also units of measure as symbols, page 642, and examples of symbols, page 36. Most dictionaries include a section on common signs and symbols.

Credits

Index

Checklist for Com
ising a Paper

Title

Clearly stated, in a phrase   11
Precise indication of paper emphasis   11
Correctly capitalized   643, no. 6
No quotation marks around the title   657, no. 6
See sample pages   40, 54, 63, 81, 89, 115, 143, 166, 186, 206, etc.

Organization

(See "Procedure" outline in each chapter.)

Introductory section (See roman numeral I in each "Procedure" outline.)
 Background information, overview of topic, or identification of subject   310, 402–403, 605–606
 Obvious lead-in or forecasting statement (key sentence, controlling statement, central idea)   310, 402–403, 605–606
Body (See roman numeral II in each "Procedure" outline.)
 Adequate development of the lead-in or forecasting statement   310, 606–607
 Topic statements of paragraphs or sections that point back to the lead-in or forecasting statement   605–606
 Carefully chosen details, specifics, and examples   11–14, 76
 Information arranged in logical sequence   138–139, 160, 199–203, 313, 320, 334–335, 344–345, 351, 608–610
Closing (See roman numeral III in each "Procedure" outline.)
 Compatible with purpose and emphasis of the paper   310, 607–608

Content

(See "Procedure" outline in each chapter.)

Suitable for the intended audience and purpose   3–4, 28–32, 75–76, 118–120, 135, 160, 184, 218–219, 603–604
Accurate   305
Clear   305
Complete   606–607
 Adequate coverage of major and minor subdivisions   606–607
 Length of paper appropriate   43, 68
 No significant points omitted   42–43
All information directly related to the topic   600
Concise   14–18, 305–307
Coherent   13–14, 610–612